LEAN PRODUCTION FOR COMPETITIVE ADVANTAGE

A Comprehensive Guide to Lean Methodologies and Management Practices

LEAN PRODUCTION FOR COMPETITIVE ADVANTAGE

A Comprehensive Guide to Lean Methodologies and Management Practices

John Nicholas

CRC Press is an imprint of the
Taylor & Francis Group, an **informa** business
A PRODUCTIVITY PRESS BOOK

Productivity Press
Taylor & Francis Group
270 Madison Avenue
New York, NY 10016

Printed in the United States of America on acid-free paper
10 9 8 7 6 5 4

International Standard Book Number: 978-1-4398-2096-4 (Hardback)

Library of Congress Cataloging-in-Publication Data

Nicholas, John M., 1945-
Lean production for competitive advantage : a comprehensive guide to lean methodologies and management practices / John Nicholas.
p. cm.
Rev. ed. of: Competitive manufacturing management, 1998.
Includes bibliographical references and index.
ISBN 978-1-4398-2096-4
1. Production management. 2. Lean manufacturing. 3. Costs, Industrial. 4. Quality control. I. Nicholas, John M., 1945- Competitive manufacturing management. II. Title.

TS155.N43 12010
658.5--dc22 2010001495

Visit the Taylor & Francis Web site at
http://www.taylorandfrancis.com

and the Productivity Press Web site at
http://www.productivitypress.com

To Frank and Emily,

Elmer and Dolores

Contents

SECTION I CONTINUOUS IMPROVEMENT, WASTE ELIMINATION, CUSTOMER-FOCUSED QUALITY

SECTION II ELEMENTS OF LEAN PRODUCTION

Preface

Around 1989, after reading Richard Schonberger's *World-Class Manufacturing: The Lessons of Simplicity Applied* (The Free Press, 1986), Robert Hall's *Zero Inventories* (Dow Jones-Irwin, 1983), and Kiyoshi Suzaki's *The New Manufacturing Challenge* (The Free Press, 1987), and talking to former students who had become practitioners and consultants in manufacturing about emerging developments, I decided to offer a course for operations management majors focused on just-in-time (JIT). At the time there were no textbooks on JIT; some trade books covered the topic, but in a style and at a level more appropriate for practicing managers, not college students. Hence, to the chagrin of my students, the required reading material came to consist of a few trade books and a packet of journal articles. Trade books tend to skip over the basics and do not include homework questions and problems, and for this I had to write my own material. Over the years I accumulated more and more of my own writings, and about the time this material covered half the topics in the course, I decided to write a textbook to put everything in one place.

The first edition of this book was published in 1998. Management concepts seem to come and go with the seasons, but the important ones remain alive even though the terminology changes. JIT is alive and thriving, though nowadays it is called lean production and the term JIT is reserved for narrower, logistical applications of lean. In this book the terms lean production and JIT mean roughly the same thing and are used interchangeably.

In and around my city of Chicago, there are many factories and ample opportunities to see both the best and the worst in manufacturing management. I have been excited and encouraged in visiting plants that have embraced the concepts and methods of lean production, including continuous improvement and emphasis on waste elimination and customer-focused quality. Managers and workers in these plants enthusiastically talk about what they are doing. Students in my class often express the same level of enthusiasm, discovering that, yes, lean production is cool—in a way. Some of them with manufacturing experience also express frustration because of problems in their workplace and management's ignorance about lean production concepts or unwillingness to change. I can relate to that, having seen plants that seem like throwbacks to 60 years ago, and talking to workers with demeaning jobs and managers out of touch with the workforce and customers. These plants represent the antithesis of lean production.

The concepts and principles described in this book have revolutionized manufacturing practice and business conduct in the same way that Henry Ford's system of mass manufacturing did almost a hundred years ago. They have done so because, simply, they work—they are effective. Because they are effective, the only companies that can compete with companies that have embraced lean principles are companies that have also embraced them and happen to make products that customers want.

Beyond effectiveness, many find lean production appealing because of the somewhat high level of responsibility and dignity it attaches to the jobs of workers on the shop floor. I was raised in a working-class family and always felt that my parents' abilities far exceeded what they could exercise in the workplace. My dad was one of the smartest and all-around most capable men I have ever known and I always thought he could build or fix anything requiring mechanical, electrical, or carpentry skills. When I think about teams of workers in lean organizations, I envision people like my parents—for certainly they are representative of many millions of workers and of the opportunities and fulfillment that lean-style production can offer them. That is not to say that lean organizations are a kind of utopia but that, on balance, workers in lean factories have more opportunity to find meaning in their jobs and get more earned respect from management than workers in other factories.

The emphasis in this book is on lean production concepts and tools. As such, it focuses on methods and procedures, although it must be said that in any organization the success of methods and procedures depends on their acceptance and implementation. Lean production is more than methods and procedures; it is a management philosophy that emphasizes relationship building and trust, and responsibility conferred to frontline workers and suppliers. Thus, successful adoption of lean production means not only adopting particular methods and procedures, but adopting an organizational culture that supports those methods and procedures. In lean companies decision-making responsibility is much more decentralized, and its locus is decidedly much lower in the organizational hierarchy than in non-lean companies. In most cases this shift in responsibility accompanies a makeover in assumptions that people at all levels of the organization, from the shop floor to company president and CEO, make about the roles and responsibilities of frontline workers and suppliers, and particularly their involvement in day-to-day decision making. If an organization is not able to successfully adopt the methods and concepts of lean production, it is likely because it is not able to adopt the true meaning of employee and supplier involvement in its culture.

There is much to learn about lean production. Beyond the basics described in this book you can learn about new developments and applications of lean production in all kinds of industries from the pages of trade journals such as *Target*, the journal of the Association for Manufacturing Excellence.

Einstein said, "I know why there are so many people who love chopping wood. In this activity one immediately sees the results." Lean production practices are that way. They represent pragmatic approaches to chopping away at waste and problems in organizations. In many organizations you can start to see results on the shop floor or in the office not too long after putting aspects of lean production into practice. Improvements on the balance sheet happen too, of course, but will appear a little later.

Audience and Use of Book

Lean Production for Competitive Advantage was written for three audiences:

1. Bachelor of business and MBA students interested in production and operations management
2. Industrial and manufacturing engineering students
3. Practicing manufacturing managers and engineers seeking a better understanding of lean production.

The book is intended for a second-level course. Students who have taken an introductory course in production and operations management should have no difficulty grasping the material.

The book is divided into an introductory chapter and three main Sections. Chapter 1 and Section I provide foundation concepts for everything that follows. Section II covers the core concepts and methods of lean production. Section III covers integrated planning and control in pull production and the supply chain. It might be difficult to cover the entire book in depth in a typical one-term college course, and the instructor must decide on which topics to focus. I believe that everything in Sections I and II should be covered in some depth. Depending on students' prior exposure to planning and control, portions of Section III might be skimmed or deleted, although I suggest that no chapter be completely skipped. Most of the topics in this book are interrelated and to gain a full understanding of lean production it is necessary to know about all of them.

Throughout this book you will see terms bolded. These are topics discussed later in the book.

Acknowledgments

In writing this book I have had the good fortune of being assisted by many bright and capable people. Drs. James Zydiak and Enrique Venta, my friends and colleagues, read many parts of this book and provided useful suggestions. Sosamma Mammen, Marco Menaguale, Marlene Abeysinghe, and Omar Saner, my graduate assistants, provided research and editorial help. In case you find the going difficult with some of what you read, you can only imagine what it was like for them many drafts earlier. On the other hand, if perchance you think the material reads with exceptional clarity, it is no doubt due to their exceptional efforts. I also wish to credit Leslie Bailyn and Diane Petrozzo for their editing and gopher support.

Thanks also to Avi Soni and Al Brouilette, two enthusiastic managers who I believe represent the best examples of people who have embraced the true meaning of lean production. I have learned a great deal about lean production from frequent visits to their plants and numerous discussions with them and their co-workers. Avi Soni championed the successful makeover to lean production in his plant and is to me the quintessential transformational leader.

I also want to acknowledge the reviewers of an earlier edition of this book whose comments and suggestions greatly improved the end product: Mary Jo Maffei, formerly of the University of Cincinnati; Behnam Malakooti, Case Western Reserve University; Unny Menon, California Polytechnic State University; George Schneller, Baruch College—CUNY; Kenneth Ramsing, University of Oregon; Joe Biggs, California Polytechnic State University; Karen Donohue, University of Pennsylvania; Vaidyanathan Jayaraman, University of Southern Mississippi; Pitu Mirchandani, University of Arizona; Byron Finch, Miami University; and George Petrakis, University of Missouri.

Thanks also to the folks at CRC/Productivity Press, especially to the late Ray O'Connell for encouraging me to sign on, and to Maura May for guiding me along.

Finally there is my wife, Sharry, who has my deepest appreciation for patiently having assumed responsibility for managing virtually every aspect of our home life so I could work undistracted.

The assistance of so many people made writing this book not only doable but enjoyable. Most of them share with me an excitement about lean production. My one wish is that you, after having read this book, come away with that same sense of excitement.

I apologize in advance for any typos and mistakes in the book. I had final say over everything, so I accept responsibility for these as well as for any other source of anguish this book might cause. For your sake I hope there aren't too many, but I do appreciate hearing from you about them.

Chapter 1

Race without a Finish Line

> What we really cannot do is deal with actual, wet water running through a pipe. That is the central problem we ought to solve some day.
>
> **—Richard Feynman**

Most people have heard of Xerox, the company that invented the paper copier and for a time one of America's best-known corporate names and employer of 100,000 people.[1] Like so many successful companies, however, it has had its share of struggles—both in maintaining market share and even staying in business. In the early 1980s, Xerox executives realized that if something radical was not done, Xerox probably would not survive to the end of the decade. The problem: Japanese companies were selling copiers for less than it cost Xerox to make them, and they had targeted Xerox to capture its market and wipe it out altogether. For Xerox the situation was dire: Its market share was less than 15%, down precipitously from the more than 90% share it had enjoyed a decade earlier. But the competition alone was not the cause: there was no quality control to speak of, overhead and inventory costs were excessive, and there were too many managers. Further, Xerox had lost touch with its customers and was not giving them the products or service they wanted. At one time Xerox had no competitors. Now it had many, and customers were flocking to them.

Xerox managers and a core group of unconventional corporate thinkers mapped out a strategy to remake the company. The strategy succeeded. Xerox doubled its production output, reduced its costs by nearly 50%, reduced its product development cycle time by nearly a year, and dramatically improved the performance of its copiers. By 1990, Xerox had become the producer of the highest quality office products in the world and had gained back market share from the Japanese. By 2004 it still retained a dominant share of the high-end color copier market. Xerox accomplished this not with government subsidies, trade barriers, or import quotas but rather by adopting new management approaches that included the tenets of **lean production** and **total quality management**.

Successful recovery is, of course, not the fate of every company that has faced tough competition from overseas. It took Xerox 7 years to reinvent itself, but many companies cannot afford such time. Global competitors make everyone vulnerable. Aeronautics is a case in point. Up until the late 1960s, the United States was the world leader in the commercial aircraft business, holding nearly 100% of the free-world share. However, in 1969 a consortium of companies from France,

Germany, the United Kingdom, and Spain began to pursue that market, and the U.S. share has steadily declined. The European share is now roughly 50%. This is not an insignificant fact given that the aeronautics industry is the single biggest manufacturing contributor to the U.S. balance of trade.

As the number of skilled contenders in a given market increases, so increases the intensity of the competition. Initially, in new and growing markets many players can survive by absorbing a portion of the growth in market size. In the global economy the market size is large, but so is the number of players. Eventually, as market size levels out or as players differentiate themselves by ability to compete, the more skilled players drive out the less skilled players.

Competitive Advantage: Better, Cheaper, Faster, More Agile

In no small part what differentiates competitiveness in industrial companies is the way that each designs and builds products. Paying attention to customers and knowing what they want is a fundamental and important beginning. However, given that those competing companies pay attention to what customers want, the key to competitiveness then becomes production capability. What differentiates winners from losers is that winners are better able to consistently provide products and services that are competitive with regard to quality (better), price (cheaper), time (faster), and response to change (agile).

To this end, companies adopt different manufacturing philosophies, strategies, and methods. These differences are part of what distinguishes lean companies from traditional companies:[2]

Making things better: Traditional manufacturers strive for quality by relying heavily on computer-aided design (CAD)/computer-aided manufacturing (CAM) to enhance product design and manufacturability. In contrast, lean firms rely more on group technology and cellular manufacturing, good condition and proper placement of equipment, smaller manufacturing units, and improvement-focused employee teams. Both kinds of manufacturers rely on statistical process control, defect reduction programs, and vendor quality programs.

Making things cheaper: Traditional manufacturers tend to rely on job enlargement programs, automation, and robotics to reduce direct labor content. In contrast, lean firms seek to achieve low cost by redesigning and simplifying products and processes, standardizing products, and reducing lead times and cycle times.

Making things faster: Traditional producers rely on robotics and flexible manufacturing systems, location of facilities, and improved labor–management relations. Lean companies emphasize continuous reduction of lead times and setup times, equipment maintenance, and broadening of workers' jobs.

Being more agile: The ability to introduce new products and to respond quickly to changing customer demands is an area where lean companies maintain a wide lead over traditional firms.[3] Traditional firms seek agility through technology, process flow improvements, quality management, and cross-functional communication improvements. Lean firms consider agility an integral part of quality and delivery capability, and a by-product of programs to improve these areas. Thus, programs aimed at producing things better and faster are also aimed at achieving manufacturing agility.

Lean producers also take a somewhat different approach to product development. In traditional firms, product ideas tend to move sequentially through functional areas (from marketing to engineering to production, and so on), whereas in leans firms they are honed in production development teams. The latter allows integration of ideas during the early design stages, takes less time, and results in a better product at lower cost.

Lean Production and Total Quality Management

Another difference in the competitive strategies of traditional manufacturers versus lean manufacturers is that the former seek improvement at discrete times through capital intensive means such as automation and new technology, whereas the latter seek improvement through small but continuous refinements in processes and procedures, and investment in human capital. Lean production and the tandem management philosophy of total quality management (TQM) emphasize *continuous* attention to product and process improvements, and involvement of frontline workers in those improvement efforts.[4] Briefly:

- Lean production is management that focuses the organization on *continuously identifying* and *removing sources of waste* so that processes are continuously improved. Lean production is also called **just-in-time** or **JIT**.
- TQM is management that focuses the organization on *knowing the needs and wants of customers*, and on building capabilities to fulfill those needs and wants.

This book is primarily about lean production, but it is also about TQM because the two philosophies are mutually dependent and, in some respects, are the same. To paraphrase Schonberger,

> Lean is a quality improvement tool because it cuts time delays between process stages so that the trail of causal evidence to quality problems does not get cluttered and cold. But TQM tools are needed as well or the rate of quality improvement will not be fast enough. To oversimplify, lean without TQM will be quick response to quality problems, but to a dwindling number of customers. TQM without lean will meet correctly-identified customer needs, but using methods that are costly and wasteful.[5]

That the philosophies and practices of lean production and TQM are overlapping is not surprising since both originated in Japan in the 1950s.[6]

Lean Production and the Production Pipeline

Think of a company as a pipeline with raw materials entering at one end and products exiting at the other.[7] The goal is to minimize throughput time, that is, to move materials (or ideas, orders, whatever) through the pipeline as fast as possible. Shorter throughput time is better because, assuming price and quality remain constant, the company can respond more quickly to changes in customer needs. The customer gets the product sooner, and the company gets the payment sooner.

But the production pipeline is seldom uniform and without obstacles. What flows out of the pipeline is limited by the biggest obstruction. Portions of the pipeline that are narrow or obstructed are in reality equivalent to the stages of the production process that take more time or where stoppages or slowdowns occur. Obstructions to smooth flow and rapid throughput are commonplace, particularly in plants that produce a variety of products with different, fluctuating demands.

To speed up the flow through the pipeline, the obstructions must be identified and eliminated. As each obstruction is eliminated, the flow speeds up, but only by as much as allowed by obstructions elsewhere in the pipeline. Identifying the obstructions, understanding them, and finding ways to eliminate them is the thrust of lean production.

The pipeline analogy gives the impression that barriers to production, once identified, can be removed once and for all. In reality that is impossible. First, there are often a large number of phases, stages, or steps, and it is difficult to identify the precise location of all obstructions. Also, the sources of obstructions keep changing—machine breakdown, parts run short, and so forth. As some obstructions are removed, new ones appear. Further, the pipeline itself and the things that flow through it are always changing. Customer orders change, so the flow rate must be adjusted to accommodate the right kind and quantity of materials. In the analogy, the pipeline diameter might have to be widened or narrowed. A pipe that is wider than necessary is wasteful, and so is a flow rate that exceeds demand. In addition, the products are changing, too, so the process must be adapted (the pipeline itself must be modified or replaced), and that introduces a whole new set of obstructions. In short, work on the pipeline is continuous.

Lean production is a way of continuously tinkering so the pipeline and the material coming out of it are the best possible. Lean production continuously seeks way to make the pipeline more adaptable to whatever materials or flow rates are desired, to match the flow of materials as closely as possible to customer demand, and to make the material coming out of the pipeline ever more acceptable to the customer.

The Lean Difference

Most organizations seek to identify and eliminate obstructions in the pipeline, but two features distinguish lean organizations. First, lean organizations greatly increase the number of people involved in identifying and eliminating obstructions. Whereas most companies assign professional staff people to diagnose and solve problems, in lean organizations everybody does it. People at all levels are trained in analysis and problem-solving techniques and given some level of responsibility to implement improvements. The general level of responsibility of workers to make and carry out decisions is much higher in lean companies than elsewhere.

A second difference is in the process employed to identify and prioritize problems and sources of waste. In lean companies the primary process is reduction of inventory. Returning to the pipeline analogy, inventory buildup inside the pipe narrows the pipe's diameter and reduces the rate of flow. When the rate of flow is slow enough, obstructions in the pipeline are hidden from view. Lean organizations reduce the size of inventory to reveal the obstructions and prioritize them. With fewer inventories, the obstructions become visible, which means they can be identified and removed. With each small reduction in inventory, additional obstructions appear, and these become the new priority for removal.

It is important to note, however, that inventory reduction per se is not the objective of lean production. Inventory reduction by itself yields a one-time benefit of freed-up capital. Unless problems in the production system, problems previously hidden or ameliorated by the inventory, are resolved, the process will not be improved and might be worsened. This is why companies that try to reduce inventories as a solution rather than as a tool to identify, prioritize, and make process improvements usually fail.

The following section is a brief history of the evolution of manufacturing and the origin of present-day lean production.

Evolution of Manufacturing

The Machine That Changed the World[8]

One of the most informative and interesting introductions to lean production is the book *The Machine that Changed the World*. The machine in the title is the automobile, and the book is about the auto industry, production methods, and global competition. The automobile is truly the machine that changed the world because twice in the 20th century the auto industry changed the way all products are made. The first change came after WWI when Henry Ford and Alfred Sloan advanced manufacturing from *craft production* to *mass production*. Whereas for centuries Europe had led the manufacturing world with the former, America swiftly became the dominant global economy with the latter. The second change came after WWII when Eiji Toyoda and Taiichi Ohno at Toyota Motor Company pioneered lean production. The economic rise of Toyota and of other companies and industries in Japan and elsewhere that adopted lean production was a consequence.

Craftsmanship Yields to Industrialization

Prior to the beginning of the Industrial Revolution, usually associated with James Watt's development of the steam engine in 1769, the emphasis in production was on skilled craftsmanship. Craft guilds promoted workmanship and manual skills using hand tools. Apprentices learned workmanship while performing preliminary tasks for a master craftsman, who performed tasks that demanded high-level skills. Craftsmen and their apprentices produced almost everything. Not until 1776 when Adam Smith wrote *The Wealth of Nations* did the notion become prominent that dividing tasks among more than one specialist could increase productivity in large-volume production.

Around 1780, Eli Whitney in the United States and Nicolas LeBlance in France independently developed the concept of the *interchangeable part*. Parts are made in batches such that any one part would fall within design tolerances and fit into an assembled product. With the implementation of the interchangeable part in production of rifles, clocks, wagons, and other products, the slow transformation that would replace hand labor and craftsmanship with mechanization and division of labor began.

Around the early 1900s Frederick Taylor introduced the idea of improving operations by studying and simplifying them. Always looking for the **one best way** to do something, he developed techniques for systemizing and improving economies of work motion, as well as a complete management philosophy that included time analysis, wage incentives, separate responsibilities for managers (planning) and workers (doing), an accounting system, and principles for running a business on a scientific basis. Taylor inspired legions of contemporaries, including Frank and Lillian Gilbreth who extended Taylor's time study to detailed analysis of motion. Their ideas about work measurement, analysis, and management contributed to the theory of *scientific management*.

One consequence of this theory was to take most of the skills and thinking away from factory workers (de-skill job) and give them to legions of managers and specialists. Factory work was divided into narrow, repetitive tasks, and workers were stripped of all control. As this happened, work on the shop floor began to lose its appeal. It is significant to note that the Japanese never wholly embraced Taylor's system of specialization and its rigid rules separating responsibilities between workers and managers. In Japan, workers in factories continued to develop broad-based skills that would allow them to rotate freely among a variety of tasks. This continuing emphasis on skill development, as well as on delegation of responsibility and the expectation that factory workers would both plan and do work tasks, resulted in a level of worker commitment virtually unseen in Western factories.[9] The consequence of this would not be recognized until half a century later.

Craft Production of Automobiles

Cars were a luxury only the rich could afford. They were initially handmade by skilled craftsmen who were knowledgeable in design principles, materials, and machine operation. Many of these craftsmen went on to form their own machine shops and become contractors to auto-assembly companies.

Few cars were actually identical since every shop produced parts according to its own gauging system. Specifications on parts were only approximations, and car assemblers would have to file down the parts so they would fit together. Lack of product uniformity did not matter much since most cars were built for individual buyers. Minor design features, location of controls, and some materials were changed to meet buyer preferences.

In the early 1900s there were hundreds of companies in Western Europe and North America producing autos, though each made fewer than a 1,000 a year, and rarely made more than 50 from a single design. Because the small shops that supplied parts lacked resources, there was very little product or process innovation. As a result, cars were poor in terms of consistency, reliability, and drivability.

Ford's Mass Production System

As a young machine-shop apprentice, Henry Ford envisioned producing an inexpensive auto. In 1903, at the age of 40, he started the Ford Motor Company and began producing the Model A. Each car was produced on a fixed assembly stand, and a single craftsman worker assembled all or a major part of it. The worker got his own parts; if a part didn't fit, he filed and pounded it until it did. Ford saw the limitations of parts being inconsistent and introduced Whitney's idea of standardized, interchangeable parts.[10] Using interchangeable parts, any worker could be trained to assemble a car and all cars would be virtually the same, so anyone could drive and repair one. Thus, Ford began to insist that all parts be produced using the same gauging system. He also had the parts delivered to the work stands so assemblers could work uninterrupted at one place all day.

By 1908 when the Model T was introduced, Ford had modified the process so that each assembler moved from car to car and did only one task on each car. Though, presumably, such specialization would increase efficiency, the problem was that faster workers would catch up with and have to wait behind slower workers. As a solution, Ford introduced in 1913 another major innovation: a *moving assembly line.* He got the idea at a slaughterhouse watching carcasses on hooks move from workstation to workstation by an overhead bicycle-chain mechanism. The line, which brought the cars past stationary workers, eliminated time wasted by workers walking, and forced slower workers to keep up with the pace of the line. Whereas before it had taken 13 hours to make a car, it now took just 1½. Ford's competition realized the productivity advantage of the combination of interchangeable parts and the moving production line—what became known as Ford's *mass production system*—and eventually companies around the world adopted that system.

While perfecting the interchangeable part, Ford was also perfecting the "interchangeable worker" by obsessively applying the teachings of Frederick Taylor. Whereas the Ford worker of 1908 gathered parts and tools, repaired tools, fitted parts together, assembled an entire car, and checked everything, the Ford worker of 1915 stood at the moving line and did one simple task. Workers caught slacking off or performing inadequately were quickly replaced. The atmosphere did not inspire workers to point out problems or offer suggestions, so problem solving was assigned to foremen, engineers, and battalions of functionally specialized workers. With time, these specialties all became narrowly subspecialized, eventually so much so that staff in one subspecialty would have trouble communicating with staff in other subspecialties.

By 1931 Ford had brought every function necessary to car production in house, the concept of vertical integration taken to the extreme. Not only did the company make all of its own parts, it also controlled procurement and processing of basic materials such as steel, glass, and rubber. Ford did this partly because he could produce parts to closer tolerances and tighter schedules than his suppliers and partly because he distrusted his suppliers.

To make parts inexpensively, machines were needed that could make them in high volume and with little downtime for changeovers. Ford eliminated this downtime by using machines dedicated to doing one task at a time. Because it is difficult and costly to modify products using dedicated machines, new products were avoided. But the rigidity of the system did not matter because from 1908 through 1927 Ford only produced the Model T. Throughout this period Ford's goal was to keep increasing Model T volume so that the cost would continuously decrease. Ford believed that mass production precedes mass consumption because high volume results in lower unit costs and allows prices to be reduced. By 1926 Ford was the world's leading manufacturer, producing half of all cars. Not until the 1950s did a car outsell the Model T in numbers; that car was the Volkswagen Beetle.

Quality control of the Model T was lax. Finished cars were rarely inspected, but they were durable and could be repaired by the average user. This concept was a major selling point. But in 1926 sales began to slip. Because so little about the car had been changed in 10 years, price had become its sole selling point. Customers were looking elsewhere for more innovative and exciting products.

Emergence of Modern Mass Production

General Motors (GM) was created when William Durant acquired several car companies. Early on this enterprise faired poorly because the companies had separate management and overlapping products. Upon becoming GM's president in 1920, Alfred Sloan instituted a new management philosophy that would complement the mass production system introduced by Ford. He reorganized GM into five car divisions covering the market continuum, low end to high end, and into parts-specialty divisions.[11] He also divided management into functional areas and originated the new functions of financial management and marketing. By 1924 GM had become a serious challenger to Ford.

The production system at Ford put workers under increasing pressure, sometimes pushing them to the extremes of endurance. By 1913, when most jobs in the system had become menial and relentless, turnover reached 380%. To stem the flow and attract better workers, in 1914 Ford doubled the daily wage to the then-phenomenal amount of $5. The pay hike worked, and most workers stayed and came to view Ford as a place of permanent employment, despite unbearable work conditions. But the workers were expendable; in a market turndown, Ford would lay them off in a moment. Not only at Ford, but also at GM and other automakers, they were considered little more than pieces to manipulate as needed. As a result the workers organized and formed a union, the United Auto Workers (UAW). By the late 1930s the UAW had gained enough strength to force the automakers into an agreement whereby seniority, not job competency, would determine which laborers were laid off. Seniority also dictated who got certain job assignments. Management in all major industries, not just automobiles, fought labor unions long and hard, and by the time the unions finally gained power, hate and distrust had swelled. Labor and management had become adversaries and would remain so for decades.

The system of mass production today is largely a result of the combined influence of Ford, Sloan, and organized labor. Starting in the 1930s the system served as the means for increasing economic gains for both employers and workers in the United States. By 1955 sales of the Big Three U.S. automakers accounted for 95% of the more than 7 million autos sold in the United States that year. That was their peak year. Afterward, their share began to dwindle as the share from imports steadily grew.

Mass Production Around the World

Economic chaos, nationalism, WWII, and strong attachment to craft production had prevented European automakers from widely adopting mass production until the 1950s. When the Europeans did adopt mass production techniques, low wage costs and innovative features gave them a burst of success in world markets. That success dampened, however, during the 1970s, by which time wages had increased and work hours had decreased to the point where European cars were no longer price competitive. In 1973 the oil embargo hit; gasoline prices worldwide soared. Even though the typical U.S. car was a gas-guzzler and the average European car was smaller and more fuel efficient, the latter was also higher priced and posed no real competitive challenge to U.S. automotive dominance. Elsewhere, however, the real challenge was forming, though it was totally unperceived. It was in Japan.

Toyoda and Ohno

The Toyoda[12] family had been in the textile business since the 1800s. It began producing cars in 1935, but the cars were crude and poorly made. In 1950 young Eiji Toyoda visited the Ford River Rouge plant to learn the methods of mass production. The plant seemed to him a miracle of modern manufacturing. Almost everything that went into an automobile—parts, components, and assemblies—was produced in this one monstrous plant. In one end went raw materials like iron ore and out the other end rolled cars, 7,000 a day. Toyoda wanted to learn how the Americans did that. (At this time, the Toyota Motor Company over its entire history had produced fewer than 2,700 cars.)

After 3 months of studying the plant, however, Toyoda concluded that Ford's system of mass production was unworkable in Japan, which was still struggling to recover from the ravages of war. Since Japan had few auto manufacturers at the time and a small auto market, Toyoda wanted to make a variety of cars in just one plant. In the United States, only one type of car could be produced in a plant. Because of strong Japanese company unions, he knew he could not readily hire and fire workers as was common in U.S. firms. Also, because of the short supply of capital, he would not be able to invest heavily in modern equipment and technology.

Returning home he called on Taiichi Ohno, his chief production engineer, to help him develop a workable system. Ohno too visited Detroit and concluded he would have to design a system that would be less costly and wasteful, but more efficient and flexible than the traditional system of mass production. The system he and others at Toyota developed, called the **Toyota Production System**, is the prototype for lean production and just-in-time manufacturing. Though developed for automobile production, the principles of the system have since been applied in all kind of industries.

Toyota Production System—Prototype for Lean Production

This section gives an overview of the features of the Toyota Production System and, in particular, what distinguished it from traditional mass production. These topics are covered in more detail throughout the book.

Reduced Setup Times

The American practice was to use hundreds of stamping presses, each for making only one or a few kinds of parts for a car. These huge presses sometimes required months to set up. Since Toyota's budget limited procurement to only a few stamping presses, each press would of necessity have to stamp out a variety of parts. To make this practical, the setup times for switching over from stamping one kind of part to another would have to be drastically reduced.

By carefully analyzing existing procedures, Toyota devised methods that slashed setup times from months to just hours. By organizing procedures, using carts, and training workers to do their own setups, the company was eventually able to get the setup time down to an amazing 3 minutes.[13] The procedures developed at Toyota can be applied to any setup in almost any workplace.

All setup practices are wasteful because they tie up labor and equipment, and add no value to the product. Despite this, the tradition, before Ohno, was to take any setup practice as a given; if the setup takes a long time, so be it. In fact (so goes traditional thinking), if a setup takes a long time, it would be necessary to produce things in large batches to justify that setup time. Since setup times were usually long, large-batch production became the norm in manufacturing.

Small Lot Production and One-Piece Flow

The traditional practice of producing things in large lots (batches) was justified not only by the high setup cost, but also by the high capital cost of high-speed dedicated machinery. Dedicated machinery is very efficient, but it is also expensive, and somehow managers feel the expense can be justified if they keep the machinery running to produce things in massive quantities, regardless of demand.

But producing things in large batches results, on average, in larger inventories because it takes longer to use up the batch, and that, in turn, results in higher holding costs. Plantwide, larger batch production also extends lead times because it ties up machines longer and reduces scheduling flexibility. The effect on lead times is negligible when demand is constant but it increases as the demand variability increases.

Large size batches also tend to have higher defect costs. Production problems and product defects often happen as a result of setup mistakes that affect the entire batch. The larger the batch, the more items affected.

The classical way to determine optimal batch size is based on the economic order quantity (EOQ) formula:

$$\text{EOQ} = \sqrt{2DS/H}$$

where D is demand, S is setup cost, and H is holding cost.

Since, traditionally, holding cost has been underestimated because it ignores the effects of batch size on quality and lead times, and since the setup cost, a direct function of setup time, is usually large, the formula (and managers' thinking) has been biased toward large batch sizes.

Once ways were found to make setups short and inexpensive, it then became possible for Toyota to economically produce a variety of things in small quantities. In other words, Toyota could produce any sized quantities, whatever demand dictated—even if demand was very small. The smallest size demand is one unit, and Toyota set the goal of being able to produce anything, one unit at a time. This is called **one-piece flow**.

Employee Involvement and Empowerment

In U.S. plants there was a specialist for doing just about everything. Other than machinists and assembly workers, though, few specialists directly added value to a product. Ohno reasoned that most tasks done by specialists could be done by assembly workers and probably done better because assemblers were more familiar with the workplace. To this end he organized his workers by forming teams, and gave them responsibility and training to do many specialized tasks. Each team had a leader who also worked as one of them on the line.

In addition to their assigned work tasks, teams were given responsibility for housekeeping and minor equipment repair. They were allowed time to meet to discuss problems and find ways to improve the process. Further, they would collect data to help diagnose problems, develop solutions and plans, and use suggestions from the specialists who were still on hand but in relatively few numbers compared to those in Detroit. The notion of workers **asking why five times** to get to problem root-causes was first introduced at Toyota. Eventually, worker responsibility was expanded to include many areas usually held by specialists, including quality inspection and rework.

Quality at the Source

Ohno saw that the traditional manufacturing practice of stationing inspectors at locations throughout and at the end of the line did little to promote product quality. Defects missed by inspectors were passed from one worker to the next. As a product was assembled, defects became progressively more embedded inside, and, thus, more difficult to detect. If detected, defective products were scrapped or sent to rework areas; if not, they were passed on to the customer. Thus, relying solely on inspectors was not a practical means by which to eliminate defects and the costs associated with making and reworking defects.

Ohno reasoned that to eliminate product defects, defects must be discovered and corrected as soon as possible, which means going to the source of defects and stopping them there. Since the workers are in the best position to discover a defect and to immediately fix it, Ohno assigned each worker responsibility for detecting defects. If the defect could not be readily fixed, any worker could halt the entire line (called **Jidoka,** or **line stop**) by pulling a cord. At first the Toyota line was frequently stopped and output was low, but over months and years of refining the process, the number of defects began to drop, sources of errors were eliminated, and the quality of parts and assembled products got better. Eventually, the quality of finished goods was so high that the need for rework was practically eliminated.

Equipment Maintenance

Manufacturing organizations laden with work in process (WIP) inventory are not much concerned about equipment maintenance because the inventory is a buffer that allows work to continue (for a while) even when equipment breaks down. Further, many organizations actually bank on key equipment breaking down and carry enough inventory between operations to cover for that eventuality. Some managers are of the belief that the most productive way to run a machine is to run it constantly (three shifts if possible) until it breaks, then fix it. This philosophy runs square in the face of waste reduction because, accordingly, inventory is held, operations are idled during equipment repair, and repair cost are higher.

Consistent with the philosophy of worker empowerment at Toyota, operators are assigned primary responsibility for basic **preventive maintenance** since they are in the best position to detect early signs of malfunction. Maintenance specialists now diagnose and fix only complex problems, train workers in maintenance, and improve the performance of equipment. The combination of maintenance specialists and workers both practicing regularly scheduled preventive maintenance curtails equipment breakdowns.

Pull Production

In traditional manufacturing plants, products are fashioned by moving batches of materials from one stage of the process to the next. At each stage an operation is performed on an entire batch. Because batch sizes and processing times vary from stage to stage and because usually there are several jobs that need work at each operation, it is difficult to synchronize the flow of material from one stage to the next. As a result, materials wait at each operation before they are processed, and typically the wait time far exceeds the time to process the batch. Plantwide the result is large amounts of WIP waiting at various stages of completion. In terms of inventory holding costs and lead times, the waste can be staggering.

To reduce these wastes Ohno developed the **pull production** method wherein the quantity of work performed at each stage of the process is dictated solely by the demand for materials from the immediate next stage. Ohno also developed a scheme called **Kanban** to coordinate the flow of materials between stages so that just as a container was used up, a full container from the previous stage would arrive to replenish it. This is actually where the term *just-in-time* originated. Production batches are kept small by using only small containers to hold materials.

Although pull production reduces waste, it is, in truth, not always an easy system to implement. Toyota took 20 years to work out the process.

Standard Work

At Ford and most modern industrialized organizations, work standards developed by engineers and specialists specified what the frontline employees were expected to do. But the standards, developed by staffers somewhat detached from the actual workplace, were often unrealistic, impractical, and provided no motivation for improvement. Ohno, during his early years at Toyota, had developed the philosophy that workers should create their own job descriptions and work steps because, he reasoned, only then would they be able to fully comprehend the details of their work, know why they had to do things that way, and be capable of pondering better ways to do it. He developed the **standard work** or **standard operations** concept wherein worker teams create the standards that define the work they currently do. The standard work also serves as the baseline from which to improve processes to better suit changes in customer demand and the work environment.

Supplier Partnerships

Ohno also recognized problems with traditional customer-supplier relationships in the U.S. Typically the manufacturer would develop detailed specifications for each product part and then contract with suppliers to make the parts through a competitive process that, usually, awarded the lowest bidder. Multiple suppliers for each part were retained, and these suppliers were routinely played one against another to keep prices down.

Ohno saw the need for a different kind of relationship wherein the manufacturer treats its suppliers as **partners** and, as such, integral elements of the production system. Essentially, suppliers are trained in ways to reduce setup times, inventories, defects, machine breakdowns, and so forth, and in return they take responsibility for delivering the best possible parts/services to the manufacturer in a timely manner. With this arrangement the manufacturer, customer, and supplier benefit.

By the early 1960s Toyota Motor Company had worked out most of the major principles of lean production. Eventually all other major Japanese auto firms adopted their own versions of lean production. In 1968 Japan passed West Germany as the number 2 producer of vehicles in the world. In 1980 Japan passed the United States to become number 1. In 2005 it produced 50% more vehicles than it did in 2001.

America's Fall from Manufacturing Grace[14]

It would be a mistake to assume that Japan's economic success resulted solely from lean production. Japan is a very communal society, and following WWII the ministries of government decided where to concentrate the nation's limited resources to best serve the country. The nation's educational system put emphasis on turning out engineers. By creating an excess of engineers, it was thought, more of them could be put on the shop floor where they could tinker with improvement. The cumulative effort of so many talented people working on so many things was incalculable. It was also the case that, deprived of opportunities in aerospace, nuclear, and other cutting-edge technologies with military potential, the cream of Japan's technical brainpower was funneled into more prosaic industries such as autos, steel, and machine tools; in the United States such industries that were having trouble attracting the best people.

While the Japanese were developing improved methods of production, American manufacturers were being distracted. Beginning in the 1950s a new breed of executives came to power. Here again the change was first seen at Ford, starting with the hiring of a group of young men known as the Whiz Kids, men who during the war had gained a reputation in the Air Corp for prowess in applying analytical techniques to managerial problems.[15] But the Whiz Kids were not car men, nor even product men. Lacking product know-how, they minimized its importance and played up their management systems, which emphasized financial criteria that (presumably) told management what to do, regardless of market, product, or industry. The group was at first a welcome relief at Ford. Henry Ford had distrusted accountants and fired most of them, so company records and procedures after the war were in total disarray. The Whiz Kids reorganized everything and imposed tight financial and accounting controls. They cleansed the system, though later people would question whether or not their kind of cleansing had been best. The issue is not the efficacy of financial management systems; such systems are an important component of modern management. What is wrong is that in the Whiz Kids' brand of management, financial controls are allowed to dominate and smother every other component of business—marketing, engineering, and manufacturing.

At Ford and at most other corporations in the United States, a powerful new bureaucracy was installed. As it assumed power it displaced the manufacturing people who had customarily dominated. Given that one of the easiest ways to bolster profit performance is to curtail capital expenditure and implement cost cuts, top management pared back on capital investment and product innovation, made cuts in facilities and labor, closed plants, and moved manufacturing overseas.

Over the next several decades, graduates of America's business schools swelled the ranks of the new elite. The best students went into finance and systems analysis, which had become the fast tracks to senior management. Rarely did good students go into manufacturing, which was viewed as a dead end. It was just as rare in big companies to find a manufacturing person on the board or as president. (Manufacturing in Japan carried higher prestige, and boards there had many men with factory experience.) As U.S. managers changed their style, they also changed their business agenda. Driving up the price of shares on the stock market became more important than making a good product with a profit.

Japanese producers slowly began to take away market share in autos, steel, and electronics. Rather than learn from it, American managers used the challenge of Japan as a threat to keep wages down. They told workers "You better take the contract we are offering you or see your jobs moved to another country."

Climbing Back

The Whiz Kids approach to management is still taught in business schools and practiced in corporations, but signs indicate the mistakes of recent decades are being recognized and corrected. There are also signs that U.S. manufacturing is on the road to becoming competitive once more. This is not to say that all is well in U.S. manufacturing or that America will regain its position of dominance in world production. Many executives still view manufacturing as but an expense that must be pared back or fully outsourced. The United States is still the most innovative country in the world, but its innovators often have a hard time getting financial backing at home and must look overseas where investors and manufacturers seem more eager to snap up promising new ideas. These manufacturers, of course, become the long-term beneficiaries of American innovativeness.

Modern Developments

To round out this brief history we note some other manufacturing developments in the last four decades. These developments can be summarized in two words: computer technology. Since the 1970s, probably the single greatest use of computers in manufacturing has been for material requirements planning (MRP) systems that link together information about parts and components that go into a finished product and generate schedules and purchase orders for assembled and procured parts. Without the computer's capability to manipulate large quantities of data, coordinated scheduling and rescheduling of thousands of parts for producing hundreds of products would be impossible. These systems are often integrated with marketing, financial planning, and supply chain management function in company-wide enterprise resource planning (ERP) systems.

Other noteworthy developments are computer-aided design, computer-aided manufacturing, flexible manufacturing systems, computer-integrated manufacturing, and electronic data interchange. Computer-aided design (CAD) enables designers to design a part or product, and test its features and compatibility with other parts and products, all with a computer. Computer-aided manufacturing (CAM) refers to software that translates design requirements into instructions for controlling production machinery. Given CAM's dependence on CAD for input, CAD/CAM is often applied as a single, integrated system.

A flexible manufacturing system (FMS) aims to achieve high-variety output at low cost. Although an FMS can be manual, the usual notion of FMS is a computer-controlled, automated system. A large FMS consists of many machines or processes, each with varying degrees of mechanization, linked by automatic transfer systems, guided vehicles, or robots, and all controlled by a central computer. With a computer regulating changes in machine settings, parts, and tools, one machine can perform numerous functions, and require very little time for changeover between parts.

The next step beyond FMS is computer-integrated manufacturing (CIM). CIM links CAD/CAM; automated material handling; robotics; and automated manufacturing planning, control, and execution into a single, integrated system.

Electronic data interchange (EDI) refers to computer-to-computer exchange of information, usually meaning between multiple companies. EDI enables quick, accurate sharing of production schedules and placing of orders between companies that are customers/suppliers to each other.

Computer technology is also responsible for another aspect of mass manufacturing: mass customization. In one scenario, a sales clerk at a clothing store enters into a computer a customer's vital statistics to create a digital blueprint of a pattern, which is transmitted to the factory and instructs a robot to cut the fabric to precise measurements.[16] In another scenario, customers phone an engineer to discuss the kind of part they need, and the engineer enters the specifications into a CAD/CAM system to design a one-of-a-kind part. Overnight, automated machines grind out the custom parts.

In many industries, technologies like these have cut labor, material, transportation, and inventory costs, and have improved manufacturing cycle time, flexibility, product quality, and customer satisfaction. In other cases, however, because the technologies were too costly, poorly implemented, prematurely made obsolete, or ill-suited for the operation, they helped little or made things worse. In a study of FMSs used in companies in the United States and Japan it was revealed that while the average U.S. FMS was used to produce 10 types of parts the Japanese FMS produced 93. Said the researcher who did the study, "With few exceptions, the flexible manufacturing systems installed in the United States showed an outstanding lack of flexibility."[17]

The Imperative

Lean production programs have been adopted by organizations everywhere. Despite variations in the ways these programs are implemented, the fundamental ingredients of all of them remain largely the same. In manufacturing, many of these elements represent a departure from traditional ways, though many of them also conform to principles of good operations and manufacturing management that have been in the books for decades. Some are technically or culturally difficult to implement, and some provoke resistance from managers and workers. As a result, lean programs require long-term commitment to implement, and few of them will show positive results in the first several months or even years of effort. The alternative to not adopting these methods, however, is potentially disastrous. Players who have successfully adopted the methods of lean production and who faithfully apply these methods to improve their processes and products have gained a substantial competitive advantage over others within the same market who continue to operate by the old rules of mass production.

The proof is in the record. It is not an overstatement to say that companies that have been challenged by producers committed to the principles of lean production have been able to meet or beat back those challengers only by committing to the same principles. Companies that were challenged but that did not become lean producers, either because they ignored the challenge or did not have time to make the transition, lost; either they chose to drop out of the market, or they were forced to drop out.

Organization of Book

This book is divided into three parts. Part I is an overview of three concepts fundamental to lean production philosophy and practice: **continuous improvement**, **elimination of waste**, and **focus on the customer**. Survival and competitiveness in manufacturing mandate continuous improvement of products and processes—forever striving to be better, faster, cheaper, and more agile. Attempts to improve a process imply the process contains waste, so part of continuous improvement is identifying waste and eliminating it. This part of the book begins with an overview of the strategic importance of improvement in competitive environments, kinds of improvement, sources of waste, and the basic lean principles for waste reduction and process improvement in manufacturing.

In the pursuit of competitive advantage, a company must target its improvement efforts, and that is where the concept of quality comes in. In the notion of **customer-focused quality**, customer-defined quality requirements are used to set priorities for all improvement and waste-reduction efforts. This concept and the related topics of **total quality management**, **statistical process control,** and **Six Sigma quality** are also covered in Part I.

Part II covers the main elements of lean production. In a typical company, waste exists in the form of product defects, inventory, overproduced items, idle workers and machines, and unnecessary motion. Through lean production practices in the factory it is relatively easy to identify the sources of these wastes and eliminate them. This part of the book presents the core practices of lean production, namely, producing in **small batch sizes**; reducing process and equipment **setup times**; **maintaining** and **improving equipment**; using the **pull system** (Kanban) for production control; organizing facilities into **focused factories** and **workcells**; **standardizing operations** on the shop floor; and eliminating manufacturing defects, the notion of **pokayoke** or mistake-proofing. The tools and methodologies of lean production are largely interrelated, and the successful implementation of one is usually predicated on substantial implementation of many of the others. In every case, the methodologies are aimed at institutionalizing the process of continuous improvement. In every case also, frontline workers play a central role. This role and real-life implementation issues about lean production are also addressed in this part of the book.

Part III reconsiders the lean production concepts from earlier chapters but in the context of a working manufacturing system. Assuming the elements of lean production are well along into implementation, the question then is, "How is production planned and scheduled to take the fullest advantage of these changes?" This part of the book takes a step back to look at lean and pull production methods as elements of a total production system, a system that forecasts demand, accumulates orders, translates forecasts and orders into plans and schedules, authorizes material procurement, and executes work tasks to ultimately yield a finished product. Particular topics of this part of the book include the concepts of **production leveling** to smooth fluctuations in production plans and schedules, **balancing capacity** and **synchronizing of operations**, and a **framework of overall planning and control** to tie together many of the topics from elsewhere in the book. Part III also addresses the relative roles of centralized systems and decentralized (shop floor) systems in lean production planning and control, and the subject of adapting MRP systems to pull production.

The quality, cost, and delivery performance of every manufacturer depends in large part on the performance of its suppliers. Without good relationships—agreements between a customer and its suppliers about the quality, cost, delivery, and service expected on materials received, a company will be hard pressed to achieve lean production ideals. The last chapter of the book deals with the concept of **supply chain management**, the changing role of the purchasing function, and application of lean concepts to improve performance and reduce waste throughout the supply chain.

Originated and developed in manufacturing, the philosophy and tools of lean production have since been widely adopted in all kinds of organizations to improve processes and eliminate waste. To illustrate lean production practices as applied to a wide variety of service and manufacturing situations, many examples will be given throughout the book.

Notes

1. D. Kearns, and D. Nadler, *Prophets in the Dark: How Xerox Reinvented Itself and Beat Back the Japanese* (New York: HarperBusiness, 1992).
2. A. Roth, A. DeMeyer, and A. Amano, 1989. International manufacturing strategies: A competitive analysis, in *Managing International Manufacturing,* ed. K. Fellows, 187–211. Amsterdam: Elsevier Science.
3. National Center for Manufacturing Sciences, *Competing in World-Class Manufacturing* (Homewood, IL: Business One Irwin, 1990), 118.
4. A related philosophy is **time-based competition** (TBC), which, obviously, puts more emphasis on the time criterion. Much of the practice of TBC overlaps with TQM and JIT practices, although an express goal of TBC, one which is not necessarily a goal of JIT or TQM, is to shorten product development times, that is to improve "agility" in terms of product redesign and innovation. See P. Smith and D. Reinertsen, *Developing Products in Half the Time* (New York: Van Nostrand Reinhold, 1991); and J. Blackburn, *Time-Based Competition* (Homewood, IL: Business One Irwin, 1991).
5. R. Schonberger, *World Class Manufacturing: The Lessons of Simplicity Applied* (New York: The Free Press, 1986), 137.
6. J. Dalgaard and S. M. Dahlgaard-Park, Lean production, six sigma quality, TQM and company culture, *The TQM Magazine* 18, no. 3 (2006): 263–281.
7. The analogy has been used before. See H. Mather, *Competitive Manufacturing* (Englewood Cliffs, NJ: Prentice Hall, 1998), 12–20; W. Sandras, *Just-in-Time: Making It Happen* (Essex Junction, VT: Oliver Wight, 1992), 7–8.
8. Material in this section is derived from two principal sources: J. Womack, D. Jones, and R. Roos, *The Machine That Changed the World* (New York: Rawson, 1990); and D. Halberstam, *The Reckoning* (New York: Avon Books, 1986).
9. See W. Lazonic, *Competitive Advantage on the Shop Floor* (Cambridge, MA: Harvard University, 1990).
10. Ford's innovative contribution to the interchangeable part was in its application and execution: he imposed on all suppliers of each part a single standard to which they were all expected to conform.
11. DuPont and Carnegie Steel also developed new forms of organization. DuPont was the first to establish corporate divisions, and Carnegie the first to vertically integrate an industry—raw materials to finished product. The point: It was U.S. companies that were making the innovations. Nothing like it was happening in Europe.
12. The family name Toyoda means "rice field" in Japanese; the word Toyota, which they chose to call the corporation, has no meaning.
13. Womack et al., *The Machine That Changed the World,* 53.
14. This section is derived from Halberstam, *The Reckoning.*
15. See J. Byrne, *The Whiz Kids* (New York: Currency Doubleday, 1993).
16. G. Rifkin, Digital blue jeans pour data and legs into customized fit, *New York Times* (Nov. 8, 1994): A1.
17. R. Jalkumar, Postindustrial manufacturing, *Harvard Business Review* Nov–Dec (1986): 69–76.

Suggested Readings

P. Dennis. *Lean Production Simplified: A Plain-Language Guide to the World's Most Powerful Production System,* 2nd ed. New York: Productivity Press, 2007.

J. K. Liker. *Becoming Lean: Inside Stories of U.S. Manufacturers.* Portland, OR: Productivity Press, 1997.

J. Liker. *The Toyota Way.* New York: McGraw-Hill, 2003.

Y. Monden. 2010. *Toyota Production System: An Integrated Approach to Just-in-Time,* 4th ed. New York: Productivity Press, 2010.

T. Ohno. *Toyota Production System: Beyond Large-Scale Production.* New York: Productivity Press, 1988.

Questions

1. Explain how each of the following is a basis for competitive advantage in manufacturing: cost, quality, time, agility.
2. Distinguish between delivery time and time to market.
3. In 10 words or less, what is the primary focus of lean production?
4. What is meant by the term *production pipeline*? What does the production pipeline have to do with lean production?
5. What features differentiate lean organizations from other organizations?
6. What is craft production?
7. What is mass production?
8. What were Henry Ford's most significant contributions to mass production?
9. Why did Eiji Toyoda and Taiichi Ohno decide against copying Detroit's system of mass production? How did that lead to the Toyota Production System?
10. Describe the principles of the Toyota Production System and how it is a departure from traditional production systems.
11. Explain some of the reasons why America's position as the world's supreme industrial power began declining in the 1960s and 1970s.
12. What is the relationship between quality and production in craft production?
13. What effect did interchangeability of parts have on product quality?
14. Why have overseas corporations been successful at capturing large shares in U.S. markets? What is the role of manufacturing?
15. What are some potential barriers against adopting industrial policies and manufacturing practices in the United States that would help make American industry more competitive?
16. A principal feature of lean production is large-scale employee involvement and employee empowerment. What do these concepts mean? Why in practice are they considered somewhat radical in U.S. organizations when they are accepted as commonplace in Japanese organizations?

Research Questions

To answer these questions, do a literature search and include material beyond this chapter.

1. Name some newsworthy, contemporary companies that are successful (high market share, high profits, leading edge products, and so on). Which of them are "turnaround companies" (rebounded after hard times)? Which of them are lean companies?
2. What are some big-name companies that in recent decades stopped producing certain products after Asian or European producers aggressively entered the market? Which of these Asian or European producers use lean practices?

I

CONTINUOUS IMPROVEMENT, WASTE ELIMINATION, CUSTOMER-FOCUSED QUALITY

The term *improvement* implies that something about a process has changed for the better. In business operations, improvement is often expressed in terms of changes in the process output or input. For example, increasing the process output–input ratio (O/I) is considered improvement, achieved by either an increased level of output for a fixed level of input or by a reduced level of input for a fixed level of output. Though more output for the same input or the same output for less input are common ways of measuring improvement, too often the view of what constitutes "input" and "output" (and, hence, improvement) is narrow and simplistic. Real improvement represents an attempt to make something better while trying to minimize cost and adverse consequences. For example, if a manufacturing process is changed to increase the rate of output, the change represents an overall improvement only as long as aspects of the output or other processes have not been degraded. If the change results in diminished product quality, increased production cost, worsened working conditions, or increased toxic wastes, the increased output may not be an improvement at all.

In the global marketplace the concept of improvement is synonymous with "never stop trying." External factors (technology, resource availability and costs, and competitors' capabilities) over which an organization has little control but which influence its competitiveness and profitability are constantly changing. Inside the organization, factors like work conditions, employee motivations and skills, and processes are changing, too. As a result, whatever was good enough yesterday will probably not be good enough tomorrow.

There is also no logical end to customers changing their expectations, neither in level or direction, nor to what competing organizations will do to try to meet those expectations. Thus, there can be no such thing as ultimate improvement. Others keeping pace with changes will surpass organizations that try to achieve ultimate improvement or are content with staying in one place. In short, survival and success mandate *continuous improvement.*

Paralleling the idea of continuous improvement is the concept of *elimination of waste*. To use a metaphor, the road to improvement is potholed with waste. Like potholes, sources of waste are everywhere and no matter what you do, they keep coming back. Any attempt to improve a process by increasing output with a fixed level of input or by maintaining the same output from reduced input implies that the process contains waste. Part of continuous improvement is the drive to continuously identify waste and eliminate it.

In the pursuit of competitive advantage, an organization must be able to target its improvement and waste-elimination efforts. There must be some scheme for putting priorities on where to expend time, effort, and resources in improvement and waste-reduction efforts. Here is where the concept of *quality* comes in. In seeking competitive advantage, *customer-focused quality* is the criterion an organization uses for prioritizing improvement and waste-reduction alternatives. Quality has a price, however, so the concept of *value* is important, too. To retain or increase a product's value, efforts devoted to increasing product quality must be accompanied by efforts to hold or reduce costs. Lower cost can result from improvement efforts directed solely at costs, though lower cost is also one of the many byproducts of eliminating waste in a process.

To avoid getting into a semantical debate about distinctions between the three concepts of improvement, elimination of waste, and customer-focused quality, let us just say they are all related, which is why they are discussed together here. The concepts and their relation to one another, to lean production and TQM, and to competitive manufacturing are topics of the chapters in Part I:

Chapter 2: Fundamentals of Continuous Improvement
Chapter 3: Valued-Added and Waste Elimination
Chapter 4: Customer-Focused Quality

Chapter 2

Fundamentals of Continuous Improvement

Even if you are on the right track, you'll get run over if you just sit there.

—Will Rogers

A rut is a grave with the ends knocked out.

—Laurence Peter

An organization's ability to survive and thrive depends on how well the organization adapts to demands imposed by a changing environment. Organizations come and go, though relatively few outlive the people who have worked in them. Organizations that survive for a century are rare. Nowadays it seems that big companies whose names were once synonymous with power and prosperity are being eclipsed at a quickening rate; their demise hastened by a failure to adapt.

The challenge of change in a business environment comes along many fronts: competitors introduce new products, industries develop new processes and technologies, and the scope of the business environment expands. One time, not so long ago, a business could feel safe if it had captured most of the market from its domestic competitors. Today no organization anywhere can feel safe just by having beaten domestic rivals. The business environment has expanded to put U.S., Asian, and European organizations in direct competition, and this competition has changed the definition of what constitutes doing a good job. Organizations everywhere are always striving to offer something new and, as a result, customer expectations are always expanding, and satisfaction tends to be a fleeting phenomenon. A company's survival and success now depend on its ability to **continuously improve** its products and services to meet and exceed customer expectations.

Continuous improvement is measured in terms of producing things better, faster, and cheaper, and being more agile. But to improve products and services it is necessary to go beyond the products and services themselves; it is necessary to examine and improve the materials and basic processes intrinsic to them. Continuous improvement is thus synonymous with **continuous process improvement.**

Lean production is, effectively, about continuous improvement, so this chapter sets the stage for everything to follow. The chapter starts by discussing continuous improvement and what it means to be *continuous*. Also discussed is the role of frontline employees in improvement efforts, and different improvement approaches such as PDCA (plan–do–check–act), value analysis, reengineering, kaizen projects, and value stream mapping. The chapter concludes with a review of the seven basic analysis and problem-solving tools.

Continuous Improvement as Tactics and Strategy[1]

A premise of continuous improvement is that processes and products can be improved without limit. Yet we know intuitively there must be limits, if only because improvements require resources, and resources are limited. Part of continuous improvement involves knowing where to direct improvement efforts so they contribute the greatest good. This means being able to identify the portions of the system that contribute most to increasing product quality and meeting customer requirements, as well as the ability to recognize when a product or process is already as good as it can be.

Incremental Improvement: Kaizen

The concept of limits to economical improvement can be described in terms of the S-curve shown in Figure 2.1. The curve shows the relationship between the effort or resources to improve something and the incremental result of that effort. The kind of improvement represented by the S-curve is called **incremental improvement**, which is the process of making something better through the accumulation of small, piecemeal improvements, one at time.

At first progress is slow, hence the early part of the curve is nearly flat. As the object under scrutiny is better understood, however, learning accelerates and improvement occurs at an accelerated rate. The idea that great improvement eventually comes from a series of small, incremental gains is the Japanese concept of ***kaizen***. Accordingly, employees throughout the organization patiently work to continually improve the processes in which they are engaged.

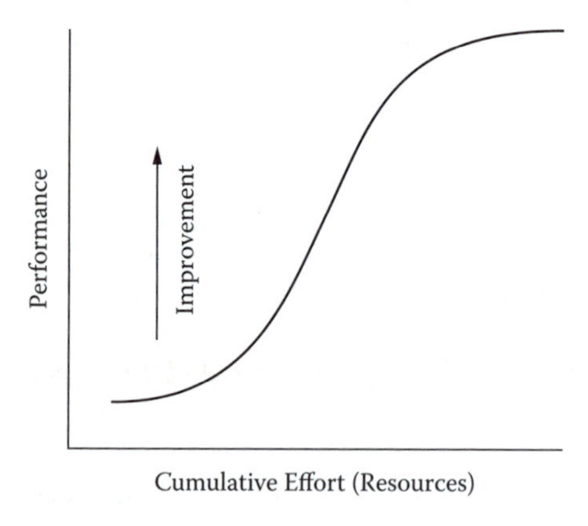

Figure 2.1 The S-curve of incremental improvement.

Case in Point: LCD Technology at RCA and Sharp

A good example of traditional U.S. and Japanese kaizen approaches to improvement is Hedrick Smith's discussion of the history of the liquid crystal display (LCD).[2] In 1968 RCA unveiled a new technological discovery, the LCD, a flat-display screen that promised to replace cumbersome cathode-ray tubes in TV sets. To transform the new technology into useful products, however, would take years of patient developmental work and considerable financial investment, neither of which RCA's corporate culture supported. In fact, RCA at the time was diversifying into areas like car rental, carpeting, and TV dinners because these provided quick returns, despite taking away funds from RCA's core business: electronics. Unlike founder David Sarnoff, who appreciated technology and pushed for innovation, RCA at this time was now run by marketing and financial people who cared more about corporate stability and short-term financial success than industry leadership. After a half-hearted effort at producing LCD wristwatches and calculators, RCA sold its LCD patents to the Japanese in 1973.

Contrast this with what happened at Sharp in Japan. Sharp, at the time a modest-sized radio and TV producer dwarfed by RCA, saw a market niche in lightweight, energy-efficient LCD pocket calculators. It put $200 million into development, and the calculators became a market success. Profits from the calculators provided Sharp a springboard into other LCD and flat-panel display applications. Ultimately, Sharp invested a billion dollars in developing LCD technology and building plants to manufacture flat-panel displays.

Today LCD technology is everywhere—in industrial gauges, clocks, phones, TVs, computers, medical imaging systems, automobile dashboards, and aircraft cockpit controls, to mention a few applications. By the 1990s, Sharp and 18 other Japanese firms controlled 95% of the world market for flat-panel displays, although more recently Korea and Taiwan vie with Japan for LCD dominance. Many U.S. manufacturers have become dependent on LCDs, and Japanese producers have opened assembly plants in the United States to avoid tariffs. Total market revenues for flat-panel displays and associated electronics by 2007 were $42 billion.

As for RCA? All it got from the LCD was what it sold the patents for: $2 million.

Incremental improvement can, over time, lead to substantial improvement, although, as the S-curve illustrates, at some point that improvement will slow. Regardless of effort, improvement gains get smaller and smaller. Only so much can be done economically to improve something, after which any residual improvement comes only at great expense. This concept of improvement has universal applicability, no matter the product or process.

What happens when the improvement threshold is reached? Does improvement cease? For an existing technology, product, or process, the answer is yes, at least for now, because the cost of further gains becomes just too great. This is important to note because it means that if further improvement is necessary for the organization to survive, then something new and perhaps radically different is necessary.

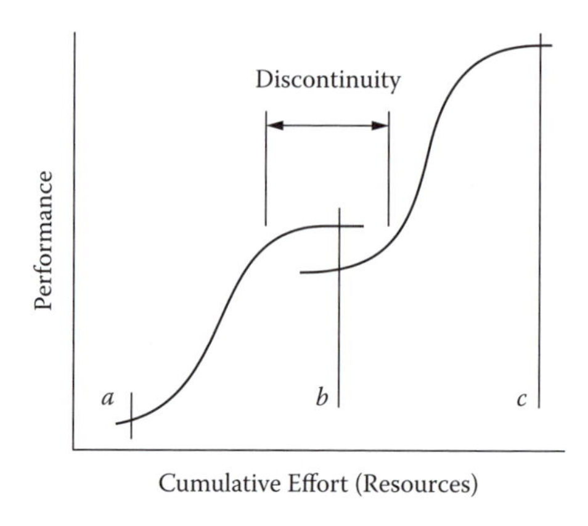

Figure 2.2 Incremental improvement: each curve; innovation improvement: jump from lower curve to higher curve.

Innovation Improvement

Once the limit is reached, higher performance is achieved only by adopting a new way or new technology (something fundamentally different and innovative, something that does not have that performance limit). This new way has its own S-curve, shown as the curve on the right in Figure 2.2. Figure 2.2 illustrates the difference between two improvement approaches: the continuous, incremental approach is represented by a single S-curve; the **innovative** approach is represented by the jump from one S-curve to the next. For example, vacuum tubes, propeller-driven airplanes, and the open-hearth method of making steel are three technologies that were incrementally improved over decades. Eventually each technology reached its limit and little could be gained through additional improvement efforts. The big leap in performance came when these technologies were eclipsed by the introduction of three entirely new technologies: semiconductors, jet engines, and continuous steel casting, respectively. In terms of performance, these new technologies far surpassed the old ones. Each became the focus of incremental improvement, and each is now well along its own S-curve, someday to be eclipsed by still newer technologies.

Figure 2.2 also suggests that although the initial change from the old to the new might require substantial initial effort (remember Sharp's investment in LCD technology), once the new way has been adopted, incremental improvements thereafter might be achieved for about the same amount of effort as was needed for the old way ($c - b = b - a$).

The leap from one curve to the next is also the way improvements happen within a given system. Figure 2.3 shows how a system—product or process—is moved along its S-curve by virtue of aggregate improvements among its components. Each component is incrementally improved and may even be replaced by new, better components. Eventually, however, simply improving components is not enough to elevate the performance of the system, so the entire system must be replaced. A form of innovation improvement called **process engineering** will be discussed later.

Making the Leap

Between the S-curves of two succeeding technologies lies a *discontinuity,* the time between crossing over from the old way to the new. Referring back to Figure 2.2, during the discontinuity the

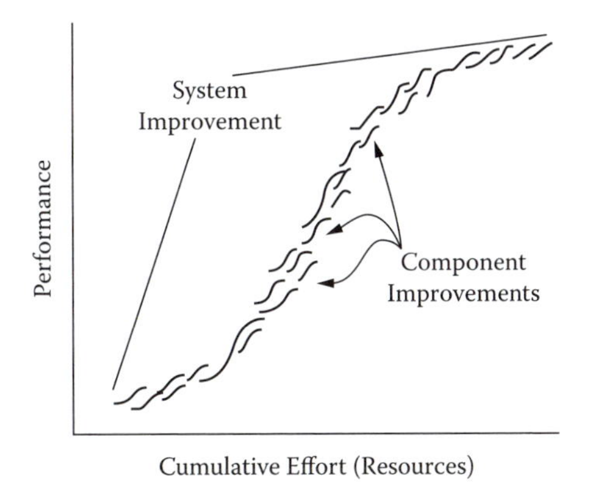

Figure 2.3 Systemwide improvement through improvements in components.

performance of the new technology is less than the old one. Any technology performs worst at the time it is first introduced; the trick is to resist ignoring it because of the initial poor showing.

Many businesses fail to appreciate a new technology's potential to make the current technology obsolete, and at the same time are so caught up with current products or processes that they fail to appreciate when the limit to incremental improvement has been reached. LCDs are one example, commercial jets are another. Early jet aircraft were inefficient and did not perform as well as the best propeller-driven airplanes, so some engine and airframe manufacturers tended to ignore jets in favor of further development of propeller technology. It only took a few years, however, for jet technology to move ahead of propeller technology, and companies not already involved in the transition were put at a serious competitive disadvantage. Boeing, for example, had always trailed Douglas, until then the world leader in commercial aircraft production. In the early 1950s Boeing made a bold move: it put all of its commercial resources into development of a jet transport, the 707. Douglas then raced to develop a jet of its own, but it had a late start. In just a few years Boeing surpassed Douglas. Douglas merged with McDonnell, and in 1996 McDonnell-Douglas merged with Boeing but the corporate name and headquarters remained Boeing's.

A similar misstep happened to Zenith, a former world leader in television, radio, and stereo manufacturing. When transistor technology was first introduced, Zenith chose to cling to the old vacuum tube and fell behind. Sony meanwhile was quick to adopt the new technology and has since become a world leader in television manufacturing. Zenith saw its small market dwindle and in the late 1990s was acquired by the South Korean conglomerate LG Electronics.

The ramification of the S-curve and its upper limit is that for real continuous improvement, there must be both innovation improvement and incremental improvement. At the corporate level this means strong, continued commitment to research and development for new products and processes; at the middle management and shop-floor worker levels, it means openness to new and bold ideas, and willingness to try new ways of doing things. That an organization is comfortably ensconced in a continuous improvement effort does not mean it will excel or even survive, particularly if its efforts involve only incremental improvement of existing products and processes. When the S-curve begins to flatten out, the technology of the particular product or process is ripe for innovation. Whoever takes first advantage of that opportunity will gain the competitive lead.

Of course, knowing which innovations to pursue and which to ignore is not an easy choice, and even large investment in promising technology is no assurance a new technology will meet expectations. For example, in the 1970s the auto manufacturer Mazda began production of the Wankel engine, which functioned on the rotary principle and was a marked change from the traditional, reciprocating engine. After only a few years Mazda ceased production because of poor market response, problems with excess pollution, and difficulties with the engine's rotary seals.

While it is not necessary to draw an actual S-curve for a product or process, it is important to be aware of a product's or process's approximate position on the curve. For each improvement effort, the degree of improvement achieved for the effort expended should be estimated. Marginal improvements that get harder to achieve could suggest that the technology is moving into the latter portion of the S-curve, in which case it could be the time to cut back on incremental efforts and concentrate on finding a whole new approach. Knowing where competitors stand on the S-curve is important too, because the farther along they are on the S-curve, the more likely they are working to make the leap to a whole new way. There is little sense in trying to incrementally improve an existing way when a competitor is well along in developing a new way that will make it obsolete.

Even after making the technological leap, it is essential to immediately begin incremental improvement to remain competitive. This is necessary because today the process called **reverse engineering** whereby one company takes apart another's invention, analyzes it, copies it, and improves it has become a high art. Though inventions and intellectual property are protected through laws and the efforts of groups such as the World Intellectual Property Organization (a United Nations agency), reverse engineering often gets around such protection. Personal computers are an example. Most components are off the shelf, and only the central processing unit (CPU) is patented. By studying the CPU, a competitor can learn its functions and design one that functions virtually the same without violating laws. The point is that simply being the originator of an idea or the first to introduce it does not guarantee that the inventing company will retain any advantage. The leader will be the company that successfully improves and commercializes the invention.

Manufacturing **process improvement** is likewise important because it is not enough to originate an idea for a product; you have to be able to control the manufacture of a product to be able to control the market for it. For example, the U.S. firm Ampex invented the VCR (video cassette recorder) but never developed the manufacturing and marketing means to support it. It sold the technology to Sony, which became a major manufacturer in a multibillion-dollar-a-year industry.

The improvement S-curve has relevance to processes at every level of the organization—individual tasks, manufacturing subprocesses and technology, and service processes. No matter what the nature of the task or process (assembling a product, responding to a customer complaint, maintaining a piece of machinery, arranging the layout of a plant or office, or changeover of equipment), the dynamics of the S-curve apply. That means virtually everything about an organization's operations can be improved through either incremental or innovative means.

Improvement as Strategy

Manufacturing strategy has been defined as creating the operating capabilities a company will need for the future. Since improvement alternatives are the stepping stones to achieving long-term organizational goals, then the continuous improvement process can be considered a strategic tool. Every technology possesses capabilities, though only certain capabilities are relevant to an organization's competitive success. Thus, any decisions about whether to continue improving an existing technology or to work on new ones (and if so, which ones) should be based on an assessment of the capabilities each will provide and the importance of those capabilities to competitiveness.

As an illustration, consider a company that is pondering ways to achieve the long-term goals of reduced production lead times and inventories. One way is to work with the existing production

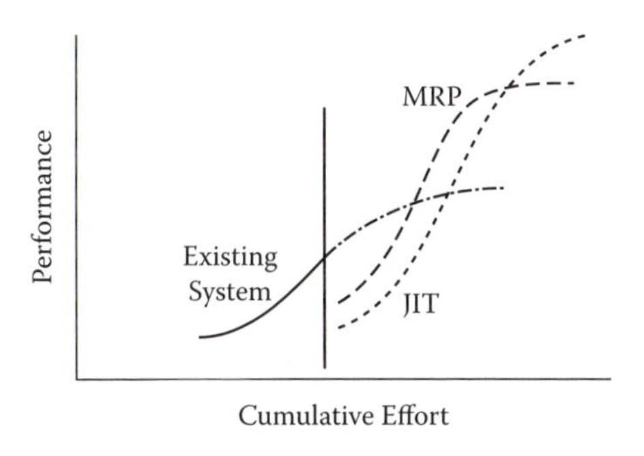

Figure 2.4 Performance improvement opportunities of alternative manufacturing strategies.

process while continuously striving to reduce batch sizes. Another is to adopt a computerized material requirements planning (MRP) system to improve production scheduling and ordering. A third is to convert to a lean pull production system. All three options will impact lead times and inventories, although the similarity stops there. The first option can be started immediately and will cost little. The other two will be costly and will not show improvement until some time later. Without fundamental changes to the production planning and control system, improvement in inventory and lead time reduction and batch size reduction will soon hit the limit; whereas for the latter two options, the opportunity for incremental improvement will likely continue for some time. This concept is illustrated in Figure 2.4.

The opportunities and capabilities afforded by each alternative must be assessed in terms beyond the immediate criteria. Reduced lead times and inventories are one set of goals, but there are certainly others, and the assessment should point out the alternative that best provides the capabilities necessary to reach most or all of those goals. Incremental improvement in MRP happens through improvements in data accuracy, shop-floor organization and control procedures, and computer software. Incremental improvement in pull production happens through improvements in setup times, equipment maintenance, quality checking, and problem solving by frontline workers. Each option builds different capabilities, and the best option is the one that develops capabilities that the firm will need to capture and maintain a unique position among competitors.

Consider a final example, the decision about whether to try to reduce costs and improve product quality from an existing manufacturing process or to cease producing the product and outsource it to a supplier. Although a reputable supplier could provide immediate improvement in terms of reduced costs and increased quality, by ceasing production the company relinquishes certain skills and capabilities associated with the product's design and manufacturing. It will be difficult or impossible to later recapture and improve upon these capabilities. If those capabilities represent a core technology of the company, that is, something unique about the company and fundamental to its competitive position, or are something that customers identify with the company, then outsourcing can pose a threat to the company's very survival.

Just as outsourcing diminishes the capability of a company to improve upon the outsourced product, it increases the capabilities of the supplier to improve both the product design and its manufacture. These days, product quality and customer service dictate close customer–supplier relations, and through such relations some or much knowledge associated with the product gets transferred to the supplier. Since the supplier already has the capability to make the product, the supplier will be the one to improve upon that capability and, potentially, also to develop a design capability as well. With both design and manufacturing capability, a determined supplier can eventually set out on

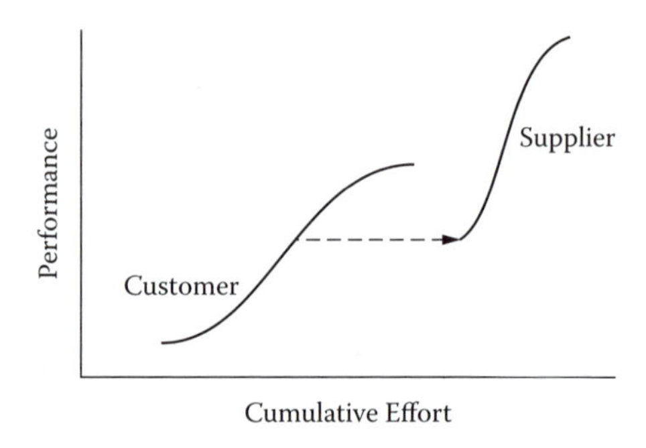

Figure 2.5 Transfer of technology from outsourcing of production.

its own and become a new, formidable competitor in the marketplace. This transfer of technology is illustrated in Figure 2.5. Such a transfer is what happened in television manufacturing where, at one time, Asian firms such as Toshiba served solely as suppliers to U.S. manufacturers. Today Asian firms not only make virtually all U.S.-brand TVs, but they also make their own brands and, of course, they are no longer simply suppliers. Outsourcing, like other ways to achieve improvement, must be viewed for the long-term improvement opportunities it affords or precludes.

Finding and Implementing Improvements

Several methodologies are used as guides to continuous improvement. This section begins with a somewhat generic methodology for identifying and diagnosing problems called the PDCA cycle. It also covers two other improvement approaches: *value analysis* for incremental improvement and *process reengineering* for innovation improvement. Later chapters will address specific improvement methods from lean philosophy.

PDCA Cycle

Shigeo Shingo, the renowned manufacturing improvement expert, has said that improvement requires a continuous cycle of *perceiving* and *thinking*.[3] A structured way to apply this cycle is **PDCA**, which stands for plan–do–check–act, illustrated in Figure 2.6. PDCA is also called the **Shewhart cycle,** after its developer A. W. Shewhart, and also the **Deming cycle**, after the person

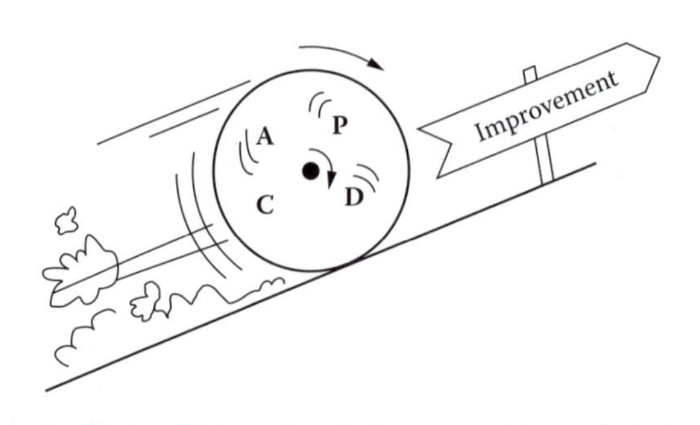

Figure 2.6 PDCA problem-solving cycle.

who popularized it, W. Edwards Deming.[4] The process can be visualized as starting with the plan step and moving clockwise through the other steps. For continuous improvement, however, it is better to visualize it as a continuous process with no start or finish and recurring: PDCAPDCA ...[5]

Here is a brief description of the steps in PDCA:

Plan Step

This step results in a plan to accomplish improvement. It has four substeps:

1. Collecting data to understand the current situation. Improvement is always instigated by dissatisfaction, the perception that a problem exists, or that things are not as good as they might be. Before seeking a solution, the existence of the problem should be substantiated and its root causes identified. This requires developing a complete understanding of the existing situation, which includes understanding not only the problem or issue, but the system, context, or environment that supports or led to the problem. Collecting data involves systematic observation, documentation, and analysis for purposes of verifying, understanding, and getting to the root cause of a problem. It requires firsthand observation at the source or location of the problem, called the *gemba*. Complete understanding of the current situation requires separating the symptoms from the causes, and getting to the root cause of the problem.
2. Defining the problem. Once the source of dissatisfaction is understood and its causes are known, the problem can be defined unambiguously. Clear definition of the problem is a necessary precursor to solving the problem.
3. Stating the goal. A goal is specified as the target of improvement. For example, the goal might be a more efficient way of doing a task, elimination of a particular waste, or a better environment. The goal should contain acceptable upper and lower limits (level of improvement), the criticality of achieving the goal (nice to have versus necessary at all costs), and a deadline. The goal must be unambiguous and agreed upon by everyone working toward it. In keeping with the philosophy of continuous improvement, the goal is only provisional. Once close at hand, the goal may be set at a higher level or a broader scope. Incremental improvement of a product or process is achieved through a stepwise approach of ever-more ambitious goals.
4. Solving the problem. Almost any improvement can be reached by various means. In solving the problem, multiple ways of achieving the goal should be considered along with the costs and benefits of each. Considered solutions should be shared with the parties affected and account for their reaction and suggestions. The cause–effect relationship between the root cause of the problem and the chosen solution should be explicit.

The result of the plan stage is a plan to accomplish improvement. The plan includes a clear definition of the problem, the goal, the solution or countermeasure to the problem, and the method of implementation. The method of implementation should describe the steps to be taken, people responsible, a schedule, expected outcome, and the budget. Creating an effective implementation plan involves the 5W1H principle, that is, *who* is going to do *what*, *where*, *when*, *why*, and *how*.

Do Step

This is the implementation of the plan. Plans that call for wide-ranging changes are typically implemented on a piecemeal or trial basis. The implementation and resulting changes are monitored and the plan modified according to circumstances.

Check Step

The implementation should be thought of as an experiment. Data are collected following implementation and analyzed to assess the results and to see to what extent goals are being accomplished. Side effects or adverse consequences are noted.

Act Step

In the final step, actions are taken based upon results from the check step. If the planned changes were successful, they are retained, standardized, and institutionalized. If the changes were successful but were implemented only in a limited way, they are expanded. If the results do not turn out as predicted, the situation is investigated again to discover why not and what new actions should be taken. Of course, even when the changes appear successful, continuous improvement mandates that eventually the plan of action be reviewed again and possibly replaced in favor of something better.

The rationale behind the PDCA cycle is improvement through learning and the scientific method. The plan stage involves not only analysis to identify root causes and a solution, but a prediction about the outcome of the improvement and how it will be measured. In the check stage, the actual outcome is compared to the prediction and any discrepancies are assessed. Starting with a prediction or hypothesis is important, because without it the PDCA cycle would be little more than trial and error. Outcomes that do not conform to the hypothesis indicate an incomplete understanding of the current situation or the theorized relationship between the problem and proposed solution.

The PDCA cycle of improvement is embodied throughout lean production systems. For example, as elaborated in the next chapter, one benefit of working to avoid overproduction is smaller work-in-process inventories. But with smaller inventories there is less to buffer against production problems such as equipment breakdowns, material shortages, erratic schedules, and poor quality, which means that when these problems occur they are likely to interrupt production. In lean production, however, the PDCA approach dictates that when such a problem occurs it is addressed with a solution (plan). The solution is implemented (do) and the results are assessed (check). Effective solutions are adopted and incorporated into the process; others, less effective, are rethought, redone, and reevaluated (act). No solution is ever considered perfect; even if it were, it couldn't remain so because the surroundings are always changing.

Five-Why Process

A procedure commonly employed in the plan stage of the PDCA cycle is the five-why process. Whenever analyzing a problem, the question why the problem exists is asked five times. The purpose of the procedure is to assure that the root causes, not merely the superficial symptoms, are corrected.

For example, suppose the observed problem is defective parts:

1. *Why* are the parts defective?
   Answer: The machines on which they are produced do not maintain the proper tolerance.
2. *Why* do the machines not maintain the proper tolerance?
   Answer: The operators of the machines are not properly trained.
3. *Why* are the operators not properly trained?
   Answer: The operators keep quitting and have to be replaced with new ones, so the operators are always novices.

4. *Why* do operators keep quitting?
   Answer: Working at the machines is repetitive, uncomfortable, and boring.
5. *Why* is the work at the machines repetitive, uncomfortable, and boring?
   Answer: The tasks in the operator's job were designed without considering their effect on human beings.

Now, assuming that answer 5 is taken to be the root cause of the problem, a solution can be suggested: Redesign the job so that it is less uncomfortable and more fulfilling for the operator.

In most situations, real problems and root causes are obscured by apparent problems. Notice in our example how different the solution might have been if the cause had been taken to be the answer to question 2, 3, or 4 instead of 5. Asking why repeatedly, possibly more than five times, directs the focus toward real causes so problems can be resolved permanently.

The five-why process is not necessarily as simple or straightforward as the example suggests. The real problem is often far afield from the problem as initially perceived, and answering each why requires considerable, thoughtful analysis.

Value Analysis/Value Engineering

Value analysis and **value engineering** are techniques for assessing the *value* content of the elements of a product or process. Value is what people are willing to pay for something. Since the price they are willing to pay for something cannot exceed the value they perceive in it, the cost of aspects of the product's manufacture should not exceed the value contributed by those elements.

The terms value analysis (VA) and value engineering (VE) are sometimes used interchangeably, but the former usually refers to ongoing improvements, especially of processes, whereas the latter refers to first-time design and engineering of a product or process.

Although all production and support activities add to a product's cost, not all add to its value. Similarly, while all components add to the cost of a product, not all contribute effectively to its functional value. VA and VE focus on the ability of product elements (product components) or process elements (steps and procedures) to add to the worth of a product from the customer's perspective.

VA and VE look at how much value something adds compared to the cost it adds. The concepts and processes associated with VA and VE are similar, though our discussion will emphasize the former. Value analysis often focuses on those activities that transform materials from one form to another. Rather than simply minimizing the labor content in a product, VA seeks to maximize the contributions (value-added effort) of people in an organization.

Value Analysis Procedure

Value analysis typically involves five steps: information gathering, analysis, creation, evaluation, and implementation.[6] (Notice the similarity with the PDCA cycle.) Ideally, the procedure is carried out by a cross-functional team from marketing, product design, manufacturing engineering, procurement, production, and finance. The steps in VA are as follows:

1. Information gathering. Data are collected about the components or elements of the system being scrutinized (a product, process, or some aspect thereof), and for each element or component its cost, requirements, and features (functions, attributes, why they are included, etc.). The worth of each element is estimated based upon the least expensive known alternative to it. Given the worth and cost of each component, components having the highest cost-to-value

ratio are identified. These components are the ones where the most could be gained by elimination or substitution with other, lower cost components. The emphasis of VA is not simply reducing cost but reducing cost for the same value or increasing value for the same cost.
2. Analysis. The chosen element or component is studied for the function it serves. The big question is "What does it do?" If the element is a step in a process, the answer should explain why the step is there. If it is a component in a product, the answer should explain what function the component serves and why it is needed. If it is a product, the answer should explain what the product does and why customers buy it. In all cases, another question is "Can it be made better or faster or cheaper?"
3. Creation. Ideas are generated for better ways of serving the same function as the element under scrutiny. Examples include simplifying the design, replacing nonstandard with standard parts, and using less expensive materials.
4. Evaluation. The alternative that seems to have the best chance of fulfilling the function, but at lower cost, is chosen as the replacement. In some cases it is discovered that the element can be eliminated because the function it serves can be served as well by other, existing elements, or that the function is no longer necessary to the purpose of the process or product.
5. Implementation. The selected alternative is adopted and the results are monitored for costs and to ensure that the function is being well served by the new element.

The following examples illustrate this approach.

Example 1: VA for Improvement of an Existing Product

A VA team devotes 80 labor hours to studying a product and learns the following:

1. Among all components of the product, one, a metal part, can readily be substituted by a plastic part. The substitution will save $0.15 per unit.
2. One of the drilled holes in the frame of the product is a holdover from an earlier model and is unnecessary. Not drilling the hole saves $1.10 in labor and overhead cost for each unit.

The team then spends 20 more labor hours making changes and performing tests on one unit to ensure that its functioning, reliability, and quality have not been reduced. They discover that by eliminating the hole, the product is somewhat strengthened—an important marketing feature that will enable the company to increase the unit sales price by $1.00.

If the product currently sells for $45 per unit, costs $22 per unit to manufacture, and sells at a rate of 50,000 units per month, the per-unit cost savings and value-added improvement will be $0.15 + $1.10 + $1.00 = $2.25. Without the change, monthly profit is ($45 – $22) × 50,000 = $1,150,000; with the change, monthly profit is [($45 – $22) + $2.25] × 50,000 = $1,262,500.

If the total salary for the team is $60/hr and the cost of the VA process is primarily the team's salary, then the VA cost is 100 hrs × $60 = $6,000. Hence, for a one-time cost of $6,000 the monthly profit will be $112,000.

Example 2: VA in Improvement of an Existing Process

Two engineers each devote 40 hours to studying the parts on a product. For the two parts shown in Figure 2.7 they discover the following:

1. Both parts are punched out one at a time on a punch press from the identical kind and gauge of sheet metal.
2. The cost of the sheet metal is $0.23 per large part, and $0.15 per small part. Metal left over from the process is sold as scrap, and $0.05 is recovered per large part, and $0.03 per small part.

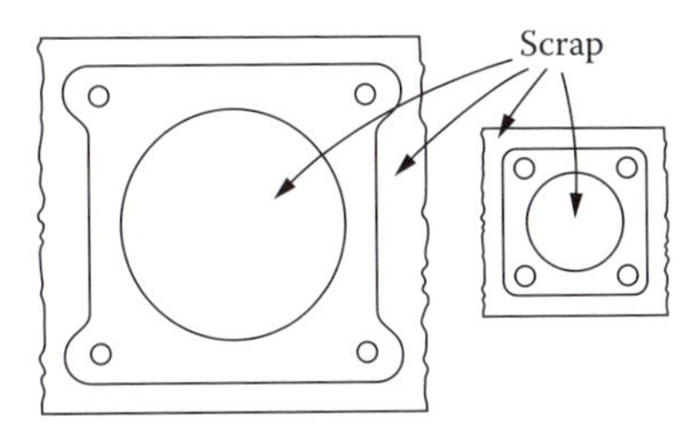

Figure 2.7 Two parts used in same product.

3. The outside dimensions of the small part indicate that it could be produced from scrap punched from the center of the large part.

The engineers each spend another 40 hours investigating how scrap from the large part can be used for the small part. Since the center piece of the large part is shaped differently than the piece from which each small part is currently formed, a new die set must be designed, which will cost \$4,000. However, in using scrap from the large part, the step of cutting sheet metal into right-size pieces for the small part can be eliminated. That step is estimated to cost about \$0.30 per part. Recovery on scrap from both large and small parts produced from a single piece of metal is estimated at \$0.04 per unit. The material cost savings per unit will thus be

$$(0.23 - 0.05) + (0.15 - 0.03) - (0.23 - .04) = 0.30 - 0.19 = 0.11.$$

If the demand for the product that uses these parts is 50,000 per year, the annual savings from material and the eliminated step will be 50,000(0.11 + 0.30) = \$20,500. If the engineering salary is \$30 per hour per engineer, then the cost of the VA study and the die-set redesign will be 80(2 × \$30) + \$4,000 = \$8,800. Thus, for a one time cost of \$8,800, the annual savings will be \$20,500.

Example 2 illustrated VA of an existing process. When VA is performed for a proposed process and is in conjunction with a new product design, it is a step in **concurrent engineering** and a part of a methodology known as **design for manufacture and assembly**, topics covered in the next chapter.

Process Reengineering

In the seminal book on the subject, Michael Hammer and James Champy define **reengineering** as "the fundamental rethinking and redesign of business processes to achieve dramatic improvement in critical contemporary measures of performance such as cost, quality, service, and speed."[7] That sounds a lot like innovation improvement, and, indeed, reengineering is probably the best contemporary example of a planned change process for achieving sweeping, innovative improvement. Process reengineering and kaizen are counterparts in improving processes—the former by quantum jumps, the latter by small, methodical steps. Unlike kaizen, reengineering never uses the existing process as the basis for improvement. Most things about the existing process (rules, procedures, structures, and systems) are discarded, and a new process is invented from scratch to replace them.[8]

Reengineering Fundamentals

Reengineering emphasizes process simplification and elimination of nonvalue-added steps. The following results of reengineered business processes are common:

- Several jobs are combined into one.
- Steps in the process are performed in natural, linear sequence.
- Workers in the process make decisions as part of their jobs.

- Work is performed where it makes the most sense in order to reduce overall process cost and time, and to improve overall performance.
- Processes have multiple versions and applications.
- Checks and controls are reduced by examining aggregate patterns rather than individual instances.

Reengineering projects are performed by cross-functional teams and with a systems perspective of the process. Unlike making small changes in an existing process where the results can be understood piecemeal, redesigning a process from scratch requires a complete understanding of the elements and dynamics of the process, which is what the systems perspective affords. Unlike kaizen projects wherein improvement suggestions originate from people within the process, most ideas for reengineering come from designers and planners outside the system. Insider involvement is important as a source of data and suggestions, and to gain commitment to the new process, but an outsider's perspective is necessary to assess how well the system performs vis-à-vis other areas of the organization.

Employee-Driven Kaizen

Opportunity for improvement is everywhere and you do not have to be a genius to find it. Still, many organizations operate as if improvements were solely the responsibility of managers, consultants, analysts, and engineers. As a result most employees do not look for places needing improvement, and even when they do see one, they do not tell anyone. When they do, few listen. This experts-only approach to improvement preconditions virtually everyone not to think about improvement.

But many times an improvement is the result of doing something very simple, and afterward people ask, "Why didn't we do this before?" As this book will show, many improvements derive from logical analysis and common sense, and once the problem has been diagnosed a solution is easy to find. Improvement does necessarily involve advanced technical thinking; in fact, such thinking might actually handicap improvement because it leads to overly complex solutions that might not work as well or be as cost effective as simpler solutions.

The most expeditious way to find improvement opportunities is to make improvement-seeking everyone's job. No matter what the job or situation, people doing the same tasks day in and day out will see alternatives that the experts overlook. When the alternatives require special analysis or technical expertise to implement, that's when the experts are called in.

One reason that Japan was able to make such significant manufacturing gains post-WWII was because Japanese workers were given responsibility for improvement, and trained and coaxed to seek out problems and resolve them. While Japanese managers and engineers were making fundamental changes to production systems (innovative improvement), factory workers were making small, continuous changes to better fit these systems to individual processes and day-to-day operations (incremental improvement)—the concept of kaizen and repetitive PDCA.

In general, workers must be given the opportunity and skills to make improvements. If they appear uncaring, disinterested, or unmotivated in seeking improvement, it is because they never before were given the opportunity, or the organization did something to turn them off. Much like the natural curiosity of children that is sometimes stifled by parents and educational systems, workers' attitudes about improvement are repressed by management practices, status barriers, and company culture.

While frontline employees should be involved in planning improvements for their own jobs, ideally they should also be allowed to assist in larger scale, strategic planning. Schonberger cites Zytec Corporation, where one-fifth of the workforce from every area of the company critiques the 5-year strategic plans and translates them into measurable monthly goals for themselves.[9]

Employee-driven improvement also requires that workers be given ownership over process data (data they themselves record, and use to monitor and improve the workplace), and be rewarded in ways commensurate with their contributions (pay, prizes, stocks, job opportunities, public recognition, or a simple thank you). The old distinction between white-collar and blue-collar workers also needs to be erased. Some companies have abolished the word *worker*, or even *employee*; they refer to frontline employees as *associates*, and managers and staff as *facilitators*. With the entire workforce looking for improvement opportunities, one might think that after a while ideas for improvement will be exhausted (the upper right of the S-curve). But in all companies products and demand keep changing and so too the manufacturing processes, and with each change comes new opportunities for improvement.

Kaizen Projects

A common approach to incremental improvement is an event called a kaizen project or kaizen blitz. Each event usually runs 2 to 5 days and is conducted by a team facilitated by an expert (person experienced in lean production and team facilitation) and led by the process owner (supervisor or manager who oversees the process). The team focuses on a particular process, its problems and wastes, which—as the term *blitz* suggests—the team "attacks."

In addition to attacking problems and wastes, a purpose of these events is to demonstrate and teach lean principles and methods. Sometimes the event is sponsored by a professional organization such as the Association for Manufacturing Excellence (AME) and is opened to outsiders interested in learning improvement methods and lean tools.

Most kaizen projects begin with a daylong kickoff meeting, starting with a presentation about the focus and scope of the project, a review of lean concepts and analysis methodology, and a tour and scrutiny of the physical facility of the process. The last point is crucial: In every kaizen event, the team collects data from direct observation and measurements of the *gemba*—the place where the work is done and improvements are to be made. Afterward the team sets measurable targets and decides on additional data it needs to further analyze the process. Over the next several days the team meets to review findings and suggest improvements.

The kaizen team is comprised of everyone directly involved in the target process (a cross-functional group of shop-floor associates, line management, and support staff) as well as staff and associates from other areas who might be able to contribute ideas or use what they learn from the event in their own areas.

The following example illustrates a kaizen project.

Case in Point: Kaizen Project at McDonnell & Miller[10]

McDonnell & Miller is the Chicago division of ITT Industries that manufactures fluid monitoring and control equipment. In the early 1980s it began adopting lean methods and philosophy, and using regular kaizen projects to identify and implement process improvements. One project was directed at a workcell and focused on reducing wastes associated inventory, transportation and material handling, and defects.

The project initiated with a 9 a.m. kickoff meeting attended by the workcell team leader; engineers from design, quality, and manufacturing who support the cell; the product line manager; a resident Black Belt (person with advanced training and experience in lean and statistical problem-solving tools); and a few team

Figure 2.8 Kaizen workshop.

leaders and support staff from other areas of the plant. The team leader opened the meeting by stating the project objectives: to understand the material and information flows in the workcell process and to change the process to enable one-piece, continuous flow production. He defined the customers of the process: the purchasers and users of the workcell's products, which is necessary to determine whether process activities are value added or nonvalue added. He then gave a slide presentation to describe the principles of value stream mapping (VSM), the tool they would use to draw out the current process, distinguish value added from waste, and conceptualize the future, ideal process. (VSM is discussed later.)

At midmorning, six operators and the supervisor from the workcell, who had already been briefed about the project, joined the meeting, and the team leader turned the meeting over to them to answer questions about details. The operators then returned to their workstations and the team leader led everyone else on a guided tour of the workcell and connected processes before and after it. To emphasize the customer's perspective, the tour began at the last step in the extended process, shipping, and ended with the first step, raw materials. The team members, clipboards and pencils in hand, questioned the operators about why this or that is necessary and what problems they face.

In the afternoon the team members returned to the conference room to discuss their findings and map out the process. As the team members described the process steps they had seen, the team leader marked the steps on little sticky notes and posted them on sheets of paper hung on the wall (Figure 2.8). After iterations of adding notes, moving them around, and connecting them with arrows, a picture emerged of the material and information flow in the workcell process.

What also emerged were many questions that required additional data to answer: cycle times, inventory levels, changeover times, machine uptimes, quality levels, and so on. The team decided it would collect the data directly from onsite observation and not use engineering or production records. Over the next few days, the team collected the data and met several more times, during which it created a detailed map of the process. It identified areas of waste on the map, developed improvement plans, and set about immediately to implement the changes. Ultimately the cell was converted to one-piece flow. This reduced the number of needed operators by half, significantly reduced inventory, and freed up workspace.

Basic Problem-Solving and Improvement Tools

Kaizen projects and other improvement efforts generally follow the PDCA procedure and rely on a set of data collection and analysis techniques called the **seven basic problem-solving (or improvement) tools**. In lean companies, everyone involved in improvement efforts is trained and skilled in using these tools—frontline workers as well as supervisors, support staff, and managers.[11] The following gives a brief description of the tools by way of an example.

Check Sheet

Improvement begins by collecting data to confirm initial perceptions and suggest a course of action. The check sheet (also called tally sheet) is a sheet where data from observations are recorded and tallied. The sheet is designed to suit a particular purpose and, hence, its content and format will vary depending on the data being collected. Figure 2.9 shows a check sheet for tracking several possible types of defects observed during the final inspection of a product.

The categories, terminology, and layout of the check sheet are carefully determined. All terms used must have clear meaning, and any that are somewhat ambiguous (such as "paint problems") must be operationally defined. In Figure 2.9, what constitutes a paint problem (e.g., the degree of paint problem) must also be defined (e.g., any perceptible paint smudging, or only smudging that shows, say, from a distance 12 inches or more).

The check sheet and its method of usage must be designed to minimize interobserver subjectivity, meaning that the results of observations recorded on the sheet would be the same, no matter who is filling in the sheet. To this end, observers are instructed about how to interpret what they observe and how to mark results on the sheet.

Histogram

The histogram graphically shows the frequency distribution of a variable. Figure 2.10, for example, shows the distribution of total daily defects recorded for 31 days of observations. The horizontal scale is divided into intervals that represent the number of defects observed in one day. The vertical scale represents the number of days, and the bars show the number of days in which a given range of defects was observed. For example, the first bar means that in two of the 31 days observed, defects numbered 0 to 4. Histograms show the spread or variability of the observations (0–39), as well as the mode or place where observations occur most frequently (15–19 defects per day).

In constructing a histogram like Figure 2.10, certain rules apply: the horizontal scale should extend over the full range of actual, observed values (in Figure 2.10, the number of daily defects observed is 0–39); the intervals must be of equal width (e.g., 5 defects); and there must not be too many or too few intervals (intervals should be sized to clearly show the distribution of the data).

Though the horizontal scale on the histogram typically utilizes numerical intervals, it can also represent non-numerical classes or categories. For example, the histogram in Figure 2.11 shows the distribution of types of defects observed over 20 days.

Figure 2.12 is a histogram showing the distribution of paint defects for each of the last 6 months. On the horizontal scale the monthly intervals have been time-ordered, and the increasing frequency indicates that the number of paint defects is on the rise.

Date Jan. 27		Product R2-D2
Shift 1		Operator/Inspector Himmelman
Defect	*Tally*	*Total*
Rough edges on body	||	2
Loose rivets on frame	||	2
Paint problems	~~||||~~ |	6
Distorted handle	||	2
Misalignment of wheels	~~||||~~ ||	7
Missing wheel pin	|	1
Cracked wheel cap	||	2
Cracked handle	|	1
Miscellaneous (specify):	|	1
Chipped trundle		
Total items inspected 60	Total defects	24
Total items rejected 9		
Special data (specify)		
Special instructions:		
Observations		

Figure 2.9 Check sheet.

Histograms usually do not suggest the causes of variation or problems. To better understand relationships in the data and find causes of problems requires other tools.

Pareto Analysis

Pareto analysis is a tool for separating the vital few from the trivial many. It is useful for deciding which of several problems to attack first. The kinds of problems that Pareto analysis seeks to identify are those relatively few problems that occur with the greatest frequency (or account for the biggest dollar loss, the greatest number of defects, the biggest headaches, etc.). A visual tool used to assist in Pareto analysis is the Pareto chart, which looks similar to a histogram, except the bars are ordered starting on the left with the bar representing the greatest frequency.

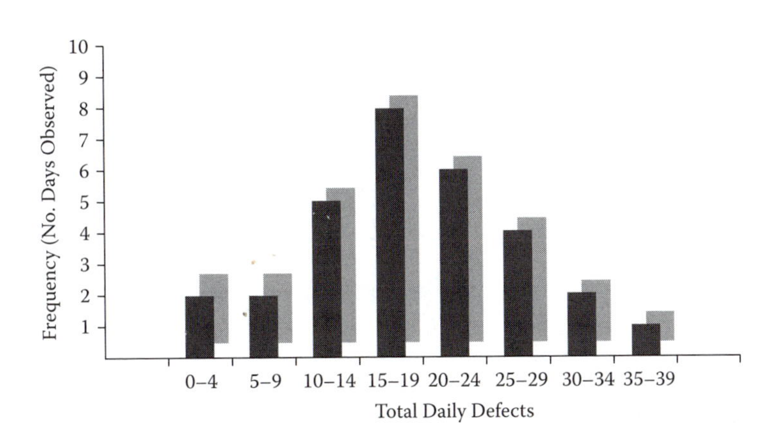

Figure 2.10 Histogram, numerical intervals.

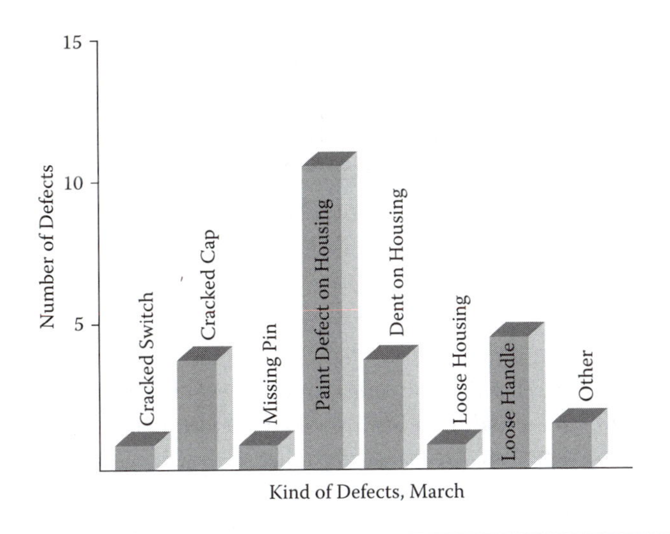

Figure 2.11 Histogram, nonnumerical intervals (classes).

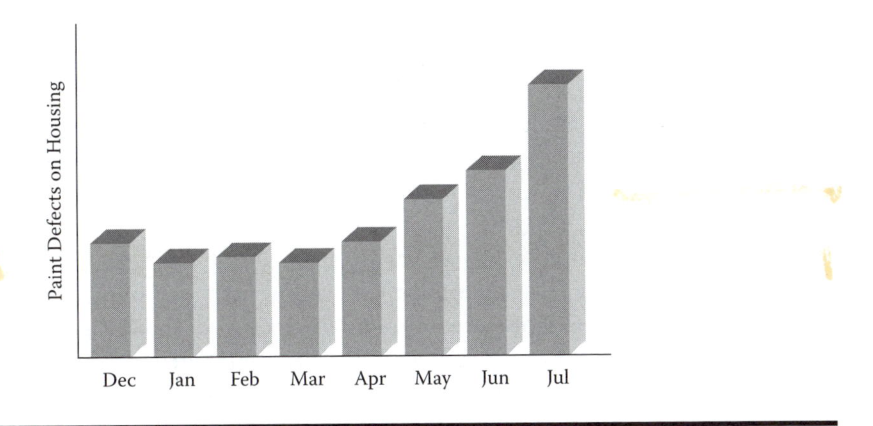

Figure 2.12 Histogram, nonnumerical, time-sequenced intervals.

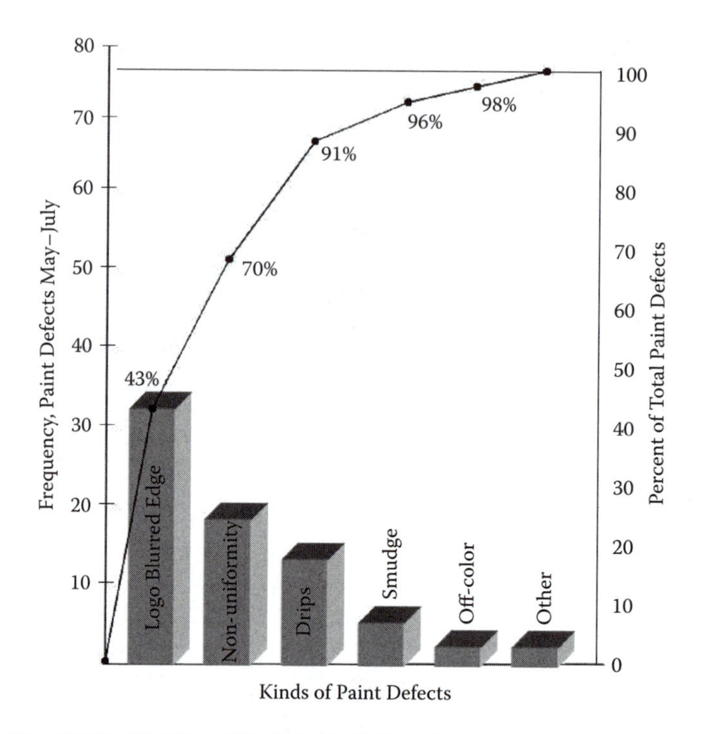

Figure 2.13 Pareto diagram with cumulative line.

Using the previous example, suppose to better understand the paint-defect problem, data are collected over several weeks of the various kinds of paint defects that occur. Figure 2.13 is the Pareto chart for types of paint defects observed over 20 consecutive shifts, and it clearly indicates that the major type of paint defect is paint blurring around the edges of the product logo. (Management thinks this is a serious problem because the logo is one of the first things customers see. If the logo is sloppy, what will customers think about how the rest of the product is made?)

The chart shows other information, too. The scale on the right of Figure 2.13 shows that, for example, of 200 paint defects, about 43% are blurred edges and 27% are nonuniform paint. The line moving diagonally across the chart shows the cumulative contribution to total paint defects by category of defects. For example, the first two categories account for 70% of all paint defects, the first three categories account for 91%, and so on.

Scatter Diagram

The scatter diagram is used to reveal possible relationships between variables. Suppose in the example that, along with defects, the plant temperature at the end of each shift was noted and also recorded on the tally sheet. The scatter diagram in Figure 2.14 shows plant temperature and number of blurry-edge defects occurring in a 40-day observation period. The plot reveals an apparent correlation between the two variables: number of defects and plant temperature rise and fall. The correlation does not say that temperature is a cause of defects or vice versa. Yet it does suggest a potential relationship that should be investigated further.

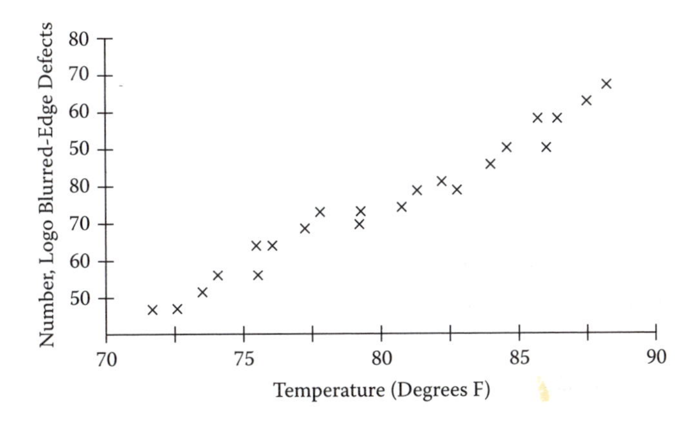

Figure 2.14 Scatter diagram.

Process Flow Chart

The process flow chart shows the steps in a process and is useful for analyzing a process to pinpoint sources of problems. Figure 2.15 shows the flow chart for the portion of the paint process that is relevant to the blurry-edge problem. It is a rather generic type of flow chart, the kind that anyone without specialized technical training might create. Many standardized conventions exist for

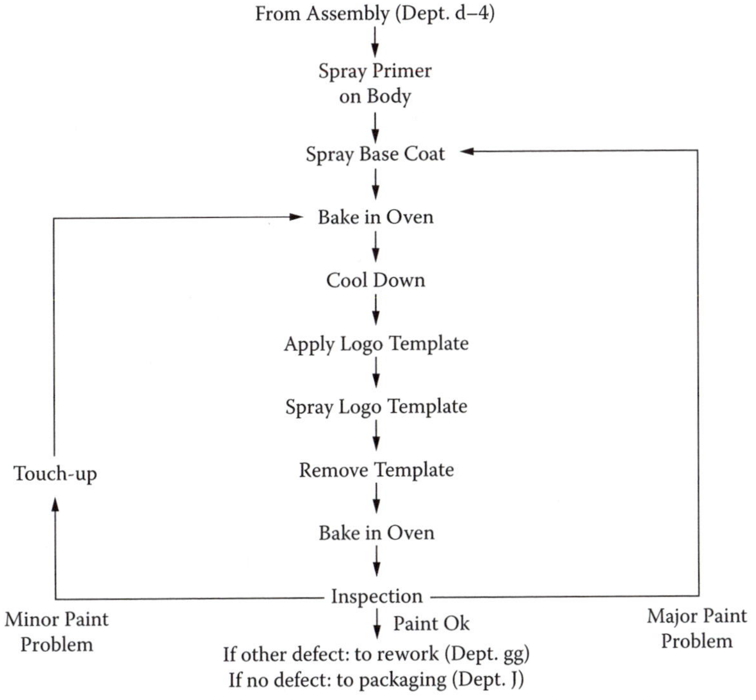

Figure 2.15 Process flowchart.

flowcharting that each result in a different-looking kind of chart. A chart created by, for example, an industrial engineer would look different than one created by a systems analyst. Since the purpose of the process flow chart is to clearly portray the steps or elements in the process and how they interrelate, the best charting technique (format and symbolism used) is whatever describes the process best. The chart should show all relevant activities, value added as well as nonvalue added. Care must be taken in deciding what is not relevant to the process and to not exclude from the chart steps or information that, in fact, are crucial for understanding the process or pinpointing causes of the problem. The amount of information and detail shown on the chart should depend on the level of detail necessary to understand the process.

Cause-and-Effect Analysis

Cause-and-effect analysis is used to identify all of the possible contributors (causes) to a given outcome (effect). A diagram called the cause-and effect diagram (or Ishikawa diagram after its originator; or fishbone diagram after its appearance) is used in the analysis. Figure 2.16 is such a diagram showing possible causes for blurred edges around the product logo. As is typical, causes are divided into the categories of manpower, materials, methods, equipment, and environment, though others can be used depending on the problem being studied.

A small team usually conducts cause-and-effect analysis for a given problem with members from different areas and levels of the organization. The team brainstorms to generate as many ideas as possible about causes for the problem. Every idea is considered, no matter how farfetched or ridiculous it might seem at first. As each idea is generated, it is categorized and recorded at the appropriate place on the diagram. To keep things organized, ideas considered as subelements of other ideas are attached at the appropriate places. For example, under the heading Materials, adhesive is shown as a subelement of template.

All ideas listed on the diagram are considered as possible root causes of the problem or as candidates for more detailed scrutiny using Pareto analysis, histograms, flow charts, and so on.

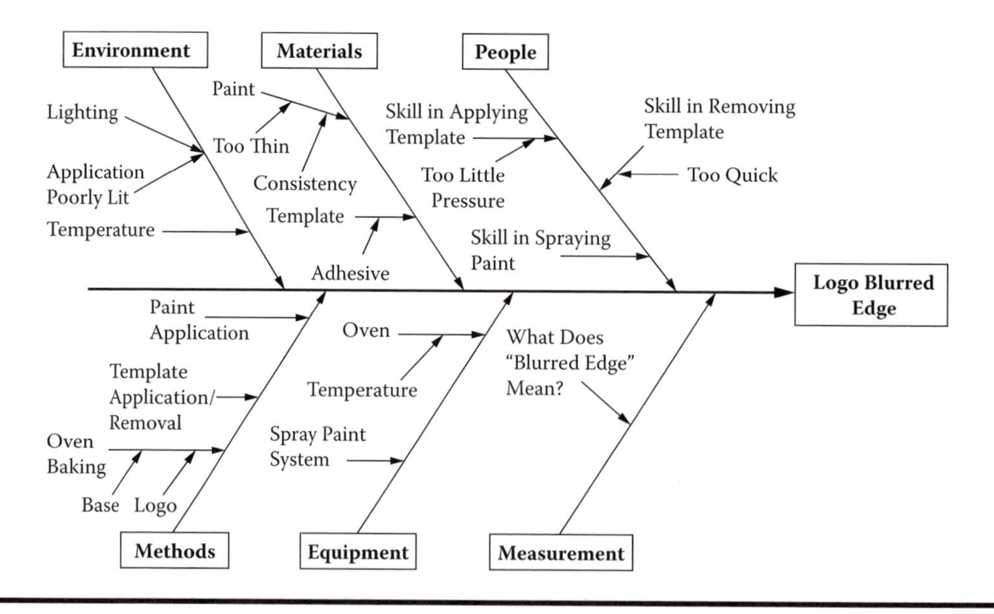

Figure 2.16 Cause-and-effect diagram.

Figure 2.17 Run diagram, two shifts.

In the blurry-edge problem, the team observes the fact that the number of defects seemed to be increasing (Figure 2.12), and that also the defects seem to increase with temperature (Figure 2.14). Let's say that the workers who applied the paint and inspected the product had been doing the same job for years, which would appear to rule out manpower or methods as potential causes. Unsure about how to proceed, the analysis team decided it needed to collect more data and use a different analysis tool.

Run Diagram[12]

A run diagram shows the results of observations taken at prescribed intervals (for example, every 10th unit, 1 unit every 10 minutes, etc.). Observations are plotted versus time to reveal any excessive or out-of-ordinary results. In the example, suppose the attribute being inspected is paint around the edge of the logo, and results are classified 0 for no blurring, 1 for slight blurring, and 2 for blurring all around. Figure 2.17 shows the results of 100% inspection for 2 shifts, each shift producing 60 units. The diagram indicates no clear pattern, which is often the case when the period of observation is short. Suppose run diagrams with 100% inspection were compiled over a 10-day period and aggregated to give average classification ratings over time. The result, shown in Figure 2.18, suggests a general increase in severity of blurred edges during the day shift and a general decrease during the night shift.

In practice the basic problem solving tools are used in combination, as needed. Returning to the example, the analysis team took the pattern in Figure 2.18 as further evidence that neither manpower nor methods were causing the blurred-edge problem. That left equipment, environment, and materials remaining as likely causes, and environment in particular because the scatter diagram had shown a possible relationship between number of defects and temperature (Figure 2.14).

The team looked again at the painting procedure (Figure 2.15) and cause-and-effect diagram (Figure 2.16), then decided to focus its attention on the materials, particularly the paint and the logo template. Closer investigation revealed the following: the template has a sticky backing to hold it in place as the paint is sprayed on, and when the template is removed, a slight amount of adhesive residue sometimes remains on the product. This was not considered earlier since the residue is difficult to see. The team discovered that the adhesive is temperature sensitive; the higher the

Figure 2.18 Ten-day average of run diagrams.

temperature of the plant, the more residue that remains on the product. When the product is put into the oven, the residue next to the logo paint slightly melts and causes the paint to bleed before it dries, resulting in a blurred edge around the logo. As plant temperature goes up, so does the amount of adhesive residue on the product and, hence, the number of blurry-edged defects. Since the average temperature inside the plant rises in the summer, so too do the number of defects. Also, plant temperature is slightly higher in the day than at night, explaining the pattern in Figure 2.18.

Why wasn't this problem noticed earlier? When the team notified the template supplier, they learned that the supplier had switched to a new adhesive in January. At the time, the temperature in the plant was low enough to cause negligible adhesive residue. Not until warmer weather did the problem show up. The supplier readily agreed to try other adhesives that would leave no residue, regardless of the temperature.

As this example suggests, finding the root cause of a problem can be, well, sticky. It also shows how the basic problem-solving tools can be a great aid in collecting and analyzing data, determining root causes, and finding solutions.

Value Stream Mapping

Besides the basic problem-solving and analysis tools, another method used in improvement efforts and especially kaizen projects is to flowchart the process using **value stream mapping (VSM)**. The VSM methodology for flowcharting uses standard icons and diagramming principles to visually display the steps in the process and the material and information flowing through it, start to finish ("dock to dock"). As the name implies, the methodology focuses on the **value stream,** which is the sequence of all activities, both value added and nonvalue added, in the creation of a particular product or service. The VSM methodology was derived from methods used by Toyota consultants in the 1980s to teach suppliers about the Toyota Production System (TPS).

VSM methodology emulates PDCA, starting with data collection to understand the current process. That understanding is used to create a process flow diagram. That diagram, called the *current state* map, is used to stimulate conjecture about opportunities for improvement and about how the process ought to look, which is the ideal or *future state.* The degree of detail shown on the map depends on the level from which the process is viewed: a view of the process from 40,000 feet will obviously show much less detail than a view from 10,000 feet of a single subprocess.

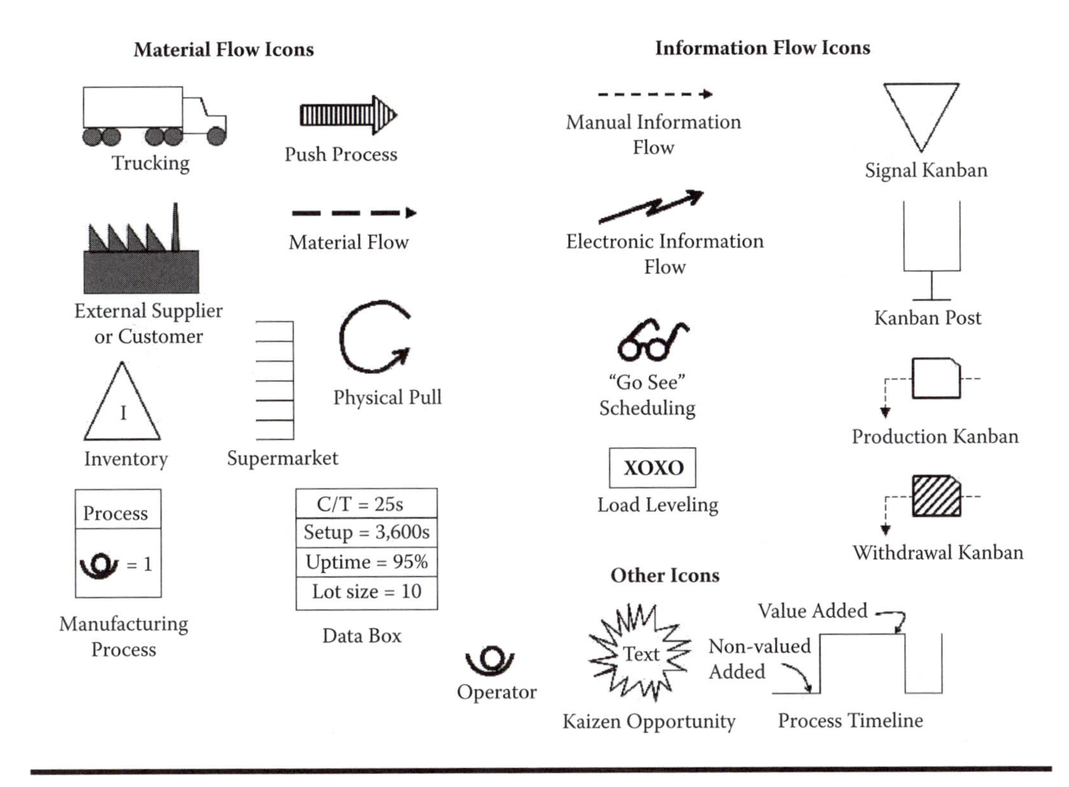

Figure 2.19 Common icons used in VSM.

VSM methodology employs some two-dozen standard icons (examples, Figure 2.19), which are fully described in books such as the one by Rother and Shook.[13] The icons represent notable features of the process such as process steps, inventory, transfer of materials, kanbans, operators, shipments, schedules, manual and electronic information flows, and truck shipments. Included on the map for each process step is information about the step such as cycle time, changeover time, uptime, scrap/defect rate, number of operators, and batch size.

Kaizen events and workshops commonly employ VSM methodology. After the kaizen team has completed a walkthrough and scrutiny of the process being analyzed (the *gemba*) and gathered essential information, the team and the leader or process owner develop a map they feel accurately represents the current process. They use the map to distinguish value-added from nonvalue-added activities. Value added is anything in the process for which the customer is willing to pay. Nonvalue added is waste to be eliminated (making allowance for nonvalue added but necessary support activities such as accounting, planning, and purchasing).

Like the other analysis and problem-solving tools, VSM provides a way to visualize and communicate facts and ideas about the process. Information necessary to create the map is collected directly from the shop floor, which leads to scrutiny of all the parts of the process and how they fit together. Collecting the data and creating the map in the workshop stimulates questions and ideas, which leads to conceptualization of the future state map. Development of the future state map is iterative: as ideas are accumulated, the current state map is modified and morphed, resulting in the future state map.

A thrust of VSM is to begin making improvements immediately. In reality, however, the time from kaizen kickoff to implementation of all planned changes for the future state might be lengthy.

In that case, intermediate future states maps are created and implemented gradually, although always to target dates. The following case illustrates VSM in a kaizen project.

Case in Point: VSM at McDonnell & Miller Company[14]

During the McDonnell & Miller kaizen event described earlier, the team created a current state map for the process. The map was posted in two places: one on a large sheet of paper hung outside the team leader's office near the workcell; the other, an electronic copy, on the computer network. Although the VSM is best developed and updated using paper and pencil, the electronic copy provided a speedy way for team members to continuously review and communicate ideas about the process.

Over several weeks the team leader added to the map additional data and suggestions he received from the team. He also added suggestions for process improvements, from which he fashioned a new map to represent the future state (Figure 2.20).

Using the future state map as a goal, the team continued to review and document various aspects of the current process, including assembly and testing procedures, tool storage, material flow, and ergonomic and safety issues. After more meetings it became apparent that one of the project's goals, continuous, one-piece workcell flow, could in fact be achieved. The future state map was finalized and intermediate implementation plans prepared. Within a year most of the plan had been implemented and the future state realized.

Consensus Building[15]

As originated at Toyota, decision making through consensus is an inherent feature of continuous improvement and PDCA. After a problem solver has prepared a plan, he seeks consensus from everyone involved with or affected by the plan. Gaining consensus helps ensure not only that the necessary perspectives have been considered, but that the plan will be properly implemented. Without consensus, plans fail for lack of understanding of the situation, resistance, or improper implementation.

Consensus does not mean that everyone completely agrees on everything; it means that everyone agrees to support the decision arrived at by the group. If parties are unable to reach consensus and a resolution is necessary, then management steps in. No opinions or suggestions are ever ignored. To parties whose ideas cannot be accommodated, reasons are explained with the hope of gaining some measure of their support.

Gaining consensus acknowledges the systems view of organizations. Accommodating a broad range of viewpoints in every plan helps avoid situations where success in one area of the organization is achieved to the detriment of other areas.

Nemawashi

Nemawashi refers to the process of circulating a plan or proposal among all the people affected by it or who must approve it. The proposal is passed back and forth among interested parties and modified to incorporate their suggestions and opinions. The interested parties include anyone who will execute, be

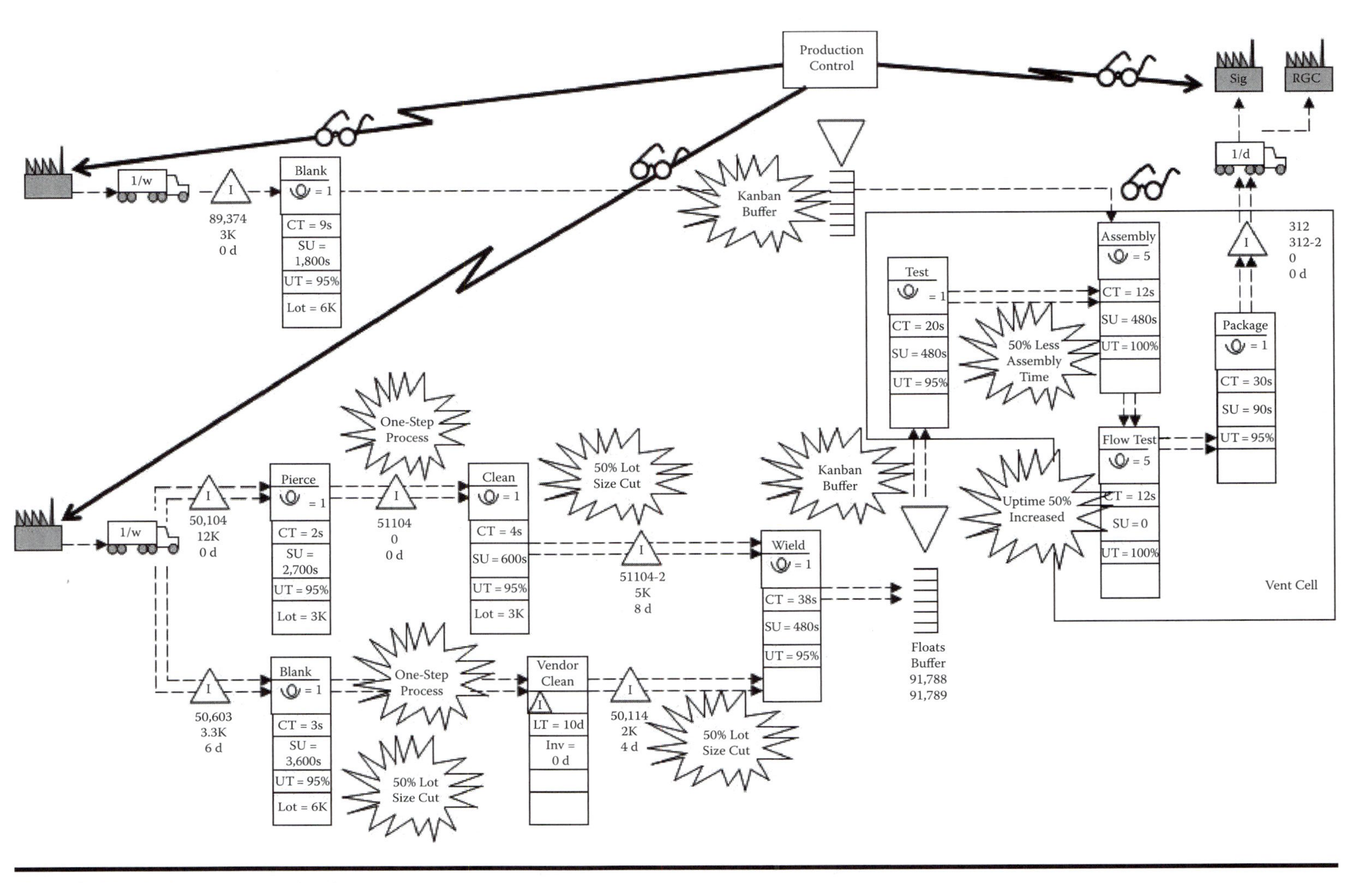

Figure 2.20 Value stream map, future state.

affected by, approve, or be able to improve upon the plan. In this way, the final formal approval is merely a formality because consensus will have been achieved and approval tacitly conveyed by everyone.

The nemawashi process is used in many ways, such as to reach decisions in the early stages of product design.[16] Each design is analyzed through study drawings, possible problems, and alternative solutions, all of which are combined in a binder. The binders are circulated among numerous people from various positions and departments, with the purpose of identifying and addressing all issues and reaching consensus on design decisions. The process is applied in similar fashion for problem solving, developing proposals, and reporting the status of ongoing programs or improvement efforts.

Pascal Dennis[17] likens this consensus seeking to playing catchball. For example, senior-level managers pass a plan or goal to the managers below them, who translate it into a plan at their level, which they toss back to the managers above them and ask "Is this what you intended?" and "Will this plan accomplish what you want?" Sometimes, then, the senior managers modify their goal or plan to accommodate the subordinates' plans. After going back and forth like this, both sides reach consensus. Next, the middle managers toss their plans to lower level managers, and the process repeats. Ultimately, the plans at every level will be interlinked with and aimed at accomplishing the goals of plans at levels above them.

A3 Report

The average person looking at nemawashi and consensus seeking in general has to wonder how anything ever gets done. Certainly all of this back-and-forth to achieve consensus must be very time consuming. In fact it could be, although as practiced at Toyota the process is handled very efficiently, in part because of the standard format used in many reports and proposals: a simple one-page document with lots of graphs and charts called an **A3**. Rather than hidden in bulky reports, everything important about the situation, analysis, and conclusions is clearly presented on a single, standardized-format page. This means that anyone looking at the report can to get to the crux of the issue rapidly.

A3 Format and Purpose

A3 is the designation for a standard 11″ × 17″ sheet of paper commonly used in Japan. An A3 report is a full report, not a memo, giving all necessary information, including data, graphs, and figures, on one side of the sheet. Toyota's widespread usage of A3 reports stems partly from Taiichi Ohno, who had a reputation for not reading reports longer than one page.

The format for every A3 is somewhat standardized, with topics listed in logical order. Upon receiving an A3 report, the reader immediately knows where to look. Unlike typical lengthy, wordy reports, there is no need to ferret out information. An A3 report shows everything, clear and simple: problem, analysis, recommended solutions, and costs and benefits. The typical A3 report includes data charts, value stream maps, fishbone and Pareto diagrams, and so on. The size, format, and content of an A3 report are designed to communicate information about problems, decisions, or plans in the simplest and most effective manner, and to serve as a tool for structured problem solving and maximizing learning.

A3 Process

The creation and use of a typical A3 report embodies PDCA and consensus seeking: a team prepares the initial A3 report stating the problem, analysis method, results, suggested solution, and cost/benefits.[18] The report is circulated among everyone who would be affected by the decision and is redone to incorporate their suggestions. In catchball fashion, the report goes back and forth until consensus is reached, of course the process being expedited because everything important—numbers, facts, opinions, and arguments—is on a single sheet of paper.

Creating an A3 requires discipline and focus. The authors must work hard to gain a deep understanding of the situation and insight about root causes and how to address them. This discipline tends to spill over and improve the efficiency of meetings, at which are people who have seen the A3 reports and, thus, are prepared, know the issues and objectives, have shared their opinions, and are ready to focus on remaining issues and make decisions.

A3 reports can be used in a variety of ways, the three most common being for problem solving; presenting a proposal; and describing the status of a plan, problem, or issue. Each of these kinds of reports corresponds to different steps of the PDCA cycle:

- A problem-solving A3 is written after the plan, do, and check steps are completed (although it must be started much earlier).
- A proposal A3 is written during the plan step but before starting the do step.
- A status A3 is written during and after completing the check and act steps.

The remainder of this section will concentrate on the problem-solving A3.

Problem-Solving A3

In addition to being a report with standard size and format, A3 also refers to a problem-solving process that mirrors the PDCA cycle described earlier.[19]

Plan

- Step 0: Problem or need perceived
- Step 1: Conduct research to understand the current situation
- Step 2: Identify the root cause
- Step 3: Devise countermeasures and visualize the future state
- Step 4: Create an implementation plan
- Step 5: Create a follow-up plan (specify how actual results will be verified against predicted outcomes)
- Step 6: Throughout steps 1 to 5 discuss with affected parties
- Step 7: Obtain approval (ensure that entire investigation and report writing process is done thoroughly and objectively)

Do

- Step 8: Execute the implementation plan

Check

- Step 9: Execute the follow-up plan; evaluate the results. If targets are met, go to step 10, otherwise return to step 0.

Act

- Step 10: Establish process standards.

The results of steps 0 through 6 are recorded on the problem-solving A3 report, the format of which embodies aspects of PDCA and the problem-solving process.[20] Figure 2.21 shows the

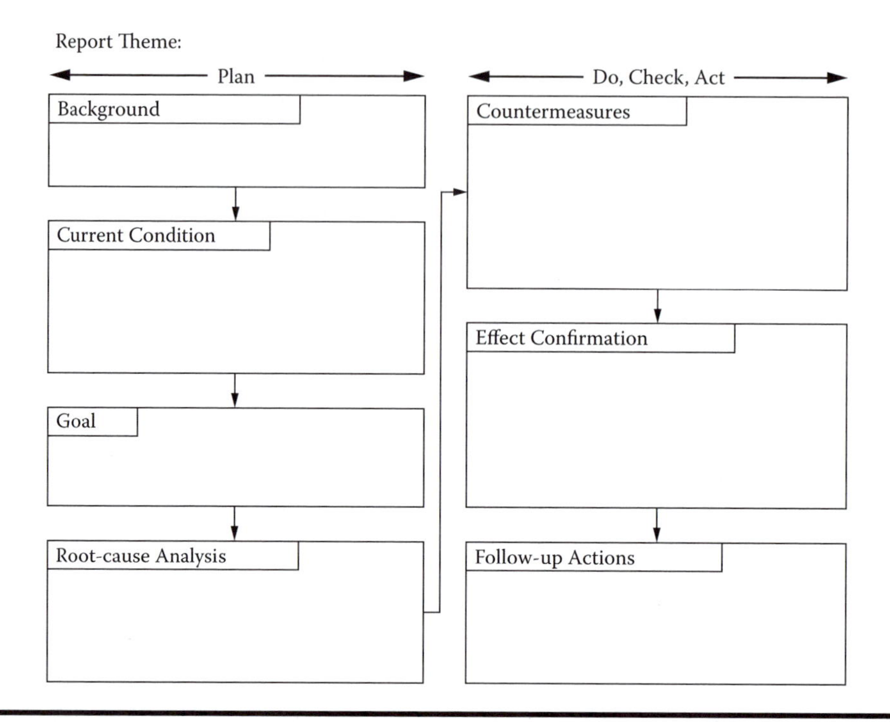

Figure 2.21 Format of typical problem-solving A3 report. (D. Sobek and A. Smalley, Understanding A3: A Critical Component of Toyota's PDCA Management System, p. 31, CRC/Productivity Press, Boca Raton, FL, 2010. Reprinted with permission.)

format of a typical problem-solving A3 report. The left side of the report is devoted to the plan aspect of the PDCA cycle, including aspects such as background and current condition, goal, and root cause analysis. The right-hand side is devoted to the do, check, and act aspects of the cycle.

As shown in Figure 2.21, the main sections of the report are as follows:

- Report title or theme: Describes the problem addressed by the report.
- Background: Provides information showing the importance or extent of the problem, sometimes by relating the problem to company goals.
- Current condition or problem statement: Describes the current condition in a way that is easy to understand; typically uses visuals (charts, graphs, tables, etc.) to highlight the importance and extent of the problem; should facilitate agreement about a course of action.
- Goal statement: Addresses what will define success after the problem is resolved and how success will be measured.
- Root cause analysis: Shows the most salient aspects of the analysis and the root cause as established through logical deduction or experimentation.
- Solutions or countermeasures: Describes solutions to counter specific root causes.
- Effect confirmation: Shows the results or effects of the countermeasures and that in fact they are effective; verifies the causal linkage between the implemented action and the observed effect.
- Follow-up actions: Specifies additional actions to be taken to ensure the improvement is sustained; addresses remaining issues to be investigated or actions to be taken in order to sustain or expand the improvement.

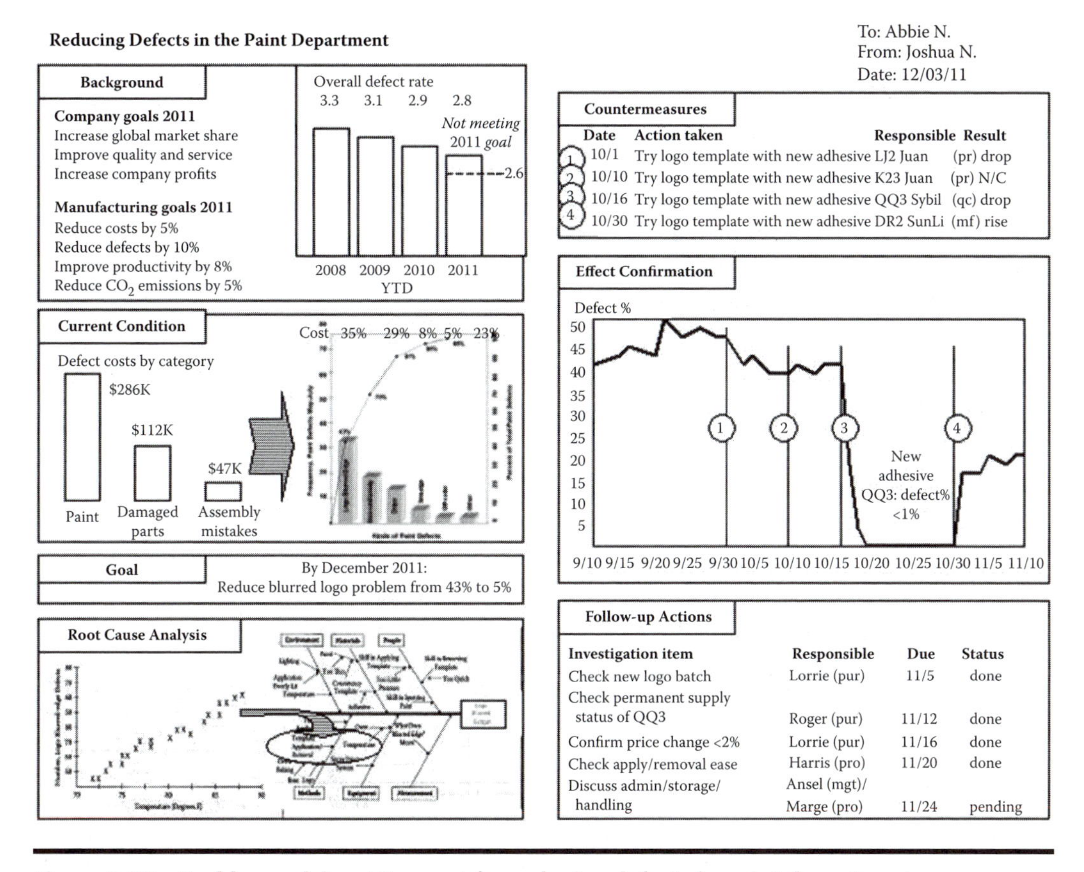

Figure 2.22 Problem-solving A3 report for reducing defects in paint department.

As an example, Figure 2.22 is the problem-solving A3 report for the blurred logo situation addressed earlier in the chapter.

Summary

Kaizen is the Japanese concept of long-term, continual improvement through small, incremental changes made one at a time. The approach takes patience, although over time many small improvements accumulate and can ultimately result in big gains. But the incremental improvement potential of every process and technology is limited either by resources or by physical, natural laws. To continue improvement beyond these limits it is necessary to innovate—to find a new, innovative way of doing things that is not subject to the same physical laws or resource constraints. Thus, continuous improvement of products and processes requires both incremental and innovation improvement.

Many methodologies and tools are used to assess and implement improvements. The PDCA cycle is a pragmatic, step-by-step approach for collecting and analyzing data, solving problems, and implementing and following up on solutions. Asking why five times is a way to get beyond symptoms and to the root causes of problems. Value analysis and value engineering are approaches to improve a product or process through questioning the value of every one of its components or steps, eliminating those with no or questionable value, and seeking value-enhancing alternatives for the remainder. Process reengineering is a form of innovation improvement wherein the old process is replaced by an entirely different one. The emphasis is on determining the most fundamental, stripped down way for a process to effectively accomplish its purpose.

Potential for improvement is everywhere, and the more eyes looking for it, the more likely that that potential will be realized. In kaizen events, frontline workers, trained and skilled in data collection and problem-solving techniques, participate with support staff and managers in resolving problems and eliminating waste in target processes. Such events often focus on improving the entire process for creating a product or service—the value stream—and employ value stream mapping methodology. Ideally, the continuous improvement draws on suggestions and gains consensus from all parties affected, a process that is facilitated by A3 problem solving and report writing.

Notes

1. For a look at these concepts as applied to product innovation, see R. Foster, *Innovation: The Attacker's Advantage* (New York: Summit Books, 1986).
2. H. Smith, *Rethinking America* (New York: Random House, 1995), 6–27.
3. See A. Robinson, *Modern Approaches to Manufacturing Improvement: The Shingo System* (Cambridge, MA: Productivity Press, 1990). Though the concept of "thinking and perceiving" is discussed on p. 88, virtually the entire first half of this book is about topics related to methods of perceiving, thinking, and PDCA as applied to manufacturing.
4. W. Edwards Deming was an American advisor to Japanese industry on quality control who introduced PDCA (which he learned from Walter Shewart) in Japan in 1954. He spoke to top executives of Japan's largest companies, many of which—including Toyota—took his advice to heart and incorporated his teachings into their business philosophies. It was not until the early 1980s that Deming's contributions to quality and management became widely recognized in the United States. Thereafter and until his death in 1993, he was a much sought-after business consultant and quality guru in the United States and elsewhere. See M. Walton, *The Deming Management Method* (New York: Perigee, 1986).
5. D. Sobek II, and A. Smalley, *Understanding A3 Thinking: A Critical Component of Toyota's PDCA System* (Boca Raton, FL: CRC/Productivity Press, 2008), xi.
6. C. Fallon, *Value Analysis to Improve Productivity* (New York: Wiley Interscience, 1971), Chapters 5–9.
7. M. Hammer, and J. Champy, *Reengineering the Corporation* (New York: HarperBusiness, 1993). See also M. Hammer and S. Stanton, *The Reengineering Revolution* (New York: HarperBusiness, 1995).
8. The term *reengineering* as applied to business processes is a misnomer since most business processes were never engineered (designed) to start with; they got to where they are in an evolutionary, sometimes arbitrary fashion. Reengineering for most business processes equates to "engineering the first time."
9. R. Schonberger, *World Class Manufacturing: The Next Decade* (New York: The Free Press, 1997), 33.
10. Adopted from J. Nicholas and A. Soni, *The Portal to Lean Production* (Boca Raton, FL: Auerbach Publications, 2006).
11. An early reference that first described all seven in one source is K. Ishikawa, *Guide to Quality Control* (Tokyo: Asia Productivity Organization, 1976).
12. Sometimes instead of the run diagram, the control chart is considered the seventh basic problem-solving tool. The subject of control charts, mentioned again in Chapter 4, requires considerable discussion and is covered in books on statistical process control.
13. M. Rother and J. Shook, *Learning to See: Value Stream Mapping to Add Value and Eliminate Muda* (Brookline, MA: Lean Enterprises Institute, 1999).

14. Thanks to Keith Kornafel, Black Belt and former team leader of the Hoffman Line, McDonnell & Miller Company, for contributions to this and the next section.
15. Portions of this section adapted from J. Liker, *The Toyota Way* (New York: McGraw Hill, 2004), 241–248; P. Dennis, *Lean Production Simplified* (New York: Productivity Press, 2002), 122–125.
16. Liker, *The Toyota Way*, 242–243.
17. Dennis, *Lean Production Simplified*, 123.
18. Liker, *The Toyota Way*, 244–248.
19. Adapted from D. Sobek and A. Smalley, *Understanding A3: A Critical Component of Toyota's PDCA Management System* (Boca Raton, FL: CRC/Productivity Press, 2010), 19–27.
20. Adapted from Sobek and Smalley, *Understanding A3*, 31–45.

Suggested Readings

J. Brown. *Value Engineering: A Blueprint.* New York: Industrial Press, 1992.

M. Imai. *Kaizen: The Key to Japan's Competitive Success.* New York: McGraw-Hill/Irwin, 1986.

S. K Kuma. *Value Engineering: A Fast Track to Profit Improvement and Business Excellence.* Oxford, U.K. Alpha Science International, Ltd., 2005.

Productivity Press Development Team. *Kaizen for the Shop Floor: A Zero-Waste Environment with Process Automation* (Shopfloor Series). New York: Productivity Press, 2002.

J. Shook. *Learning to See: Value Stream Mapping to Add Value and Eliminate MUDA.* Brookline, MA: Lean Enterprise Institute, 1999.

D.H. Stamatis. *Six Sigma and Beyond: Problem Solving and Basic Mathematics.* Boca Raton, FL: CRC Press, 2001.

J. Womack, and D. Jones. *Lean Thinking: Banish Waste and Create Wealth in Your Corporation*, 2nd ed. New York: Free Press, 2003.

Questions

1. Explain why change is an essential process for businesses to be competitive. In what ways must businesses keep changing?
2. Compare and contrast incremental improvement with innovation improvement.
3. Explain the concept of the S-curve as it relates to continuous improvement.
4. Give examples of product and process technologies not mentioned in the chapter where kaizen led to competitive advantage.
5. Give examples of new product and process technologies not mentioned in the chapter that made the older technologies they replaced obsolete, and that led to competitive advantage.
6. Explain the role frontline workers play in continuous improvement.
7. Describe the steps in the PDCA (Deming or Shewhart) cycle.
8. What is the meaning of "ask why five times"?
9. What is value analysis and value engineering? How are they related to continuous improvement?
10. What is reengineering? Where is reengineering in the context of the S-curve?
11. Describe the sequence of events and what occurs in a typical kaizen event (kaizen blitz).
12. What are the seven basic problem-solving tools? Describe each.
13. Describe value stream mapping. What is its focus and purpose?
14. Describe the consensus-building process. What is *nemawashi*?
15. What are the significant features of an A3 report? How does creating an A3 report fit into the PDCA cycle of improvement?

PROBLEMS

1. Analysis of the cost of a product reveals the following sources:

Materials	40%
Direct labor	10%
Overhead and administration	50%
	100%

You are considering ways to reduce the cost of this product. Where should you begin?

2. Following is the number of seconds customers have to wait for a service representative. Create a histogram of the data using 4 seconds as the range of each interval (0–3, 4–7, etc.). What does the data indicate?

7 8 2 8 4 10 5 7 7 15 21 8 18 14 5 15 22 10 6 10

3. A tally of customer complaints shows the following:

Shipping errors	966
Billing errors	2,070
Delivery errors	540
Ambiguous charges	9,880
Delivery delays	7,430

Construct a Pareto diagram. What do you conclude?

4. A group of machine operators suspect that the machine speed affects the defect rate. To test this they keep track of the number of defects in same-size batches produced at different machine speeds. The results are:

Machine Speed (rpm)	*Number of Defects*	*Machine Speed (rpm)*	*Number of Defects*
1,900	10	2,300	9
2,450	17	2,550	16
1,800	12	2,150	6
1,850	14	1,950	7
2,000	6	2,100	6
2,350	15	2,400	12
2,200	7	2,250	7

Plot a scatter diagram. What does the diagram indicate?

5. The following numbers are from a tally sheet:

Total deliveries observed	1,860
Total deliveries with problems	204
Delivery problems	
1. Late delivery	120
2. Early delivery	12
3. Shipment batch too large	57
4. Shipment batch too small	56
5. Excessive defects in shipment	13
6. Wrong items delivered	4
	262

a. Create a histogram showing relative frequency of problems.
b. Modify the histogram into a Pareto diagram.
c. Why is the sum of the delivery problems, 262, less than the number of deliveries with problems, 204? In addition to the listing of problems shown, what other information about the problems listed should have been gathered? Design a tally sheet to collect information about the delivery problems that would be more useful than the simple listing shown for this problem.
d. If you wanted to find solutions to delivery problems, where would you begin? What additional data would you collect?

6. Draw a flow diagram for each of the following processes:
   a. Withdrawing money from an ATM machine.
   b. Programming a DVR to record a one-time broadcast.
   c. An entire day spent downhill skiing (include subprocesses like buying lift tickets, renting equipment, going uphill/downhill, returning equipment, etc.).
   d. Any process with which you are familiar.

7. In each of the processes in problem 6, which steps would you concentrate on to improve the overall process. Explain your criteria for improvement, and suggest what you would do to the steps.

8. Consider the following situations:

   - You are late to work (or school, a meeting, etc.)
   - You are painting a ceiling and the paint is dripping on your face.
   - Your average grocery bill is twice your neighbor's.
   - You make coffee every morning, but it tastes lousy.
   - A person who doesn't know you but is an important business contact does not return your phone calls.
   - You buy a new appliance, and it won't work.

   a. Draw a cause-and-effect (fishbone) diagram.
   b. Explain which of the causes from the diagram you would look at first to solve the problem.

9. Select a simple product (e.g., pen, corkscrew, tape dispenser, etc.). Analyze its components and consider for each what it does (function), what it should do, and how it might be modified or replaced to do the same thing for less cost.
10. In a customer service department, the process for handling complaints is as follows:

    Representatives list complaints over a 12-hour period. The complaint lists are collected from each phone representative at the end of the day, and early the next morning are reviewed by the manager, who sorts them by severity. Severity is determined by the customer's demand for service (immediate or not), and by the nature of the complaint, that is, whether it is (1) a problem that requires an immediate solution and follow-up call, (2) a problem that requires solution and follow-up, but can be delayed, or (3) a situation that can be handled with a letter.

    If the complaint regards a technical problem, the manager decides which technical specialist is best qualified to handle it, then forwards the complaint to that person. If the complaint is informational and does not require a technical solution, the manager directs the complaint to a person who prepares a letter thanking the caller for the compliant. In all cases, if the caller demands an immediate response (whether the complaint is technical or informational), a copy of the complaint is made and sent to a person who calls the complainer that day or as soon as possible. Before she calls back the complainer, if the complaint is about a technical problem, she calls the technical person assigned the problem to determine if a solution is close at hand, and if not, how long it will take. She then calls the customer.

    After a technical specialist solves the problem, he phones the customer. For serious problems that require replacement parts, before calling the customer, the specialist contacts the company warranty specialist to determine whether parts should be sent to the customer free or for a charge, and how much the charge would be. For all complaints that involve a technical problem, a letter is sent two weeks later asking the customer about whether the problem was solved, and the customer's satisfaction with follow-up calls and service.

    a. Draw the flowchart for this process.
    b. Suggest opportunities to improve the quality of service.

11. The president of Zemco Plastics Company reads in a business journal that his company's closest competitor is spending about three times as much as Zemco on R&D for a plastic that both companies currently produce. From information in the article, he concludes that, despite this, the competitor apparently has not gained any discernable technological or profit advantage over his company in that particular plastic. What might Zemco's president conclude about the plastic?

12. Cylo Electronics has two production divisions. In terms of sales volume, cost, and profitability, the two are currently about equal, though the product lines and production processes of each are very different, and Division A has been operating for about 10 years, whereas Division B, only about 4 years. The CEO of Cylo notices that over the last few years productivity improvement at Division A has been poor, and several product development projects came in very much over budget and over schedule. Division B, in contrast, has been able to continuously reduce its production costs and has been successful in all of its developmental projects. In hopes of improving the situation at Division A, the CEO is considering transferring several product and process designers and engineers from Division B to Division A. Comment.

Chapter 3

Value Added and Waste Elimination

> The only things that evolve by themselves in an organization are disorder, friction, and malperformance.
>
> —**Peter Drucker**

> If it doesn't add value, it's waste.
>
> —**Henry Ford**

A cornerstone of continuous improvement is the concept of **value added** and that anything in an organization or process that is not value added is considered **waste**. A goal of lean production is to identify and eliminate waste. This is done through adherence to a set of guiding **lean principles**.

The first two lean principles, **simplification** and **cleanliness and organization**, are the most rudimentary of improvement approaches. Simplification implies elimination of nonessentials; cleanliness and organization imply thoroughness and attention to detail. The third principle, **visibility**, ensures that frontline workers have the right information and stay informed through simple observation; they know goals and status in operations by just "seeing" it. The fourth principle, **cycle timing**, is the idea that production output should be uniform yet closely coincide with demand. Lean methods seek to achieve the seemingly inconsistent goals of meeting fluctuating demand while maintaining uniform production output. **Agility**, the fifth principle, is a manufacturer's ability to switch over products and processes as customers and markets dictate. The sixth principle, **measurement**, is fundamental to PDCA (plan–do–check–act) and is the means by which improvement is gauged. It is also part of every value-added, waste-elimination, and improvement effort. The final principle, **variability reduction**, refers to continuing efforts to reduce process variation, a prime source of waste and contributor to poor quality, cost, and time performance.

This chapter covers the concepts of value added and waste, kinds of wastes, and the aforementioned guiding principles behind lean production. It concludes with a discussion of lean production philosophy, perceived limitations, and implementation issues.

Value-Added Focus

Value added is the concept that every activity and element of a system (materials, humans, time, space, or energy) should add value to the output of the system. As such, it provides perspective for determining what needs improvement in business processes.

The value-added concept also relates to the earlier discussion about what constitutes real improvement. In particular, improvement should be applied only to valued-added and necessary nonvalue-added activities (concepts described next). If you try to improve some aspect of a process that is unnecessary in the first place, then you are wasting your time.

Necessary and Unnecessary Activities

Combining materials to form a product or performing a service for a customer are examples of value-added activities. These activities directly add value to the output, whether product or service. In contrast, a task such as processing a purchase order is a nonvalue-added activity, even though it is necessary for doing value-added activities. The value-added concept classifies activities as either value added or nonvalue added. Within the latter category, the absolutely necessary activities are separated out, and all the others are candidates for elimination.

Distinguishing necessary, nonvalue-added activities from the unnecessary, wasteful ones is tricky because unnecessary activities in organizations often seem necessary. Purchasing-type tasks are necessary because they procure the materials needed by value-added activities for transformation into the final output. Activities such as inspecting incoming parts for defects or counting materials in inventory also seem necessary; inspection prevents defective parts from going into the product (a valuable endeavor), and counting ensures that inventories are kept at the right levels (also valuable). The fact that an activity fulfills a valuable purpose, however, should not be confused with its adding value. For inspection and counting, alternatives exist that would obviate the need for either of them. For example, by requiring vendors to deliver only zero-defect parts, the need for incoming inspection is eliminated. By using production procedures that limit inventory levels, the need for counting is reduced or eliminated. Many such valuable purposes can be fulfilled in different ways without necessitating the preservation of a nonvalue-added activity.

In summary, the value-added concept says to distinguish value-added from nonvalue-added activities. Among the latter activities, seek out the ones that are unnecessary and try to eliminate them. The remaining activities, which are the value-added and necessary nonvalue-added ones, then become the focus for improvement.

Support Organization

To distinguish value-added from nonvalue-added activities, it is useful to think of an organization as comprised of two organizations: one, the **production organization**, makes the product or provides the service; the other, the **support organization**, assists and supports the production organization but does little that qualifies as value added. In common parlance, the production organization is called the **line** (or **frontline**) and the support organization is called the **staff**. In actuality, the two blend together without regard for work function, job level, or job category, and a person's job might readily involve doing things in both organizations.

In many firms, the support organization accounts for a significant proportion of total organizational costs in the form of overhead. Often this cost exceeds the cost of the production organization by a wide margin.

The number of activities within the support organization can be quite large and most of them, though necessary, are nonvalue added. The purchasing, inspection, and inventory counting tasks cited earlier are examples. The following categories give an idea of the expanse of these activities:[1]

- Planning, control, and accounting activities: Forecasting, production planning and scheduling, purchasing, master scheduling, requirements planning, production control, customer order processing, order tracking and expediting, responding to customer inquiries, and all associated data entry, bookkeeping, data processing, and follow-up on errors.
- Logistical activities: All ordering, execution, and confirmation of materials movement within an organization, including everything associated with receiving, shipping, work orders and expediting, as well as data entry and processing, and follow-up on errors.
- Quality activities: All quality-related work such as definition of customer requirements, assurances that necessary activities have occurred, defect prevention, quality monitoring, and follow-up on defects, mistakes, or complaints.
- Change activities: All revisions and updates to other activities, including, for example, customer orders, product designs, and planning and control systems. Change transactions have a multiplier effect: a change in a product design usually requires changes in material requirements, bills of material, and product routings, each of which requires data entry and processing activities to implement them.

Support activities can often be eliminated by simplifying products and processes, eliminating defects at the source, improving steps to remove mistakes and duplication of effort, and improving product design and production planning to reduce the number of changes. Lean management practices that eliminate the need for support activities will be discussed throughout this book.

Employee Involvement

While the value-added approach seems simple in concept, it can be difficult to apply. Within a particular job or task, there may be only subtle differences between necessary and unnecessary activities. The person best qualified to make the distinction, then, is the one most familiar with the task—the person doing it. Frontline workers usually know what is essential and what is not, and given the opportunity, they will share that knowledge. Worker involvement is thus fundamental to improvement efforts.

On the other hand, few people will point out unnecessary portions of their work if they think it will jeopardize their job or continued employment. While the value-added process requires that workers scrutinize their jobs and suggest ways to remove unnecessary activities, the process must not threaten workers' job security. Getting employees to participate when the covert goal is to eliminate jobs will have only short-lived success and probably preclude any future, meaningful participation from the employees.

Historical Note: Gilbreths and the One Best Way

Among the earliest proponents of eliminating waste were the husband–wife team of Frank Gilbreth (1868–1924) and Lillian Gilbreth (1878–1972). They thought that the role of management is to find the simplest, easiest way to do the job. Their philosophy of "work smarter, not harder" meant that every task should be carefully studied and all wasted motion eliminated to arrive at the *one best way* to do the job. During their lifetimes, the Gilbreths applied this philosophy to almost every conceivable kind of work. The Gilbreths owned a construction company that specialized in speed building. By applying the one-best-way philosophy to the task of bricklaying, they reduced the number of basic motions for laying a brick from eight to six. At a time when bricklayers were laying about 500 bricks a day, the bricklayers for the Gilbreths averaged 2,600 a day.

Though important contributors to management thought, the Gilbreths could hardly be included among the founders of modern lean production since the latter goes far beyond their way of thinking and in some instances stands counter to it. The Gilbreths were interested primarily in one particular kind of waste, waste of motion. As we will discuss later, there are many other sources of waste. Also, in any process there is no such thing as one best way. If there ever were, it would only be temporary and later superseded by a still better way (the idea of continuous improvement). Also, lean production stands by the premise that seeking out waste and finding better ways is the responsibility of everyone, not just managers.

Sources of Waste

When all the obvious sources of waste have been removed, continuous improvement efforts switch to searching for the hidden sources.

Toyota's Seven Wastes

One contribution of Toyota Motor Company to modern manufacturing is its strong advocacy of eliminating waste (called *muda*) as a strategy for continuous improvement. Toyota defines waste as anything other than the minimum amount of materials, equipment, parts, space, or time that are essential to add value to the product. Though sources of waste vary within and across organizations, the similarities are great. The following sources of waste, the mudas identified by Toyota and first described by Taiichi Ohno, are universal in most every organization:[2]

- Waste from producing defects
- Waste in transportation
- Waste from inventory
- Waste from overproduction
- Waste of waiting time
- Waste in processing
- Waste of motion

Producing Defects

Defects in any product or service are a major source of waste. Consider defects not remedied by the producer and discovered by a customer. The costs of these defects include warranty or reparation expenses assumed by the producer, aggravation of customers, and loss of existing and potential customers who hear about the defects.

Product defects are ideally detected and remedied before products go to a customer; however, detecting and fixing defects are themselves wasteful and costly activities. The simple expectation that defects will occur requires that producers devote time and resources to inspecting items and sorting out defectives. Defective products accrue additional labor and material expenses related to disassembly and rework. For those items that must be scrapped, all of the labor, material, and resource expense of producing them is wasted. Products with minor defects might be usable, but must be sold as "seconds" and at reduced prices. Product defects delay production and increase production lead times. Multiply the efforts to find and correct defects by how often they must be done to eliminate defects and the result is a large amount of wasted labor, material, and other resources. This is all unnecessary if everything associated with products or services was **done right the first time**.

Transportation and Material Handling

In many organizations, the items (people, products, parts, supplies, etc.) being processed or serviced must be moved from one location to another over several stages. Two factors determine the distance through which items must be moved and the transportation means (conveyors, carts, forklifts, overhead cranes, etc.) to move them: the **layout of the facility** (the location of machines, desks, departments, reception areas, shipping and receiving docks, and so on) and the **routing sequence of operations** to produce or service the items. For example, Figure 3.1 shows the layout and routing sequencing for processing three items (products 1, 2, 3). The facility could be a factory or an office that processes many kinds of items. Letters inside the facility represent equipment to perform the operations. Notice in the figure the overall distance through which the items must move in the course of the process. In many organizations this distance can total miles, and the time involved is very large. Since typically no work is performed on items while they are being moved, time spent en route is wasted. All equipment and labor involved in moving and tracking the items is costly and wasteful, too.

Figure 3.2 shows an alternative equipment layout, part of which is devoted solely to the three items. By rearranging the layout and putting equipment for sequential operations close together, the distance through which the items move is a fraction of the previous distance. The time to move the items and the cost of systems to move and keep track of them, as well as space required for the processes, are been substantially reduced.

Inventory

Toyota calls inventory the root of all evil.[3] That is a strong statement meant to imply that wastes stemming from inventory go far beyond the items held in stock. Inventory represents items waiting

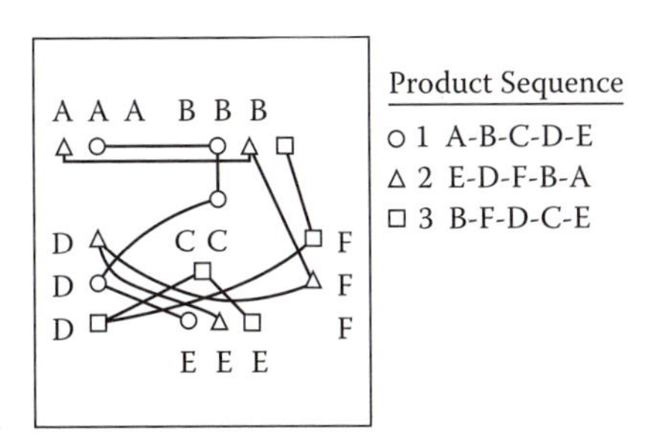

Figure 3.1 Routings for three products.

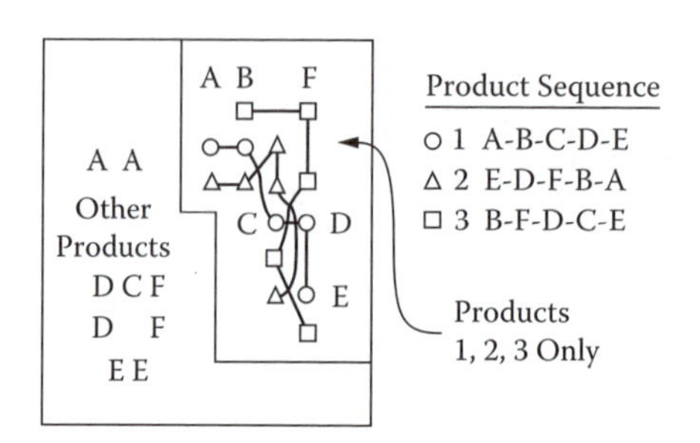

Figure 3.2 Alternate layout.

for something to happen, a waste in that there are costs associated with items waiting and lost time since no value is being added to them. Inventory holding costs increase with the size of the inventory since it costs more to hold more. Holding costs include the charges for storage space, paperwork and handling, insurance, security, and pilferage. Since the capital needed to acquire and produce the items in inventory cannot be used elsewhere, there is an opportunity cost as well. If the inventory comprises items procured with borrowed funds, there is also an interest expense. The sum holding costs among all inventories carried by an organization throughout a year can be sizable.

Inventory is also considered evil because it covers up other kinds of wastes and encourages, or allows, wasteful practices. Inventory has been called a just-in-case management philosophy, meaning that managers use inventory as a hedge against things that might go wrong. While they recognize the costs of inventory, they also think of inventory as necessary to overcome other kinds of problems. What they fail to see are alternatives for dealing with these problems. Three such scenarios follow.

1. Inventory is carried so that material flow will be uninterrupted in the event of equipment breakdowns or delivery delays. Preventive maintenance programs and close customer–supplier working relationships can eliminate most equipment breakdowns and delivery delays, which would obviate the need for protective inventory.
2. Inventory is carried to cover defects in materials and finished products. Making suppliers responsible for the quality of their products and improving product quality through better product design and production processes can eliminate defects at the start, which would make inventories to cover defects unnecessary.
3. Large inventories result from large production runs, which managers say are necessary because of time-consuming and costly production setups. If production setup methods were improved and the setup costs were reduced, then small batch production would be economical. The by-product of smaller production runs is smaller inventories.

We can use the analogy of a ship on water to clarify the point. As Figure 3.3 illustrates, a high water level makes it unlikely that a ship will encounter the rocks below. When the water level is lowered, the rocks begin to be exposed and care must be taken to guide the ship around them. Inventory is analogous to water level: high inventory covers up problems in the system and allows management to cruise without fixing them. As inventory is lowered, problems in the system (poor forecasting, poor maintenance, long setups, poor product design and quality control, etc.) are exposed, and management has to resolve them in order for the system to work. In lean production inventory reduction is not an end in itself; it is a device for exposing problems and wasteful practices in the production system.

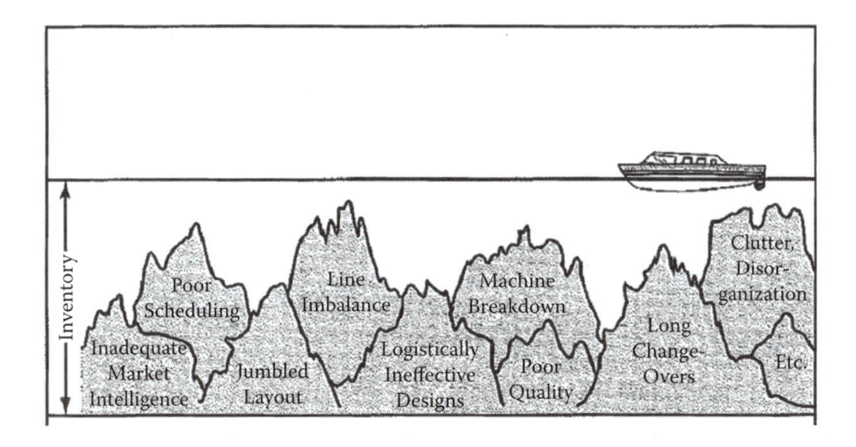

Figure 3.3 Inventory as a way of avoiding problems (when the inventory tide goes out, the skipper must carefully navigate between the rocks or find ways to eliminate them).

Overproduction

Companies sometimes produce more than they have sold or might sell because they want to build inventories (for reasons given earlier) or because they want to keep their equipment and facilities running (to achieve high-level resource utilization). Whatever the reason, making products for which there is no demand is wasteful. If demand does not materialize, then at some time the items will have to be discarded or disposed at reduced price. In the meantime they are held in stock where they accrue all the costs and wastes associated with inventory.

Waste from overproduction is difficult to identify, and unless you compare what is produced with what is sold and shipped, nothing appears wrong. In organizations that habitually overproduce, everyone is busy, and when everyone is busy no one has time to scrutinize what is happening or see what is wrong.

Waiting

Unlike waste of overproduction, waste of waiting is easy to identify. It takes many forms, including waiting for orders, parts, materials, items from preceding processes, or for equipment repairs. It also occurs in automated processes, as when an operator loads and turns on an automatic machine, then watches and waits until the machine is finished.

Some companies pride themselves in minimizing waste of waiting with a policy of keeping workers busy and machines running, regardless of demand. In other words, they overproduce. This practice replaces one waste (waiting for demand) with a worse waste (overproduction) since shutting down machines and idling workers on occasion is less costly in terms of material, equipment, and overhead than producing inventories for which there are no orders.

Processing

A process may itself contain steps that are ineffective or unnecessary. Take, for example, a product that goes through two steps: cutting, then filing to remove burrs along the cut edge (Figure 3.4a). This process might be altered to reduce wasted time and steps. Automatic filing of the edge is more efficient than manual filing (Figure 3.4b); still better is periodic maintenance or replacement of the

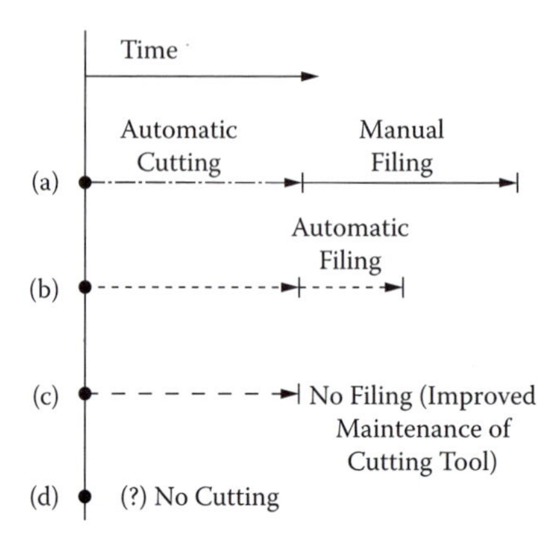

Figure 3.4 Paring waste from a cut and burr-removal process.

cutting tool so it gives a smooth edge that does not need filing (Figure 3.4c). The item might even be redesigned so the cutting operation is eliminated (Figure 3.4d).

Taking advantage of natural forces such as gravity can eliminate processing waste. Instead of having a worker remove a finished part from a machine and put it in a bin, the part can be disengaged automatically from the machine and fall down a chute into the bin.

Motion

People in work settings often confuse being in motion with working. In reality motion and work are not the same. For definitional purposes, **work** is considered a particular kind of motion that either adds value or is necessary to add value. A person in constant motion throughout the day (i.e., a busy person) may in actuality be doing little work. Motion that is not necessary to do the work is considered waste. A useful concept for identifying waste of motion is **work content**, or the proportion of all motion in a job that is actually considered useful work:

$$\text{Work content} = \text{Work/Motion}$$

For example, a job that takes 10 minutes but involves 6 minutes of work and 4 minutes to pick and place tools and materials has 60% work content.

For any job the goal should be to achieve work content near 100%. This is attained by eliminating wasteful motions, not by increasing work. Among wasteful motions in jobs, the most common are searching, selecting, picking up, transporting, loading, repositioning, and unloading. These motions take precious time and increase the cost, but do not add value.

Although it is common for companies to increase output by increasing the number of workers or number of hours worked, a better way is to attack wasted motion. For example, suppose the work content of a busy worker is 50% (half of the worker's time is spent on wasteful motion) and his daily output is 10 units. To double the output, the company might put on an additional worker or ask the current worker to double his hours. An alternative, however, would be to examine the worker's job and try to eliminate all the wasted motion. This, in effect, would double the work content to 100% and double the worker's output. The worker would still be busy, but he would be busy doing only useful, value-added activities.

Canon's Nine Wastes

Toyota's seven wastes emphasize factory wastes. Canon Corp. uses a broader classification scheme of nine wastes that could be applied even to service companies:

- Waste caused by work-in-process
- Waste caused by defects
- Waste in equipment
- Waste in expense
- Waste in indirect labor
- Waste in planning
- Waste in human resources
- Waste in operations
- Waste in startup

Classifying wastes is useful because it is easier to focus on particular wastes than to try to attack everything at once. A company can begin eliminating waste by using the seven (or nine) waste categories, then tailor the categories to better suit its purposes and programs.

Lean to Green

Efforts to eliminate waste benefit not only the company but the environment as well. Generally, reduced waste and improved operations within an organization spill over to reduce environmental damages associated with doing business. Waste reduction efforts result not only in less landfill waste from reduced use of raw materials, but also less air and water pollution from reduced manufacturing by-products, and lowered carbon emissions from reduced energy usage. So-called **lean to green programs** specifically aim to reduce both production-related wastes and the environmental and consumption wastes to which they are linked.

Although our focus is primarily on waste in manufacturing, waste reduction efforts to achieve farther-reaching cost and environmental ramifications must of necessity extend beyond this perspective. **Total waste reduction** involves a total product life-cycle way of thinking, starting with a product's design and ending with its disposition at the end of its operational life.[4] It means designing, manufacturing, and distributing products, and, in general, doing business in ways that utilize less materials and less energy, and a high proportion of recycled materials. The focus is on recycling resources rather than extracting and discarding them after use, and reducing the total resources needed. A design philosophy called **design for environment** (DFE) emphasizes environmental consideration in product and process design.[5] DFE principles include

- Minimize usage of hazardous and bulky materials as well as materials that involve energy-intensive methods of production
- Maximize usage of materials that are recyclable and environmentally friendly
- Design products for ease of repair and reuse so they are not readily discarded
- Design products for ease of disassembly after disposal

Design for disassembly considers how the product will be torn apart at the end of its useful life. The matter of disassembly is becoming more important in design as interest grows in separating out components for reuse and recycling.[6]

Many organizations have developed schemes for recycling scrap and for becoming recyclers of materials they use in their products. BMW and other automakers are designing automobiles wherein entire subassemblies can be separated from the vehicle and recycled. In 2007 Xerox manufacturing operations generated 21,000 metric tons of nonhazardous solid waste (consisting of paper, wood pallets, plastics, corrugated cardboard, scrap metal, waste toner, batteries and lamps, and trash) of which 81% was reused or recycled.[7] Recycling efforts at its 10 largest U.S. and European sites have reduced waste by more than 25,000 tons.[8]

In lean to green programs, fundamental wastes (such as Toyota's seven) are translated into environmental waste categories such as energy and carbon footprint, water, and raw materials. For example, overproduction and defects result in overconsumption that affects all categories of environmental waste. Excess finished goods inventory results in energy waste (warehouse heating and lighting) and material waste (packaging).[9] The lean to green philosophy will grow in importance as organizations accept responsibility for reducing consumption of all resources and curbing global warming.

Lean Principles

Lean principles are a set of beliefs and assumptions that drive operational decisions and actions about products and processes. They address general issues about what a company should do in terms of product and process improvements. In general they are worthwhile prescriptions for the conduct of any organization or business, lean or otherwise.

Simplification

In virtually any work situation, an action to reduce waste will result in a simplification of whatever existed before. The converse is also true; taking action to simplify something usually results in a reduction of waste. Ideas about how to simplify have been around for a long time; consider, for example, the principles of motion economy developed by Frank and Lillian Gilbreth in the early 1900s. Traditional methods like time-and-motion analysis have long been used to improve efficiency of tasks and processes. In general, however, simplification efforts extend far beyond removing wasted motion or redesigning predefined tasks. They also extend to simplification of products and services, as well as the overall processes and individual procedures involved in providing these things. Often the best approach is to simplify both the product and the process simultaneously.

Product, Process, and Procedure Simplification

Simplification means accomplishing the same ends but in a less complex, more basic way or with fewer inputs. For any system, whether product or process, this means critically scrutinizing the components or elements of the system with an eye toward combining, streamlining, or eliminating them. Simplification also means cutting out or cutting down on features that do not add value. Consider some examples.

Example 1: Product/Process Simplification

The component in Figure 3.5a is assembled from three kinds of parts—A, B, and C. Part C is a casting (a part formed by pouring molten metal into a mold, solidifying it by cooling, then popping it out like an ice cube from a tray), and the decision was made to alter it into the casting, D, shown

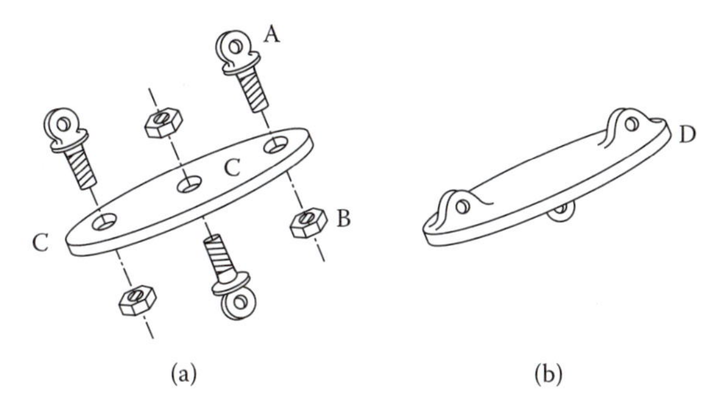

Figure 3.5 Product/process simplification.

in Figure 3.5b. Though the new casting is slightly more expensive to produce than the original, it eliminates the need for Parts A and B. As a result, the assembly operation is eliminated, and since the number of parts is reduced from three to one, costs associated with parts procurement and storage (ordering, inventory, bill of materials, inspection, etc.) are also reduced. Further, if the bolts in the original assembly had to be aligned, every assembly risked an alignment error. Redesign eliminates the possibility for this kind of error.

Example 2: Process Simplification

A part must be processed on two automatic machines, A and B, each run by an operator. Figure 3.6a shows two timelines, one for each machine–operator. As soon as the operation at A is completed, the part is transferred to B, as indicated by the wiggly line. The operators load and unload the machines but otherwise are idle while the machines are running. The machine cycle time (CT) for operation A is slightly longer than for B, so the worker at B periodically has to wait for parts from A. Figure 3.6b shows the process simplified by moving the machines closer together so one operator can run both. (The wiggly line shows points where the operator goes from machine A to machine B with the part, then returns to machine A.) The operator experiences no idle time, though there is still a small amount of machine idle time at B because of the disparity in machine CTs.

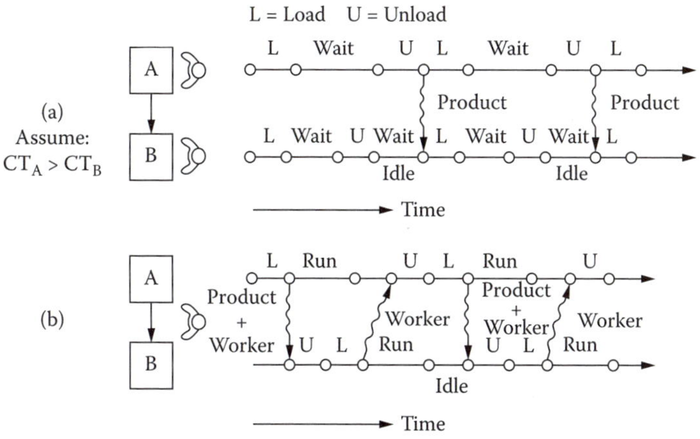

L = Load U = Unload

(a) Assume: $CT_A > CT_B$

A: L Wait U L Wait U L

Product Product

B: L Wait U Wait L Wait U Wait L

Idle Idle

Time

(b)

A: L Run U L Run U

Product + Worker

Worker Product + Worker Worker

B: U L Run U L Run

Idle

Time

Figure 3.6 Process simplification.

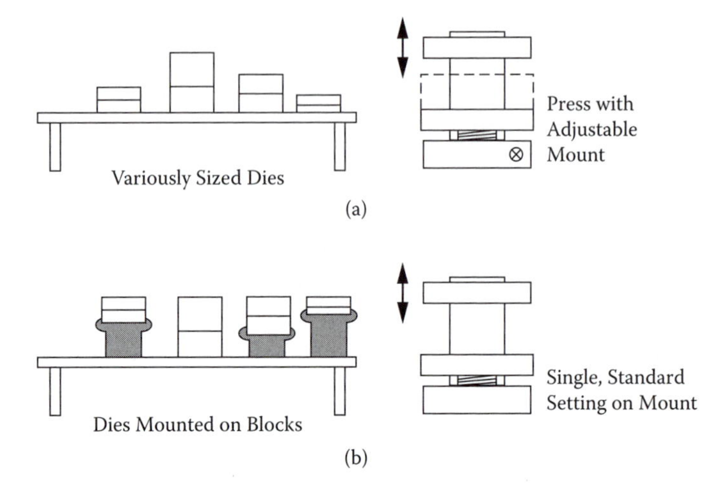

Figure 3.7 Procedural simplification.

Example 3: Procedure Simplification

Stamped metal parts are made with a press using heavy metal forms called dies. A metal sheet is put between the upper and lower matching faces of the die (called "male" and "female" faces) and the two dies are stamped together to form a part. A typical press can be used to stamp a variety of metal parts simply by changing dies. The dies, however, are of different sizes and shapes, and getting them to precisely fit the press requires delicate, time-consuming adjustment (Figure 3.7a).

A way to simplify the installation procedure is to make a block or fixture for each die to sit upon so that the combined height of the die and its fixture is the same for all dies (Figure 3.7b). When the die and its fixture are installed together, no adjustment for height is necessary, saving much of the time required to set up the dies. This is an example of *setup reduction*, the topic of Chapter 6.

Example 4: Product/Procedure Simplification

Two different molded plastic parts (A and B) with pegs on the bottom are glued into holes on a board as shown in Figure 3.8a. The parts look similar so assemblers sometimes glue them into the wrong holes. Also, the parts should be installed with a certain orientation, but assemblers sometimes put them in pointing the wrong way. The solution is to change the shapes and sizes of the pegs and holes so the parts can only be inserted in the right place and facing the right way (Figure 3.8b). In addition, each part is molded in a different color plastic (one black, one white),

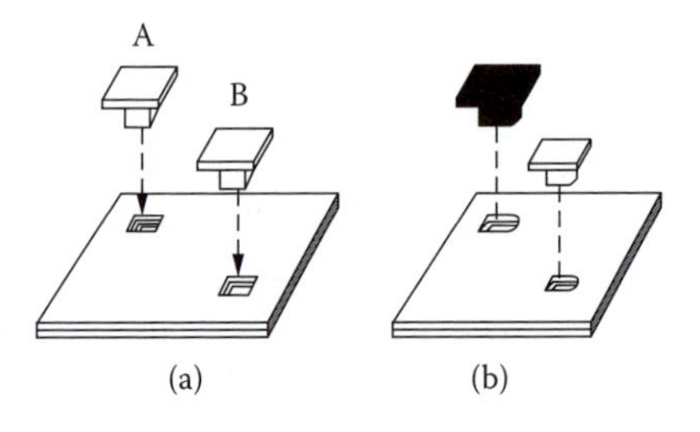

Figure 3.8 Product/procedure simplification.

which somewhat precludes an assembler from even starting to put a part in the wrong place. These modifications eliminate any possibility of assembly error and, hence, the need for inspection. They also save assembly time since workers need not scrutinize the parts or think about their orientation in each installation. This is an example of *mistake-proofing,* which is covered in Chapter 12.

Concurrent Engineering

Examples 1 and 4 raise an important point worth brief comment here. To make some kinds of improvements requires coordinated effort among different functional areas. The improvements in these cases could only happen if manufacturing and product design people talked to one another. Designers need to know how the product will be made if they are to design the product so it can be made simply and well. Such knowledge comes from working closely with the people in manufacturing. Yet in many organizations people in different departments (or buildings, professions, wage and salary grades, etc.) tend not even to speak to one another. Instead of having a dialogue, they use a linear process where designers develop the product, then hand the design over to the manufacturing group, which has to wrestle with the problem of how to make it. The product design in Figure 3.8a give assemblers headaches because it inevitably leads to mistakes, no matter how careful they are. The simple changes in Figure 3.8b eliminate mistakes, even if the assemblers get careless.

To do something right means considering all the parties affected (manufacturing, product design, procurement, finance, marketing, and so on), and incorporating those considerations into the product design before the production process gets underway. This concept of a multifunctional team designing and developing a product while also thinking about how it will be made and designing the process to manufacture the product is called **concurrent engineering**.

Cleanliness and Organization

Facilities in many organizations are dirty, cluttered, and disorganized. This is wasteful because it makes doing work more difficult and often results in poor quality work. Time is wasted looking for misplaced or lost tools and materials; equipment problems are camouflaged by grime and clutter; movement from one place to another is difficult; obsolete and discontinued materials are mixed up with current, needed materials; tools are bent or broken; and gauges and equipment are damaged and out of calibration. In general, messy facilities show an uncaring attitude about the workplace, but worse they foster similar attitudes about work and the finished product. Disorganization and clutter as shown in Figure 3.9 suggest a lack of discipline and the likelihood that all kinds of waste prevail in other ways throughout the organization. For these reasons it makes sense to begin the continuous improvement process by cleaning and organizing the facility.

Improvement Kickoff

Making housekeeping the responsibility of everyone is a way to ease workers into the improvement process and to prepare them for greater responsibility later on (Figure 3.10). To this end, staffers and frontline workers should get together to clean and organize their workplace together. Using the facility as a focus of attention will help workers begin to develop the right attitudes and work habits. The facility itself becomes a symbol of the new order, and getting it cleaned up is a way to introduce the problem-solving skills described in the previous chapter.

Figure 3.9 Scrap and parts clutter.

Figure 3.10 Proud worker and his work area.

More than just getting the facility cleaned up and organized, the emphasis must be on keeping it that way; for this employees will have to exercise continuous discipline, caring, and attention to detail.

The Five Ss

One way to instigate and maintain ongoing, workplace improvement is a program called five S, referring to Japanese names for five dimensions of workplace organization. The five Ss and their English equivalents are

- *Seiri*: Proper arrangement (sort everything; toss out anything not needed)
- *Seition*: Orderliness (specify a place for everything; designate location by number, color coding, name; put everything in its place)
- *Seiso*: Cleanliness (wash, clean, or paint everything so abnormal or problematic situations become obvious)
- *Seiketsu*: Neatness (create procedures or principles for maintaining the first three Ss)
- *Shitsuke*: Self-discipline (take responsibility for maintaining a clean, organized workplace)

(Some companies add a sixth S for safety.)

The five Ss are implemented through frequent inspection and grading of each work area and department. At Canon this is done using a check sheet similar to Figure 3.11.[10] At some companies five S committees do the inspections and grading; at others the workers or supervisors themselves do them. Problem areas are photographed and the plant or work area is expected to come up with a solution and plan. Work areas that show good housekeeping practices are awarded recognition. Results of evaluations are posted to foster responsibility and pride.

The five S process helps change attitudes, and employees tend to conform to workplace rules that previously were difficult to enforce (such as keeping parts and tools in the right place). As a result, performance measures such as number of accidents, equipment breakdowns, and defect rates tend to improve.

Benefits

Cleanliness and organization are important because without them opportunities for improvement and sources of problems are often obscured. Specifically, a clean workplace makes it easier to see cracks, missing parts, or leaks on equipment; reduces the chance of products being contaminated; improves work safety and reduces the chance for accidents; and makes it easier to see product defects. Further, a safer, nicer place to work improves morale. Likewise, keeping equipment, processes, and procedures organized makes it easier to find things (tools, parts, materials); makes it easier to assess processes and procedures. and pinpoint trouble spots; and makes it easier to move from place to place. Figure 3.12 depicts an exemplary workplace.

It is worth repeating that housekeeping is a good starting point and way to develop and reinforce the work habits, attitudes, and skills important for waste reduction, continuous improvement, and lean production. This is not the same, however, as saying that having a clean, organized place is an indication the organization will work well. Cleanliness and organization are only background and mean nothing unless there is also continuous effort to move forward and make improvements in the areas that give the firm a competitive advantage.

5S Inspection Sheet		*Evaluation Rank*			*Rank A: Perfect score Rank B: 1–2 problems Rank C: 3 or more problems*
	Item	*A*	*B*	*C*	*Comments*
Proper Arrangement	(Sort out unnecessary items)				
	Are things posted on bulletin board uniformly?				
	Have all unnecessary items been removed?				
	Is it clear why unauthorized items are present?				
	Are passageways and work areas clearly outlined?				
	Are hoses and cords properly arranged?				
Good Order	(A place for everything and everything in its place)				
	Is everything kept in its own place?				
	Are things put away after use?				
	Are work areas uncluttered?				
	Is everything fastened down that needs to be?				
	Are shelves, tables, and cleaning implements orderly?				
Cleanliness	(Prevent problems by keeping things clean)				
	Is clothing neat and clean?				
	Are exhaust and ventilation adequate?				
	Are work areas clean?				
	Are machinery, equipment, fixtures, and drains kept clean?				
	Are the white and green lines clean and unbroken?				
Cleanup	(After-work maintenance and cleanup)				
	Is the area free of trash and dust?				
	Have all machines and equipment been cleaned?				
	Has the floor been cleaned?				
	Are cleanup responsibilities assigned?				
	Are trash cans empty?				
Discipline	(Maintaining good habits at Canon)				
	Is everyone dressed according to regulations?				
	Are smoking areas observed?				
	Are private belongings put away?				
	Does everyone refrain from eating and drinking in the workplace?				
	Does everyone avoid private conversations during work time?				
	Rank totals				

Figure 3.11 Five S inspection sheet. (Japan Management Association, Canon Production System: Creative Involvement of the Total Workforce, Productivity Press, Inc., Portland, OR, 1987; English translation. Reprinted with permission.)

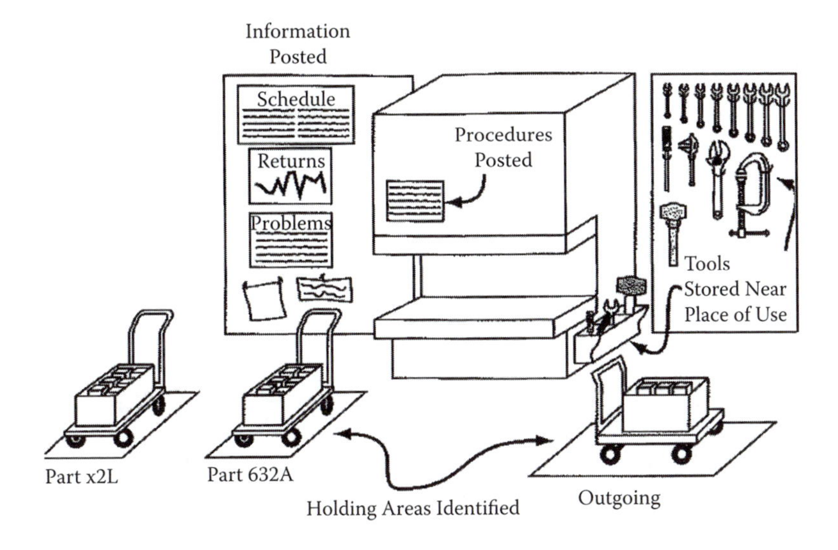

Figure 3.12 An organized workplace.

Case in Point: Stretch the Web[11]

At many companies, five S inspection results are plotted on a spider-web diagram (Figure 3.13, lower left). The ratings are posted next to each work area and employees are expected to stretch the web—to maintain perfect or near-perfect scores in all categories.

Sometimes, the rating for each category is based on the average rating of several subcategories. Schonberger describes a small company where every Friday the plant manager made rounds to rate every employee on six categories of criteria, many of them related to housekeeping and organization. One category, Straighten, includes the subcategories of unnecessary things around table, cleanup of trash and waste, uncompleted things in the right place, and proper notices on the wall. Some of these subcategories might seem picayune but, Schonberger emphasizes, in combination they add up to a system that stamps out sloppiness and mistakes and promotes attention to details. Even the vigilance of the plant manager is monitored. A special audit team rates him on a regular basis so that, in effect, he too must try to maximize the rating on his own spider web.

Visibility

Visibility means knowing what has been, must be, or should be done by seeing it. In traditional manufacturing organizations, information is the bevy of a privileged few. The masses of frontline workers get little information and are allowed to see only what management approves. Much information that would be important and useful to them, they never get to see.

5 S Approach: *Performance Summary Score Chart*

Minimum Score in each area is 15. Anything less requires immediate action to resolve.
FF Leaders are responsible to follow up and report progress results to the
Operations Manager until the cause(s) of the unacceptable scores are fixed

Summary Audit Score						
Section	Sort	Straighten	Shine	Standardize	Sustain	Total
Actual Score	25	25	24	23	24	121
Goal	25	25	25	25	25	125
Opportunity to improve	0	0	1	2	1	4

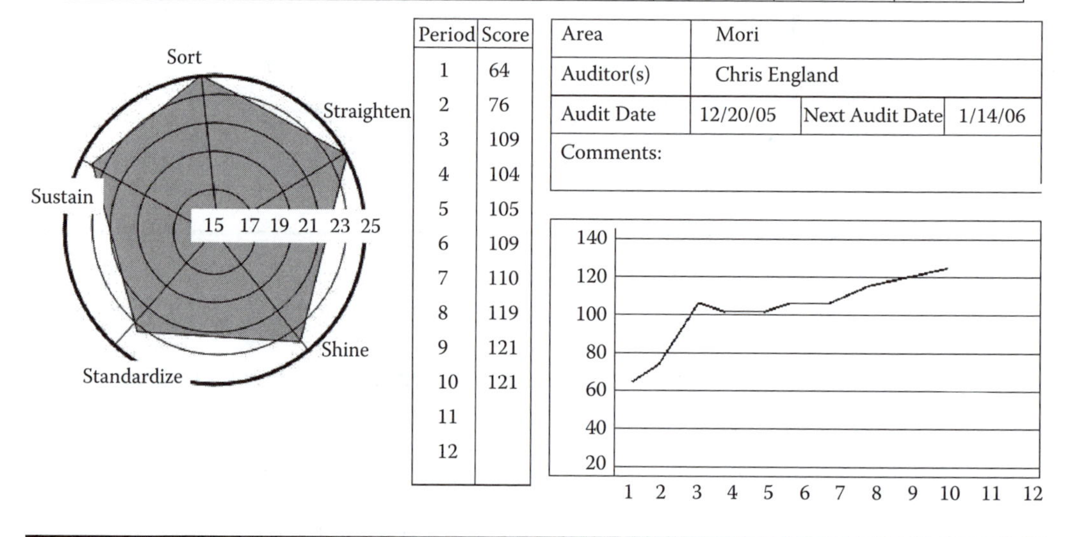

Period	Score
1	64
2	76
3	109
4	104
5	105
6	109
7	110
8	119
9	121
10	121
11	
12	

Area	Mori		
Auditor(s)	Chris England		
Audit Date	12/20/05	Next Audit Date	1/14/06
Comments:			

Figure 3.13 Process check sheet.

The essence of visibility is to redirect and redefine information so it is visible to workers on the frontline, and immediately so, whenever they need it. An example is a daily production schedule, prominently posted on the wall of a production area. Everyone, not just the supervisor, sees what the day's work is to be. Throughout the day, production results are displayed so everyone knows who is on schedule and who is not.

Another example is charts posted in work areas showing instructions, production standards, and goals. At each department or workstation, charts are posted to show workers what should or must be done. Often the workers themselves contribute to setting the standards and targets listed on the charts. This involvement helps gain buy-in from workers for the prescriptions on the charts and helps ensure that targets are achievable. Special achievements, quality awards, notices of recognition, and photos of workers and teams are also posted so people can show off their successes and accomplishments.

The overall status of the shop-floor operations is also readily visible. Signal lights on machines and overhead displays indicate the functioning of workstations and equipment. Workers switch on these lights to show that they are experiencing problems; short on parts; or that a setup, tool change, or quality check is needed.

Workplace organization and cleanliness contributes to visibility. When everything has a designated place, it is easy to see that everything is in its place. When equipment is kept clean, it is easier to spot abnormalities, malfunctions, or problem symptoms such as cracks, leaks, or lose fittings and fixtures.

The very layout of the factory also contributes to visibility. Equipment, workstations, and stock areas are situated so workers easily see what has been or needs to be done. Arranging a product line into a U-shape allows workers at the start of the line to see what is happening at the end of the line and vice versa. Putting subassembly areas next to final assembly enables everyone to see the final products into which their parts are going, which contributes to higher morale and fewer defects. Putting inventory stock areas on the shop floor next to workstations allows workers to easily monitor inventory levels, anticipate when they will need to order more parts, or to see when they need to produce more parts.

The lean concept of pull production, discussed in Chapter 8, is a manifestation of the principle of visibility. It relies on workers' observation of simple cards, containers, or inventory stock levels to tell them when to initiate or stop some action, or change what they are doing.

Reports can always be fudged but seeing physical evidence tells the truth about the quantity and quality of items supplied and produced, how well supply matches demand, and how well actions are conforming with plans. Visibility creates immediate feedback so workers know what to do without getting orders from supervisors or pronouncements from management. Visibility in practice is called *visual management* or *visual control,* more examples of which will be cited later in this book.

Cycle Timing[12]

The time interval that elapses between occurrences of something is called the cycle time. Cycle time can connote different things, depending on the way it is used. It can be the time

- Between placing job orders or performing different jobs
- Between preparation of business plans or accounting statements
- For a machine or person to perform a single operation
- Between completion of units for an individual operation or an entire process

In all cases, the concept of cycle timing suggests a regularity of timing or rhythm. Regularity of timing in manufacturing is beneficial because it reduces production uncertainty and permits managers and workers to better anticipate and prepare for the future. Manufacturing regularity ensures that products are produced at a steady rate, which means that materials flowing through the process are consumed at a steady rate. This directly or indirectly benefits virtually every activity associated with producing, moving, and monitoring material, including material handling and procurement, preventive maintenance, quality inspection, machine setup, and personnel scheduling. (Note that the emphasis is on maintaining a steady, predictable production rate and not on increasing that rate. The production rate should be whatever the demand requires: If the demand is low, the rate of production should also be low.) Regularity stimulates standardization and simplification. Workers who set up and operate the same machine at about the same time every day are stimulated to develop standardized, simplified procedures for that setup.

Agility

Chester Richards[13] describes the agile manufacturer in terms of the "OODA loop" concept in which opponents in a competitive situation *observe* the situation (absorb information), *orient* themselves (put the information in context), *decide* (select a course of action), and *act* (carry out the action). The agile manufacturer is a producer that has the ability to observe changes in its environment, quickly reorient itself, and accordingly make timely and effective decisions.

Through rapid execution of the OODA loop, companies can quickly increase product variety and steal customers away from competitors. Honda, for example, had lost market share to rival Yamaha for 15 years. In the early 1980s it began a counterattack and in an 18-month period replaced its entire product line twice and introduced three times as many model changes as Yamaha. Honda rapidly surpassed Yamaha and remains today (with Indian joint venture Hero Honda) the world's leading motorcycle producer.

Agile manufacturing is not exactly the same as flexible manufacturing, which is the ability to produce different products using the same line, machine, or process, and to switch over to meet planned changes in demand. It goes beyond that. It is the ability to economically switch back and forth among various products, produce any of them in almost any quantity, and do so quickly in response to unplanned changes in demand.

Automation is not necessarily the way to agility. In fact, automation can actually hamper agility since most automated systems are limited in terms of what they can produce. Nonetheless, as automated systems and robotics improve and evolve to incorporate artificial intelligence, these systems themselves will become more agile.

Case in Point: Agility at Prince Castle[14]

Prince Castle is a leading manufacturer and supplier of foodservice equipment for national brand franchises. It makes a variety of products, from electronic cooking timers and condiment dispensers, to tomato slicers and deep fat fryer cleaning equipment. In the past the company was primarily an assembler; that is, it produced virtually all of its products by assembling parts and components that were provided by suppliers.

Demand in the foodservice equipment business involves periodic surges. When a national chain introduces a new product, say a sandwich with a new kind of sauce, it needs a special portion-control dispenser for the sauce. Before the product can be launched, every restaurant in the chain (possibly many thousands) must have the dispenser. Since new products are carefully guarded secrets, even long-time suppliers like Prince Castle might receive only a few months or weeks notice. To ensure that it could meet such demand surges, Prince Castle had to carry ample inventory of finished goods and expensive components.

Further contributing to Prince Castle's inventory is international growth of American fast-food chains. Equipment used in different countries often requires different components to meet different safety requirements and standards, and Prince Castle had to carry all of those components.

To reduce its reliance on inventory and increase its agility, Prince Castle decided to switch strategies: instead of being just an assembler, it would become a full-scale integrated manufacturer, that is, it would not only assemble parts, but also make most of those parts in house. This strategy was a success: it reduced inventory, which freed up cash and space that Prince Castle used to acquire and install a computer numerical control-flexible manufacturing system and other machines needed to produce a multitude of parts, panels, and even fasteners.

Prince Castle also moved to a three-shift operation. Under normal demand conditions, the first shift is fully staffed and the others run on skeleton crews. The rationale for maintaining skeleton crews is that in the event of a demand surge (new product launch), each shift has an experienced supervisor and core group of workers who can train and oversee temporary workers hired from an agency. Full-time workers on every shift are multiskilled and able to work at different workstations and operations, as needed.

The overall result of the strategy change was to cut inventories in half, recapture capital costs, and bring specialty products from concept to the customer in substantially less time. Customers know they can count on Prince Castle to fill an order of virtually any size for virtually any product, even on short notice.

Measurement

Another lean principle is measurement. For whatever we seek to improve or wastes to eliminate, measurement is necessary to know exactly where we are, where we have been, and where we are going. It is fundamental to the PDCA cycle in both the plan stage (collect data) and the check stage (collect data). Any area for which improvement is sought must be initially measured to establish a baseline against which to gauge progress. Once an improvement plan has been approved and implemented, data are collected to assess the extent of progress made and to determine what more needs to be done. Thereafter, data are collected and results announced to ensure progress is maintained and inspire further improvement.

Grass Roots Measurement

Lean organizations utilize measurements in ways unlike many traditional measurement systems. To begin with, measurement results are not used for performance appraisal of personnel or work groups. Also, the specific types of data to be collected are not mandated by upper management. Finally, the measurement results are not entered into centralized databases for purposes of control. Instead, data are collected to help groups and individuals assess problems, find solutions, and track progress. Workers on the shop floor decide what needs to be measured and how it should be measured. Also, as much as possible, they collect the data themselves, which means they must first be trained and become skilled in data collection and analysis methods. Involving workers in tracking performance is one way to develop the norms of vigilance and attention to detail.

An unspoken rule in organizations is "If it's measured, it's important," and the corollary is "If it's not measured, it's not important." Of course, in reality many things get measured that never amount to anything (e.g., data are collected, results filed, then no one ever looks at them). When people discover this, both the rule and the corollary no longer apply. Nonetheless, on average, if you give an employee two tasks and measure only one of them, the one measured will get more attention. Measurement thus helps to establish priorities and focus attention on the areas most in need of improvement.

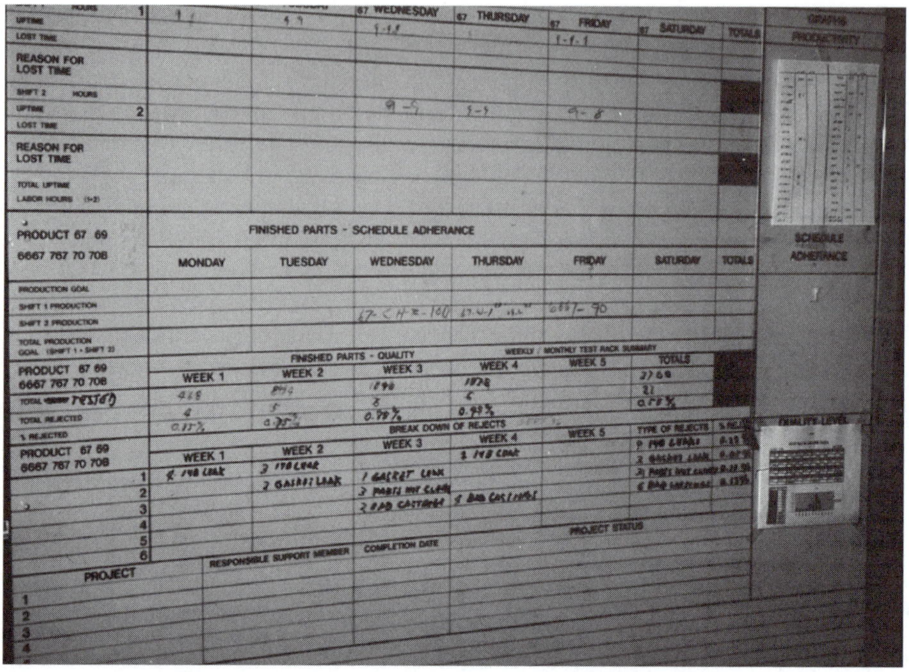

Figure 3.14 Schedule adherence charts maintained by shop-floor workers.

Visual Management: Information Post-Its

Workers retain most of the measurement results they collect for their own usage. One way to gauge an organization's dedication to continuous improvement is to look at how much measurement is going on and how the results are shared among the people who need it. In organizations serious about improvement, information about performance is posted on the walls, next to machines, and wherever else it is convenient or expeditious to display it. Information about levels of waste, quality, productivity, and service is posted along with goals so workers can readily see trends and gaps between goals and practice (Figure 3.14). This is an aspect of visual management.

Management must be careful not to abuse the results of grass roots measurement. The purpose of these measures is to serve PDCA, and management must ensure that the data are used for that purpose. Data collected for any other reasons, such as worker performance appraisal, must be kept separate from these data. If workers find that their measurements are being used for purposes of appraisal or manipulation by management, they will soon learn to misrepresent findings or measure only those things for which they can readily show progress (neglecting everything else).

Getting to the Bottom Line

Eventually it will be necessary to translate the results of improvement efforts into broader measures of organizational health or competitiveness. To assure that continuous improvement and waste reduction efforts become permanent, measures of improvement must ultimately be expressed in terms consequential to everyone, including shareholders. Canon Corporation, for example, utilizes a system that expresses waste reduction efforts in terms of profit and loss.

Case in Point: Waste-Cost Estimating at Canon[15]

Every year Canon Corporation estimates the cost of waste by aggregating the estimates of waste costs from all areas across the corporation. It uses the concept waste elimination profit (WEP), where

WEP = Degree of improvement over the previous year, in dollars

= (Prior waste rate – Current waste rate) × Current production

Suppose in one year the production output is $100 million. If the cost of waste from product defects is $1 million, the waste rate for defects would be 1%. Suppose in the next year the cost of waste for defects is still $1 million, but production output increases to $125 million. The waste rate for this year would be 0.8%. From one year to the next, the total savings (or earned profit) of reducing the defect waste rate is

WEP(defects) = (1% – 0.8%) × $125 million = $250,000

That is, $250,000 is the cost savings (accrued profit) attributed to waste reduction efforts focusing on product defects.

The same kind of analysis is performed for all of Canon's previously mentioned nine wastes, and the WEPs for all nine are combined to get total WEP. This figure is divided by the number of personnel in each section, division, factory, or department, to get WEP per person. One reason for Canon's success at waste elimination has been its ability to tie improvement measures to traditional financial accounting measures. Each factory is required to produce an annual waste elimination plan with high WEP goals. Though the immediate aim of WEP is to improve operations, the result is typically to increase value added and reduce fixed costs. Value added results from reducing wastes from defects, equipment, planning, and operations; and lower fixed costs result from reducing wastes from WIP, expenses, and indirect labor. Combining higher value added with lower fixed costs reduces the breakeven point and improves product profitability.

Variation Reduction

Variation (variability) represents the amount by which something differs from some nominal value (standard, target, or expected). Whereas variability in some things adds surprise, spice, and novelty to life, in a manufacturing process it is a curse manifested by waste and poor quality. Nonetheless, variability runs through every aspect of manufacturing: workers' skills, motivation, abilities, and attendance; equipment and process operating capabilities, cycles times, setup times, and reliability; quality and scheduled delivery of raw materials and components; batch sizes; and in other innumerable aspects of production.

Because of the persistence of variability, product and process requirements are often defined in terms of a **tolerance range** with upper and lower specification limits, and a **target** value at the midpoint. Performance is considered acceptable as long as it lies within the range. But even if every component meets its individual specification, however, that will not ensure that a higher-level product or process, one formed by a combination of components, will meet its requirement. When one component that is at the extreme of its tolerance range is mated with another component that is at the opposite extreme, the result is a bad fit. This is called **tolerance stackup**, and it is observable in products by gaps between parts that should fit snugly or tight-fitting parts that should fit loosely. Either way, it's a sign of poor quality.

Variability also has a large effect on production costs and lead times, what Hopp and Spearman refer to as the "corrupting influence" of variability on system performance. They note that in a steady-state system, variability always acts to increase average cycle times and WIP levels, and that this effect on WIP and cycle times increases the earlier in the routing sequence (in a sequential process) that the variability appears (that is, variability at an early stage of a process has a greater influence on overall WIP and lead times than variability at later stages).[16]

In a meticulously designed product or process, that is, one where target values everywhere have been set to optimize system performance (or, say, to provide total customer satisfaction), any amount of deviation from any of the target values will result in less than optimal performance. Genichi Taguchi, a quality expert, expressed this less-than-optimal result as a "loss" to the manufacturer and the customer.[17] Thus the closer a product or process comes to meeting its target value, the lower the cost and the better the overall performance of the system as experienced by the producer and customer. In lean production, variation reduction is enforced through methods such as standardization of procedures for operations, machine setup, and preventive maintenance, and by using leveled, regularized production schedules.

Lean Principles beyond Manufacturing

Manufacturing is not the exclusive province of waste and, as illustrated throughout this book, the principles of lean production apply to all kinds of organizations. The aforementioned principles are universally applicable, though many organizations choose to add to them or create their own. The following case in point illustrates this.

Case in Point: Lean Principles at Virginia Mason Medical Center[18]

Hospitals are complex organizations with numerous, interacting processes that are rife with waste in the form of inefficiency and mistakes. Virginia Mason Medical Center is among a growing legion of healthcare organizations that have adopted lean production to address these wastes. Borrowing from the Toyota Production System, it has created its own Virginia Mason Production System (VMPS) and relies on the following five principles to guide its waste reduction and defect elimination efforts.

- *Value*: Whatever the customer is willing to pay for, or that changes the form, fit, or function of a product or service.
- *Value stream*: The sequence of steps to provide a product or service.
- *Flow*: The movement of items or people through the value stream without delays or waiting.
- *Pull*: Upstream steps of the value stream only produce or respond to demand or requests from downstream steps.

- *Perfection*: Eliminating waste so all activities in the value stream only create value.

At Virginia Mason, as at many hospitals, nurses were overburdened with nonvalue-added work and rework that prevented them from attending to patients. This was a big problem at the 27-bed acuity care unit, which suffered from random assignment of patients, disorganized and inadequate supplies, lack of information regarding patient and nurse status, no routine surveillance of patients, and poor financial performance due to incidental overtime from nurses missing breaks and lunches. To rectify the situation, lean principles of the VMPS were applied to the unit. Changes implemented at the unit included several lean methods and concepts (described later in this book), such as the following:

- *One-piece flow*: Morning rounds scheduled such that all work for one patient is completed before moving to the next patient.
- *Workcells*: Patient bed areas arranged in U-shaped clusters to minimize walk distance.
- *Standard work*: Standard procedures created to ensure nurses address specific patient needs before leaving the room and going to the next patient.
- *Point of use*: Materials and supplies located where needed for usage without searching.
- *Kanban*: Visual system to flag needed replenishment of materials and supplies, ensure just-in-time resupply, and control inventory.

Improvements in the unit attributed to the changes included the following waste reductions: 85% in staff walking distance, 50% in time for nurses to complete morning work for patients, 85% in time spent searching and gathering equipment and supplies. Also, patient complaints regarding responses to call lights dropped to 0, and nursing hours per patient day dropped from 9 to 8.36. Nurses had more time to attend to patients so they missed fewer breaks and lunches, which decreased overtime.

Virginia Mason is one of numerous examples of lean concepts applied to the service sector. More will be described later in this book.

The Meaning of Lean Production

In this chapter lean production has been referred to as a management *philosophy* principled on simplicity, organization, visibility, agility, and so on. Lean production also represents a collection of *methods* for small-batch production, setup time reduction, maintenance, pull production, and the like (discussed in Part II of the book), as well as a *system* for production planning and control (discussed in Part III). Referring to different concepts like this using the same moniker can be confusing. Even in the literature, there is some confusion and debate about what lean production really is—philosophy, methods, or a system.[19] Continuing the debate here is pointless, since lean production can be all of these things, depending on what aspect of it you choose to look at.

Lean production began as JIT, just-in-time, a technique used by Toyota for the purpose of controlling production and reducing inventory. From there, it evolved to include techniques for setups, maintenance, worker participation, supplier relations, and even product development. When JIT was introduced to the rest of the world in the 1970s, it began as a manufacturing technique centered on the kanban method and pull production. Over time, lean production has evolved into a complete management philosophy driven by waste reduction and continuous improvement. Organizations that adopt lean production seem to go through the same kind of evolution: they start using lean methods to improve shop-floor control, then they adopt the broader principles of lean philosophy for organization-wide management.[20] Today most managers no longer think of lean as just a set of manufacturing methods, but as a philosophy for managing a company and doing business.

Implementation Barriers

Articles abound about companies that tried to implement lean production and failed. Much of this commentary points to one fact: lean philosophy requires changes in traditional attitudes about business relationships, work management, and shop floor practices, and it requires long-term commitment to quality and waste reduction.

Attitudes

In traditional plants, frontline workers are given limited, singular responsibility for assembling parts or running machines. Contrast that to workers in lean plants who also have responsibility for continuous improvement. In lean plants, broadened worker involvement results from a transfer of responsibilities from support staff, supervisors, and managers to frontline workers. The transfer is successful only when staff, supervisors, and managers accept the proposition that workers are capable of handling more responsibility. The workers, too, must wholly embrace this concept.

Lean is a team-oriented philosophy. Teams of workers, cross-functional staff, suppliers, and customers are involved in projects to reduce waste, design and improve products and processes, reduce setup times, and improve supplier/customer relations. Every area of the organization is affected. Managers unwilling to break down traditional walls between areas, and with workers and suppliers will be incapable of implementing lean production.

Time Commitment

Lean programs take time to show benefits, especially financial benefits. In one study, the time required for lean programs to yield significant benefits and deterioration of management support were cited as two major barriers to implementation.[21] In another study, among companies implementing JIT 59% said they were still 6 months to 3 years from full implementation; 26% said they were more than 3 years away.[22]

In the short run, the benefits of improved quality, agility, and cycle time are transparent, and, financially, lean production shows up solely as a cost. In terms of payback period, lean can appear risky, not because of the capital expense involved but because of the time and expense in training programs as well as improvement and waste-reduction projects. In lean companies, chronic problems such as machine breakdowns, production mistakes, and poor organization are eliminated through steady, continual effort, but that takes time.

Quality Commitment[23]

A prerequisite for lean success is commitment to quality. Among 1,035 organizations surveyed in a study of lean practices, the largest percentage, 85%, were practicing total quality control. Commitment to quality requires changes in policies affecting procurement, production, product design, problem troubleshooting, and relationships with suppliers. For example, instead of just low price, the criteria for choosing suppliers must include guarantees for high-quality products in terms of specifications, delivery dates, and delivery quantities. Failure to find suppliers that can meet these requirements or that are willing to adapt to them can be a major obstacle to implementing lean production.

Quality must be designed into the product and the production process, and that, in turn, requires adopting a new product/process design methodology. Also, frontline workers must be given time to troubleshoot and resolve quality problems at the source.

Misunderstanding Lean Production

Because lean production is associated with inventory reduction, some people believe that the primary focus of lean is on inventory. It is not. The lean prescription is to reduce inventory slowly, identify problems, then change policies and practices to remove the problems; having done so, reduce inventory a little bit more, and so on. But confusing means with ends, companies try to reduce inventories without resolving the problems. When the production system comes to a screeching halt, as it surely will, they blame lean production.

A further misunderstanding about lean is that it is a physical system to be implemented. The most common mental picture of lean production for many people is a system called pull production or kanban, topics of Chapter 8. Implementing lean, they believe, equates to tearing out whatever production system is in place and installing a pull process with kanban cards and small containers. While it is true that a physical pull system with kanban control is a feature of lean production, such a system does not represent the entirety of lean, nor is it a necessary component of lean. A lean program that focuses on long-term commitment and cultural change devoted to waste reduction and customer-oriented quality can succeed even without a pull system.

Social Impact of Lean

Beyond the conduct of business, lean production is having a broader social impact. The philosophy and practice of lean production returns to workers a degree of dignity, recognition, meaning, and pride of work that is often absent at the lowest levels of organizational hierarchies. Lean philosophy says that the center of wisdom does not reside solely at the top of an organization; rather, it is distributed throughout, though often it is only wisdom in potential form, and it needs developing to be realized. The ultimate waste recognized by lean production, then, is the wasted potential of most workers.

First Things First

A quick review of this chapter says what lean is really about—simple, commonsense ideas, largely instituted at the shop-floor level by the frontline workers themselves. It also says what lean is not about—automation, robotics, and computerized manufacturing systems designed and installed by engineers and technical specialists. Both lean production and automation are important to manufacturing competitiveness, but whereas the latter is important in varying degrees to some or many manufacturers depending on industry, the former is important to virtually all of them.

Lean production is also a matter of priorities. If an organization decides to automate, it had better first be well down the road to having eliminated much of the waste in its production system, and to having improved its product and process designs and procedures for manual systems. Automating something that is wasteful simply casts the waste into cement, making it harder and more costly to remove it later. General Motors spent $40 billion on factory automation, and afterward was still plagued with the same quality and productivity problems it faced before. Continuous improvement and elimination of waste is a movement back to the basics. Having mastered the basics, a manufacturer will be in a better position to take advantage of the benefits offered by computer-integrated manufacturing systems (CIMSs), electronic data interchange systems (EDISs), flexible manufacturing systems (FMSs), and the like.

Learn as You Go

Lean production involves so many aspects and is so all-encompassing that manager and workers can be intimidated into avoiding it. Although it is so broad and has so many aspects, however, it does not have to be implemented all at once. In fact, rushing headlong into lean production and pushing for change everywhere is likely to be too much, too soon, and to cause increased resistance. On the other hand, introducing it slowly allows problems to be identified and resolved in a more orderly way and resistance to atrophy for lack of cause. Given that management understands the principles of lean production, acknowledges its long-term commitment and the possible delayed payoffs, a company can then proceed to implement lean in any of numerous ways. McDonnell & Miller and Strombecker are examples.

Case in Point: Lean Production at McDonnell & Miller[24]

McDonnell & Miller has been implementing lean methods for the last 20 years. The plant is divided into focused factories, each with its own minimanagement team. The focused factories, in turn, are divided into workcells, each run and managed by a team of operators. Pull production has been introduced in many areas of the plant but not everywhere. Frontline workers are considered associates. They not only operate machines and do product quality inspections, they also do job scheduling and rudimentary equipment maintenance. When touring the shop floor, a visitor can easily detect the pride workers take in their jobs and the company as they explain their responsibilities and contributions to improvements. Quality of design, statistical process control, close supplier relations, preventive maintenance, and other key aspects of lean production are entrenched practices at McDonnell & Miller.

Case in Point: Lean Production at Strombecker Inc.[25]

Strombecker manufactures a variety of toys under the Tootsietoy logo. It became a lean company starting in 1994 under tutelage of consultants from Toyota. In many ways Strombecker is not as far along in lean production as McDonnell & Miller, though in some ways it is considerably ahead. Pull production and kanban have been implemented on most major lines, and Strombecker has developed its own pull system for materials procurement. It is experimenting with workcells, defect mistake-proofing, and standard operations, though it has yet to develop a strong preventive maintenance program and has had difficulty in getting suppliers to subscribe to lean concepts.

Perhaps the big surprise about Strombecker is that it has undertaken so much change without empowering its frontline workforce. That was not a conscious choice, but rather a result of the fact that most of the company's shop employees speak little or no English. This creates problems training workers so they can assume broader responsibility. Strombecker realizes the limiting effect this has had on quality and process improvement and is committed to increasing worker participation. To that end it has begun an English as a second language program. Though not to the extent of McDonnell & Miller, Strombecker frontline workers are treated as associates, and a visitor to the plant with a question is often introduced to a worker who can give an explanation.

Summary

The key concepts in lean production are value added and waste elimination. The value-added concept says to identify those activities in a process that directly increase the value of an item and consider all the rest wasteful and candidates for elimination. Only activities that add value, or that do not but are necessary, should be retained. These become the focus of continuous improvement.

One way to distinguish value-added from nonvalue-added activities is to look at common wastes in organizations such as transportation, inventory, waiting, and setup. These add to cost but not to value, so focusing on them is a logical first step in waste elimination. Once the obvious, prominent sources of waste have been removed, the next step is to pursue wastes that have been hidden. Focus on waste reduction not only improves end-item quality and organizational performance, it is also good for the environment. Broadly construed, less waste in production translates directly into conserved resources, recycled and reused materials, and less pollution.

Decisions and actions in lean organizations are guided by a set of principles:

- Simplification: Given multiple ways to achieve the identical result, simpler is better.
- Cleanliness and organization: A clean, organized workplace promotes a disciplined attitude about work and products, reduces waste, and helps pinpoint incipient problems.
- Visibility: Visible information available to everyone who needs it enables people to do their jobs better, motivates them to do the right thing, and eliminates non-value-added activities.
- Cycle timing: Regularity and recurrence of workplace patterns reduces uncertainty, increases learning, and permits better planning and action toward meeting customer demand.
- Agility: Changing customer demand is a fact of life; companies must be able to quickly adapt to such changes without relying on inventory and other wasteful means.
- Measurement: Improvement and waste elimination efforts at any level of the company depend on people using data to assess where they are now, where they should be going, and how well they are doing.
- Variation reduction: Reducing the variability by which a process deviates from standards, goals, or expectations results in less process waste and improved process performance.

Lean production is a management philosophy that addresses not only production practices but also the roles and responsibilities of managers, support staff, frontline workers, and suppliers; their relationship to one another and to customers; and broader issues about the conduct of business. The principles of lean production apply to all organizations—large and small, services and manufacturing. Problems with implementing lean production stem largely from lack of commitment, resistance to change, and misunderstanding about what lean production really means. Lean production is a move back to basics, and there can be little argument that the principles behind lean make good business sense.

Notes

1. T. Vollman, W. Berry, and D. Whybark, *Manufacturing Planning and Control Systems*, 3rd ed. (Homewood, IL: Irwin), 72–73.
2. T. Ohno, *Toyota Production System—Beyond Management of Large-Scale Production* [in Japanese] (Tokyo: Diamond Publishing, 1978).
3. It has also been called the "flower" of all evil, meaning that inventory is the result of mismanagement of production. See R. A. Iman, Inventory is the *flower* of all evil, *Production and Inventory Management Journal* 34, no. 4 (1993): 41–45.
4. T. E. Graedel, and B. R. Allenby, *Industrial Ecology* (Englewood Cliffs, NJ: Prentice Hall, 1995).
5. B. Patton, 1994. Design for environment: A management perspective, in *Industrial Ecology and Global Change*, ed. R. Socolow, C. Andrews, F. Berkhout, and V. Thomas (Cambridge: Cambridge University Press, 1994); M. E. Henstock, *Design for Recyclability* (London: Institute of Metals, 1988).
6. See P. Dewhurst, Product design and manufacture: Design for disassembly. *Industrial Engineering* 25, no. 5 (1993): 26–28.
7. Xerox Web site, December 2008, http://www.xerox.com/ Static_HTML/citizenshipreport/2008/nurturing-page9-8.html.
8. Xerox spreading its recycling philosophy, *APICS—The Performance Advantage* August (1992): 13.
9. I. Young, Beyond lean toward green, *Target* 25, no. 3 (2009): 19–26.
10. Japan Management Association, *Canon Production System* (Cambridge, MA: Productivity Press, 1987), Chapter 6.
11. R. Schonberger, *World Class Manufacturing* (New York: Free Press, 1986), Chapter 2.
12. For elaboration, see R. Hall, *Attaining Manufacturing Excellence* (Burr Ridge, IL: Dow-Jones Irwin, 1987) , 98–103.
13. C. Richards, Agile manufacturing: beyond lean? *Production and Inventory Management Journal*, 37, no. 2 (1996): 60–64.
14. Conversation with William Kinney, Sr. V.P. of Sales and Marketing, and John Esselburn, Sr. V.P. of Operations, Prince Castle Incorporated.
15. Japan Management Association, *Canon Production System*, 23–25.
16. W. Hopp and M. Spearman, *Factory Physics* (Chicago: Richard D. Irwin, 1996), 282–288.
17. See P. J. Ross, *Taguchi Techniques for Quality Engineering*, 2nd ed. (New York: McGraw-Hill, 1995).
18. D. Nelson-Peterson and C. Leppa, Creating an environment for caring using lean principles of the Virginia Mason Production System, *Journal of Nursing Administration* 17, no. 6 (2007): 287–294.
19. R. Vokurka and R. Davis, Just-in-time: the evolution of a philosophy, *Production and Inventory Management* 37, no. 2 (1996): 56–58.
20. G. Plenart, Three differing concepts of JIT, *Production and Inventory Management* 31, no. 2 (1990): 1–2.
21. K. Crawford, J. Blackstone, and J. Cox, A study of JIT implementation and operating problems, *International Journal of Production Research*, 26, September (1998): 1565–1566.
22. Touche Ross Logistics Consulting Services. Implementing Just In Time Logistics, *National Survey on Progress, Obstacles, and Results* (1998): 5.
23. National Center for Manufacturing Sciences, *Competing in world-class manufacturing* (Homewood, IL: Business One Irwin, 1990), 227–228.

24. Conversation with Avi Soni, Manager of Manufacturing Engineering, ITT McDonnell & Miller.
25. Conversation with Vern Shadowen, Manager of Plant Operations, and Al Brouilette, Vice President, Strombecker Inc.

Suggested Reading

B. Carriera. *Lean Manufacturing That Works: Powerful Tools for Dramatically Reducing Waste and Maximizing Profits*. New York: AMACOM, 2004.

T. J. Goldsby, and R. Martichenko. *Lean Six Sigma Logistics: Strategic Development to Operational Success*. Fort Lauderdale, FL: J. Ross Publishing, 2005.

P. J. Gordon. *Lean and Green: Profit for Your Workplace and the Environment*. San Francisco, CA: Berrett-Koehler Publishers, 2001.

J. Persse. *Process Improvement Essentials: CMMI, Six SIGMA, and ISO, 9001*. Sebastopol, CA: O'Reilly Media, 2006.

P. Posey. *Seeing is Believing: How the New Art of Visual Management Can Boost Performance Throughout Your Organization*. New York: AMACOM, 2004.

S. D. Young, and S. O'Byrne. *EVA and Value-Based Management: A Practical Guide to Implementation*. New York: McGraw-Hill, 2000.

Questions

1. What is the distinction between value-added and nonvalue-added activities in a process? Give examples of each (other than from this chapter).
2. Distinguish between necessary and unnecessary nonvalue-added activities. Explain how you differentiate them. Give examples (other than from this chapter). Can a nonvalue-added activity in some places be considered necessary, in other places unnecessary? Explain.
3. Describe typical support activities. Are they value-added or nonvalue-added, necessary or unnecessary? Explain.
4. What role do frontline workers play in distinguishing value-added from the nonvalue-added activities in a process?
5. What are Toyota's seven wastes? What are Canon's nine wastes? What are categories of waste that neither list includes?
6. Suggest ways to reduce or eliminate each of the wastes listed in question 5 (Toyota's, Canon's, and your own).
7. Explain the following lean principles and give an example of each:
   a. Simplification
   b. Cleanliness and organization
   c. Visibility
   d. Cycle timing
   e. Agility
   f. Measurement
   g. Variation reduction

8. Explain the various interpretations of the term *lean production*. In its broadest interpretation, what is lean production?
9. A hospital administrator turns down his assistant's request to attend a lean production seminar, saying, "We don't make things here. It is not relevant to us." Comment.

PROBLEMS

1. Refer to the processes in Problem 6 in the previous chapter. In each of the process flow diagrams:
   a. Classify the activities (steps) as value-added or nonvalue-added, necessary or unnecessary. Explain how you make the distinction in each.
   b. Based on the way you classified activities in (a), what would you do to reduce waste in each process? Would the process be improved in terms of time, cost, quality, agility? Explain.

2. Refer to Examples 1 through 4 in this chapter. Each of these illustrates "improvements" through product and/or process simplification. For everything, however, there are at least two sides, and decisions about simplification should always consider the tradeoffs. Consider in each of the four examples potential drawbacks associated with the proposed simplification in terms of time, cost, quality, or agility. Discuss the improvement–drawback tradeoff.

Chapter 4

Customer-Focused Quality

> Quality ... you know what it is, yet you don't know what it is ... But for all practical purposes it really *does* exist ... Why else would people pay fortunes for some things and throw others in the trash pile?
>
> **—Robert M. Pirsig,**
> *Zen and the Art of Motorcycle Maintenance*

What does quality have to do with lean production? In a word: everything. Every product that has a physical flaw or fails to meet requirements or customer needs represents multiple wastes and nonvalue-added costs. For starters, a flawed product discovered by the producer must be reworked or discarded. To discover the flaws, the producer has to incur costs for inspection. A flawed product that gets into the hands of customers causes the customer aggravation and the producer additional costs for returns, repairs, warranties, and declining business.

A lean producer must have a scrupulous quality assurance methodology to ensure that customer needs and requirements are incorporated into the product design, and that the manufacturing process is able to produce products that faithfully conform to that design in each and every unit produced. The methodology must apply rigorous data collection, problem solving, and statistical analysis tools to build a process capable of matching production closely to design requirements and containing no defects.

This chapter argues that to achieve high quality levels a company must adopt a *total quality management* approach. The chapter gives an overview of this approach and briefly describes two quality improvement and control methods, *Six Sigma* and *statistical process control.*

Quality Defined

Pirsig's quandary, and ours too, is that the concept of quality does not lend itself to easy description or simple measurement. Any definition of quality is dependent on the person doing the defining—the customer or the producer—and even within those categories definitions vary. A principle of the **total quality approach** is that any definition of quality must start with the customer's perspective. That perspective represents what is termed **customer-focused quality**, which is the way customers view or feel about a product. Obviously, the product that customers see and use is influenced by the way the product is designed, the way it is manufactured, and the services the customer receives after buying the product. These factors represent actions or procedures taken by the producer to translate customer opinions and requirements into a product design, and to monitor and control materials and manufacturing so the final product conforms to customer requirements.

Customer's Perspective

Broadly speaking, the term *customer* refers to the recipient of the output of any process. In manufacturing, the customer can be a company using materials or parts to produce a product, a person on a production line working on the product, a retailer selling the product, or a person who purchases the product. W. Edwards Deming considered the person who purchases the product as the most important customer. The customer's perspective of quality, called **fitness for use**, is how well the product compares to what the customer expects from it.

David Garvin identifies eight dimensions of quality:[1]

1. Performance: Includes operating characteristics such as speed, comfort, and ease of use; most products have multiple performance features, and customer preferences determine the relative importance of each (e.g., high acceleration versus high gas mileage).
2. Features: Extras, add ons, or gimmicks that enable a customer to somewhat customize a product.
3. Reliability: The likelihood that the product will perform as expected (not malfunction) within a given time period.
4. Conformance: The degree to which the product satisfies or conforms to preestablished standards.
5. Durability: The length of time, or extent of use, before the product deteriorates and must be replaced; durability is a function of the product's operating environment and reliability.
6. Serviceability: The speed, ease, and convenience of getting or making repairs, and the courtesy and competency of repair people.
7. Aesthetics: The look, feel, taste, sound, or smell of the product based on personal taste; though subjective, some aesthetic judgments tend to be universal.
8. Perceived value: Subjective opinions about the product based on images or attitudes formed by advertising and/or the reputation of the producer.

Besides these is the dimension for *service quality*, for example, the availability of the product, the support offered by the manufacturer after the product is purchased, and a sales and support staff that is knowledgeable and courteous.

A product does not need to be rated highly by customers on all dimensions, only on those they think important. Customers weigh the dimensions against the cost of the product, and are willing to pay more for higher quality on certain individual dimensions or on a greater number of dimensions. The connection between quality and cost is the concept of *value*, the extent to which customers feel they paid a good price for the quality they received in return.

Producer's Perspective

Once a company has determined the customer's expectations and quality requirements, it must translate them into a product design and, then, into a manufacturing process that can make the product according to the design requirements. These two aspects of the producer's perspective are quality of design and quality of conformance.

Quality of Design

The term **quality of design** represents the ability of a product as designed to satisfy or exceed customer requirements. Although quality of design initially focuses on customer requirements, eventually it must also take into account product demand, availability of materials and parts, and the capability of the manufacturer to produce the product. Related matters such as production costs, profitability, and the actions of competitors also come into play when determining what aspects of quality (say, from among Garvin's eight dimensions) to emphasize in the design.

Quality of Conformance

Given that the product design does in fact satisfy customer requirements, it is then incumbent on the manufacturing function to produce the product so it remains faithful to that design. **Quality of conformance** is the term used to connote that the manufactured product consistently upholds the requirements as set in the product design. Emphasis in quality of conformance is on two things: defect detection and defect prevention, where the term *defect* implies a deviation from the design requirements or an inability to meet some fitness-for-use criteria.

- **Defect detection** refers to inspection, test, and analysis of products using (typically) statistical sampling procedures to determine the presence of defects and to draw conclusions about the quality of an overall process or batch. Defect detection helps ensure that the product going to the customer is OK, but it does nothing to improve the quality of the product.
- **Defect prevention** includes monitoring and controlling process variation, again using statistical procedures and identifying causes of variation that, left alone, could lead to defects. The goal is to prevent defects. For some processes, defects caused by errors or inadvertent mistakes can be virtually eliminated through **error-proofing** or **mistake-proofing** procedures. It should be clear that under the total quality and lean production philosophies, the emphasis is overwhelmingly on defect prevention, not defect detection.

No company can achieve the levels of quality expected by customers without paying attention to both quality of design and quality of conformance. A product must be designed to meet the customer's requirements, yet it must also be designed to fit the capabilities of the production process. After the product is designed to do both, the production process must be adjusted and controlled to ensure that the output continuously conforms to the requirements of the design. To excel at both quality of design and quality of conformance requires the efforts of people from all functions and levels of the organization.

Total Quality Management

Total quality management or **TQM** refers to a management approach to focus all functions and levels of an organization on quality and continuous improvement. The term *total quality* implies quality not just for customers of the final product, but for the organization's internal customers as well. Further, it means total participation and commitment from everyone in the company.

Like continuous process improvement, TQM is not a program to achieve specific, static goals, but is instead a process committed to continuous quality improvement. Continuous quality improvement is essential to surviving and thriving in a changing, competitive world:

- The competition is not sitting still. No matter what level of quality standard a company sets for itself, the competition will eventually meet and exceed that standard.
- Customer expectations are continuously increasing. Brand loyalty for its own sake is a thing of the past. Customers seek out whatever producers can best satisfy their requirements.
- No level of quality can be sustained on its own. Without continuous effort, the inclination is to fall back to former, lower-level quality standards.

Though many companies recognize the strategic importance of continuous quality improvement, many others do not, and as a result the acronym TQM has been overused and misused. Many self-proclaimed TQM producers have actually done little in the way of improving their processes for quality assurance and quality control; they merely use TQM as a label to create the image that they think quality is important and to suggest to customers that their products are high quality.

TQM Integrative Framework

TQM builds quality into the product through attention to quality improvement at every organizational function, from marketing and design to manufacturing to customer service. Figure 4.1 is a framework illustrating the responsibilities of functional units in TQM and their relationship to the primary tasks of TQM. For example, in Figure 4.1 **quality of performance/service** refers to the extent that the finished product measures up to the customer's expectations and requirements. How well the product does that, of course, depends on how well the company executes the quality-of-design and quality-of-conformance tasks. If a product fails to meet customer requirements, the fault is with the product's design, its manufacture, or both. This suggests the importance of departments and teams responsible for quality of design and quality of conformance receiving frequent, accurate feedback about the product's performance (represented in Figure 4.1 by arrows a and b). Responsibilities for TQM are spread across all the functional areas as discussed in the following sections.

Marketing, Sales, and Finance

Marketing and sales are the company's primary points of contact with the customer; as such, they provide customer-related information essential for developing new products and upgrading existing products. Through consumer research, marketing determines customers' wants and needs, defines customer quality characteristics and requirements, and determines what customers are willing to pay. Through consumer research, marketing determines customers' wants and needs, defines customer quality characteristics and requirements, and determines how much customers are willing to pay. The sales force provides customer opinions about current products and customer suggestions for future products.

Traditionally, finance and accounting monitor and guide an organization's financial performance; in TQM, this role is expanded. Measures of financial performance are integrated with measures of operational performance to provide a balanced view of how well the company is meeting customer-defined requirements.[2] Finance and accounting provide the performance

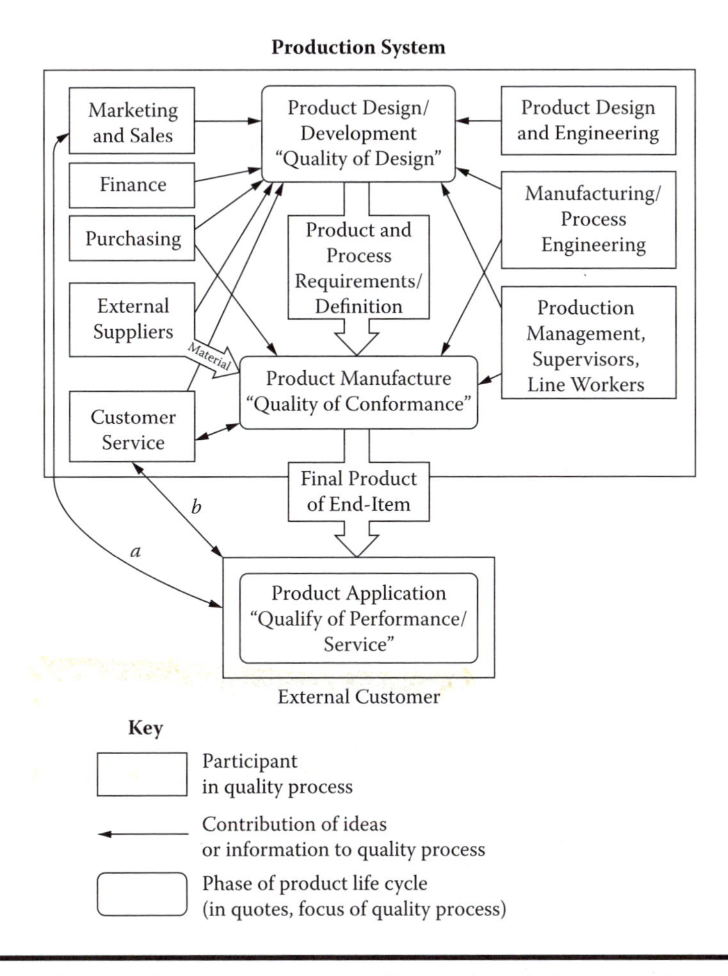

Figure 4.1 TQM framework: Participants in quality product design, manufacture, and service.

information necessary to support continuous improvement efforts and make competitive operational decisions.

Product Design and Manufacturing Design

Product designers and engineers work with people in marketing and sales to translate customer needs, wants, and expectations into product requirements and a physical design. Since a product is only as good as the process that produces it, product designers must also work with manufacturing engineers to ensure the product design accounts not only for the customer's requirements, but also for the company's processes, equipment, and labor skills that will convert the design into an actual product.

In the past, products were often designed without consideration as to how they would be made. Product designers would conceive and design a product in detail, then "toss it over the wall" to manufacturing (Figure 4.2), the wall being the metaphorical barrier that traditionally separated the design and manufacturing functions (and that separated all the other functions, too). The consequence was that sometimes manufacturing could not make the product to conform to the

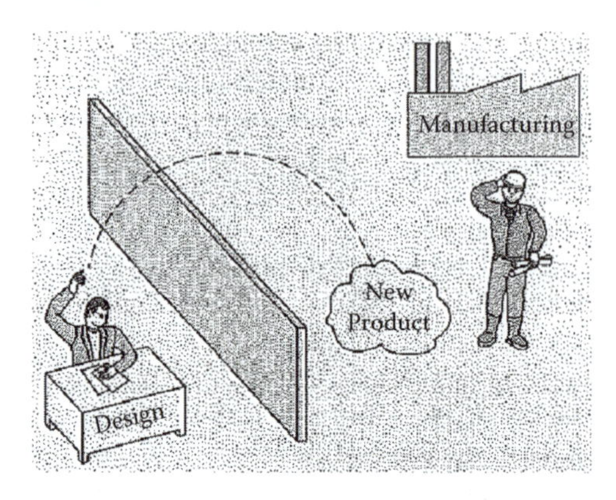

Figure 4.2 Over-the-wall product design.

design (the result: high defect rates), or it could make it to conform, but only through extraordinary effort (the result: high cost). The point is, issues relating to the production process (workplace equipment, technology, and workforce) must be addressed during product design. Among the approaches for integrating customer requirements and production-process issues into the product design are **concurrent engineering** and **design for manufacture and assembly (DFMA)**.

Unlike the sequential process illustrated in Figure 4.2, in concurrent engineering the functional areas of design and manufacturing work as a team to concurrently create a product design that will satisfy the customer and address issues regarding the product's manufacture. The product is designed so as to be manufacturable (readily produced and assembled well within the company's capability), which means that it can be produced with high quality and for low cost. To ensure all aspects of the product life cycle are addressed in the product design, the concurrent engineering team commonly also includes representatives from sales, finance, and purchasing, as well as customers and suppliers.

DFMA considers features in the product's design that will simplify its manufacture and assembly, and, usually, simplify the manufacturing process, reduce manufacturing costs, and improve manufacturing quality. DFMA includes methodologies and tools to make sure that the product's design includes considerations such as materials use, material forming and shaping procedures, machining processes, material handling, and machine and tool changeover procedures. Among the DFMA guidelines intended to improve manufacturability are the following: minimize the number of parts in a product, eliminate screws and fasteners, eliminate the need for special tools, and use standard parts. DFMA also includes guidelines for assessing procedures to assemble parts and determining the best method and sequence of assembly.

The emphasis of these integrated approaches is to *design in* quality and to *design out* defects, an aspect of **quality at the source**. Such emphasis is crucial to product improvement since as much as 80% of all defects result from poor design.[3] More time and effort devoted to quality in the early stages of design result in less time and effort in dealing with quality-related problems during production and following sale.

Purchasing and Suppliers

A high-quality product is achievable only if it has high-quality parts, and for most products at least half the parts come from suppliers. A role of purchasing in TQM is to ensure that purchased parts and materials meet quality requirements and that suppliers meet delivery and service requirements and guarantee quality parts. Purchasing agents participate in product design by providing design engineers with information about the availability, technology, and costs of parts and materials that go into products.

The purchasing function also serves as the intermediary between the manufacturer and its suppliers. Lean organizations form *customer–supplier partnerships* wherein a supplier's managers and engineers participate in the customer's product design process and offer recommendations concerning the parts they will be supplying. In turn, suppliers are provided with training and support to help them better satisfy the customer's requirements. Partnerships are discussed more in Chapter 16.

Production Management and Frontline Workers

The production function must be managed so the product design is executed and the product is made according to expectations. Production management is responsible for planning and controlling the materials, equipment, and frontline workforce so that variations in the process, such as material defects, equipment breakdowns, order backlogs, and other problems that contribute to poor quality, are minimized. Whereas firms have traditionally relied upon quality-assurance inspectors to spot defects and monitor processes, TQM organizations give most of that responsibility to frontline workers. This is another aspect of quality at the source—putting quality responsibility with the people in the best position to readily identify and quickly remedy defects in production. Supervisors and frontline workers actively participate in problem-solving sessions and are integral to the quality improvement process.

In theory, if parts and raw materials, processes, and operations all meet specifications, then the finished product will meet specifications, too. Modern customer–supplier partnerships and guaranteed incoming quality have largely eliminated the need for the customer to perform acceptance inspection of procured parts and materials, and suppliers and freight carriers working with customers have eliminated most of the defects originating from packaging, delivery, and warehousing. As a result, virtually all of the quality emphasis in the production function is now on controlling the manufacturing process using **statistical process control (SPC)**. SPC is discussed later.

Customer Service

A high quality product includes customer services to ensure the customer gets adequate information, support, and assistance to install, operate, and maintain the product. Customers expect the producer to be responsive to their complaints and warranty claims, and willing to replace or repair products that do not meet promises or expectations. Customer service is an important aspect of quality of performance/service.

Customer service, like marketing and sales, is a source of information about customers' opinions and suggestions. This information is channeled back to the design, production, and procurement areas so that sources of existing quality problems are identified and eliminated, and suggestions for improvements are integrated into new products.

Within the TQM framework, lean customer-focused organizations improve and sustain quality using a variety of tools and methodologies. Besides the problem-solving tools and kaizen approaches described in Chapter 2, two main quality-focused methodologies are Six Sigma quality programs and statistical process control.

Six Sigma

Motorola first used the term *Six Sigma quality* in the 1980s to represent its efforts applying Japanese quality techniques to manufacturing. Subscribing to the maxim that quality products require quality parts, it began offering training courses in quality techniques to its parts suppliers and other interested companies. Although focused originally on manufacturing, Motorola later expanded its Six Sigma concept to other functions such as distribution, marketing, and order processing. In 1988 Motorola achieved recognition for its quality efforts by winning the Malcolm Baldrige National Quality Award. Today, quality in some products at Motorola has improved so much that defect rates are now measured in parts per billion.

Over time the Six Sigma concept spread throughout industry, and in 1995 its popularity soared when General Electric corporate management adopted six sigma as a companywide strategic initiative. Such initiatives take a long time to bear fruit, but the results can be dazzling. GE claims that for a $400 million investment, mostly in training, it reaped $1.2 billion in benefits.[4] Beyond profits, other reasons companies undertake Six Sigma programs include becoming more competitive, exceeding customer requirements, and being certified to supply Six Sigma business customers. Six Sigma programs provide direction and priority in quality improvement and are a natural partner to continuous process improvement in lean production. Six Sigma efforts emphasize customer focus, variability reduction, product and service performance, financial performance, and ability to meet quality requirements. To every lean effort, Six Sigma contributes better ability to achieve customer expectations, and product and service performance goals.

Statistical Interpretation

Where does the term *six sigma* come from? Most people believe that 99% quality is pretty good—that is, until they understand the ramifications. Based on current demand figures, 99% quality means that, on average, we would expect the following number of errors:[5]

- 5,000 surgical operations performed incorrectly each week.
- 200,000 drug prescriptions filled incorrectly each year.
- Electrical service interrupted 15 minutes each day.

If you are in need of surgery, a prescription, or electricity, it is doubtful you will consider this as "good" quality. How much better is 99.9%? Ten times better, but still leaving lots of room for problems: 500 incorrect operations, 2,000 incorrect prescriptions, and 1.5 minutes without electricity. Not so good either. Well then, what is good quality? According to the Six Sigma standard, good quality is 99.99966% quality, which translates into 3.4 errors per million. Six Sigma quality is more than a thousand times better than 99% quality.

The word *sigma* refers to standard deviation. The output of every process is variable, and the variability in many processes conforms to a normal distribution. If the mean of the distribution lies close to the desired or target value, then the smaller the standard deviation, the higher the percentage of the output that will be close to the target and the smaller the percentage that will be so far away as to be considered defective. Figure 4.3 illustrates this with two example processes. Both processes have the same mean, but the standard deviation of process 1 is much smaller than for process 2. If the shaded regions represent unacceptable items (defective) because they are too far from the mean, then process 2 will have many more unacceptable items than process 1. In a

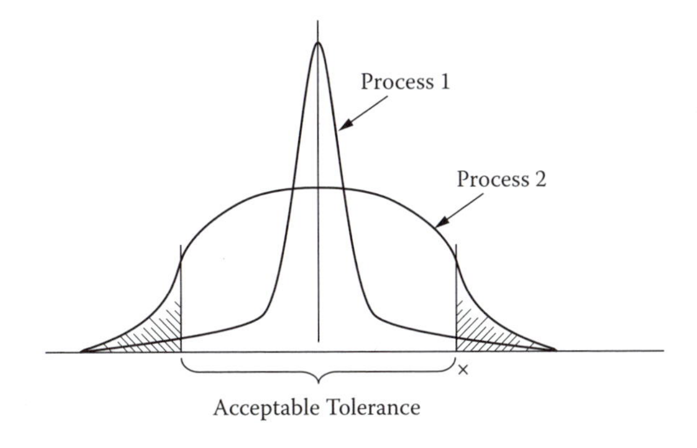

Figure 4.3 Distribution of output of two processes.

Six Sigma quality process, the unacceptable items are located 6 standard deviations away from the mean. In such a process, the number of defects in the shaded regions will number only 3.4 per million occurrences.

For example, suppose a process makes machined shafts and that the maximum allowable diameter of the shafts is 2.5 cm. Any larger diameter is considered unacceptable. Suppose x in Figure 4.3 represents 2.5 cm, so anything to the right of x is unacceptable. The shaft can be made in either process 1 or process 2, but the number of unacceptable items for process 2 is much greater than for process 1. The reason for the difference is because the standard deviation in process 1 is much smaller than the standard deviation in process 2. In statistical terms, this says that x falls more standard deviations away from the mean for process 1 than it does for process 2. In a Six Sigma process the standard deviation is small enough that results beyond x (unacceptables) are 6 standard deviations away from the mean. The number of items beyond ±6 standard deviations is 3.4 per million.[6]

Six Sigma quality represents a goal that in reality might be very difficult to achieve. In a multistep process, even if Six Sigma quality were attained in each of the steps, the total process output might not be six sigma. If a process requires five steps in sequence, and each step has attained 99.99966% (six sigma) quality, the yield of the process will be (99.99966%) which is only 99.998%. Despite the difficulty, however, committing the effort to achieve Six Sigma quality is a worthy goal sometimes made necessary to equal or exceed the best competitors.

DMAIC Improvement Process

The Six Sigma philosophy is enacted by employee teams in quality improvement and waste reduction projects. These projects follow a methodology developed by GE and widely used by companies around the world. The methodology is a called DMAIC, which represents the following five steps:[7]

1. Define (D): Define the problem, the customer of the problem, and critical-to-quality attributes or CTQs (the criteria the customer considers most important).
2. Measure (M): Identify the processes that influence the CTQs and measure their performance.
3. Analyze (A): Determine the causes of problems or poor performance in the process; determine key factors causing large or erratic process variation.

4. Improve (I): Confirm the impact of the key factors on the CTQs. Determine methods for measuring variation, the maximum acceptable range of variation, and methods to make the process acceptable.
5. Control (C). Employ the methods to ensure that the process stays within the acceptable range.

The thrust of DMAIC is process improvement through identification and reduction of process variation (steps A and I), which, as represented in Figure 4.3, is a main contributor to defects and poor quality.

Although the methodology was originally created for manufacturing processes, companies now apply it to all kinds of processes—whatever most impacts their ability to meet customer product and service requirements.

Belts and Certification

Companies with Six Sigma programs engage in ongoing projects similar to the kaizen projects described in Chapter 2. The projects are done by cross-functional teams of frontline workers and support staff, many of whom are trained in the basic data collection and problem solving tools. The projects follow the steps or DMAIC and are facilitated by "belts," experts certified in the DMAIC process and trained in data collection, statistical analysis, and problem solving, and experienced in Six Sigma projects. Most Six Sigma programs incorporate different level belts to account for different level qualifications. The belt levels originally defined in GE's program and used today by many other companies are

- Green Belts: Trained in data collection and analysis tools and Six Sigma methodology, plus additionally trained by the more qualified Black Belts.
- Black Belts: Promoted to assume full-time responsibility for leading and consulting with improvement teams throughout the organization, its customers, and its suppliers organizations, and to train and mentor Green Belts.
- Master Black Belts: Promoted to take on responsibility for setting qualities strategies and deployment methods, and to train and mentor Black Belts.
- Champions: Business leaders trained in Six Sigma tools and responsible for promoting and leading the companywide Six Sigma program.

Statistical Process Control (SPC)

Statistical process control (SPC) is a methodology for establishing and maintaining high-quality output. Although not derived from lean production or TQM per se, it is nonetheless essential to both. SPC includes a set of tools and principles for determining if a process is stable, for monitoring a process for possible changes in behavior, and for assessing whether a process is capable of meeting production requirements. This section gives a brief, nonmathematical overview of SPC concepts and tools.

Described earlier and illustrated in Figure 4.3, a notable feature of every process is variability. Variability is inherent in all outputs and processes, although if the variability is small enough it is inconsequential. High-quality products originate from processes wherein the variability is so small that only a very tiny proportion of the output fails to meet requirements. Thus, one key to achieving a high-quality process is to manipulate aspects of the process so that the variability is very small.

Control Chart

Assuming production engineers have gotten everything to work such that the process conforms to specifications (the average of the process is close to the target value and the variability [standard deviation] is very small), after that they have to ensure that the process doesn't change. If, say, the current manufacturing process meets requirements 99.99% of the time, and managers consider that good enough, then some way is needed to monitor and control the process to make sure it stays that way. The assumption here is that as long as the process does not change, it is OK; if, however, the process changes, it might not be OK and something will have to be done to get the process back to where it was. The problem is every process eventually changes. The producer needs to know exactly when a change has occurred or is about to occur so the cause can be found and necessary corrections made to get the process back to the desired state.

The tool used to monitor a process for potential change is the **control chart**. The significant features of the control chart, shown in Figure 4.4, are a center line and upper and lower control limits, all of which are computed from sample data and statistical formulae. Once the values for these lines are set, the control chart is used to monitor the process by taking periodic samples of the process output and plotting the results on the chart.

For example, an assembly worker (Figure 4.5) performs tests every hour on a sample of five units of process output. The results of the tests are averaged and plotted on a control chart like Figure 4.4. The test results will vary slightly from sample to sample due to random variation in the process, but as long as they remain within the control limits and exhibit random variation (not too many points in succession trending upward or downward, nor too many lying above or below the center line), the process is considered in control, which means the variation from sample to sample is random and the process has not changed. If otherwise—points lie outside the control limits or in succession exhibit nonrandom behavior—the process will be suspected of having changed. In that case the worker will stop the process to investigate what might have happened.

There are no guarantees. Although the variation might in fact be entirely random, sometimes points fall outside the control limits or appear nonrandom, leading the worker to mistakenly conclude that the process has changed, even though it hasn't.

A "good" process is one where both the *mean* and the *variability* of the process in combination result in a very high percentage of process output meeting product or process requirements. Thus,

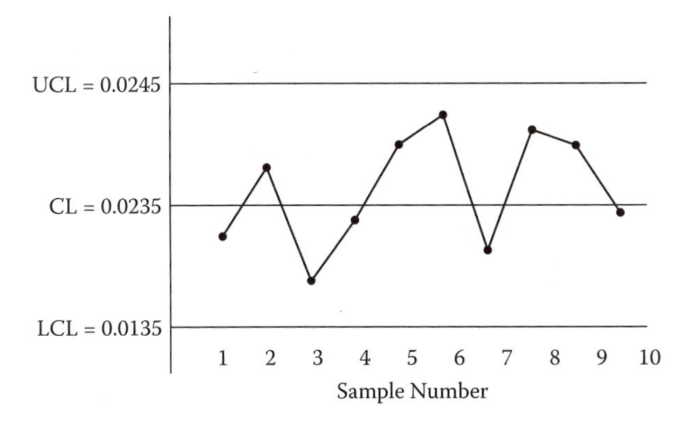

Figure 4.4 Example control chart.

Figure 4.5 Assembler inspecting output.

two kinds of control charts are used together: one to track the process mean and the other to track the process range or standard deviation (two ways to measure variability). If either chart indicates a potential change in the process, the process is stopped and investigated for the cause.

Process Stability

Although every process has variability and, thus, every unit produced is slightly different, a large number of the units when combined create a distribution for which the mean and standard deviation can be computed. The mean and standard deviation are important parameters for monitoring a process because they describe the overall behavior of the process and how well it compares to requirements. Besides for use in constructing control charts, they enable us to determine, for instance, that only 8 parts per million will be defective.

But the mean and standard deviation of a process can be determined only if the process is **stable**. A stable process is one where fundamental features of the process are repetitive and unchanging. In a stable process the mean and standard deviation are unchanging, that is, they will remain the same until something happens in the process that changes them. In the production of circular rods, for example, the average diameter of the rods produced will be the same today as it was yesterday—given that nothing has happened to the process to move the average up or down. The same applies to the standard deviation of the diameter. Even though the diameters of individual rods will vary slightly, in a stable process the standard deviation of the rods remains the same.

In contrast, in an **unstable** process the mean and/or standard deviation of the process are constantly changing due to changes in factors of production such as workers, equipment settings, process procedures, or material qualities. As long as they keep changing, it is impossible for the process to produce output that is on average constant and uniform. The point is, unless a process is stable, the process mean and variability cannot be determined, and the process cannot be monitored for quality.

Thus, first things first. Before employing control charts, it is important first to ensure that the process is stable. If it is not, the sources of instability (procedures, equipment, labor, material, worker skills, etc.) must be identified and remedied.

Process Capability

Besides stability the other desideratum of quality is process capability. A process that is behaving as it should is called **capable**, which implies that with rare exceptions it produces output that conforms to specification requirements. A six sigma process is considered exceptionally capable, but a three sigma process only marginally capable. Process capability is specified by a **capability index**, which roughly indicates how well the process fits within the product specification limits (i.e., the range between the lower specification limit and the upper specification limit). In Figure 4.3, process 1 has higher capability than process 2 because more of its output is acceptable.

Thus, again, first things first. Before applying control charts, the process capability must be determined. If the capability is good, then control charts should be applied to monitor the process and keep it as is. But if the capability is poor, the process should first be improved. For the latter case, the procedure works like this: determine the process capability and whatever fixes are necessary to improve it, institute the fixes, then again determine the capability, and repeat. Once the process capability is considered good enough, control charts can be instituted to help assure it stays that way.

Nonstatistical Process Control

Lean production contains many practices that assist in process control, a good example being visual management. With visual management, everything is out in the open for everyone to see. Materials are arranged in bins on the floor or in racks; instructions for standard operations, setup, and preventive maintenance are posted at machines and workstations; and charts showing quality goals, requirements, and progress are posted in work areas. This makes it easier to know and enforce standards and requirements, and to spot deviations that contribute to defects and nonconformities.

Workplace discipline and organization also benefit process control. Most everything that flows through a lean process—raw materials, parts, components, and assemblies—follows the first-in, first-out priority rule. The rule is reinforced with posted standard procedures ("always put material at the rear and take material from the front") and suitable equipment (e.g., racks with tilted shelves so parts slide to the front, dairy-case like, when the front item is withdrawn). Items move in a stream that prevents the bad ones from being intermingled with the good.

Employee Involvement and Quality Ownership

Frontline Worker Responsibility

One feature about lean companies—whether or not they label their quality initiatives TQM or Six Sigma—is that virtually everyone in the company feels some degree of responsibility for quality. Of particular importance are the people who directly add value to products and services—the frontline operators and assemblers on the shop floor and the service workers who interact daily with customers. If you ask any of them "Who here is responsible for quality?" you will hear back "I am" or "our team is." Accepting personal responsibility for quality, again, is a precept of quality at the source, the concept that if you want to eliminate a defect, you have to go to the source of the defect.

This is not to say that frontline workers are the source of defects, but rather that they are as close to the source as is humanly possible. Being within the process, they are able to observe defects and mistakes as they occur and, often, to enact small fixes that eliminate the causes. Some companies still rely on inspectors, but that is an outmoded and ineffective way to handle defects. Inspectors see only a sampling of the output, and they easily miss mistakes that are hidden or covered up. When an inspector discovers a defect, often she is at a loss to explain the cause because she discovered it too late or far away from where the defect originated. It is one thing to discover a defect but quite another to determine its source. When every employee feels responsible for the quality of his or her every action, the origin of a defect is more readily identified and soon fixed, and the likelihood a defect will survive and emerge from the process is significantly decreased.

Some managers think that delegating primary responsibility for quality to frontline employees is risky or unproductive. They see many potential drawbacks (reduced production capacity, lengthened cycle times, employee inability to be competent inspectors and problem solvers) but few benefits. Yes, DMAIC and kaizen projects take time, and yes, cycle times do lengthen, at least initially, although not as much as expected. But scrap, rework, and other wastes associated with defects plummet, and, ultimately, overall cycle times quicken and capacity rebounds. On balance, overall rates of production go up and costs go down, even allowing for the time and cost of training.

In lean companies, continuous improvement in quality is the result of an ongoing offensive to identify problems, understand root causes, and eliminate them. Shouldering the brunt of responsibility for that offensive are the frontline workers.

Process Orientation

Improvement teams in lean TQM organizations are organized around processes, not functions. Each team is comprised of workers and support staff linked by a common process. One example is a group of workers who serve each other as suppliers and customers. The team includes workers from different areas of the plant that each produce or use one another's parts and subassemblies. Another example is a customer–vendor team. Workers from both the customer plant and the vendor plant meet periodically to discuss problems about quality, delivery, and cost and to jointly decide on how to resolve them. Finally there are the kaizen and DMAIC teams, mentioned earlier. These too are process oriented and rely substantially on participation from the workers and support staffers who know the most about the process. In all cases, teams like this raise the consciousness level about the customer perspective for both internal and external customers, and facilitate cross-functional problem solving.

Quality Training and Education

Says Kaoru Ishikawa, one of the early promoters of Japanese quality improvement programs: Quality control (QC) begins with education and ends with education. To promote quality with participation by all, QC education must be available to all employees, from president to assembly-line workers.[8]

Perhaps obvious is that associated with responsibility for quality is the ability to collect and analyze data, identify and solve problems, and to work effectively in teams. Thus, a major component of all quality improvement programs is training frontline workers in these skills.

The best kind of training is **just-in-time style training**, that is, training and education that can be applied immediately, as needed, and on the job.[9] Simply training masses of workers (an unfortunate common practice in many organizations) is not effective. Far better is an ongoing series of short training sessions (a few hours a week, at most) during which workers are taught only what they can apply soon (and not overloaded with information they cannot). The sooner they can apply learned skills, the more effective the learning and the longer it will be retained.

Of course, managers and supervisors are trained too, not only in the tools of analysis and problem solving, but also in how to manage and supervise empowered workers. Many of them are lacking in coaching and facilitating skills, and need training and reinforcement in skills related to participative management, active listening, and collective decision making.

Usage of the key concepts and tools for quality is enforced from one level of the organization to the next by managers at each level requiring their subordinates to use quality tools and concepts. For example, a vice president requires that his managers use Pareto charts and cause-and-effect analysis during presentations, and also requires that those managers, in turn, require the same from their direct reports. Everyone, from frontline workers to executives, shares in the usage of common methods for quality improvement and problem solving.

Implementing TQM

The ways of implementing TQM are almost as varied as the number of companies adopting TQM. One theme, however, pervades them all: they are intended to change organizational culture and to drive customer focus, continuous improvement, and employee involvement at every level and job throughout the organization. In terms of the way they go about doing this, several themes are common:

- Top management sets a company vision or broad fundamental objective.
- The vision is broken into narrower, more focused, and shorter range objectives and plans at every level of the organization.
- The vision and objectives are set high and focus largely on the external customer.
- Objectives are developed by teams at every level so as to involve most of the employee population.

Also important as mentioned is widespread training to make sure employees develop the necessary problem solving and improvement skills relevant to their jobs, and frequent use of prespecified metrics to assess process improvement and progress toward meeting targets, which are set as means to provide concrete direction and motivation for improvement efforts.

Barriers to Successful TQM[10]

The barriers to implementing TQM somewhat mirror those for lean production because both TQM and lean involve what are, for many companies, dramatic changes in ways they set goals and try to achieve them. Additionally, both TQM and lean necessitate a cultural transformation that shifts the locus of responsibility somewhat toward the frontline workers. Thus, the two main reasons for lean failure and TQM failure are the same: lack of long-term management commitment and leadership, and lack of empowerment of frontline workers.

The literature gives other reasons for TQM failure:[11]

- Misdirected focus on the trivial many problems facing a company rather than the critical few
- Emphasis on internal processes to the neglect of external (customer-focused) results
- Emphasis on quick fixes and low-level reforms
- Training that is largely irrelevant and lacks focus
- Lack of cross-disciplinary, cross-functional efforts

Managers often see improvement efforts as solutions to particular problems (e.g., lean to make certain operations less wasteful, TQM to improve one product's quality) rather than as stepping stones to cultural change and improved competitiveness.[12] Thus, they give little forethought as to whether the allocation of scarce resources to correct particular problems or weaknesses is in the best long-range interest of the company.

Lack of appropriate measures, performance reporting, and reward systems also contribute to failure.[13] The cost of quality is inadequately measured; focus is on visible costs such as scrap and warranty, while invisible costs such as lost sales and customer defections are ignored. Financial measures encourage short-term performance at the expense of long-term improvement. The traditional cost accounting system does not assess quality costs and improvement benefits, and these things are not included in formal management reports. Reward systems and formal methods of recognition are not tied to quality or improvement metrics, so there is no monetary or promotional incentive for employees to change.

In short, TQM like lean production is a long-term, lasting commitment to continuous improvement. Both require a willingness to change and sustained devotion to see the change through.

TQM and Lean Production

Lean production, TQM, and Six Sigma all have similar roots in the continuous improvement and waste elimination movement that began in Japan in the 1950s. Thus, they considerably overlap and, some argue, are essentially the same.[14] Yet there are distinctions, and practicing lean should not be equated to practicing TQM. Lean is a process broadly aimed at increasing value added and eliminating waste, and includes techniques geared toward those things. But eliminating waste cannot be achieved solely through efforts directed at manufacturing. Process improvement and waste reduction require a companywide, integrated effort that includes all functions—marketing, sales, finance, product engineering, purchasing, customer service, accounting—as well as manufacturing, which is the emphasis of TQM and its companywide commitment to quality. While lean production seeks improvement through reduction of waste, TQM seeks improvements that matter most to customers. As an example, manufacturers used to think that inspection to sort out defects was an improvement because it reduced bad products going to the customer. It did reduce waste and it was an improvement. Nonetheless, since sorting out defects did not involve finding the root causes of defects (and allowed the causes to stay in the system), the improvement was not fundamental. Fundamental improvement means eliminating the sources of problems, not just the symptoms, and it requires changes in functions besides manufacturing, including product design and purchasing. Lean production in theory seeks to eliminate wastes that might mask problems, but TQM provides the mechanisms for finding the root causes of waste so they can be eliminated.

Because lean systems run with little slack or inventory, the production process must be reliable, stable, and predictable. The biggest detractor from reliability and stability in processes is variability in production schedules, procedures, materials, machine functioning, and worker skills. TQM and in particular Six Sigma provide the procedures and tools for identifying and eliminating the sources of excess variability. Since companies are always changing products and processes, and since the manufacturing environment (both internal and external to the company) is always changing, any source of variability, once identified and removed, will soon be replaced by others. Using TQM and Six Sigma, the new sources of variability will be identified and dealt with before they cause problems.

The joint application of lean production principles with TQM and Six Sigma methodology has become so common that a term has been coined for it: **lean Six Sigma**.

Summary

Quality can be defined from two perspectives: that of the customer and that of the producer. The customer-focused quality approach starts by taking the customer's perspective, then works backward to translate customer requirements and needs into the producer's perspective, which are considerations about product design and manufacturing that will result in a product or service that meets or exceeds customer requirements. Orienting the design and manufacture of products to meet these requirements is referred to as quality of design and quality of conformance, respectively.

Total quality management is an organization-wide management approach committed to quality and continuous improvement. It is also a process in which an organization is continually aware of the changing needs and expectations of customers, of challenges posed by the competition, and of threats and opportunities in the business environment. TQM is also a management framework for integrating the efforts of sales, marketing, finance, engineering, manufacturing, purchasing, suppliers, and customer service around determining what customers want and need, and delivering the products and services that satisfy them. It is a concerted, continuing effort to broaden and deepen worker involvement in quality and improvement efforts and to give frontline and support workers the skills, training, education, and support necessary for them to meaningfully contribute to those efforts.

Any organization adopting lean production must also be involved in TQM initiatives and employ quality of design methods such as concurrent engineering and DFMA, and the process control methods of SPC. Whereas lean production provides techniques and philosophy for process improvement through waste-reduction efforts, TQM provides methods, teamwork, organization-wide participation, and customer-focus so that these improvement efforts address fundamental problems and result in better products and services from the customer's perspective.

Notes

1. D. Garvin, 1987. Competing on the eight dimensions of quality, *Harvard Business Review* November-December (1987): 101–109.
2. See C. L. McNair, *World-Class Accounting and Finance* (Homewood, IL: Business One Irwin, 1993).
3. B. Flynn, Managing for quality in the U.S. and in Japan, *Interfaces* 22, no. 5 (1992): 69–80.
4. K. Henderson and J. Evans, Successful implementation of six sigma: Benchmarking at General Electric Company, *Benchmarking: An International Journal* 7, no. 4 (2000): 278.
5. M. Harry, 1997. *The Nature of Six Sigma Quality* (Schaumburg, IL: Motorola University Press, 1997), 1–2.

6. The strict interpretation of Six Sigma is somewhat more involved than described here. Though the example assumes the population mean coincides with the target value, in reality it is difficult to get the process mean to coincide exactly with a given target value. Thus, the defect rate of 3.4 per million allows that the population mean can be as far away as 1.5 standard deviations from the target value. Nonstatisticians need not worry about this detail; the point remains: Six Sigma means high quality and a defect rate of practically zero.
7. Notice the similarity of DMAIC to the general improvement process of PDCA (plan–do–check–act).
8. K. Ishikawa, *What is Total Quality Control: The Japanese Way* [trans. D. Lu] (Englewood Cliffs, NJ: Prentice-Hall, 1985), 37.
9. D. Muther and L. Lytle, Quality education requirement, in *Total Quality: An Executive's Guide for the 1990s,* ed. Ernst & Young Quality Improvement Consulting Group (Homewood, IL: Business One Irwin, 1990), Chapter 7, 105–107.
10. For example: D. Garvin, 1988. *Managing Quality: The Strategic and Competitive Edge* (New York: The Free Press, 1988); G. Bounds, L. Yorks, M. Adams, and G. Ranney, *Beyond Total Quality Management* (New York: McGraw-Hill, 1994); Ernst & Young Quality Improvement Consulting Group, *Total Quality: An Executive's Guide for the 1990s* (Homewood, IL: Dow-Jones Irwin, 1990); S. George and A. Weimerskirch, *Total Quality Management: Strategies and Techniques Proven at Today's Most Successful Companies* (New York: John Wiley & Sons, 1994).
11. R. Chang, When TQM goes nowhere, *Training and Development Journal* 47 (1993): 22–29.
12. R. Hayes and G. Pisano, Beyond world calls: the new manufacturing strategy, *Harvard Business Review* January–February (1994): 77–86.
13. L. Tatikonda and R. Tatikonda, Top ten reasons your TQM effort is failing to improve profit, *Production and Inventory Management* 37, no. 3 (1996): 5–9.
14. J. Dalhgaard and S. Dahlgaard-Park, Lean production, six sigma quality, TQM and company culture, *The TQM Magazine* 18, no. 3 (2006): 263–281.

Suggested Readings

R. Amsden. *SPC Simplified: Practical Steps to Quality*. Portland, OR: Productivity Press, 1998.

R. Connors, and T. Smith. *How Did That Happen?: Holding People Accountable for Results the Positive, Principled Way.* New York: Portfolio Hardcover, 2009.

J. Liker, and D. Meier. *Toyota Talent: Developing Your People the Toyota Way.* New York: McGraw-Hill, 2007.

T. Stapenhurst. *Mastering Statistical Process Control: A Handbook for Performance Improvement Using SPC Cases.* Burlington, MA: Butterworth-Heinemann, 2005.

G. Taylor. *Lean Six Sigma Service Excellence: A Guide to Green Belt Certification and Bottom Line Improvement.* Fort Lauderdale, FL: J. Ross Publishing, 2008.

Questions

1. What is meant by the customer's perspective of quality? The producer's perspective? How do these perspectives relate to customer-focused quality?
2. What are the eight dimensions of quality as identified by Garvin? What are other possible dimensions of quality? Select three products and discuss how the dimensions apply to each.
3. What is the meaning of quality of design? Quality of conformance? Quality of performance and service?
4. In what ways is employee involvement and employee ownership in the quality concept cultivated in TQM organizations?
5. What is concurrent engineering?

6. What is design for manufacture and assembly (DFMA)? What is the relationship between DFMA and concurrent engineering?
7. What is meant by Six Sigma? What is the statistical interpretation of Six Sigma?
8. Describe the steps in the DMAIC procedure.
9. Describe SPC. What is the main purpose of SPC?
10. What is a control chart? What are its main features and how is it used?
11. What does it mean for a process to be stable?
12. What does it mean for a process to be capable?
13. Many large corporations have demonstrated their commitment to TQM by requiring that every employee attend seminars on quality concepts, and know the tools for problem solving and SPC. Everyone belongs to a problem-solving group, and groups are expected to meet once a week. Comment on this approach to TQM.
14. What is quality at the source? What role do frontline workers have in quality at the source?
15. What are the difficulties in implementing TQM? What distinguishes companies that say they are TQM companies from those that are TQM companies?
16. Why be a TQM and lean company? Why not be lean only?

ELEMENTS OF LEAN PRODUCTION

Consider again the wastes listed in Chapter 2:

- Defects
- Waiting time
- Transportation
- Processing
- Inventory
- Motion
- Overproduction

In a typical manufacturing plant it is relatively easy to find these wastes by just walking around. It is also easy to identify many of the *contributors* or *sources* of these wastes by observing management directives, shop-floor practices, and working conditions.

Among the primary sources of wastes are:

- Large lot-size production
- Inefficient setup procedures and long changeover times
- Poor performance and breakdown of equipment
- Poor layout of equipment for the processes required
- Inefficient procedures and lack of performance standards
- Poor shop-floor coordination and control

Some of these sources stem from outdated or incorrect notions about relationships between production costs, quality, efficiency, and demand. For example, when equipment changeovers are costly and time consuming, large lot-size production is considered as one way to drive costs down, despite the fact that it also contributes to increased inventories, which drives other costs up.

In some cases, waste comes from neglect. For example:

- Lengthy process and equipment changeovers and setups exist because no one has tried to find ways to do them better and faster.

- Machines run poorly and are unreliable because no one attends to them unless they malfunction or break down.
- Processes include wasteful and inefficient steps and procedures because no one has investigated them and prescribed "improved" procedures or the standards to which they should be held.
- Materials in the factory have to be moved long distances and then wait in queues between operations because the location of equipment and workstations is arbitrarily or unrelated to the process flow.

Sometimes the source of waste is simply that the system of production planning and control is inadequate to meet requirements of the production process and customer demand.

Outdated and incorrect notions, neglect, errors in judgment, and inadequate systems for planning and control are all somewhat interlinked; as a result, any effort to eliminate waste and improve processes in manufacturing must address all of them. For example, start with the premise (argued in Chapter 5) that small-lot production is generally preferable to large-lot production: If, then, you want to reduce the average size of production batches, you will probably first have to reduce the time and cost of production changeovers, possibly dramatically. And to be able to economically process those many small batches through a multistep series of operations, you will probably also have to rearrange the operations so they are closer together. Small lot-size production reduces inventory, but that leaves less stock to buffer against equipment problems and machine-induced defects, so you will also have to improve equipment reliability and the quality of the output.

Because all these approaches deal directly with the sources of manufacturing waste, including waste from defects and quality problems, they are considered the fundamental elements of lean production. This part of the book covers these approaches:

Chapter 5: Small Lot Production
Chapter 6: Setup-Time Reduction
Chapter 7: Maintaining and Improving Equipment
Chapter 8: Pull Production Systems
Chapter 9: Focused Factories and Group Technology
Chapter 10: Workcells and Cellular Manufacturing
Chapter 11: Standard Operations
Chapter 12: Quality at the Source and Mistake-Proofing

Chapter 5

Small Lot Production

Little strokes fell great oaks.

—Benjamin Franklin

Most everything manufactured is produced, procured, and transported in lots. A **lot** is a batch of something; in fact, the terms *lot* and *batch* are interchangeable. The smallest lot-size alternative is one unit, for example, pick up one lobster from the market, make one pizza from scratch, or form one metal part on a press. The alternative is multiple-unit lots—pick up 10 lobsters at once, make 10 pizzas, or shape 1,000 parts one after another. Which alternative is best depends on many factors including time, cost, and demand. Determining the right lot size is called **lot sizing**.

Traditional manufacturing has had an affinity for large lot sizes. It just seems more practical to do things in large quantities. Sometimes it is more practical, though by no means always. Many of the costs associated with lot sizing are not initially obvious. Besides cost, lot size has a major impact on production throughput and lead time, manufacturing flexibility, product quality, and manufacturing wastes such as inventory, waiting time, and transportation. In short, lot sizing is important because it impacts manufacturing speed, cost, quality, and agility. As this chapter will illustrate, a manufacturer and its customers potentially have much to gain by using **small lot sizes**.

The chapter reviews the factors relevant to lot sizing decisions and traditional approaches for lot sizing. Benefits of small lot production are discussed, as well as situations that merit large lot sizes. Practical and economical issues of using small lots in procurement and shipping and the role of lot-size reduction in continuous improvement are also covered.

Lot Size Basics

The effort required of going to the market, mixing together ingredients to make pizza, or preparing a machine to make a kind of part is called **setup**. The desire to minimize the time and cost of setup leads to working with larger lots—like buying multiple lobsters at once to save trips to the market, making multiple pizzas to save time on mixing ingredients, or manufacturing many parts at once to save time and cost associated with setting up a machine. Intuition says it is more efficient to do things in large lots.

But unless demand is large enough to quickly consume all the units in a large lot, many of them have to be stored. Excess lobsters and pizzas crowd the freezer, making it hard to put in or take out other things. If we eat 5 lobsters a year and start with 10 in the freezer, half will remain a year later. They might be edible, but how will they taste? Large-sized lots offer savings in some ways but they increase costs in others.

Dollar Costs Associated with Lots[1]

The two principle dollar costs associated with lot sizing are setup costs and holding costs. In the lobster case, setup cost is the cost of going to the market (time, fuel, etc.), and holding cost is the cost of storing the lobsters (space occupied, taste deterioration, etc.). Formally, the costs are defined as follows:

Setup cost, S, is the cost of preparing to make a batch or of ordering a batch. When applied to manufacturing, S is the cost of changeover from making one kind of item to another kind. It may include the cost of lost production while a machine or process is being changed and the cost of scrap incurred while adjusting the machine. When applied to purchasing, S is called the **order cost** and is the cost of placing and receiving an order. For purposes of lot-sizing analysis it makes no difference whether S represents setup cost or order cost. S is a fixed cost, independent of the size of the batch produced or ordered.

Holding cost, H, is the cost of holding a unit in inventory for a given time period. H includes expenses such as storage (rent, lease, mortgage, utilities, maintenance, etc.), tracking and monitoring of inventory, damage and pilferage, insurance, interest on money to produce or procure the items in inventory, and the opportunity cost of money tied up in inventory (and not available for use elsewhere). Usually items that are more valuable or costly to produce are also more costly to hold; that is, H is a percentage of the value of the item held in stock:

$$H = rP$$

where

P = **Unit production cost** or **unit procurement cost**, the cost of manufacturing (materials, labor, overhead) or purchasing one unit

r = Percentage based upon rates for borrowing, insuring, investing (opportunity), and so on

Lot Sizing and Setup Reduction

Small batch production and delivery is a main feature of lean production. Yet stand in a typical U.S. factory and look around. What you see too often are crates, boxes, bins, and pallets of materials and assemblies stacked floor to ceiling. This is a consequence of U.S. managers' preference for large batch production and procurement, which stems, in part, from the tendency to view material ordering, handling, and setup as immutable, fixed activities.[2] If you think the time and cost of a setup activity are fixed, then that time and cost will dictate your thinking about lot sizes: If S is large, then the lot size must also be large; if S small, then the lot size can be small, too. Traditionally, however, S has been large.

This concept of starting with S, assuming it is fixed, then determining the lot size as a function of S, is in sharp contrast to the lean approach to lot sizing. Lean production views all activities associated with setup, ordering, and handling of materials as nonvalue-added, wasteful things that should be minimized and eliminated. Thus, not only is S treated as not being fixed, but the explicit goal is to reduce S to zero, if possible. Lean production emphasizes **setup reduction** as another form of continuous improvement; it is the continuous reduction of the time and cost of setup, ordering, and handling of materials. As the time and cost of these activities is made smaller, production and procurement in small lots is easier to justify, even in high-demand situations. The next chapter discusses setup procedures and the topic of setup-time reduction.

Kind of Lots

A lot is a quantity of items purchased, produced, or transported. When a lot is the quantity of items manufactured as the result of a single setup it is referred to as a **production** or **process batch**. When it is the quantity of materials purchased from a supplier it is called a **purchase** or **order quantity**. A lot moved or transferred from one operation or workstation to another is called a **transfer batch**. Finally, a lot shipped between supplier and customer is called a **delivery quantity**.

As examples of these concepts, consider a machine that is set up to produce 100 units. As every 10 units are completed, they are moved to the next operation. In this case the size of the process batch is 100 units and the size of the transfer batch is 10 units. Similarly, suppose a purchase order is sent to a supplier for 500 units, and the supplier fills the order by making 10 deliveries of 50 units each. In that case the purchase order quantity is 500 units, and the delivery quantity is 50 units.

Lot Sizing

Deciding the appropriate sized lots is important to good manufacturing management. Regardless of the kind of manufacturing scheduling and control system used, whether a material requirements planning (MRP)-type push system or a lean-type pull system, lot sizing has a major impact on the efficiency, cost, and flexibility of production.

Organizations use different guidelines and procedures for determining lot sizes, from simple rule-of-thumb methods (order enough to last the month) to analytical procedures. This section reviews some common analytical models for lot sizing of process batches and purchase order quantities. Lot sizing of transfer batches is treated later.

Process and Purchase Batches

Following are four traditional lot-sizing approaches. We first explore the impact of these models on dollar costs, and then discuss the nondollar cost ramifications.

Lot-for-Lot

In **lot-for-lot** (LFL) lot sizing, the size of the lot or batch corresponds exactly to the amount required (ordered or forecast) during a particular time period. If, for example, total customer orders for 4 successive weeks are 600, 10, 120, and 200 units, respectively, then the product would be manufactured over the 4 weeks (or as close to then as possible) in process batches of lot sizes 600, 10, 120, and 200, respectively.

	Week			
	1	2	3	4
Demand	600	10	120	200
Lot size	600	10	120	200

If each batch is shipped immediately after production, holding costs will be zero because each lot will have been sent immediately and never held in stock.

An advantage of LFL lot sizing is that it generally works well in both pull-production systems and push-production systems (both discussed in later chapters).

LFL also works well whether demand is independent or dependent. **Independent demand** means the demand for an item is generated exogenous to the production system, that is, demand is customer or market driven. For lot-sizing purposes, independent demand is determined from either a forecast, actual sales, or a combination of both. **Dependent demand** means the demand for an item is generated internal to the production system, usually as a function of the demand for a higher-level item. For example, the demand for headlights and steering wheels in automobile production is a function of the number of cars produced (1,000 cars require 2,000 headlights and 1000 steering wheels). Headlights and steering wheels are dependent-demand items; the autos themselves are independent-demand items. Dependent-demand items can be procured from suppliers (headlights) or produced internally (fenders and trunk lids).

LFL dictates production or procurement whenever there is demand (whether independent or dependent), even if the demand is small. Thus, demand that occurs in frequent, discrete amounts will require frequent setups or orders (assuming one setup or order for every discrete amount) and, as a consequence, high costs from setups or ordering.

Period Order Quantity

The **period order quantity** (POQ) method reduces the number of setups or orders by restricting the frequency of orders. The POQ method starts with a predetermined **order frequency,** and then determines the lot size according to the demand occurring between successive orders. As an example, suppose in the earlier situation the order frequency is 2 weeks, that is, one setup or purchase order happens only once every 2 weeks. If orders are placed in weeks 1 and 3, then the sizes of the two lots would be:

	Week			
	1	2	3	4
Demand	600	10	120	200
Lot size	610		320	

While the POQ method results in fewer numbers of setups or orders than LFL (in our example, half), it also results in inventory being carried. In this example, 10 units are carried from week 1 to week 2 and 200 from week 3 to week 4. Compared to LFL, POQ involves lower costs for setup, but that is traded off with higher costs for carrying inventory. The next two lot-sizing methods balance setup and carrying costs and give the optimal tradeoff.

Economic Order Quantity

The **economic order quantity** (EOQ) model gives the lot size that most economically satisfies demand for a given time period; that is, it minimizes the sum of setup costs and carrying costs over a specified time period. The model was developed in 1915 and for 70 years was considered by academics as the fundamental method for determining lot size (managers, however, had learned from experience not to trust its results).

As suggested in the LFL and POQ examples, there is an inherent tradeoff in inventory systems between setup costs and carrying costs; as one goes up, the other goes down. If a few large lots are produced, setups will be few and overall setup costs low, but it will take longer to deplete the lots, so carrying costs will be high. If many small lots are produced, the carrying costs will be low because each small lot will be used up quickly, but the overall cost for setup will be high because of the many setups. This tradeoff of costs as a function of lot size Q is shown in Figure 5.1. The EOQ model gives the lot size Q that minimizes the sum of the two costs:

$$\min\ [\text{TC} = \text{Total setup cost} + \text{Total carrying cost}]$$

where TC represents total cost over a given time period (day, week, month, or year). In general, total annual cost is

$$\text{TC} = (SD/Q) + (HQ/2)$$

The lot value, Q, that minimizes this cost is

$$Q = EOQ = \sqrt{2DS/H}$$

where

EOQ = Optimal (least cost) lot size

D = Average demand for a year

S = Setup (or order) cost, as defined earlier

H = Unit holding cost per year, rP, as defined earlier

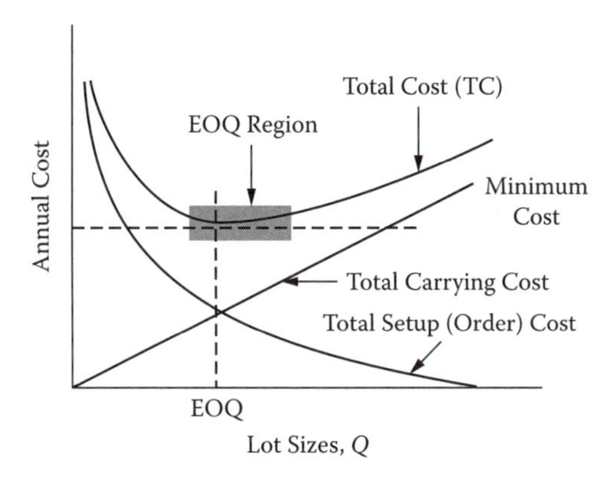

Figure 5.1 Inventory costs and lot sizes.

(See books on introductory production and inventory management for derivation of the TC and EOQ formulae.)

The model holds for the following assumptions:

1. Demand is constant and continuous.
2. Demand is independent.
3. Setup (order) cost, S, is fixed, regardless of lot size.
4. Unit carrying cost, H, is constant; total carrying cost is a linear function of lot size.
5. The entire lot is produced or delivered all at once.
6. Unit purchase price or manufacturing cost, P, is fixed, regardless of lot size (no quantity discounts or production economies).
7. Stockouts (or subsequent backorders) do not occur.

Example 1: Application of the EOQ Model

An item has an annual demand of 10,000 units, annual unit carrying cost rate of 20%, order cost of \$150, and unit purchase cost of \$5.

The annual unit carrying cost is $H = rP = 0.20(\$5) = \$1$. Thus, the lot size is

$$EOQ = \sqrt{2(10{,}000)(\$150)/\$1} = 1732$$

Given $Q = EOQ$, the total annual cost of ordering and carrying is

$$10{,}000(\$150)/1{,}732 + \$1(1{,}732)/2 = \$866 + \$866 = \$1{,}732$$

We can also determine the average purchase **order (or production) cycle** time, which is the time between orders: $Q/D = 1{,}732/10{,}000 = 0.173$ yr = 63 days. This says that, assuming constant demand, we will order a batch of 1,732 units about every 63 days.

Notice in Figure 5.1 that in the region near EOQ the TC curve is somewhat flat and, as a result, values of Q in the vicinity of EOQ all have relatively low cost. The previous example states that Q should be 1,732 to minimize cost; however, because of the flatness of the curve, lot sizes of, say, between 1,500 and 2,500 could be used instead and cost not much more that the minimum. (The exact effect is determined by substituting these values for Q in the TC formula).

Economic Manufacturing Quantity

The previous EOQ model assumes that the entire lot is delivered all at one time. For example, when a lot size of Q is ordered from a supplier, that amount, Q, is shipped in one delivery. When the lot arrives, the inventory increases by the amount Q. Now, although all-at-once delivery can happen for a batch of procured items, the situation is different when the batch represents items that are being manufactured. When items are manufactured one at a time, unit by unit, say at the **production rate** p, the inventory will grow gradually as each completed unit is moved into inventory. To account for this gradual increase in inventory the EOQ model is modified, and the optimal production lot size is called the **economic manufacturing quantity (EMQ)**. The shape of the TC curve for the EMQ situation is similar to that in Figure 5.1; thus, the procedure to determine the minimum-cost lot size is the same as for EOQ. The result is

$$Q = EMQ = \sqrt{2DS/H[1-(D/p)]}$$

With the exception that items are put into stock at the rate of p instead of all at once, the assumptions of the EMQ model are the same as for the EOQ model.

Example 2: Application of the EMQ Model

A manufactured item has an annual demand of 10,000 units. Production cost is $5 per unit; annual unit carrying cost rate is 5%. The production rate is 500 units a week.

Converting to an annual figure, p becomes

$$p = 500 \times 52 \text{ weeks} = 26{,}000.$$

Thus, $1 - (D/p) = 0.616$, so

$$EMQ = \sqrt{2(10{,}000)(\$150)/\$1(0.616)} = 2207$$

According to this model, production batches should be 2,207 units. This lot size will satisfy demand for $Q/D = 2{,}207/10{,}000 = 0.2207$ yr = 11.5 weeks, so a new batch of 2,207 units must be run every 11.5 weeks.

The EMQ model is more applicable than the EOQ model for determining the lot size of production batches, though, obviously, EMQ is but a variant of EOQ that allows for a different assumption. Still other variants of EOQ have been developed to account for other assumptions such as variable P (quantity discounts), multiple-item orders, back orders, and planned shortages.[3]

EOQ-Based Methods: Discussion

Even before lean production came along and started people thinking about the waste of inventory, the validity of EOQ methods for making lot-sizing decisions was questioned. Though the logic of the models is correct, the assumptions inherent in them delimit their applicability.[4]

One problem is demand. Virtually no industry has continuous, fixed demand, and many industries have seasonal or erratic demand. Practically speaking, with each change in D, the EOQ must be recomputed. The result could be wide swings in lot sizes over time, which would complicate planning, ordering, and scheduling of materials and production. When demand fluctuates slightly and has no trend, then an average value for D can be used. When it fluctuates much, EOQ-derived lot sizes based on average demand will result in either excess inventory or stockouts.

A second problem with using EOQ models lies in the costs. Much has been written about determining setup, order, and carrying costs, though the bottom line is that it takes considerable accounting finesse to be able to do it accurately. Few companies are able to.

Consider the difficulty in determining carrying costs, H. Even when all dollar costs in H are accounted for, H will still understate the true cost of carrying inventory because it fails to account for the detrimental effects of inventory on production quality and lead time. Generally, the larger the inventory carried, the more detrimental the effect. A similar difficulty arises with setup cost, S. Even when all dollar costs in S are accounted for, S is still potentially too low because it does not account for the effect of setup on idling the operation, machine, or process. If the operation is a bottleneck (i.e., scheduled work exceeds capacity), the cost of the setup is equivalent to the cost of all throughput lost as a result of the operation being idled because of the setup.[5] If, however, the operation is a nonbottleneck, the setup has no effect on throughput because the operation already has excess capacity. EOQ models do not distinguish bottleneck from nonbottleneck operations. These issues will be revisited later.

This is not to say that EOQ lot sizing is nowhere applicable. There are situations where EOQ-based lot sizing works, though they tend to be in nonmanufacturing situations like distribution and retailing where inventories of different items are independent, batches of materials do not flow between operations, and issues of production lead time and flexibility are irrelevant. In those cases EOQ is useful for determining the appropriate magnitude of orders, even when H and S are not precisely known. If EOQ says order 523, that probably means the lot size should be between, say, 200 and 800, though not 50 or 5,000. The flatness of the TC curve in the region near $Q = EOQ$ implies if you order "around 523" you will be near the minimum cost.

Transfer Batches

Many of the problems with large process batches come from the fact that a large batch takes a long time to produce, which ties up the machine or operation, prevents it from doing other jobs, and causes work-in-progress (WIP) inventory to build up both in front of and behind the operation. This increases the time other jobs must wait at the operation, which increases production lead times for all jobs. One way to reduce production lead time and WIP for process batches of any size is to use *lot splitting* and move items in smaller size **transfer batches**. Say, for example, 1,000 parts are being machined at an operation; as they are completed, they are put in containers that each hold 100 parts, and, upon being filled, each container is moved to the next operation. Thus, the transfer batch size is 100 units, and 10 transfer batches are required to move the entire process batch of 1,000 units.

In general, using transfer batches that are smaller than the process batch reduces production lead times because items finished at an operation wait less time before they are moved to the next operation. As an illustration, assume 1,000 units are to be processed through operation A, then operation B, and finally operation C. Each operation requires 1.0 min/unit processing time and has a process batch size of 1,000 units. The left side of Figure 5.2 shows the resulting effect of using a transfer batch size of 1,000. The right part of the figure shows the effects of using transfer batch sizes of 200, 500, and 500 units at operations A, B, and C, respectively.

The top part of Figure 5.2 represents accumulated WIP output for each operation, and the bottom represents elapsed time. The dashed triangles on the right represent the combined inventory at operations A, B, and C. The difference in the size between the triangles on the left and those on the right reveals how the smaller transfer batches result in smaller WIP. Using the smaller transfer batches also reduces the production lead time from 3,000 minutes to 1,700 minutes.

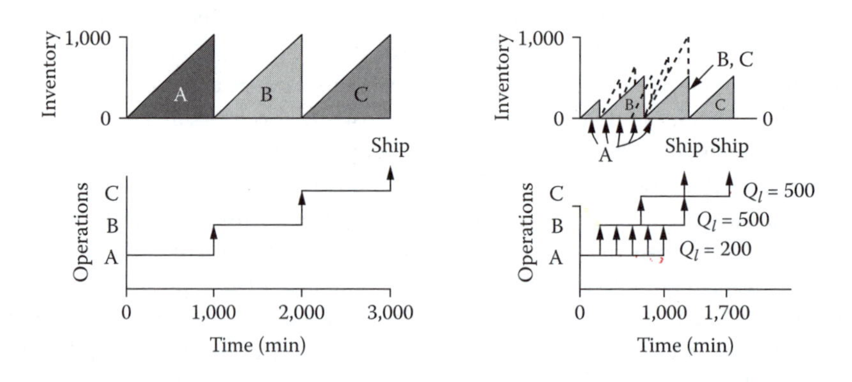

Figure 5.2 Left: Transfer batch = process batch = 1,000. Right: Transfer batches are 200, 500, 500, respectively at A, B, C, and process batches = 1,000.

There is a caveat: Using small transfer batches provides the benefits exemplified only if the transfer batches can be processed as soon as they arrive at the next operation. In Figure 5.2, the transfer batches arriving from operations A and B are processed immediately at operations B and C, respectively, because (we assume) there is no preexisting WIP ahead of those operations and the setup time is zero. The greater the WIP in front of operations or the longer the setup times, the smaller the benefit of using small transfer batches. To prevent large WIP buildup in front of operations, the setup times and process batch sizes at operations throughout the plant must be small. Reducing the setup times or process lot sizes at just a few operations will have little overall affect; it speeds up the flow of jobs through some operations, only to enable them to arrive sooner at subsequent operations (where they will have to wait longer until they can be processed).

In general, the smaller the transfer batch, the larger the number of transfers (moves) required. Therefore, any initiative to reduce transfer batch sizes must be accompanied by a program to reduce the time and cost of material handling.

As the previous example illustrated, transferring material between operations in batches that are smaller in size than the process batch can significantly influence production lead times and WIP quantities. Reducing the lot size of process batches has a similar influence, as explained next.

Lot Size Reduction

Effect of Lot Size Reduction of Competitive Criteria

To illustrate the effect of reducing the lot size of process batches, consider two products, X and Y. Assume that both are processed in sequence through three operations—A, B, and C; the setup times to changeover between products is negligible, and the daily production rates at the operations are as follows:

	Operation (units/day)		
Product	*A*	*B*	*C*
X	1,000	2,000	1,000
Y	2,000	2,000	2,000

Now, assume that 8,000 units each of products X and Y must be produced, X first, then Y. Figure 5.3 shows the schedule when both the process and transfer batch size is 8,000 units for

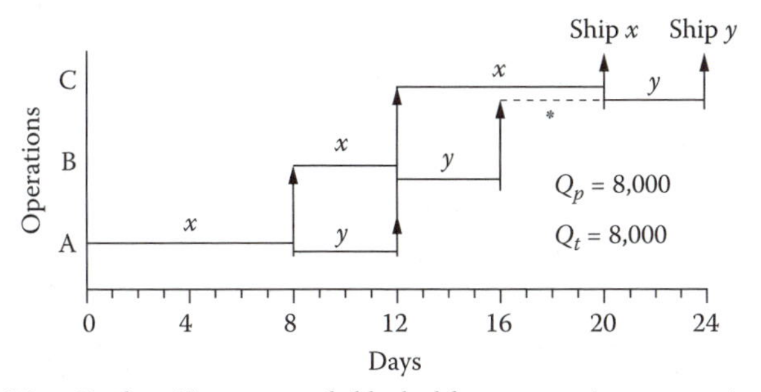

Figure 5.3 Schedule with process batches = transfer batches = 8,000.

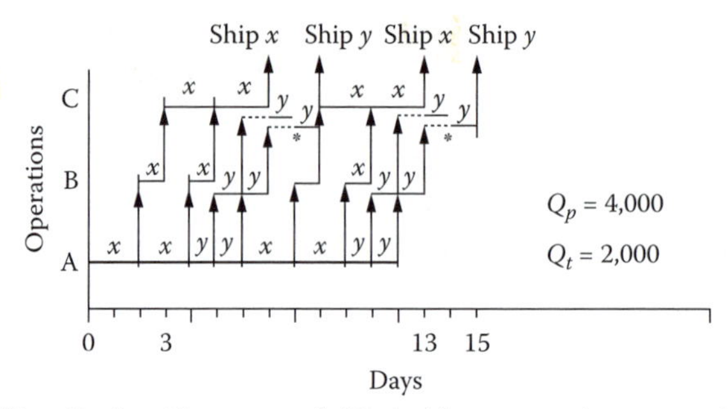

Figure 5.4 Schedule with process batches = 4,000; transfer batches = 2,000.

both products. Figure 5.4 shows the schedule when the process batch size is reduced to 4,000 units and the transfer batch size is reduced to 2,000 units. In the figures, batches of X and Y are represented by x and y, respectively. Comparing Figures 5.3 and 5.4, we can draw some general conclusions about the effects of smaller process lot size on the speed, cost, quality, and flexibility of production.

Lead Time

Production lead time (start to finish) for product X is reduced from 20 days to 13 days; for product Y it is reduced from 24 days to 15 days. With the smaller batches, product X finishes 7 days earlier, and product Y finishes 9 days earlier. In general, the smaller the size of a job, the less time the job ties up an operation, and the less other jobs at the same operation have to wait to be processed. Conversely, the larger the process batch size, the longer the job ties up an operation, the longer other jobs have to wait, and the more difficult it is to hold jobs to schedules. Even small variability in the per-unit processing time for a large batch can result in a significant job delay, and a delay in any one job at an operation causes delays in every other job scheduled to move through that operation later. In turn, the delays of these jobs at one operation invalidate their production schedules, and not only at the one operation, but all other operations the jobs are scheduled to visit later.

Note the preceding examples assume that operations are never busy with jobs other than products X and Y. This assumption, however, does not invalidate the conclusions about using smaller lot sizes. If, in the example, batches of products X and Y had to wait on other jobs upon arriving at subsequent operations, the effect would be to add the same wait time in both Figure 5.3 and Figure 5.4. Obviously, the elapsed time in Figure 5.4 would still be shorter.

Besides reducing the total lead time, breaking a large batch into smaller batches may have another benefit: lower lead time variability. Suppose, for example, that the usual batch size is 200 units and that it requires a process time with an average of t_L = 200 minutes, with a standard deviation of σ_L = 40 minutes. Assume the batch can be split into four batches of 50 units, each requiring a process time of t_S = 50 minutes, with σ_S= 10 minutes. If the four batches are independent, then the total process time average for the four batches remains 4(50) = 200 minutes, but the standard deviation for the combined four 50-unit batches is $\sqrt{4\sigma_s^2} = \sqrt{4(100)} = 20$, which is only half the standard deviation of the 200-unit batch.[6] This example is only an illustration of the potential of how dividing big jobs into small, uniform jobs can reduce the time variability of the overall process.[7]

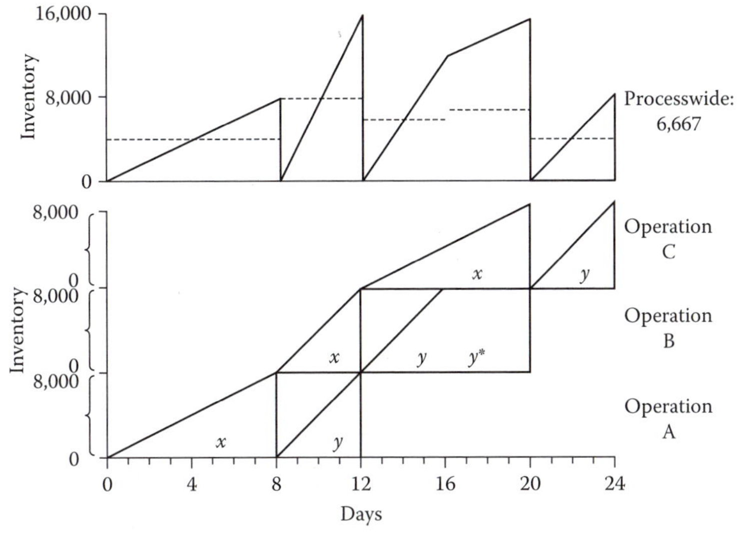

Figure 5.5 Top: Inventory level, process wide. Bottom: Inventory level of completed units following each operation when process batches = transfer batches = 8,000.

Carrying Cost

For a given production volume, WIP is proportional to production lead time. The longer items are at an operation or in the system, whether in process, in transit, or waiting, the greater the accrued costs of carrying those items. Figure 5.5 and Figure 5.6 show the accumulated output inventory processwide and for each operation. The dashed lines for the processwide inventory represent average inventory: for the larger batches, the average is 6,667; for the smaller batches, it is 4,800. Besides having smaller average inventory, the smaller batch scenario retains that inventory for just 15 days, while the larger batch scenario carries its inventory for 24 days, which is significant, especially if items have a short shelf-life or are expensive to stock.

The cost advantage of smaller inventories goes beyond dollars. With small inventories, items are easier to keep track of so inventory records can be kept more accurate. It is easy to count items and see overstocks or pending shortages. Small inventories free up floor space and make it possible to store materials next to the machines and operations where they are needed, called **point of use**. Point-of-use storage eliminates the double handling and double accounting that occur when materials are put into stockrooms and then later withdrawn and moved to the places where needed. With point of use, materials are moved directly from the receiving area to operations on the shop floor.

Setup and Handling Cost

In the example, the smaller-batch case requires twice as many setups and four times as many material transfers as the larger-batch case (Figure 5.3 versus Figure 5.4), which overall equates to a doubling of setup cost and a quadrupling of handling costs. It should thus be clear that if process and transfer lot sizes are to be reduced, the time and cost associated with setups and material handling

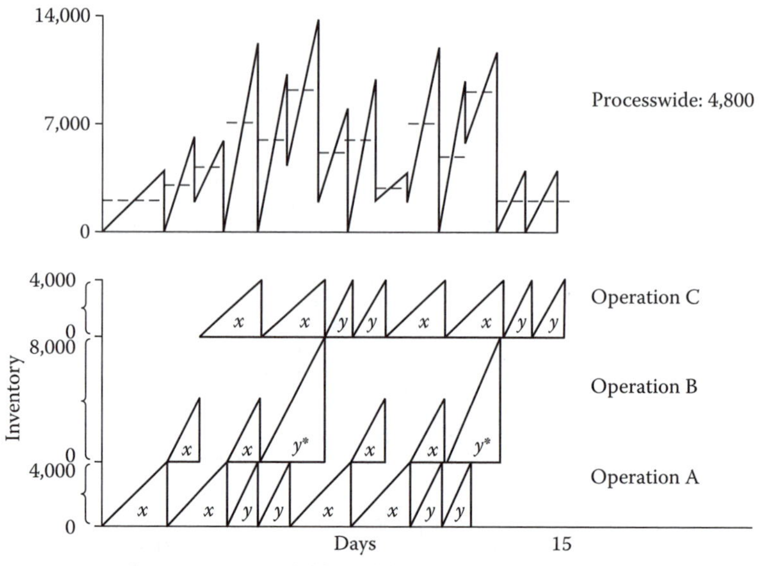

Figure 5.6 (Top) Inventory level, process-wide; (Bottom) Inventory levels of completed units following each operation using process batches = 4,000, transfer batches = 2,000.

must also be reduced to make small lot production practical and economically feasible. Fortunately, reducing setup and handling costs is not necessarily as difficult as might be presumed. Ways of reducing costs and wastes associated with setups and material handling are addressed later in this chapter and elsewhere throughout the book.

Quality

Defects are often batch specific; they result from an incorrect setup or method unique to the production batch. As a result, the larger the batch, the more units affected. In the example, 8,000 units would potentially be affected in the large-batch case, versus only 4,000 units in the small-batch case.

Often a defect is not discovered until well after it is created. With larger batches it might not be discovered until much, much later, and by that time the cause of the defect might be difficult or impossible to find. Suppose a defect is introduced in product X at operation A but is not discovered until operation C. Referring to Figure 5.3, the defect will not be discovered until 12 days after it occurred, at the earliest. With the reduced-size batches (Figure 5.4), it would be discovered in just 3 days. With the smaller WIP inventory, the reduced-batch scenario (Figure 5.5 versus Figure 5.6) also contributes to fewer defects since items wait less on the shop floor and are subject to less damage and deterioration. With smaller lot sizes, engineering changes and redesign improvements can be introduced sooner and incrementally, instead of waiting until the entire current batch is completed. For products where the required changes are released after production has begun, smaller process batches will result in fewer items needing rework.

Case in Point: Effect of Smaller Lot Sizes on Quality[8]

Inman surveyed 114 manufacturing firms to assess the relation between reduction in lot sizes and reduction in scrap and rework. Taking into account both in-house and vendor lot sizes, he found roughly a 1:1 relationship between lot-size reduction and quality improvement (suggesting, for example, that a 50% reduction in lot size corresponds to a 50% decrease in scrap and rework). Most of the firms he surveyed, however, had also implemented quality programs (supplier certification, statistical process control, kaizen teams), and some had reduced their scrap and defect rates even when they did not reduce lot sizes. Inman concluded that lot-size reduction does improve quality (through reduced scrap and rework), though probably the improvement is less than proportionate to the lot-size reduction.

Flexibility

Large WIP inventory ahead of every operation reduces the ability of a process to adjust to changes in products or demand. Whatever the nature of the change, it takes longer for large batch operations to respond because the change cannot be implemented until after the preexisting WIP is processed first. With large batch production, the only way to accommodate additional jobs on short notice is through measures such as expediting, overtime, or subcontracting. Such measures can be costly, dispiriting to workers, and disruptive to work schedules. With smaller sized batches, it is easier to change job schedules and to insert new jobs with less effect on the schedules of other jobs.

To illustrate, suppose in the prior example a special order for 4,000 units of product *Z* arrives. Product Z requires processing times identical to product Y on operations A, B, and C. Assume products X and Y must be completed by days 20 and 24, respectively. If product Z were introduced into the schedule in Figure 5.3, it could not be started at operation A until day 12 at the earliest (else products X and Y would miss their due dates) and would not be finished at operation C until day 26. Specifically, product *Z* would be processed as follows:

1. Start job Z at operation A on day 12; finish on day 14 and transfer it to operation B; wait.
2. Start at operation B on day 16; finish on day 18 and transfer it to operation C; wait.
3. Start at operation C on day 24; finish on day 26.

Now, if instead product Z were introduced into the schedule in Figure 5.4, it could be started at operation A as early as time 0 (before the first batch of product X) or at any time later between any of the process batches for products X and Y, because with process batches of 4,000 and transfer batches of 1,000, the product Z order would take 5 days to process, meaning it could be completed in day 5 at the earliest and day 17 at the latest. Regardless where in the schedule in Figure 5.4 product Z is inserted, products X and Y will still be completed before their respective due dates of day 20 and day 24.[9]

In general, large batch sizes make it difficult to make necessary changes to production schedules and priorities. They contribute to larger WIP and longer lead times, which require all jobs to be released to production far in advance of their due dates. If a customer's order arrives inside the lead time, then it will not be filled by the required due dates. Orders representing forecast demand must be forecast farther into the future, which decreases the accuracy of the forecast.

Case in Point: Lot-Size Reduction in a Pathology Laboratory[10]

Lot-size reduction can improve the performance of any almost kind of multistage process, an example being a pathology laboratory that processes human tissue samples for medical diagnosis. The processing involves a sequence of 7 steps (Figure 5.7). The steps are virtually identical for every sample, although the time at each step varies, depending on the size of the sample. When a tissue sample arrives, a pathologist gross-examines it, cuts it into sections, and places the sections into a small cassette. The samples are then accumulated for 12 hours into a batch, and at night the batch is put in the fixation machine. Small tissue samples need as little as 3 hours to fixate; large samples need up to 12 hours. Since, however, the samples are batched, all samples spend 12 hours in the fixation machine. In the morning the samples are moved to the histology area where they continue to be batch processed: all samples are first waxed, then all are chilled, and then all are sliced, placed on slides, and stained. The samples are then ready for examination by a pathologist. Since a sample takes 24 to 48 hours to complete the process, a diagnosis takes 1 or 2 days.

The alternative adopted was single-piece flow (batch size = 1). The laboratory was rearranged to enable efficient flow of tissue samples, and the cassettes were redesigned to hold individual samples instead of batches. Every evening the fixation machine is run 12 hours for large samples and any small samples that accumulated in the afternoon. In addition, the machine is run every morning on small samples that arrived overnight. Aside from the fixation machine step, all steps are performed sample by sample. In the previous method, every sample was accumulated into a 12-hour batch and then fixated

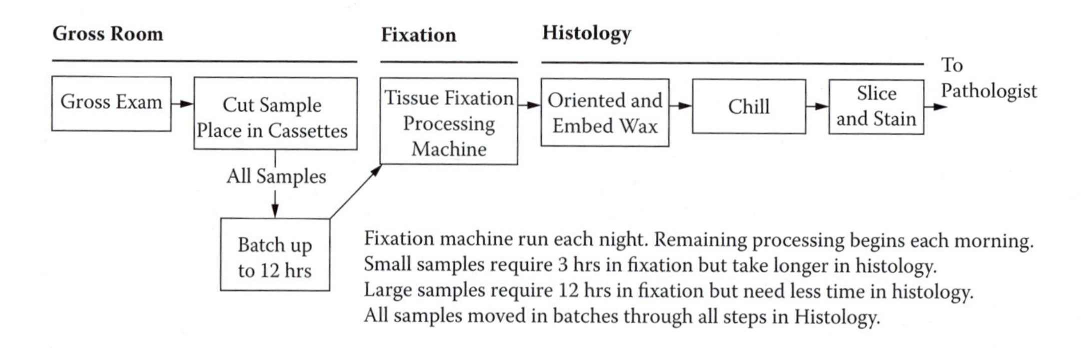

Figure 5.7 Steps in processing human tissue samples.

for 12 hours. Eliminating these steps alone cut 24 hours from the processing time. Now, virtually every sample is ready for diagnosis within a day.

In most cases, reduction of lot sizes brings improvements in cost, lead times, quality, and plant flexibility, however there are situations that call for larger lot sizes, or where constraints make small lot sizes infeasible or impractical.

Case for Larger Process Batches

In the past (before people started looking at ways to reduce setup times), the size of process batches was based on whatever would minimize the number of setups. Less time consumed by setups allows more time remaining for production. To guarantee ample time for production, it was reasoned, setups should be spaced far apart so that process batches could be large. That rationale is still valid when the time required for setup is lengthy, especially when the setup is at a bottleneck or almost-bottleneck operation.

A **bottleneck** operation is one where the production capacity is less than the demand placed upon it; it is a resource for which scheduled work exceeds the work capacity. It is overloaded. An almost-bottleneck operation is one where capacity is just barely adequate and could readily become a bottleneck if additional jobs are imposed on it or current jobs are rescheduled carelessly.[11]

At bottlenecks, larger-sized process batches might be desirable because use of larger batches involves fewer setups, which allows more time for processing and increases throughput. To create larger-sized batches, several different jobs that all require the same kind of setup can be combined into one large batch, ignoring priorities or due dates on the jobs.

At almost-bottlenecks, process batches can be smaller than at bottlenecks, though the feasible minimal size depends on the resulting number of setups and the time remaining for production. In choosing the process batch size, care must be taken so the number of resulting setups does not turn an almost-bottleneck into a sure bottleneck.

Minimal Lot Size

The minimum possible lot size for a given demand is constrained by the production capacity and setup time. Suppose the following: an operation makes five different parts and runs 40 hours a week; after subtracting the time necessary for production and maintenance, 7.5 hours a week remain for performing setups. If each setup takes 0.5 hours, then

$$\text{Number of possible setups} = \text{Time available for setups/Time to perform each setup}$$
$$= 7.5 \text{ hrs}/0.5 \text{ hr} = 15 \text{ setups a week}$$

This means that, on average, there is time to do 15 setups per week for five parts, or 3 setups per part, per week.

To satisfy demand for the given number of setups,

$$\text{Minimal process lot size} = \text{Average demand/Number of setups}$$

So, if the average weekly demand for each part is, say, 1,500 units, then

$$\text{Minimal process lot size} = 1{,}500/3 = 500 \text{ units}$$

The time to perform each setup is the **internal setup time**, which is the time the machine or operation must be stopped in order to perform the setup. Another factor that could affect lot size is the **external setup time**, which is time devoted to the setup that could be performed while the operation is running (e.g., fetching a tool needed to change over a machine for the next job while the machine is working on the current job). If, in the example, 5 hours of external setup are required in addition to the 0.5 hour for internal setup, then the total time per setup is 5.5 hours. Thus, during the 40-hour week the maximum number of setups that can be done is 40/5.5 = 7.2, or 7 setups.[12] For five parts, that allows 7/5 = 1.4 setups per part per week. Therefore, the minimal theoretical lot size for weekly demand of 1,500 is

$$1{,}500/1.4 = 1{,}071$$

The external setup time constrains the minimal process lot size only when it mandates a smaller number of setups than does the internal setup time alone. In our example, if the external setup time were only 1 hour, there would be enough time in the 40-hour workweek for 40/1.5 = 26.6 setups. But since the internal setup time of 0.5 hours allows a maximum of only 15 setups, the internal setup time is the constraining factor, and the external setup time can be ignored.

Besides setup time, product demand and production rate are the other major determinants of minimal process lot size. Smaller demand can be satisfied for a fixed number of setups with smaller lot sizes. Processes with faster production rates require less time for processing, which allows more time for setups and permits smaller process lot sizes. In other words, lot size can be varied dynamically depending on (or to take advantage of) the immediate circumstance.

A further consideration in lot size determination is delivery size. In the earlier examples the average weekly demand was 1,500 units. Suppose weekly shipments are made in quantities that average 1,500. If the computed batch size is 1,071, then to meet the weekly shipment would require two setups (produce 1,071, then produce 429). However, since the maximum feasible number of setups is only 1.4, on average only one setup will be performed a week, not two. Thus, to meet the shipment size of 1,500 the process batch size should be 1,500, not 1,071.

Small Buffer Stock

Buffer stock (also called **safety stock**) is the excess inventory held between stages of a process to avoid running short. The two reasons for carrying buffer stock are uncertainty about demand and uncertainty about lead time.

If you know exactly what the demand will be, then you can plan to make or order exactly what you need. On the other hand, if you are uncertain about demand, then to be safe you order, make, or stock more than you think you will need. The excess is the buffer stock.

The same rationale applies to lead time. If you know exactly the time to produce or take delivery of something, then you can schedule orders such that items will be completed or arrive exactly when needed. If, however, you are uncertain about the lead time, then to be safe you keep some extra on the side so you do not run short if the lead time is longer than expected.

Buffer stocks everywhere in the plant contribute to large inventories, with all the time, cost, quality, and agility consequences mentioned before. Since the purpose of buffer stock is to protect the process against variability, to reduce buffer stock it is necessary to reduce variability.

Demand Variability

One way to reduce demand variability is to level the production schedule, that is, to produce the same quantity of an item in each time period. With a somewhat uniform schedule, there is no (or less) variability and, hence, no (or less) need for buffer stocks. When demand cannot be leveled, then production should mimic demand, that is, stop-go, stop-go, producing as close to demand as possible. To do that, you need very agile production capability. (Chapter 13 discusses production leveling and ways to reduce dependency on buffer stock when leveling is not possible.)

Lead Time Variability

Variability in internal (process) lead times stems from variability in equipment functioning, setup times, worker absenteeism and skill level, material and product defects, and innumerable other sources. Variability in external lead times stems from unreliable delivery. To reduce buffer stocks while protecting against lead time variability it is again necessary to eliminate the sources of variability. This can be achieved via preventive maintenance of equipment, setup improvement, production quality control, standardized operations, and contracting with very dependable suppliers. (These topics are covered in the next six chapters on lean production and in Chapter 16 on supply chain management.)

Facilitating Small Lot Sizes

There are many ways to make small-lot production practical and economical, though they require changes in shop floor practices and procedures and, often, in manufacturing planning and control systems as well as management philosophy. These ways are covered in various chapters throughout this book. This section gives an overview of issues relating to the feasibility and practicality of small lot sizing as applied to process, purchase, transfer, and shipping batches.

Process Batches

The key factor in the feasibility of small lot process batches is setup time. *Ceteris paribus*: shorter setup times enable smaller production batches. Though the topic of setup has been mentioned before, the topic of setup reduction is so important to lean production that the entire next chapter is devoted to it.

Purchase Quantities

Reducing the size of purchase quantities requires reducing the cost of placement and processing of orders. Order costs are analogous to setup costs in that they are independent of the size of the purchased quantity. One way to reduce order costs is to move some (or much) of the responsibility for purchasing from the purchasing department to the production department and even to the shop floor, where supervisors and frontline workers place the orders (this topic is discussed later in the chapters on workcells, pull production, and shop-floor control). This reduces purchasing overhead costs and order lead times.

Another way to reduce order costs is to reduce the number of suppliers and to develop standard agreements and trusting relationships with them. With fewer suppliers, the overhead cost associated with purchasing is less; remaining suppliers get a larger share of the business and, hence, are more committed to providing better service, including more frequent, smaller-quantity deliveries. Further, with supplier partnerships, suppliers take on much of the responsibility for the quality of incoming materials, which reduces or eliminates costs associated with the customer having to perform incoming inspection.

Additionally, suppliers and customers in a partnership work together to determine ways to mutually reduce ordering, transportation, and receiving costs. The two parties can agree on standardized containers so that incoming items can be moved directly into the customer's point of use rather than being unpacked and transferred to different containers. Hewlett-Packard, for example, worked with one supplier to develop dual-purpose packaging: the package in which incoming parts are received becomes the package for the finished product. In many cases the vendors also take responsibility for determining the customer's order requirements. With each delivery, the driver checks the customer's stock level and determines the size of the order quantities for subsequent deliveries. Paperwork is reduced through blanket purchase orders covering a given time period, and the customer is billed once at the end of the period. (This subject of customer–supplier partnerships is covered in more detail in Chapter 16.)

Transfer Batches

The economic feasibility of small transfer batches depends on the cost of material handling: as the cost goes down, smaller transfer batches become more economical. The cost of material handling is principally a function of three things: (1) distance over which materials are moved, (2) number of steps or transactions involved in the move, and (3) complexity or sophistication of the required material handling system.

Distance is reduced by locating workstations and machines close together and in sequence so as to eliminate backtracking and minimize cross-traffic and congestion of material flow. This is one advantage of using focused facilities and workcells (described in Chapters 9 and 10). With workcells, the need for expensive conveyance systems, forklifts, and tracking systems is reduced or eliminated. Heavy, bulky items still require mechanical handling systems, but shortened distances keep the cost of such systems to a minimum.

Delivery and Shipping Batches

To make small sized deliveries from suppliers practical, the per-unit transportation cost must be reduced. One way to do this is to use smaller trucks and vans for short-distance deliveries or to use large trucks, but visit more customers on each trip. For example, instead of a large truck making one big delivery a week to each of three customers, the same truck makes three trips a week, on each trip visiting all three customers. On each trip, customers would get one-third of the average weekly quantity. Of course, the practicality and cost benefit of doing this depends on the relative proximity of customers to each other and to the supplier. (This subject is covered in depth in Chapter 16.)

As more companies become just-in-time suppliers, more trucking firms and for that matter the entire logistics industry are assuming greater responsibility for working out the details of delivery routes and schedules, and for coordinating deliveries between groups of suppliers and groups of customers so that frequent deliveries of small batch quantities are economical. Many companies contract with third-party distributors to work out the logistics for them. When transportation costs stay the same or even increase slightly, smaller shipments result in overall savings from smaller inventories of incoming and finished materials, higher-quality materials, and better responsiveness to short-term demand.

Continuous Improvement

The premise of this chapter is that in general smaller lot sizes (process, transfer, and buffer stock) lead to better manufacturing performance and greater customer satisfaction. Yet, as explained, there are valid reasons for using large lot sizes and carrying large buffer stocks. Thus, if a company sets out to significantly reduce lot sizes and buffer stocks, it will not be able to do it all at once. Nor should it try. In lean production, reduction of lot sizes and buffer stocks is part of the continuous improvement process. It is, like PDCA, a process. The idea is to reduce lot sizes and buffers a little bit at a time and see what happens. As soon as the reduction begins to cause a problem, the next step is to find the source of the problem, prescribe a solution, and implement it. When the source of the problem has been eliminated, the lot sizes and buffers can again be reduced until new obstacles arise, and so on.

Some problem sources can be eliminated right away. For example, a machine that is unreliable might require only periodic, scheduled recalibration; in that case, the source of uncertainty can be removed as can all the buffers surrounding the machine. Other problem sources will be much more difficult to resolve, for example, a supplier that is habitually late, not willing to change, and for which there is no alternative supplier. Even with difficult problems, however, the principle is the same: keep trying no matter how long it takes; eventually a way to eliminate the source of the problem will appear (a different, more reliable supplier will be located, or else the production of the item will be moved in house). Like continuous improvement, reduction of lot sizes and buffer stocks should proceed methodically and perpetually.

Summary

Lot sizing affects manufacturing competitive advantage because it influences the cost, quality, lead time, and flexibility of production. Traditionally, U.S. managers have favored larger lot sizes, primarily because of the large expense associated with setting up production, placing orders, and making deliveries. Large size lots are also the result of using large setup expenses in EOQ-based models, which give the economic optimum, but only in terms of dollar costs. The drawback of large lot sizes is that even when they minimize dollar costs, they lead to greater nondollar costs associated with increased production lead times, hidden defects, and reduced scheduling flexibility.

Lean production acknowledges the problems and wastes connected with using large lot sizes. Small lot-size production is achieved by giving more purchasing responsibility to frontline workers, by reorganizing the facility layout to reduce material transfer distances and cost, by relocating materials to the point of use, and by working with suppliers to find ways to increase the frequency of deliveries, reduce the need for incoming inspection, and reduce the costs of handling, purchasing, and transportation. Even when process lot sizes must be somewhat large, using small-size transfer batches can increase speed and flexibility.

To the extent that large lot sizes and buffer stocks are maintained to absorb process variability and allow continued production in the face of problems, lot-size and buffer-stock reduction is a method of continuous improvement. It is a way to reveal the sources of variability and problems, and to force their removal.

The minimal lot size of a process batch depends not just on costs, but also on demand, production capacity, and setup times. Sometimes these factors in combination mandate larger lot sizes, particularly at bottleneck or almost-bottleneck operations. However, for a given demand and production rate, the major determinant of process lot size is the setup time. Changeovers and equipment setups that are long and inefficient are the major barrier to small lot-size production. Like many other wastes on the factory floor, however, setups are inefficient and time-consuming not because they have to be, but because seldom does anyone look carefully for ways to improve them. The next chapter describes a methodology for analyzing setups and the procedures for reducing setup time.

Notes

1. There are many costs besides dollar costs associated with lot sizing. These costs ultimately affect the quality of output and functioning of the organization (and so the bottom line), but they are difficult or impossible to quantify. These nondollar costs are discussed later.
2. That is, fixed cost has been confused with fixed procedures. Indeed, setup and ordering are fixed-cost activities (*S* remains constant regardless of the size of the lot); however, they are not fixed procedures in the sense that the steps involved in setup or ordering cannot be changed. They can be changed.
3. See J. Evans, D. Anderson, D. Sweeney, and T. Williams, *Applied Production and Operations Management*, 3rd ed. (St. Paul, MN: West Publishing, 1990), Chapter 13.
4. G. Woolsey, A requiem for the EOQ: an editorial, *Production and Inventory Management Journal* 29, no. 3 (1988): 68–72.
5. M. Umble and M. Srikanth, *Synchronous Manufacturing* (Cincinnati, OH: South-Western Publishing, 1990), 113–114.
6. Assuming the batches are independent, the standard deviation of the times for a group of batches is the squared root of the sum of the variances of the times for each of the batches.
7. Assumptions: the small batches are processed independently of each other, and the ratio of the average time to standard deviation, t/σ, is the same for every small batch as it is for the large batch (i.e., average time for a small batch t_S is one-fourth the large batch t_L, and standard deviation for a small batch σ_S is also one-fourth the large batch σ_L). See W. Hopp and M. Spearman, *Factory Physics* (Chicago: Irwin, 1996), 258–260.
8. R. A. Inman, The impact of lot-size reduction on quality, *Production and Inventory Management Journal* 35, no. 1 (1994): 5–7.
9. The example assumes any savings in lead times have not been filled by other jobs. A principle in lean production is to never schedule for full capacity. Some excess capacity is always available to allow for disruptions, problem-solving activities or, as in this case, special requirements.
10. N. Grunden, *The Pittsburgh Way to Efficient Healthcare: Improving Patient Care Using Toyota Based Methods* (New York: Healthcare Performance/Productivity Press, 2008).
11. The subject of bottleneck scheduling includes considerations much beyond production lot sizing and is a topic of synchronous manufacturing and the theory of constraints. See E. Goldratt and J. Cox, *The Goal*, rev. ed. (Croton-on-Hudson, NY: North River Press, 1987); E. Goldratt and R. Fox, *The Race* (Croton-on-Hudson, NY: North River Press, 1986); Umble and Srikanth, *Synchronous Manufacturing*.
12. Since most of the setup time in this example is external time, the full 40-hour workweek can be used to determine maximum number of setups.

Questions

1. What factors are included in order and setup costs? What about holding and carrying costs? Give examples. Why is it difficult to attach a precise dollar value to these costs?
2. How does the batch size affect the average inventory?
3. Explain the tradeoff between setup cost and carrying cost.
4. Comment on the following statement: The EOQ and EMQ models are not appropriate for determining the batch size of items processed through a sequence of multiple operations, but they might be appropriate for determining the batch size of items fully produced at one operation or procured from a supplier.
5. What factors and costs do the EOQ and EMQ models ignore?
6. How does the process batch size affect quality?
7. Discuss the effects of reducing the size of process batches.
8. Discuss the effects of reducing the size of transfer batches.
9. In general, at a bottleneck operation should the process batch size be large or small? Explain. (Assume a setup between each batch, and that larger process batches can be formed by combining multiple job orders for identical or similar items.) Should the transfer batches at a bottleneck be large or small? Explain.
10. If the setup times between products at a bottleneck operation are negligible, should job orders for identical or similar items be combined to increase the process batch size or should the jobs be processed separately? Explain.
11. Suppose an operation has excess capacity (i.e., it is not a bottleneck); however the setup time between batches is not insignificant. Should process batches be large or small? What determines the size of a process batch?
12. Discuss the reasons for carrying buffer stock. Give examples illustrating why buffer stock is carried between stages of a production process, and why it is carried for incoming materials.
13. Discuss how the following costs can be reduced:
    a. Order placement and processing
    b. Material handling
    c. Shipping and delivery

PROBLEMS

1. A product has the following weekly demands:

	Week							
	1	*2*	*3*	*4*	*5*	*6*	*7*	*8*
Demand	20	70	100	190	50	20	80	120

Assume no initial inventory
Carrying cost = \$2.00/unit/week
Setup cost = \$130
Lead time = 1 week

Using the table, determine the lot sizes and calculate the total cost using:

a. Lot-for-lot

b. Two-week period order quantity (assume "inventory" is the amount remaining after the first week of every 2-week order period).

2. A manufacturer buys cardboard boxes from a supplier. The annual demand is 36,000 boxes and is uniformly distributed. The boxes cost $4 each. The estimated order cost is $6, and the carrying cost rate is 30% per year.
   a. What are the EOQ, and the annual order and carrying cost?
   b. How many times a year are orders placed, and what is the average time, in weeks, between orders?
   c. Using the answer from (b), if you round the average time between orders to the nearest week, what should the order quantity be? Would you recommend using this order quantity and time interval?

3. Suppose for problem 2 the actual demand turns out to be 72,000 boxes instead of 36,000 boxes. If you had used the EOQ from the previous problem, what would the annual order and carrying cost be? What percent larger is this cost than the cost estimated in (a)? What can you conclude about the cost of an incorrect demand estimate?

4. Referring to problem 2, again, suppose the box supplier is located close to the manufacturer's plant. For any quantity ordered from the manufacturer, the supplier fills it by making daily deliveries of up to 200 boxes per day, for as many days as it takes to fill the order. Both the supplier and the manufacturer use a 5-day workweek.
   a. What are the economic order quantity and the annual order and carrying cost? (Hint: Use EMQ.)
   b. What is the manufacturer's annual savings in carrying cost by using this system instead of the one in problem 2?
   c. What is the average time, in weeks, between orders?
   d. Suppose the manufacturer places orders at 2-week intervals. What should the order quantity be? How many days will it take the supplier to fill the order?

5. A machining area produces part QR for use in an adjacent assembly area. Estimated annual demand for the part is 20,000 units. The value of the part is $50 per unit. The annual carrying and handling cost rate is estimated to be 16%. The plant operates 250 days a year. The assembly area uses one part QR for each product, and it produces 100 products per day. When producing part QR, the machining area can produce 200 units per day. The cost of ordering and setup for part QR is $200.
   a. What is the economic order quantity?
   b. Suppose the assembly area places an order whenever the on-hand amount of part QR reaches a certain level, the reorder point. If it takes the machining area 2 weeks to begin filling an order, and if the assembly area wants to maintain a minimum, or safety stock of 200 units, at what on-hand quantity should an order be placed (i.e., what is the reorder point)?
   c. If part QR were ordered from a supplier for the same costs, and the supplier delivered the entire order all at once, what would the order quantity be?

6. Do the analysis illustrated in Figures 5.3 and 5.4, except assume processing of product Y must precede processing of product X. What effect does this have on the shipping dates?

7. A product moves in sequence through five operations: V, W, X, Y, and Z. The processing time at each operation (min/unit) is 12, 18, 10, 24, and 12. Use a chart like Figure 5.2 to show the flow of the product through the five operations and the inventory accumulation after each operation. Determine the total production lead time and average inventory. Assume the production quantity and process batch size is 100. Use transfer batch sizes of (a) 100, (b) 50, and (c) 25.

8. For problem 7, assume the inventory carrying cost is $5/unit/day, and the material handling cost is $10/transfer. Determine the total carrying and total transfer costs for each of the three transfer batch sizes. Assume carrying cost is based on average inventory.

9. An operation is used to machine 10 kinds of parts. The total weekly requirement for all 10 parts is 900. Each part requires 20 seconds machine time. Assume the machine is to be scheduled for no more than 90% of its total available time. A normal workweek is 40 hours/week, but 4 hours a week are reserved for normal machine preventive and repair maintenance. If the setup for each kind of part is 1 hour:
   a. What is the maximum number of setups per week?
   b. What is the smallest allowable lot size, assuming all batches are the same size?
   c. To reduce the lot size in (b) by 50%, what must the setup time be reduced to?

10. Three parts are routed through the same operation. The daily production volume and processing times are shown in the following table. The operation is available 400 min/day, and the setup time between parts production is 10 minutes. If the amount of time each day devoted to setup is to be equally allocated among the three parts, what are the minimum lot sizes of the three parts?

Part	*Processing Time (sec/unit)*	*Average Daily Volume (units)*
GB	10	1,050
QED	7	550
RBW	13	300

11. Four products use the same machine. Processing times, daily production volumes, and setup times for the products are shown below. Assume the machine is run for two shifts, and is available for 800 min/day. The products are to be produced in a sequence that is repeated throughout the day until the required volume is filled. How many times a day can the sequence be repeated, and what is the resulting lot size of each product?

Product	*Processing Time (sec/unit)*	*Average Daily Volume (units)*	*Setup Time (minutes)*
B	20	450	15
G	15	800	18
R	10	600	6
H	25	250	10

Chapter 6

Setup-Time Reduction

It isn't that they can't see the solution. It's that they can't see the problem.

—G. K. Chesterton

The impossible is often the untried.

—Jim Goodwin

One challenge facing modern-day production is to meet growing demand for ever-more diversified products. With increased product diversification, product order sizes tend to get smaller and product life cycles tend to get shorter. To meet demand, companies must be able to profitably produce in small quantities and make frequent product/model changeovers. Yet even for industries that produce few kinds of products in large volumes, there is something to gain from small batch production in terms of costs, quality, lead times, and agility. Whether a company makes to order or makes to stock, today's markets and competition call for production methods that require doing more setups. The lowly setup now occupies a central place in competitive manufacturing.

This chapter reviews the ways that companies have traditionally dealt with setups. It then discusses techniques for minimizing the time and effort of setup, and the organizational and procedural aspects of conducting setup reduction projects.

Improve Setups? Why Bother?

Traditional Approaches

Companies have traditionally sought to keep the number of setups to a minimum. After all, a setup operation takes time, costs money, and produces nothing. It is a nonvalue-added activity.

Setup time is time spent in preparation to do a job. In manufacturing, setup is the elapsed time between producing the last unit of one lot and the first good unit of the next. It includes time to replace fixtures and attachments on a machine and to adjust the machine so it produces parts that meet specifications. During much of the setup time the machine produces nothing. During the rest of the setup time the machine is running and producing parts, but until the machine is fully adjusted, the parts are nonconforming and must be scrapped or reworked.

Most traditional setups require a specialized knowledge of machines, tools, fixtures, and materials, and special skills for changing over and adjusting equipment to meet product requirements. As a result, companies assign setup operations to skilled workers with the title "setup person" or "setup engineer." When setup workers are scarce, equipment sits idle until they arrive. As machines are being changed over, machine operators sometimes help with the setup or they work at other machines (if they are cross-trained), though often they simply wait until the setup is done. Of course, restricting setup operations to a relative few workers like this limits the number of setups to only what those workers have time to do.

The smaller the difference between products, the smaller the difference in the processes and operations that produce them. Companies can, thus, reduce the number and types of setups by making products that are largely the same. Reducing product variety makes sense if a company can survive by making just a relatively few, similar products, though for many companies that will never be the case. At one extreme are companies that offer high-volume, standardized products, but each in numerous models with endless options; at the other extreme are companies that produce small-volume, virtually unique products, made to order. Either way, total production volume depends on the ability to offer variety, even though the unit volume for each model–option combination for each customer order might be small.

The number of setups required to process a list of jobs can be reduced by scheduling the jobs in a sequence so that all jobs with similar or identical setups are produced back to back. This method uses similarity of setup as the criterion for work scheduling. It ignores other scheduling priorities such as due dates, and results in jobs being finished earlier or later than needed (and results in either excess WIP or shortages, respectively). Scheduling jobs this way also precludes any chance to smooth workflow though multiple operations, since jobs waiting at each operation will be preempted by a job that arrives and does not require a change of setup.

Increasing the production lot size can mitigate the relative effect of setup time on each production unit. The larger the lot size, the smaller the effect of the setup on each unit. As an example, suppose setup time for an operation is 4 hours and unit processing time is 1 minute. If the lot size is 100 units, the unit operation time is

$$[4(60) + 100]/100 = 3.4 \text{ min/unit}$$

If the lot size is 1,000 units, the unit operation time is only

$$[4(60) + 1{,}000]/1{,}000 = 1.24 \text{ min/unit}$$

The 1,000-unit batch yields a 63% reduction in unit operation time over the 100-unit batch. Generally, the longer the setup time, the greater the impact of producing larger lots on reducing unit operation time.

Find Another Way

The aforementioned ways of dealing with setups are acceptable if you start with the premise that setup time and cost is immutable, inflexible, and unimprovable. But if your goal is to minimize production costs and maximize quality and customer service, then all of the aforementioned ways have major drawbacks. The traditional ways hamper product diversity, quality, and production flexibility for the sake of one thing: minimum number of setups. If, instead, you consider setup time as variable, flexible, and improvable, then you will find another way; you will seek to change the setup procedure.

Suppose, using the previous example, you want to produce in small batches, say, batches of 100 units. Also say you don't want the Operation time/Unit(Setup + Processing time) to exceed 1.24 minutes. The solution is to reduce setup time. If the processing time is 1 min/unit, then the setup time must be reduced from 4 hours to

$$x = 1.24(100) - 100$$

$$x = 24 \text{ minutes}$$

You can achieve this reduction by reducing the number of setup steps and making the setup procedure so direct that operators can do it themselves. You might think this example (reducing setup time 90% from 4 hours to just 24 minutes) is unrealistic. In fact, it's very realistic. Improvements of this magnitude happen all the time in companies where setup-time reduction is a priority. Despite infinite variety in the kind of setup activities, they all share commonalities that enables use of a common methodology for analyzing setup procedures, simplifying them, and reducing setup time.

Benefits of Simplified Setups

Simplifying setup procedures and reducing setup times provide the following benefits:

1. Quality. As a rule, people make fewer mistakes when they follow simpler procedures. That applies to setup procedures, too. A mistake during setup has potential to cause defects in every unit in a batch. Simplifying a setup procedure can thus improve product quality.
2. Costs. When changeover time is small, batches can be produced on a daily basis, which, in turn, eliminates WIP and finished goods inventory investment. Simpler setups reduce the required labor hours and skill level for setups, and eliminate scrap produced during the setup. As a result, setup related costs are reduced.
3. Flexibility. With quick setups, manufacturing has more flexibility to adjust to changing products and changing levels of demand or to changes in demand for different products.
4. Worker utilization. Simple setups do not require special skills and can be done when needed by the equipment operators, which in turn reduces their idle time. This gives setup specialists more time to devote to working on technically difficult setups and on ways to improve setup procedures. For instance, there is a clear benefit when an operator can do a setup in just 10 minutes that used to take a specialist 90 minutes.
5. Capacity and lead times. Shorter setup times increase production capacity. With short changeover times, make-to-order production becomes possible even in a traditional make-to-stock business. Also, production lead time is reduced because of the combination of smaller lot sizes and less time waiting for setup.
6. Process variability. Because each setup is itself a process with several discrete steps, it exhibits time variability.[1] The variability stems from looking for tools and fixtures necessary for the setups, tearing down the old configuration, inserting new fixtures, and adjusting parts to line up properly. To the extent that all of these steps might be ill defined, they result in large variability in the actual setup time, which contributes to variability in available production capacity. Simpler, standardized setup procedures that are clearly defined reduce setup variability and process variability.

As an example of the benefits of reduced setup time, consider the following: a 40-hour per week operation produces 5 different parts, each with weekly demand of 240 units and a processing time of 1 minute per unit. If the changeover time between parts is 4 hours, then half the workweek (20 hours) is spent on changeovers. During a 5-day week, each day must be wholly devoted to producing one of the parts—4 hours for setup, 4 hours for production.

Now, suppose the setup procedure is simplified and standardized, and the time is reduced to 24 minutes. It will then be possible every day to produce one day's demand, 48 units, of every part. Total setup time each day is 5(24) = 120 minutes, and total processing time is 5(48)(1 min/unit) = 240 minutes. Average inventory per part is reduced from 120 units to 24 units. Besides that, since each day's setup and production takes only 6 hours (120 + 240 = 360 minutes), 2 hours remain in each 8-hour workday for producing other parts, problem solving, equipment maintenance, and so on. In addition, the quality of the parts is better because the simplified, standardized setup technique results in fewer mistakes and less setup variability.

Setup: A Case in Neglect

Although setup reduction is important, before lead production companies didn't bother with it. Harmon and Peterson offer the following reasons:[2]

1. Not until recently were full-time teams given responsibility for making improvements in machines, tools, and fixtures. Setup reduction takes dedicated effort from people who know the operations and equipment best: the setup people and machine operators. In many plants these people are never asked to do analysis, contribute suggestions, or take responsibility for anything beyond their rather narrow job descriptions.
2. Managers prefer to buy new equipment rather than improve existing equipment. Setup reduction projects focus on existing operations, machines, and tools, so the primary motivation must be to improve the things you have now, not to replace them.
3. Engineers asked to come up with setup improvements often offer suggestions that are somewhat complex and, thus, rejected as too costly or impractical. In fact, most setup improvements can be achieved by relatively simple means and low cost.
4. Setup reduction requires the skills of machinists and toolmakers, who are often too busy fixing broken machines or preparing tools for new products.
5. Reducing setup on just a few machines or processes has little impact, and individual setup reduction projects are hard to justify. Setup reduction cannot be achieved everywhere, all at once; it happens machine by machine, and it takes a while to see the benefits.

It is interesting to note that companies often go to extremes to design products so their customers can easily set up and use them, yet at the same time do nothing to improve the setup and operation of their own production equipment.[3]

Setup-Reduction Methodology

Shingo and SMED

Probably the foremost authority on setup reduction is Shigeo Shingo.[4] Shingo, over many years working as a consultant to Toyota and other Japanese manufacturers, developed a methodology to analyze and reduce the changeover time for dies on huge body-molding presses. Using the methodology, which he called **SMED** for *single-minute exchange of dies*, Shingo was able to achieve astonishing improvements; for example, he reduced the setup time on a 1,000-ton press from 4 hours to just 3 minutes. Although developed for metal-working processes in the automotive industry, SMED, it turns out, can be universally applied to changeovers and setups in all kinds of processes and industries like woodworking, plastics and electronics, pharmaceuticals, food processing, chemicals, and even hospital procedures.

In any particular situation, most of the experience gained in reducing setup times at a few operations or machines can be readily transferred to setup-reduction projects at other operations and machines. Setup procedures are infinitely varied, depending on the equipment and operation involved, yet because all setup procedures consist of similar kinds of steps, they can be dealt with similarly.

The following types of steps are common to most industrial setup procedures:

- Type 1: Retrieving, preparing, and checking materials, tools, and so forth before the setup; cleaning the machine and workstation, and checking and returning tools, materials, and so forth when the setup or operation is completed.
- Type 2: Removing tools, parts, and so forth after completion of the last batch; mounting tools, parts, and so forth prior to the next batch.
- Type 3: Measuring, setting, and calibrating the machine, tools, fixtures, and parts to perform the operation.
- Type 4: Producing a test piece after the initial setting, measuring the piece, adjusting the machine, then producing another test piece, and so on, until the operation meets production requirements.

By studying, classifying, and organizing steps such as these, it is often possible to reduce the total setup time through a combination of eliminating unnecessary steps, improving necessary steps, and doing some steps in parallel rather than in sequence. The SMED approach, described next, is a four-stage methodology for doing just this.

SMED Methodology for Setup Reduction

Stage 1: Identify Internal and External Steps

An **internal step** is a step that must be performed while the machine or operation is stopped; internal setup time is downtime. An **external step** is a step that can be performed while it is running. Referring to the types of steps listed earlier, most Type 1 steps are external, while most Type 2, 3, and 4 are internal.

The primary focus in setup time reduction is not on total setup time (Internal + External) nor on setup labor time, but on internal time alone. While reducing total setup time and labor hours is desirable, it is only of secondary importance.

Learning the setup steps and classifying them requires an actual study of each setup procedure. The study is performed by detailed observation of the procedure and may involve stopwatch analysis of the steps, worker interviews, and video recording the operation. During the analysis certain steps might be identified as obsolete or no longer practical for the current application. Such steps are classified as unnecessary and eliminated from the procedure. The results of the study are recorded on a worksheet like the one shown in Figure 6.1.

In the average factory, no distinction is made between internal and external steps; everything is treated as if it must be internal. For example, although material, jigs, and tools for the next job

Setup Worksheet

Operation:	Total Setup Time:	Elapsed Setup Time:
10-t press	80 minutes	65 minutes

Step Number	*Step*	*Internal/External*	*Time (min)*		*Performed By*
1	Check in at operation, go to die storage	E		5	Setup person
2	Transfer new die	E		8	Setup person
3	Remove old die	I	10		Setup person
4	Return old die to storage	E		10	Setup person
5	Get new material	E		15	Operator
6	Attach new die	I	12		Setup person
7	Adjust machine	I	20		Setup person
			42	38	

Figure 6.1 Setup worksheet.

could be brought to a machine while the current job is running, those things are not done until after the current job is completed and the machine stopped. The machine then sits idly while the operator or setup person fetches the tools and fixtures to do the setup. This is illustrated in Figure 6.2 for the seven setup steps listed in Figure 6.1.

In Figure 6.2, note that every step, whether potentially classified as internal (solid lines) or external (dashed lines), is done after the machine is stopped. Notice that except for step 5, the operator is not involved and is idle during the setup procedure (zs).

Stage 2: Convert Internal Steps to External

The initial principal objective of setup improvement is to reduce internal setup time. The more setup steps, decisions, adjustments, whatever, that can be done on external time, the better. To that end, wherever possible, setup steps formerly done while the operation was stopped (on internal

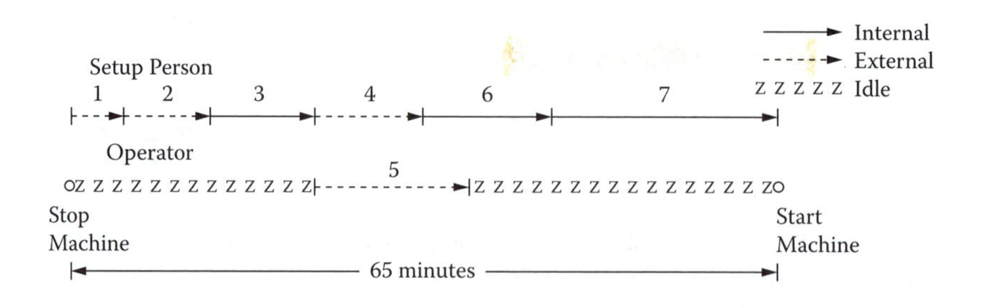

Figure 6.2 Setup procedure: no distinction between internal and external setup steps (times from Figure 6.1).

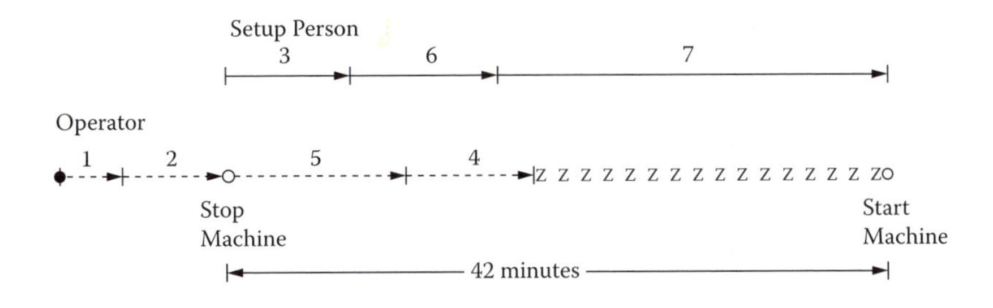

Figure 6.3 Setup procedure: internal and external steps performed separately.

time) are now done while the operation is running (on external time). This usually results in a dramatic reduction in internal setup time, frequently as much as 50%.

Figure 6.3 shows the setup time for the procedure in Figure 6.1 when the operator does the external steps, some while the machine is running. (Assumed here is that the operator can leave the machine to do steps 1 and 2.) In this case, two of the external steps, 4 and 5, are done while the machine is stopped, but that is because the operator would otherwise be idle and because doing them while the machine is stopped does not affect the elapsed setup time anyway. Elapsed setup time is affected only by the internal setup time, 42 minutes.

All internal setup steps should be reexamined closely to determine if any of them could be reclassified as external. For example, in producing metal castings, the casting mold must first be raised to a certain temperature, otherwise the castings are defective. The usual way to raise temperature is to attach the mold to the machine, then inject it with molten metal. This raises the temperature of the mold, though until the temperature gets high enough the castings produced are defective and must be melted and remolded. Heating up the mold can be converted into an external step by using gas or electricity to preheat the mold before attaching it to the machine. The mold would then produce good castings immediately. One company found that preheating the molds using the same oven that melts the metal saved 30 minutes of internal setup time. The only cost to the company was the expense of building a special rack to hold the hot molds.[5]

Stage 3: Improve All Aspects of the Setup Operation

Converting internal steps to external steps reduces setup time, although usually not enough to be in the single-minute range (less than 10 minutes). It also does not reduce the labor or material cost of the setup. (In the previous example of preheating molds there was a cost savings since melting and remolding of parts was eliminated; ordinarily however, simply converting an internal step to an external step alters the elapsed setup time but not the cost.) As long as the average total internal and external setup time exceeds the average run time, the setup time constrains the number of lots (and the minimum size of lots). Setup-reduction efforts must then focus on both internal and external steps, or on whichever most constrains the minimum lot size for a particular operation.

The emphasis in the setup in Figure 6.1 should be on decreasing internal task time, because as it stands that setup reduces by 42 minutes the time available at the operation for processing. With the operator doing the external tasks, a minimum of 13 minutes is needed for tasks 1 and 2 while the machine is running. If the machine run time is less than 13 minutes, then the time to perform tasks 1 and 2 must also be reduced to less than 13 minutes.

Reducing setup time focuses on simplifying and standardizing procedures on existing equipment. Although procuring new equipment that requires little or no setup is an option, it is often less costly and more effective to improve the setup on the existing equipment.

Setup reduction should reduce setup time to the point where setup is no longer an issue in lot-size determination, that is, to the point where the cost associated with setup is minuscule compared to inventory carrying costs. To minimize lot sizes, setup times must be small enough such that virtually any small lot size—whatever is necessary to meet demand, smooth the production flow, or meet other requirements—is practical to produce. For that to happen, the rule of thumb is that a setup should take less than 10 minutes and involve no more than a single-touch procedure. Shingo has dubbed the latter part of this goal **OTED** for *one-touch exchange of dies.*

The setup procedure should be simplified enough so that eventually machine operators can do it themselves. This deskilling of setup procedures takes setup out of the hands of a few skilled specialists and enables changeovers to be performed as needed, without schedules. As discussed in later chapters, this is one requirement for pull production.

Stage 4: Abolish Setup

Beyond OTED comes the ultimate setup improvement: complete abolishment of the setup. Here are some ways to eliminate setups:

1. Reduce or eliminate differences between parts. Fewer or no differences in parts means fewer or no changeovers to manufacture the parts. This is a product-design approach to eliminating setups. For each new product the designer raises the question "Are we currently producing parts for other products that could be used in this product?" or "How can I design this product to minimize the number of new parts (without sacrificing product functionality and customer appeal)?" For existing products, the designer asks, "What existing different parts can be standardized and used on all or many of our products?" Answers for such questions are addressed in a design methodology called **design for manufacture**.
2. Make multiple kinds of parts in one production step; for example, form two kinds of parts from the single stroke of a press (rather than forming the two parts sequentially, with a setup in between).
3. Dedicate machines to making just one item. If only one item is ever made on a machine, then the machine never needs changeover. Obviously, this approach is practical only when machines are relatively inexpensive compared to the costs of setups or when the number of different kinds of items produced is small.

It should be noted that to abolish a setup, it is not necessary to have gone through the other three steps. The alternatives for abolishing setups are very doable and it might be possible to do them right away.

The goal of setup reduction is to maximize the transfer of setup responsibility to operators, to minimize machine downtime, and to abolish setup; it is not to eliminate the jobs of setup specialists. Specialists are still needed but to do different things such as to standardize setup procedures; to modify procedures, machines, tools, and fixtures to improve setups; and to perform difficult, first-of-a-kind or one-and-only setups.

Techniques for Setup Reduction[6]

There are many techniques for setup reduction. This section reviews common techniques with wide-ranging applications. Techniques beyond these, those that apply only to a specific setup situation, are developed as a result of detailed analysis and brainstorming as part of setup reduction projects, which are described later.

This section makes frequent mention of the terms *machines*, *fixtures*, and *tools*; they are defined as follows:

- A machine is the piece of equipment that is fundamental to the operation; it is the one constant in a changeover or setup in that it is always there regardless of the item to be produced. Setup involves doing something on or to the machine (setup) so that the machine can produce a different item. A food processor is an example, a machine with which you can do many different things. One way to set up a food processor is to alter the setting (blend, chop, stir); another way is to change fixtures.
- A fixture is a device attached to a machine to adapt the machine to a particular purpose. Fixtures include dies, nozzles, blades, drill bits, cutting heads, and extensions. An example is the cutting blade on a food processor; if you want your food cut with a particular characteristic, you select and attach the appropriate blade.
- A tool is a device for adjusting fixtures and machines or for attaching fixtures to a machine. Screwdrivers and wrenches are common tools. One way to reduce the time and skill required for setup is to minimize the need for special tools. (For example, food processors are designed so the setup can be done without tools; just press a button to slip different blades in and out.)

The following subsections describe procedures and techniques for setup improvement according to four stages of the SMED methodology: separate internal and external setup activities, improve internal setup, improve external setup, and abolish setup.

Separate Internal and External Activities

Checklists

In many companies, prescribed standard setup procedures are applied only to the most common or most frequently done setups. For setups that are infrequent, the procedures are not prescribed and tend to vary, depending on who is doing the setup. The problem is that an operation's output depends on the setup, so different setup procedures will result in variation in the same item from one batch to another.

For every setup on every machine there should be a reference checklist. The checklist should provide all necessary information about the setup: the setup steps and their sequence; the fixtures, tools, dies, and other items required; numerical values of all settings, dimensions, and measurements for the machine, tools, fixtures, and materials; and specifications for the product. One purpose of the checklist is to make explicit the logic of the setup procedure and aspects of it that might need improvement. Another is to ensure that no steps, parts, tools, or requirements are overlooked in the setup procedure.

Every time the setup is done, a worker checks off steps on the checklist to ensure that everything is done correctly. When the number of different possible setups for a machine is not large, the procedures for all of them can be posted at the machine. Infrequently used checklists can be accessed and displayed on a nearby computer monitor. The checklist should be reviewed prior to each setup so that as soon as the machine or operation stops, the internal steps in the procedure can begin without hesitation. Checklists are maintained by operators and others responsible for setups, and are revised whenever the procedure is modified.

Equipment Checks and Repairs

Machines, tools, and parts must be in top working condition or else they will cause setup delays. Often the fact that equipment is malfunctioning or inadequately maintained is not discovered until setup begins, in which case repairs must be hurriedly ordered or the setup must be replaced by a different one. Equipment performance should be routinely checked as part of the external setup, and these checks should be included among the steps listed in the setup checklist. All fixtures and tools should have assigned storage locations. Tools or fixtures that are damaged during an operation should not be returned to the location without first being repaired, cleaned, and checked.

Setup Schedules

Often equipment and workers sit idly because the people and special tools and parts needed for the setup are not available. Setup operation should be scheduled in advance so that machines, tools, parts, materials, and workers will be ready when needed. The schedule shows the time when internal setup steps begin, which enables workers to determine when to begin external setup so it can be completed on time. Daily setup schedules are prepared by departmental supervisors and take into account current job orders received from production schedulers. They might have to be revised on short notice if additional, unanticipated jobs or changes in job priority are inserted into the production schedule. In pull production systems, where work centers do not have daily production schedules, the supervisor or workcell team leader prepares setup schedules based on current or anticipated orders arriving from downstream operations.

Later in this book the concepts of level production and mixed-model production are discussed. In those production methods, setups occur in a repetitive pattern, which greatly simplifies setup scheduling.

Improve Internal Setups

Parallel Setup Tasks

Using multiple workers to do setup tasks simultaneously can sometimes reduce the internal setup. As Figure 6.4 shows, setup on large machines sometimes requires the setup person to make

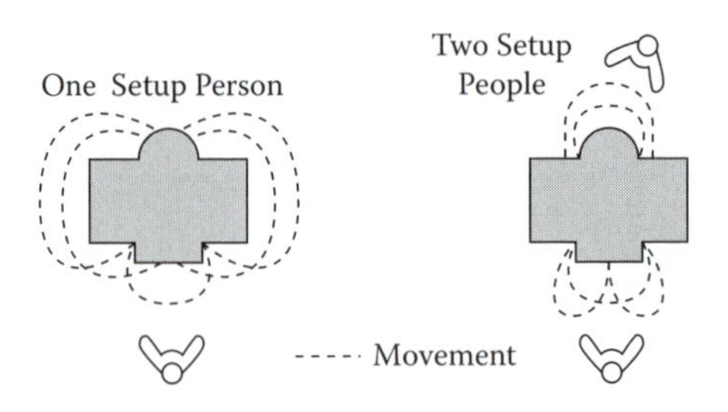

Figure 6.4 Parallel setup operations.

adjustments at both the front and the back of the machine. Assigning multiple people to the setup eliminates much of the back-and-forth walking. The tradeoff between the cost of adding extra workers and the time saved must be considered; if, for instance, two workers are assigned to a setup, ideally the internal setup time should be reduced by at least 50%.

A multiple-person setup is a team effort, which means that the entire team must be assembled for each setup. This is a potential source of delay if workers are scattered throughout the plant. Any setup that requires a team should thus be considered a temporary solution to be ultimately substituted by a procedure that requires only one worker.

Workers doing parallel setup steps often need to coordinate actions and signal each other about when to wait and when to go. Sometimes shouting back and forth works, though in a noisy factory buzzers and lights are better. For safety, foolproof mechanisms must be installed to prevent one worker from doing a step or starting an operation that might injure other workers.

When an operator is to become the primary setup person, he or she must receive the necessary training; even for the operator as in Figure 6.3, training is necessary. Usually setup specialists do the setup training. It is also necessary to upgrade the job description and pay system for operators to account for the additional responsibility and skill level of the job.

Attachment Devices

Much internal setup time is wasted attaching and securing fixtures and materials to machines. Any attachment method that requires more than one tool, one person, or one motion is a good target for improvement. The variety of devices used to simplify attachment is limitless, and there seems no end to the innovation and creativity of setup teams in designing them. This section covers but a handful of these.

Bolts are widely used for attachment, but they are very setup inefficient. Setup time generally increases with the number of bolts needed and their lengths. The number of bolts in a setup should always be questioned since often it far exceeds what is necessary to safely secure a part. All nuts and bolts used should be the same size so only one tool is needed to tighten/loosen them. In fact, just **standardizing** the kind of fasteners used in the setup can greatly reduce the setup time because the setup person doesn't have to figure out for each fastener which tool to use, search for and pick out the tool, and then repeat the process if the wrong tool was picked.

The longer the bolt, the more turns needed to tighten it. In reality, however, it is only the last turn that tightens the bolt and only the first turn that loosens it. Given that reality, setup time can be shortened by the use of innovative one-turn type bolts. Figure 6.5 shows the following four types of one-turn devices:

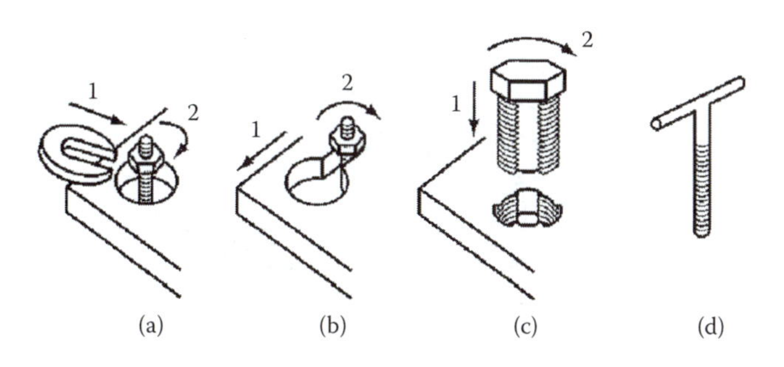

Figure 6.5 Fast bolt-type attachment devices.

1. U-shaped washers. The fixture to be attached has a hole bigger than the nut and bolt. The hole on the fixture is placed over the bolt and a U-shaped washer is slipped beneath the nut, which is tightened with one turn. To remove the fixture, the nut is loosened with one turn and the washer slipped off.
2. Pear-shaped holes. To attach a fixture, the wide part of the hole on the fixture is placed over the nut and bolt, the fixture slipped so the nut covers the narrow end of the hole, and the nut tightened with one turn. To remove the fixture, the nut is loosened with one turn, and the fixture slipped so the nut is at the wide end of the hole. (Wall telephones, small appliances, and smoke detectors often have pear-shaped holes in back for easy mounting.)
3. Split-thread bolts. The bolt has grooves cut along three sides so as to divide it into three sections. The hole for the bolt also has grooves cut to correspond to ridges of threads on the bolt. By aligning the nonthreaded parts of the bolt and hole, the bolt can be slipped all the way into the hole. The bolt is tightened with just one turn-third turn.
4. T- or L-shaped heads. When high torque is not required, bolts with hex-heads can be replaced with bolts with T- or L-shaped heads (and hex nuts replaced with wing nuts) for easy tightening by hand. The special head eliminates the need for a tool. When combined with one-turn features, tightening time is reduced to almost zero. When the setup requires many bolts and high-torque turns, power tools should be used.

Standard-size holders and pins can also hold parts and fixtures in position. Figure 6.6 shows an example. The locking pin is pulled outward and the fixture (a die in Figure 6.6) is slipped between the holders until it reaches the stop. This step aligns the hole on the fixture with the locking pin. When the pin is released, the spring pulls it back into the hole in the fixture. (Spring locking pins are also used to hold home screens and storm windows in place.)

Other simple means of attachment are one-motion devices such as **clamps**. Clamps are especially useful for securing the item to be machined because unlike bolts they do not require a hole in the item. The clamp remains affixed to the machine while the fixture or part is inserted or removed. Two clamping devices are shown in Figure 6.7. With the first clamp, the part or fixture is secured by tightening a one-turn bolt that pressures the clamp onto the fixture. Clamps like this can be used to secure only certain standard-sized parts or fixtures; other clamps, like the second clamp in Figure 6.7, are flexible and can secure parts or fixtures of different thicknesses. After slipping the part or fixture under the clamp, the handle is raised. This turns the cam, pushes the clamp down, and secures the part. A pin or other device locks

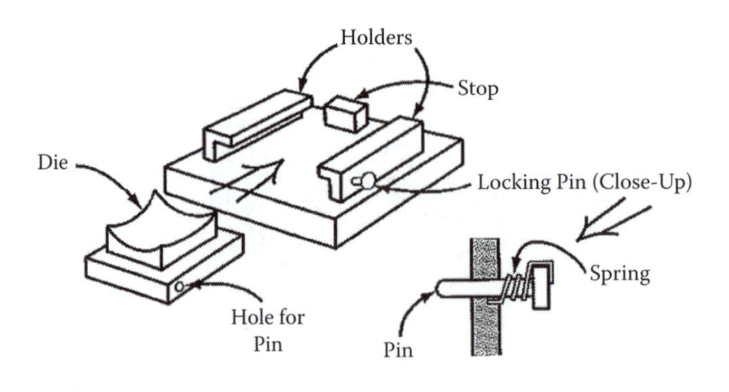

Figure 6.6 Attachment with fixed holders and pins.

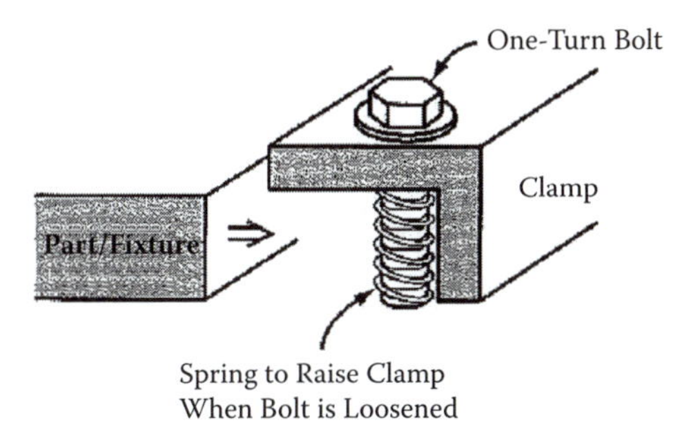

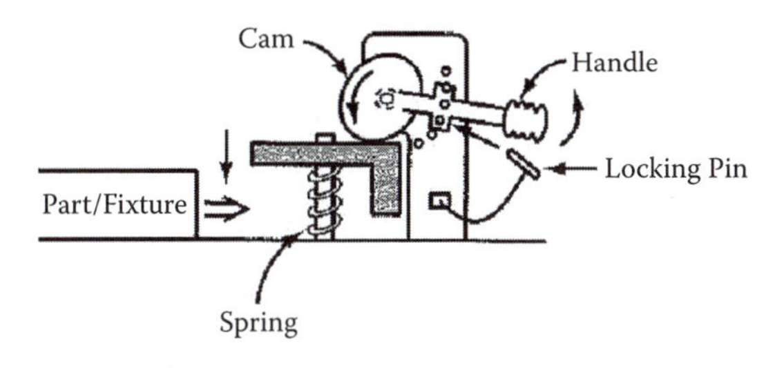

Figure 6.7 Examples of simple clamping devices.

the handle in place. The purpose of the spring is to raise the clamp when the locking pin is removed.

The clamping devices shown are simple and inexpensive, though more sophisticated (and costly) automated or hydraulic clamping systems can also be used. The best kind of device is whatever provides the simplest, least expensive, and least time-consuming solution.

Eliminate Adjustments

Most setups involve a period of adjustment. Following the installation and initial setting of the fixture or part is a lengthy trial-and-error process that involves producing sample parts, measuring the parts, and readjustment. This process occurs no matter how skilled the setup person. The run–measure–adjust cycle, repeated until every setting is correct, is often the most time-consuming portion of the setup procedure. Until the setting is right, the sample parts produced are often defective and must be scrapped or redone. Figure 6.8 illustrates how eliminating the run–measure–adjust cycle shortens the setup procedure.

The best method for reducing the adjustment depends on the kind of adjustment. There are three kinds of adjustment:[7]

1. Mounting parts and fixtures on a machine
2. Setting the parts and fixtures in the correct position
3. Setting the right combination of speed, pressure, feed rate, temperature, etc., so a part meets specifications

Mounting parts and fixtures on a machine can be simplified by using standardized fasteners, holders, and clamps. Using **shims** and **cassette-type holders** greatly simplifies the mounting of

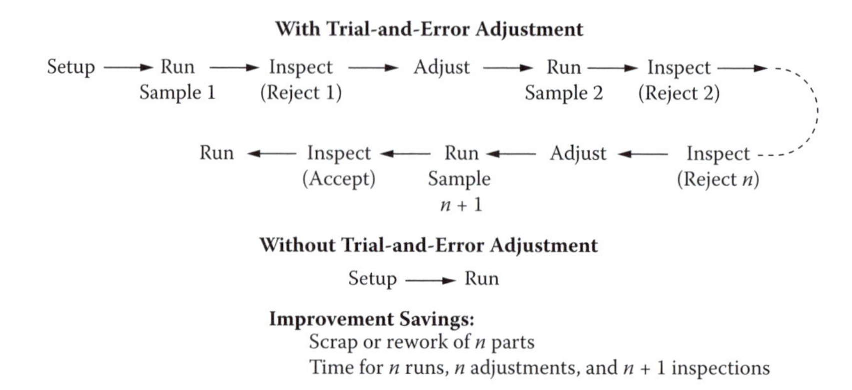

Figure 6.8 Affect of eliminating of trial-and-error adjustment.

varisized tools and fixtures into fixed-position holders and clamps (Figure 6.9). With the latter, the fixture/die goes into a holder, and the holder slips into a fixed-position clamp or other attachment. No adjustment is needed. Variable or sliding shims can be used to mount varisized pieces. For example, office duplicating machines with multiple-page feeders have sliding ledges that are adjustable to the paper size. Any shims, inserts, or cassettes necessary for the setup should be specified in the setup checklist.

Quick-change procedures apply to all sizes of machines or fixtures, including heavy dies and machine presses. A **die** is a fixture like a mold for shaping metal or other malleable material. Most dies have two parts, called male and female. On a typical machine press, one die is mounted on a stationary surface called the **bed**, the other die on a movable surface called the **ram**. When a sheet of metal is placed between the ram and the bed, the ram is lowered and the dies are pressed together, forming or cutting the metal into the desired shape.

Figure 6.10 shows a way of securing dies of different heights onto a press. The usual way to deal with variable-height dies is to raise or lower the bed, or adjust the stroke of the ram so the male and female dies match up at the right height when shut. Raising and lowering the bed and the ram stroke is time consuming, and often several sample pieces must be produced before the ram/bed setting is correct.

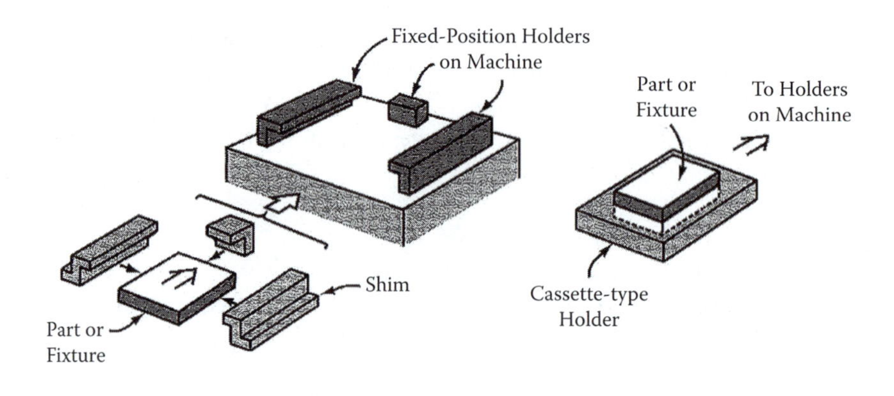

Figure 6.9 Use of shims and cassette-type holders with fixed-position holders on machine.

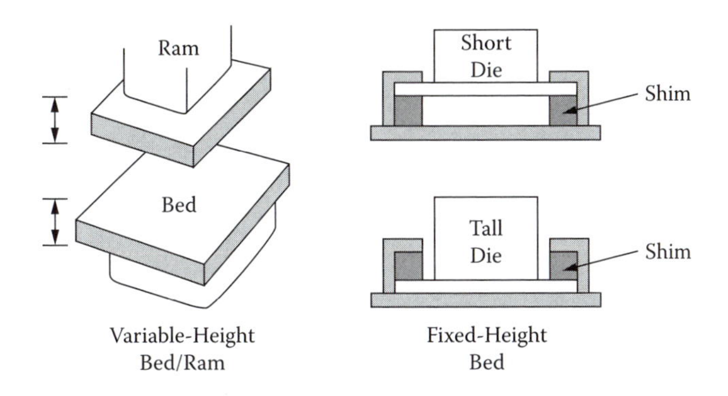

Figure 6.10 Accommodating variable-height dies without making adjustments.

A better way is to keep the bed or ram stroke fixed, and to accompany each die with right-size shims so the die and shims combined always give the same height without any adjustment.

Machines and fixtures typically have adjustable mechanisms (cranks, handles, levers, switches) that must be turned or positioned to a specific setting. For just one setup, it might be necessary to perform repeated adjustments on numerous such mechanisms. An obvious way to avoid adjustment is to determine the correct settings for each kind of setup once and for all, and record them in the setup checklist.

Adjustable mechanisms do not always have calibrated setting scales, or the scales are dirty or otherwise illegible. Figure 6.11 shows two examples, a crank and a variable lever. With the unscaled mechanisms on the left it would be difficult to precisely set the machine without using trial and error. With scales provided, as on the right, the machine can readily be set to values specified in the setup checklist, for example "set lever to 5.0; open crank three full turns and six-tenths of a turn."

Improve External Setups

Storage

Everything needed for a setup (fixtures, dies, tools, raw materials, and documentation) should be stored as close as possible to the place where the setup is done. This can save a great deal of external setup time. The more dedicated the stored items, the more setup time can be reduced.

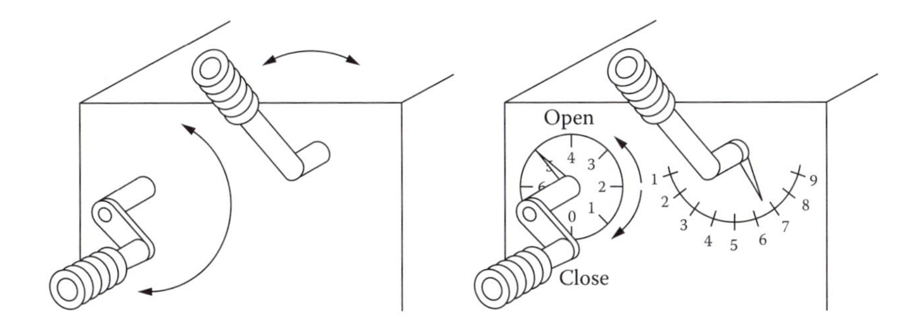

Figure 6.11 Examples of methods for showing settings on adjustment devices.

Of course, dedicating equipment to one kind of setup precludes it from use in other setups, which reduces the flexibility of operations elsewhere. Thus, dedicating equipment often means procuring more equipment, and that can be expensive. As a result, decisions about dedicating equipment must weigh the benefit of reduced setup time at one operation against the cost of purchasing additional equipment or of reducing flexibility at other operations. Items used frequently at one machine or workcell should, ideally, be kept at the machine or cell; items used less frequently can be shared and should be located centrally to all places where they are needed.

Everything about the storage location should be reviewed in terms of effect on setup time. For example, so that heavy tools and fixtures do not have to be moved up or down, the racks on which they are stored should be the same height as the carts used to transport them. (And the height of the carts, in turn, should be the same as the height of the location on the machine where they will be mounted or attached.) It is better yet to dedicate one cart to each fixture so fixtures do not have to be moved between carts and the storage rack.

When the number of different fixtures and tools in use is large, cost and space limitations might dictate that they be stored together. Storage should be arranged so everything can be found and moved easily. Tools and fixtures can be painted different colors or marked with large labels. Storage places on the floor, shelves on a rack, drawers, hangers, and so on can be painted in colors or labeled to correspond to the items stored in them. That way items are easy to find and hard to misplace.

Setup Kits and Carts

Most setup operations require an array of items such as fixtures, tools, clamps, bolts, and so on. In many shops each of these items are stored at a different place. When it is time to do a setup, someone has to go to the different locations, sort through the items, and pick the ones needed. Tools are often kept in a tool crib and must be checked out; at busy times, workers have to wait in line.

Much of the time spent walking to get, waiting for, and picking out items can be eliminated by gathering all the items needed for a specific setup procedure, putting them in a setup kit, and keeping the kit near the machine or workstation where it will be used. Examples are shown in Figure 6.12. Everything needed for a specific setup operation is in the kit. The kit should also include a list of its contents and be partitioned so each item has a designated place (and so it will be obvious when something is missing). Hospitals regularly do this, using surgical procedure kits prepared in advance and containing all the instruments necessary to perform a particular procedure.

When tools cannot be dedicated to a kit, or when kits cannot be kept at machines, then the setup tools or kits can be put on carts, and the carts partitioned with shelves and dividers so items can be quickly found. There should be a designated place for everything on the cart, and everything should be in its place. This is important not only to speed up the setup, but to keep track of tools and to hold workers accountable for the tools.

Material Handling

Equipment used for transporting tools and fixtures for setup should be dedicated to that purpose (again, taking into account the matter of flexibility of use and the cost of procuring additional handling equipment). Hand or motorized carts, pallets, bins, and overhead cranes employed in setups should not be used for other purposes like transporting WIP, raw materials, or machinery.

Figure 6.12 Examples of setup tool/fixture kits.

Fixed-place machines should be arranged with enough space around them so as to avoid impeding the movement of materials handling equipment.

Equipment should be customized to facilitate setup. Three cases are shown in Figure 6.13. For example, carts with the right height and rollers on top make it is easy to roll items off the cart and onto a machine, and vice versa. On the left of Figure 6.13 a fixture is being moved from a cart onto a machine. Presumably, the fixture from the previous job was moved onto another cart. If a

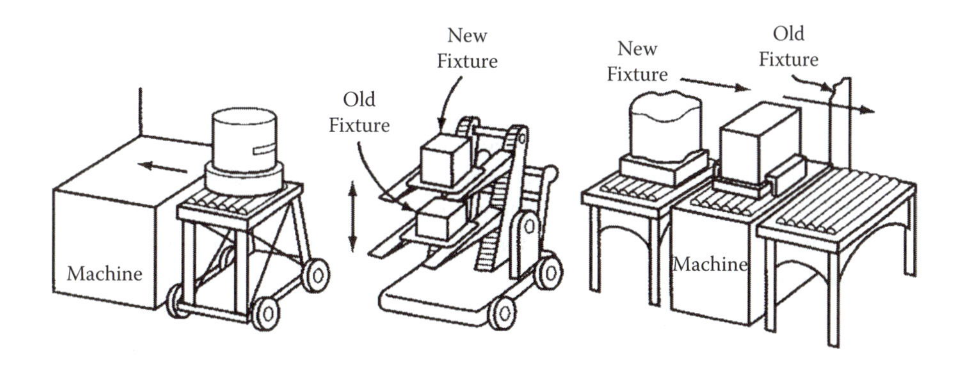

Figure 6.13 Examples of material-handling equipment.

large-enough cart were used, then only one cart would be needed: the previous fixture could be slid onto the cart next to the new fixture, then the new fixture slid onto the machine.

Figure 6.13, center, shows a manual or motorized cart with forklifts to handle fixtures for different height machines. The cart shows two sets of forks, one set to hold the new fixture while the old fixture is being slid off the machine onto the second set. The forklift is raised or lowered using manual, hydraulic, or electrical force.

When fixtures are heavy and must be transported with a forklift or overhead hoist, holding tables with rollers adjacent to the machine (Figure 6.13, right) enable workers to slide one fixture off the machine and another onto it without a forklift or hoist. The scheduled use of the forklift or hoist to transport the fixture to/from the machine does not have to coincide exactly with the scheduled time of setup; as long as the forklift or hoist can deliver the next fixture to the holding table any time prior to the setup, no time is wasted waiting for it.

As much as possible, need for forklift trucks and special hoists to move fixtures for setups should be eliminated. They take up space, often must be scheduled in advance, and cannot be dedicated to setup operations.

Abolish the Setup

One way to eliminate setup is to use the same, standardized parts and components for different products. The smaller and fewer the differences between things to be produced, the smaller the changeover between them. Through the cooperation of people in marketing, product design, and production, parts proliferation can be stopped, and the amount and difficulty of setup required to produce different parts can be reduced. Parts should be differentiated only as necessary to add to product performance, functionality, or appeal as required by customers.

Application of group technology and production by part families (discussed in Chapter 9) can also reduce setup times. William Sandras gives an example where the production of 22 different shafts was sequenced in a cell such that the first shaft required a complete setup, but the subsequent 21 required setups only one-third or less the time of the first one. Essentially, only one full setup was necessary; the other 21 were adjustments.[8]

Another way to eliminate setups is to produce multiple kinds of things with each operation. For example, every automobile has right- and left-side doors and fenders. Instead of producing each side separately in a press with a changeover in between, produce both right and left sides at once. Dies for both left- and right-side door panels are installed in the press and with a single stroke both parts are produced. Similarly, both the hood and the trunk can be formed at once. In general, simultaneous production of n different items by a single machine reduces the number of setups on the machine to $1/n$ the number if they were produced individually.

Simultaneous production and, consequentially, fewer setups can also be achieved by using multiple, dedicated machines. Instead of using a single, expensive machine that can produce many kinds of parts, use several less expensive machines, each dedicated to a single part (e.g., a machine that makes only right door panels or only left door panels). When machines are devoted to one part (or similar kinds of parts), changeover between parts is eliminated or is reduced to trivial steps. Group technology coding can help determine which parts to assign to which machines to eliminate or reduce setup times.

Projects for reducing setup time, discussed in the next section, can themselves be time consuming and costly. Especially when the existing setup procedure cannot practically be reduced, but where further improvement is needed, the only alternative might be to procure additional equipment and abolish the setup.

Setup-Reduction Projects[9]

The goal of setup-reduction projects is to reduce individual setup times so that small lot size production is feasible. Beyond that, setup-reduction projects are a good place to initiate lean production. Like workplace organization and cleanup projects, setup reduction requires the involvement of frontline workers. Setup-reduction projects are highly visible, and early setup successes help to gain the buy-in from workers and supervisors necessary to implement other lean ideas such as pull production and total productive maintenance.

Scope of Project

Each setup-reduction project must be undertaken in the context of a larger, ongoing setup-reduction program. This is because a few, standalone setup improvements will have little effect. In general, reducing setup times at an operation increases the operation's throughput. However, unless the operation is a bottleneck that has starved downstream operations for work, increasing the operation's throughput might only serve to speed up the growth of WIP waiting at downstream operations. Thus, the eventual focus of setup reduction should be plantwide.

Setup reduction has the most immediate impact at bottleneck and near-bottleneck operations because there the setup time delimits production capacity. As setup times are reduced, more time is available for production. Eventually if the setup reduction is great enough, small batch production becomes possible, even at operations that were formerly bottlenecks.

The alternative to reducing setup times at bottleneck operations is to increase capacity either through replicating the operations, outsourcing production, or using overtime. However, if setup times consume a large portion of existing capacity, then it is usually more effective to reduce the setup times rather than to increase the capacity. In general, a principle for selecting setup projects is to initially focus on operations that spend the greatest proportion of operation time on setups, then strive at these operations to achieve the maximum possible reduction in setup time.[10]

A setup project focuses on the setup of a given machine or operation to produce a particular part or product. Every piece of equipment and operation must be treated individually. Because of even small variations in seemingly identical equipment, products, and operations, every setup procedure is different. Even if exactly the same model machine is used in many places, the setup at each must be studied separately.

To be considered as a candidate for setup reduction, there must be some need to improve the particular setup operation. Need is identified analytically from computer models or empirically from shop-floor information by using as criteria the product–machine combinations that have the highest utilization rates, longest setup times, highest amounts of waiting WIP, largest batch sizes, and largest product diversity.

Candidate projects should focus on machines and products with a future, that is, products and machines that will continue to be important and are not anticipated to change soon or be eliminated. The best setup reduction projects help to improve manufacturing flexibility. For all the different items produced at an operation, a good project will result in an improved setup between any ordered pair of the items. For example, if a machine can produce three items, X, Y, and Z, the setup improvement should affect the six possible setup combinations (X to Y, Y to X, X to Z, and so on). Improved setup between just one possible pair and not the others does little to improve flexibility.

A final, important criterion for selecting early setup projects is high success potential. Successful projects generate interest, enthusiasm, and motivation to continue; more difficult and risky projects should be saved until later after confidence levels have increased. This holds true for virtually all projects to implement aspects of lean production.

These considerations result in a list of candidate projects for setup improvement. The list is based on information from setup people, operators, supervisors, production planners, schedulers, and engineers. To decide which projects should have priority and be started first, the list should be given to shop-floor supervisors.

Setup Reduction Team

Once the machine–product combination has been decided, the project team is selected. The team must include people who know the most about the machine and the existing setup, who will be affected by changes, and who have ideas for improvements; usually, these are the setup people and machine operators. Their participation in developing new procedures is important because they must be motivated to follow them. Also, setup people and operators tend to propose good ideas that are simpler and less costly to implement than ideas from engineers and other staff personnel.

The team should also include others, for example, tool and die makers, or industrial and manufacturing engineers, depending on the analytical and technical expertise required for the project. The team is informed about the project's goals and focus, and is trained in setup-reduction methodology, methods analysis, and setup-reduction techniques. Workers with prior experience in setup reduction should help do the training.

Ready, Get Set, Shoot!

The team documents features of the current setup operation using technical reports and blueprints for the product or part, the machine, and tools and fixtures used in the setup. They then prepare a plan for the major information-gathering event of the project, video recording the setup. The purpose of video recording the setup is to get a detailed picture of the current setup procedure. To that end, the recording is scheduled within normal production hours and normal production schedules.

In case the operators and setup people are not members of the setup reduction team, they should be informed in advance about the purpose of the video recording so they are aware that they personally are not being evaluated. The recorded procedure must be identical to the usual modus operandi, and the people in the setup must do everything they normally do. The entire setup operation—from the time the last part of the last batch is produced until the time the first good part of the next batch is produced—should be recorded, even if the elapsed time takes hours; throughout, the time display on the camcorder should be turned on. While one person operates the camera, another takes notes. Neither interrupts the operator or setup person with questions.

Analysis of Video Recording

The video recording is closely reviewed, and the individual, elemental (micro)steps and their elapsed times are identified. The duration of the elemental steps might range from just a few seconds to several minutes. It is important to identify microsteps because the smaller the steps, the easier it is to define their purpose and decide how to simplify, eliminate, or transfer them to external time.

The elemental steps are clustered into setup categories. For example, the initial 20 or so steps might all be clustered into "removing old fixture"; the next 30 steps, "obtaining and preparing new fixture"; the next 50 steps, "attaching new fixture and material"; and so on. Steps are clustering like this because it is easier for the team to discuss six or so procedural categories than, say, 430 elemental steps. The sum of the elapsed times of elemental steps in each category is the total time for that category; the sum of the category times is the total setup time.

Generating and Selecting Ideas

Throughout analysis of the video recording as well as in subsequent meetings and discussions, the team develops ideas about how steps can be eliminated, simplified, or transferred to external time, and from setup person to operator. The ideas are shared with (and advice sought from) operators and setup people, engineers, production planners, supervisors, and vendors of the machines, fixtures, and parts/materials involved. The impact of these ideas and suggestions on reducing times for elemental steps and categories are estimated. Not every idea can be adopted, so the biggest-bang-for-buck (Pareto) approach is used. Ideas that will provide the greatest ratio of internal setup time savings to the cost of implementation are considered first, then those with less savings per cost are considered next. The same is done for savings in total setup time. This concept of marginal improvement in setup time versus cost of improvement is shown in Figure 6.14. As illustrated, initial improvements usually result in big time savings at relatively low cost, while subsequent improvements are smaller and come with a higher cost.

Pareto analysis is also used to separate the categories of steps that take the most time, and, within each category, the microsteps that take the most time. The most time-consuming microsteps of the most time-consuming categories are the first targets of elimination or time reduction.

The proposed new procedure and its impact are discussed with people who have relevant expertise, will be affected by it, or have valid opinions. The team then presents the new procedure and the cost–benefit analysis to management. After getting the approval from management, the new setup procedures are implemented. A follow-up study is performed to assess the impact of the new procedures on setup times and determine if further changes are needed.

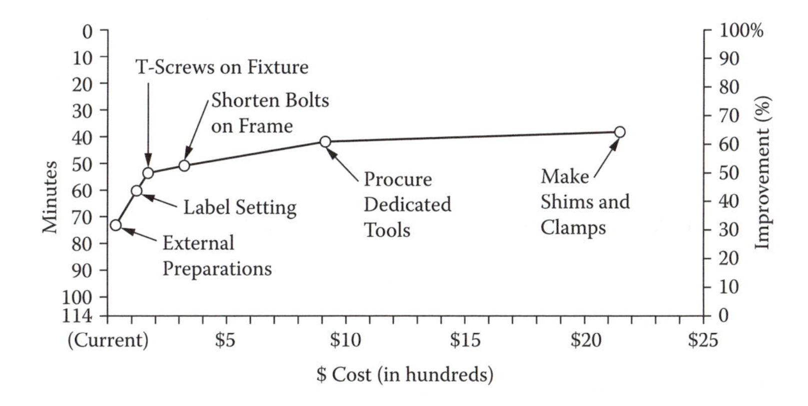

Figure 6.14 Setup time improvement versus implementation cost.

There are hundreds of published examples of the results of setup reduction projects. A few follow as cases in point.[11]

Case in Point: Setup-Time Reduction at Plastics Design

At Plastic Design Inc., a 110-ton injection molding press took 135 minutes to changeover. A team video recorded the setup procedure, then made several improvements. First, they externalized setup tasks and created two carts; one cart holds all tools with silhouette locations, setup procedures, and quality criteria; and the other cart holds hoses and fittings for changeover. Within 3 days the team reduced the setup time to 16 minutes. Aside from the team members' time, the setup reduction project cost virtually nothing.

Case in Point: Setup-Time Reduction at CSSI

At Connecticut Spring and Stamping Inc. (CSSI), 200- and 300-ton metal stamping presses were producing lot sizes of 200 to 50,000 pieces, which is 4 to 8 weeks of customer demand. A team analyzed the setup of the presses, then implemented standardized die positioning techniques such as quick clamps, locator pins, and shut-height blocks. They also externalized many tasks, put a parts storage area next to the presses, and developed an Indy 500 pit crew concept for team setup. Within only a few days, the team had cut changeover time by 50%.

In another area of the CSSI plant, a team investigated changeovers on three medium coiler machines for eight high-volume parts. They created a standard work setup procedure and posted it in the work area. The procedure cut changeover steps from 108 to 28 and reduced changeover time from 3 hours to 15 minutes. The reduced time enabled all eight parts to be run daily and cut inventory enough so that parts could be stored at point of use in the work area. Intermediate parts storage and movement from the warehouse have been eliminated.

In the foot-press area of the plant, average setup time on six presses used to be 40 minutes. After video recording and timing the procedure, the team eliminated many adjustment steps by using standardized bases with stop positions and quick clamping. New setup time was reduced to 6.8 minutes.

Continuous Improvement

Setup times should be tracked, charted, and posted as evidence that management considers performance in setup important. This tracking and charting also prevents backsliding and motivates workers to continue to find additional, simple, low-cost ways to continue to improve the setup. The setup reduction project should not just result in a cost-effective procedure to reduce setup time, but in a procedure that workers readily follow without supervision.

For each setup project, the initial goal should be to reduce setup time by 50%. Usually the resulting time will still exceed an hour, so the next goal would be to reduce it to, say, 30 minutes, then, later, to less than 10 minutes. Less than 10 minutes might seem extreme, but short setup time is essential for flexibility, particularly in processes that use pull and mixed-model production. To achieve the series of goals of ever-smaller setup times requires a succession of setup-reduction projects. Given that the average factory has many machines, it is obvious that a setup reduction campaign must be continuous.

Summary

Simplified setup and reduced setup time permit reduced lot-size production and result in increased production capacity, flexibility, and resource utilization, as well as improved product quality and customer satisfaction. Although setup procedures vary widely with type of equipment and equipment application, the methodology developed by Shingo, called SMED, can be applied to virtually any setup procedure. The principal objective of improvement is to reduce the internal setup time, that is, the setup time during which the machine or operation must be stopped. SMED has four stages:

1. Analyze steps in the setup procedure to distinguish which are internal and which are external.
2. Convert as many internal steps as possible to external steps.
3. Improve all steps to the extent that virtually any size lot is practical to produce.
4. Seek ways to eliminate the setup.

Setup reduction is implemented setup by setup, starting with operations and setups showing the greatest need, which are usually instances where setup time constrains production capacity and flexibility. The setup-reduction project is conducted by a small team of operators and setup people, assisted by engineers, toolmakers, and other specialists. The team systematically analyzes the existing procedure, develops ideas for improvement, then uses cost–benefit analysis to cull ideas that enable the largest marginal time reduction for given implementation cost.

Although the primary goal of setup reduction is to increase flexibility and make small lot production practical, the process of reducing setups is itself a useful way to begin increasing frontline workers' participation in problem solving and decision making, which is necessary for most other aspects of lean production such as preventive maintenance, pull production, pokayoke, and quality at the source, discussed in upcoming chapters.

Notes

1. S. Melnyk and T. Christensen, Understanding the nature of setups, *APICS—The Performance Advantage* (March 1997): 77–78.
2. R. Harmon and L. Peterson, *Reinventing the Factory* (New York: The Free Press, 1990) , 181–182.
3. R. Hall, *Zero Inventories* (Homewood, IL: Dow-Jones Irwin, 1983), 109.
4. S. Shingo, *A Revolution in Manufacturing: The SMED System* [trans. A. Dillon] (Cambridge, MA: Productivity Press, 1985).
5. A. Robinson, *Modern Approaches to Manufacturing Improvement* (Cambridge, MA: Productivity Press, 1990), 320–321.
6. For more setup-reduction techniques and examples see Shingo, *A Revolution in Manufacturing: The SMED System*; Harmon and Peterson, *Reinventing the Factory*, 186–200; and Hall, *Zero Inventories*, 84–111.

7. H. Steudel and P. Desruelle, *Manufacturing in the Nineties* (New York: Van Nostrand Reinhold, 1992), 200–202.
8. S. Sandras, *Just-in-Time: Making It Happen* (Essex Junction, VT: Oliver Wight Publications, 1989), 116.
9. Portions of this section are adopted from Steudel and Desruelle, *Manufacturing in the Nineties*, 179–191.
10. These conclusions are based on propositions developed by C. Hahn, D. Bragg, and D. Shin, Impact of setup variable on capacity and inventory decisions, *Academy of Management Review* 35, no. 9 (September 1989): 91–103.
11. From G. Galsworth and L. Tonkin, Invasion of the Kaizen Blitzers, *Target* 11, no. 2 (March/April 1995): 30–36.

Suggested Reading

K. Arai, and K. Sekine. *Kaizen for Quick Changeover: Going Beyond SMED.* New York: Productivity Press, 2006.

Questions

1. Describe ways that companies have traditionally dealt with equipment setup. What are the advantages of each? From a lean perspective, what are the drawbacks?
2. Discuss the cost, quality, time, and flexibility ramifications of reducing setup times.
3. What is the difference between internal and external setup procedures? Give examples of both.
4. Suppose that in the setup procedure shown in Figure 6.1 step 5 takes 25 minutes and step 6 takes 15 minutes. How long should the setup procedure take?
5. List and discuss the steps in Shingo's SMED methodology.
6. How does the external setup time constrain the minimum process lot size?
7. What is the purpose of a setup checklist? How is it used? How is it created?
8. Give industry and household examples from your experience of devices that simplify installation and attachment, and that eliminate trial-and-error adjustment. Give examples from your experience of situations where trial-and-error adjustment could be eliminated by simple means.
9. On what operations should the first setup reduction efforts focus?
10. Who should participate in the setup reduction team?

PROBLEMS

1. As part of a lean initiative, a manufacturer has reduced the setup cost for machining a component from $200 to $12.5. As a result, the new economic lot size is 10 units.
   a. The former economic lot size had been much larger. What was it? (Hint: Look at EOQ in Chapter 5.)
   b. If the goal is to reduce the economic lot size to one unit, what must the setup cost be reduced to?
2. A setup reduction team has been working on a punch-press machine that currently takes 75 minutes for internal setup. Each part requires 1.5 minutes on the machine. The machine is not a bottleneck, and the motivation for reducing setup time is to reduce the unit cost of the part and to enable the machine operator to perform the setup. The labor rate for the operator is $12.50 per hour. What must the setup time be reduced to for the lot size to be 10 units and the average labor cost per unit to be $0.75?

3. In problem 2, the setup team has set the target production lot size to five units. The production rate on the machine is 40 units/hour, the demand for the part is 10 units/hour, and the carrying cost of the part is $4.25 per unit per year. Assume annual demand is 8,000 units. What must the setup time be reduced to for this part? (Hint: Look at EMQ in Chapter 5.)

4. The official setup time for a machine is 40 minutes. Suppose a workday is 420 minutes, the machine must process 200 parts a day, and the processing time is 0.5 min/unit.
   a. What is the minimum batch size?
   b. Suppose the operation is run for one day using the batch size from (a), and at the end of the workday only 133 units have been produced. Assuming the processing time per unit was 0.5 minutes, and no stoppages except for setup, how much time was used for setup? What was the average time per setup? Explain how the actual setup time could be so much longer than the official setup time.

5. Changeover of fixtures on a machine takes 35 minutes. Following the changeover, an average of four parts must be produced and discarded before the correct setting is achieved. For each part the processing time is 2 minutes and the material cost is $1.50. The labor cost for a skilled setup person to do the changeover and adjustment is $15.50 per hour. The labor cost of the machine operator is $10.50 per hour. Currently, the part is produced in batches of 50 units. Suppose the setup procedure could be simplified so that the operator would be able to do it and not have to make any adjustment in the setting. What must the fixture changeover time be reduced to such that a batch size of 20 units could be produced for the same average per-unit cost as the present cost? Assume the per-unit cost includes labor and material costs of setup and production; also, assume the machine's excess capacity is ample and is not a consideration.

6. Don Knotts, the new assistant manager of manufacturing, has sent a proposal for a setup reduction project on the DMK-020 press to his boss, Wilie Fox. The DMK press is not currently a bottleneck, but it is used on a large number of products and typically is run at near-capacity levels. Fox says, "Before you start anything like this, I want to see some cost figures." Knotts estimates that the setup time could be reduced 50% by analyzing procedures, separating internal and external procedures, and simplifying internal procedures so the operator can do them himself. With the reduced setup time, the press would be able to produce in batches of 75 units (average size) instead of the current 450 units. The combination of reduced setup time and the operator performing the setup would reduce the labor cost of each setup by about $40. Knotts also estimates it will take a team of three workers—a machinist, the operator, and himself—about a day for the project. Knotts gives his cost analysis to Wilie, whose responds, "You propose to spend a day on a project that will save only $40! Compared to the cost of your team, the savings is insignificant. Besides, the person who does the setup now will still be on the payroll, so you're not going to save anything, really." If you were Knotts, what would your response to Wilie be?

Chapter 7

Maintaining and Improving Equipment

> Everything falls apart, count on that. Then I get to try to put it back together again. Everything that comes together sooner or later falls apart.
>
> **—Dog's Eye View, "Everything Falls Apart"**

A U.S. steelmaker and a Japanese steelmaker were negotiating to start a joint venture. The Japanese firm sent a delegation of executives to tour the U.S. plant. One executive in the delegation had a small hammer, and as the group walked through the plant, he would sometimes gently tap bolts on the machines they passed. This slightly unnerved the U.S. managers conducting the tour. They had no idea what he was doing, but of course they dared not ask and reveal their ignorance. Throughout the tour the question nagged them: Why is he doing that?

This chapter is about the executive's concern, explained later, and the reason for the hammer: it is about equipment **maintenance**. Like small batch sizes and short setups, equipment maintenance is a key element of lean production and is fundamental to competitive manufacturing. This chapter explains reasons why and gives a preview of total productive maintenance (TPM), which is the philosophy that makes equipment part of a company's competitive capability. Later sections describe concepts and measures associated with equipment effectiveness, the role of preventive maintenance and TPM in increasing equipment effectiveness, and issues associated with implementing TPM.

Equipment Maintenance

Throughout most of the last five decades manufacturing was seen by many U.S. managers as a low-status, low-glamour function; somewhere below it was maintenance—the no-status, no-glamour function. Maintenance departments in most U.S. companies were understaffed and underfunded. Maintenance was viewed as an overhead cash pit, and certainly never as a way to increase profits or competitiveness. Its role was largely one thing: breakdown repair.

Breakdown Repair

Breakdown repair is the practice of caring for equipment when, and only when, it breaks (if it ain't broke, don't fix it). If you only tend to equipment when it breaks, and if you have a great deal of equipment, then you will always be tending to breakdowns. Breakdown repair is playing catch-up; you are constantly trying to find where the breakdown occurred, remedying the breakdown, then shuffling to make up for lost production time. In companies where the breakdown rate is high, the maintenance staff works overtime to try to keep pace, though being understaffed, they cannot. Repairs are done hastily, sometimes poorly. Vital production time is lost as equipment and workers sit idle.

Breakdown repair (also referred to as repair maintenance) is the worst kind of maintenance. Like inspecting for product defects, it focuses on problems after they have occurred, not on diagnosing the problems to keep them from happening again. Equipment breakdown and malfunction directly contribute to wastes such as waiting, inventory, and product defects. Manufacturers in a competitive environment need to move beyond breakdown repair.

Equipment Problems and Competitiveness

Equipment problems have a direct effect on production cost, quality, and schedules (Figure 7.1). With each breakdown, one or more operations are idled and scheduled completion times delayed. If a machine needs a new part, it will be idled for as long as it takes to get the part, which could be days or weeks. Meanwhile, production must be transferred to other machines or covered by overtime, which disrupts the schedules of other operations and increases costs. The machine might malfunction for a while before it finally breaks and produce defective parts in the meantime that will need to be scrapped or reworked.

One way to enable the rest of the process to continue in case a machine breaks down is to carry inventories of whatever that machine produces. Subsequent operations can continue working for as long as the inventory lasts. To speed up repairs, inventories of spare parts are also carried. But the more chronic the machine's problems, the more inventory needed to cover for them. Further, while the machine is broken down, materials accumulate ahead of the machine as work in process (WIP), which results in still more inventory.

Machine Problems	*Possible Immediate Effects*	*Ultimate Cost/ Consequences*
Malfunction	Machine deterioration	Shortened machine life
	Machine inefficiency	High repair cost
	Output variability	Scrap and rework
Breakdown	Safety hazards	Injuries
	Idled workers	Inventories
	Idled facilities	High production cost
		Schedule delays

Figure 7.1 Consequences of equipment problems.

It is evident that by reducing equipment problems, a company can reduce inventories, scheduling disruptions, defects, and costs associated with these problems. At the same time, it can improve safety and reduce injuries from equipment malfunction.

Preventive Maintenance

Preventive maintenance (PM) is the practice of tending to equipment so it will not break down, and will operate according to requirements. It entails understanding and maintaining all the physical elements of manufacturing—machine components, equipment, and systems—so they consistently perform at the required levels.

Although reliable, well-functioning equipment is a good thing in general, it is a prerequisite for lean production and customer-focused quality. A hallmark of lean production is the ability to function with little inventory, and a lean process simply does not have enough WIP inventory to buffer against chronic equipment problems. Equipment that performs erratically or breaks down will bring the entire process to a halt.

A hallmark of customer-focused quality is the ability to hit product or service targets, which means no significant deviation from the targets. In manufacturing that happens only when equipment is reliable and well functioning. Regardless of what else a manufacturer is doing to improve product quality, few things will have as much impact as improved equipment functioning. For many companies PM is but a stepping stone to another, higher level of equipment maintenance, TPM.

Total Productive Maintenance[1]

Preventing equipment breakdowns is good, but even better is squeezing the ultimate potential from equipment. That is the purpose of **total productive maintenance** (TPM). The ultimate potential of a piece of equipment depends on its unique function and operating environment and, in particular, on how well the equipment meets requirements for availability, efficiency, and quality. A goal of TPM is to upgrade equipment so it performs better and requires less maintenance than when it was new.

Equipment responsibility in TPM is spread throughout many departments such as production, engineering, and maintenance, and to a range of people including equipment operators and shop workers. In TPM, operators perform basic equipment repairs and PM; meanwhile, teams of maintenance staff, engineers, machinists, and operators redesign and reconfigure equipment to make it more reliable, easier to maintain, and better performing. TPM is another never-ending facet of continuous improvement in manufacturing.

Benefits of TPM

TPM aims for greater manufacturing competitiveness through improved equipment effectiveness. By tailoring equipment to better suit a particular production environment and making it better than new, TPM increases production capacity and process reliability, and reduces the costs of lost production time, defects, repairs, shortened equipment life, and inventory. Since it involves everyone in the process, TPM also contributes to improvements in safety, morale, and pollution. The following TPM results were reported by Seeiich Nakajima:[2]

- Productivity: Breakdowns were reduced by 98% (from 1,000 to 20 times/month).
- Quality: Defect rate was reduced by 65% (from 0.23% to 0.08%).
- Cost: Labor cost was reduced by 30%; maintenance cost was reduced by 15% to 30%; energy consumption was reduced by 30%.

- Delivery: Inventory turnover increased by 200% (from 3 to 6 times per month).
- Morale: Improvement ideas increased by 127% (from 36.8 to 83.6 ideas/person per year).
- Safety: No accidents occurred.
- Environment: No pollution was created.

Managers are often unaware of the improvement potential of better maintenance because they do not know the costs of equipment-related problems. Records of the causes and frequency of breakdowns, amount of downtime, kinds and costs of repairs, and product defects from equipment are often inaccurate or nonexistent. TPM starts by assessing the state of equipment effectiveness and its impact on production.

Equipment Effectiveness

Equipment effectiveness (EE) refers to the multitude of ways equipment influences productivity, costs, and quality. The term can be used in reference to individual pieces of equipment, or, when aggregated, to all equipment in a plant or a company. High average equipment effectiveness indicates a plant or company has minimized or eliminated wastes that stem from equipment-related problems.

Equipment Losses

Think for a moment about the equipment-related sources of waste discussed before—inventory, waiting, defects, and so on. Besides these are wastes associated with machine operation, idling, and repairs. Nakajima[3] has categorized these equipment wastes into what he terms the **six big losses**:

1. Downtime from equipment setup and adjustments.
2. Downtime from sporadic or chronic equipment breakdowns.

These two losses affect the availability of a piece of equipment to perform work. The more time equipment is stopped for setups or repairs, the less time that remains for it to do work.

3. Idling and minor stoppages (equipment is running, but parts flowing into it periodically jam, or parts flowing from it are momentarily blocked by the next machine being down).
4. Reduced speed of operation (equipment is running, but at a reduced speed because it is worn out or needs adjustment).

These two losses in combination affect equipment efficiency. Equipment that is periodically interrupted by shortages or that produces at a rate less than its standard capability takes longer to do the work.

5. Defects caused by variability in equipment performance.

Equipment that is worn out or nearly broken increases process variability and causes defects. The result is reduced quality in terms of output that is nonconforming and must be reworked or scrapped.

6. Reduced yield caused by nonoptimal operation.

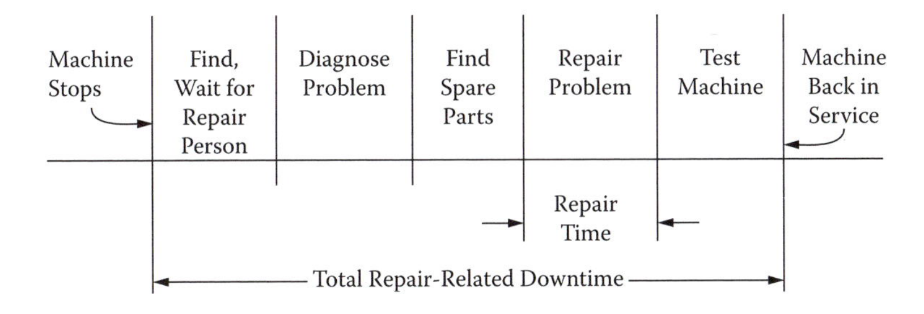

Figure 7.2 Total downtime for repair verses repair time. (Adopted from J. Moubray, *Reliability-Centered Maintenance*, New York, Industrial Press, 1992, p. 64.)

Every time a machine is stopped because of setups, breakdowns, or minor interruptions and then restarted, it takes a while for the machine to reach its normal operating conditions (speed, temperature, etc.). Until then, it produces more slowly or causes a greater proportion of defective output.

The following sections discuss the criteria for measuring the impact of these losses on manufacturing productivity.

Maintainability

Maintainability is the effort and cost of performing maintenance. It is affected by, for example, the ease of access to equipment for maintenance, the skill level required to do the maintenance, and the availability and convenience of getting spare parts and service. One measure of maintainability is **mean time to repair** (MTTR). High MTTR is an indication of low maintainability. MTTR is the average time a machine is down:

$$\text{MTTR} = \Sigma(\text{Downtime for repair})/\text{Number of repairs}$$

where downtime for repair includes time waiting for repairs, time spent doing repairs, and time spent testing and getting equipment ready to resume operation.

In some organizations, repair time is used as the downtime for repairs; however, they are not the same, as indicated in Figure 7.2.

Reliability

Reliability is the probability that equipment will perform properly under normal operating circumstances. One measure for reliability (R) is the probability of successful performance, or

$$R = \text{Number of successes}/\text{Number of repetitions}$$

where number of repetitions is the number of times the equipment does something. For example, if a machine produces 1,000 parts of which 960 are good, then the machine is 96% reliable. Similarly, if a machine used to test circuit boards for defects works 99% of the time (it misses 1% of all defective boards), then the machine is 99% reliable.

The opposite of successes is defects or failures. Failure here simply means that equipment performance is not satisfactory. It can mean that equipment is malfunctioning in some aspect or is completely broken down.

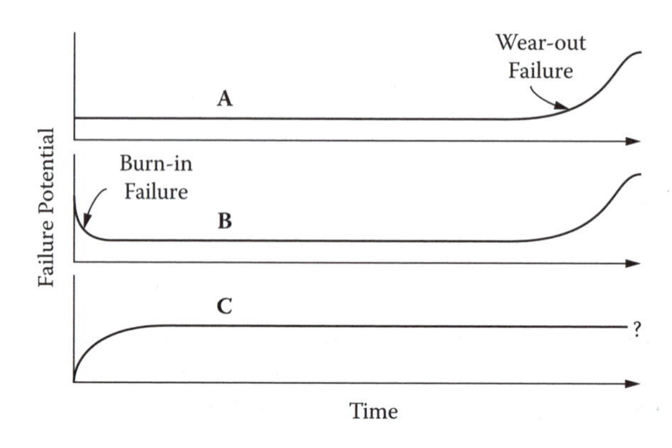

Figure 7.3 Three common failure patterns.

Failure Pattern

The more reliable equipment is, the less likely it will fail. The likelihood of equipment failing is shown with the **failure pattern**. The appearance of the failure pattern depends on the piece of equipment as well as its usage and environment. Figure 7.3 shows an example of three common failure patterns.[4]

In pattern A, the failure potential (probability of future failure given that so far failure has not occurred) is a function of age. The older the item, the more likely it will fail. In pattern B the failure potential is the so-called bathtub shape. Items having this pattern go through an early burn-in period in which the failure potential (infant mortality) is relatively high. This pattern is common for electronic components. Manufacturers of such components operate or run them for a while before shipping or provide customers with a 90-day warranty to account for infant mortality. Items that survive burn-in then have a failure pattern similar to pattern A.

Some things continue to function well regardless of age. This is represented by the constant, uniform failure potential in pattern C. Although eventually all things fail, with this pattern the point of increasing failure potential is unknown. For an item with this pattern, there is often little statistical information because managers get nervous and replace the item before it ever fails. Even if theoretically an item has a failure potential like pattern A, insufficient information about the right-hand part of the curve will default it to pattern C.

The following example illustrates how the failure potential curve is interpreted and developed from empirical data.

Example 1: Failure Potential Curve

Suppose we start with 110 components and observe the time periods within which they fail. We observe the following:

	Time Period												
	0	*1*	*2*	*3*	*4*	*5*	*6*	*7*	*8*	*9*	*10*	*11*	*12*
No. failures	0	1	1	1	1	1	1	1	2	15	46	30	10
No. survivors	110	109	108	107	106	105	104	103	101	86	40	10	0

Now, the failure potential is a **conditional probability**. It specifies for an item that has survived up until a given time period the probability that it will then fail during that time period. It is computed as the number of items that fail in a period, divided by the number of items that started the period. For example, using the above numbers and looking only at period 11, the number of observed failures is 30 and the number of items started is 40, so the failure potential in period 11 is 75%.

Performing the same computation for all periods and rounding to one decimal, gives

	Time Period											
	1	*2*	*3*	*4*	*5*	*6*	*7*	*8*	*9*	*10*	*11*	*12*
% Failure potential	.91	.92	.93	.93	.94	.95	.96	1.94	14.9	53.5	75.0	100

Notice the failure potential in this case roughly conform to the pattern A in Figure 7.3.

Implications of failure patterns for scheduled, preventive maintenance are discussed later.

Mean Time between Failure

A measure related to reliability is the **mean time between failure** (MTBF). For equipment that can be repaired, MTBF represents the average time between failures. For equipment that cannot be repaired, it is the average time to the first failure. The higher the MTBF for a piece of equipment, the greater its reliability. If we assume a constant failure rate, such as pattern C, then

$$\text{MTBF} = \text{Total running time/Number of failures}$$

The MTBF can be used to estimate the reliability of an item. In this case, the reliability, $R(T)$, is defined as the probability that the item will not fail before time, T:

$$R(T) = e^{-\lambda T}$$

where

$0 < R(T) < 1.0$

e = Natural logarithm base (about 2.718)

T = Specified time

λ = Failure rate = 1/MTBF

Following are examples to illustrate these concepts.

Example 2: MTBF

Twenty machines are operated for 100 hours. One machine fails in 60 hours and another fails in 70 hours. What is the MTBF?

Eighteen of the machines ran for 100 hours, while 2 others ran for 60 hours and 70 hours each. Thus, the total running time is 18(100) + 60 + 70 = 1,930, and

$$\text{MTBF} = 1{,}930/2 = 965 \text{ hours/failure}$$

Example 3: Reliability for a Given Time of Operation

What is the reliability of the same machine from Example 2 at 500 hours? At 900 hours?

$$\lambda = 1/965 = .0010362 \text{ failure/hour}$$

$$R(500) = e^{-.0010362(500)} = 0.596$$

$$R(900) = e^{-.0010362(900)} = 0.394$$

Thus, there is nearly a 60% probability that the machine will run 500 hours (without failure), and a nearly 40% probability that it will run for 900 hours.

Suppose the machine's performance is entirely dependent on one particular component. Each time the component is replaced, the machine's reliability returns to 100%. How often should the component be replaced so the machine's reliability is never less than 90%?

$$R(T) = e^{-\lambda T}$$

$$0.90 = e^{-.0010362(T)}$$

Transposing the formula, $T = -1{,}000 \ln(0.90)$, so $T = 109.2$. The component should be replaced every 109.2 hours.

Results from this kind of analysis are not necessarily intuitive. For example, if we wanted to increase the reliability to 95%, the computation gives 51.3, which indicates we must replace the component every 51.3 hours, despite the fact that among the 20 machines tested, none failed in less than 60 hours, and 18 were still running at 100 hours.

Availability

Availability is the proportion of time that equipment is actually available to perform work out of the time it should be available.

Availability and Downtime for Repair

One measure of availability (A) is

$$A = MTBF/(MTBF + MTTR)$$

As the formula suggests, availability can be increased through a combination of increasing MTBF, decreasing downtime for repair, MTTR, or both. This is illustrated in Figure 7.4. Strategies for lengthening MTBF and shortening MTTR are addressed later.

Availability and All Downtime

The previous formula rightfully shows that by improving MTBF and MTTR, availability is improved too; however, as a formula for determining availability it is misleading because, while it considers repair-related sources of downtime (MTTR), it ignores nonrepair sources of downtime. As such, it overstates the actual equipment availability. A more accurate measure of availability, because it includes both repair and nonrepair sources of downtime, is the following:

$$A = \text{Actual running time/Planned running time}$$

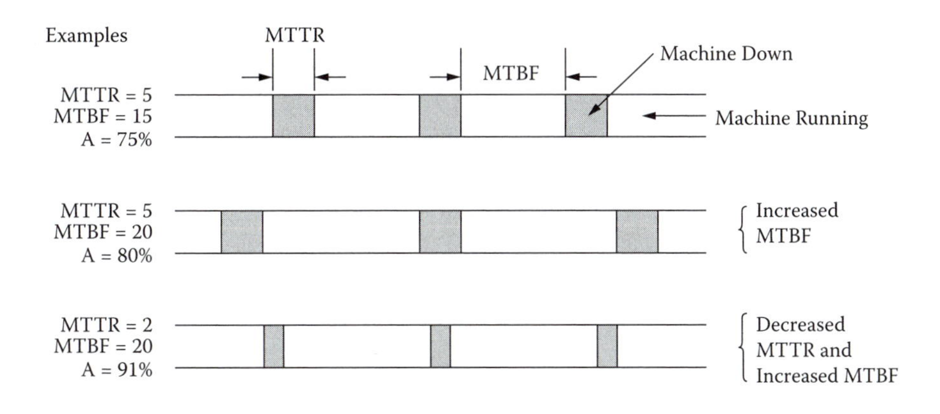

Figure 7.4 Impact of increasing MTBF and decreasing MTTR on availability.

where

Planned running time = (Total plant time – Planned downtime)
Actual running time = (Planned running time – All other downtime)

In these definitions, planned downtime includes all planned nonworking time: meals, rest breaks, meetings, and scheduled PM. No workers or equipment are considered idled by planned downtime.

All other downtime is all the other time the machine is down (for internal setup, equipment breakdown [MTTR], etc.) and during which workers and equipment are considered idled.

Suppose a plant runs two-shift (16-hour) workdays, and during each shift 2 hours are for planned downtime. The **planned running time** is thus

$$16 - 2(2) = 12 \text{ hours}$$

To determine the **actual running time**, subtract from this the time the machine is down due to setups (die changes, adjustments), equipment failures, and so on. If the machine is stopped each day an average of 110 minutes for setups and 75 minutes for breakdowns and repairs, then the actual running time is

$$12(60) - (110 + 75) = 535 \text{ minutes}$$

The availability of the equipment is thus

$$A = \text{Actual running time/Planned running time} = 535/12(60) = 0.7431$$

This measure of availability is preferable to the other because it provides incentive to reduce both MTTR and internal setup time. Though setup time might be independent of MTTR, the two are often interrelated to the extent that poorly maintained equipment is harder to change over and adjust.

Repair Downtime Variability

The aforementioned measures of availability are average measures. They indicate nothing about the impact of unscheduled downtime of process variability, which is an important matter for making equipment decisions. Suppose, for example, you have two machines, one with MTBF of 100 hours and MTTR of 10 hours, the other with MTBF of 10 hours and MTTR of 1 hour. Though both have availability of approximately 0.91, they should not be considered equivalent. Assuming breakdowns occur randomly, the first machine requires that a minimum of 10 hours of WIP buffer stock be carried to protect the process from shutdown, whereas the second requires a minimum of only 1 hour WIP. Considering only the cost of buffer stock, the second machine is preferable. In general and all else equal, machines with more frequent but less time-consuming breakdowns are better in terms of process stability and inventory cost than machines with less frequent, more severe breakdowns (logic somewhat similar to preferring smaller batches and short setups over larger batches and long setups).

Accurate assessment of availability calls for high data integrity of equipment records. Equipment records must show all time that equipment is down for internal setups and failures. In many companies records are inaccurate because they show only major breakdowns; breakdowns lasting only 10 or 20 minutes are not even recorded, even when they occur frequently throughout the day. A chart should be kept at each machine for the operator to record every instance of machine downtime.

Data for the next two equipment measures, efficiency and quality, must be collected by a team of skilled observers (engineers or trained shop people). To ensure that the data accurately measure machine performance, they should be collected over a period of many days or instances of machine operation.

Efficiency

Efficiency measures how well a machine performs while it is running. It requires answering two questions:

1. The machine is running, but is it producing output (i.e., what is the rate efficiency)?
2. The machine is producing output, but is it producing output at the right speed (i.e., what is the speed efficiency)?

Rate Efficiency

Suppose parts moving down a chute that feeds into a machine periodically jam. Every time parts jam, the flow of parts into the machine (which is running) is disrupted (Figure 7.5). The machine

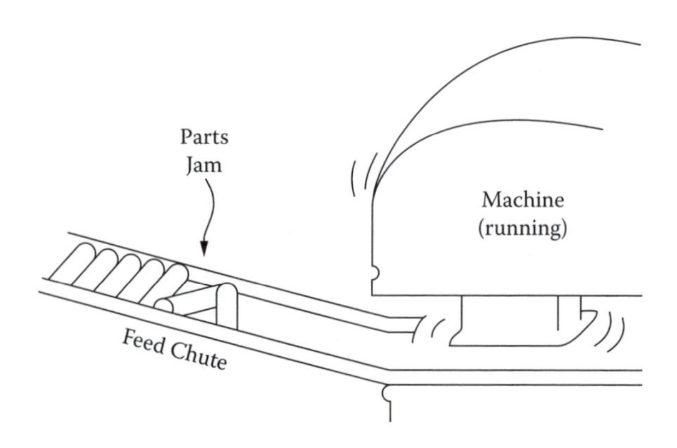

Figure 7.5 Minor disruption from parts jamming.

operator has to dislodge the parts to resume flow. In the course of a day these disruptions add up and reduce machine output considerably. They contribute to reduced yield loss, the last of Nakajima's six losses mentioned previously. Suppose this is the same machine as in the previous example, which runs 535 minutes a day. Also suppose the average daily throughput of the machine is observed to be 830 units, and actual cycle time is observed to be 0.6 min/part. In theory, it should take 830 × 0.6 = 498 minutes per day to produce these parts. However, since the machine is running for 535 minutes a day, its **rate efficiency**, RE, is

RE = (Actual production volume × Actual cycle time)/Actual running time

= (830 × 0.6)/535 = 498/535 = 0.9308

Thus, the machine is processing parts only 93.08% of the time it is running. Parts jamming in the chute plus other interruptions are consuming 6.92% of the machine's running time.

Speed Efficiency

Suppose the machine is designed to produce two parts per minute, the equivalent to a 0.5 min/part cycle time. Since in actuality the machine's cycle time was observed to be 0.6 min/part, then the **speed efficiency**, SE, of the machine is

SE = Design cycle time/Actual cycle time = 0.5/0.6 = 0.8333

The machine is running at 83.33% of its rated speed.

RE and SE are independent measures of efficiency. Multiplied they give the overall **performance efficiency**, PE, of the machine:[5]

PE = RE × SE

For the example,

PE = 0.9308 × 0.8333 = 0.7756

Sometimes it is difficult or impossible to compute SE because the design cycle time is unknown or because the machine makes different parts with different cycle time. In that case SE is ignored and PE is computed solely as a function of lost time from interruptions (i.e., machine is running but not doing work):[6]

PE = (Actual running time – Time for interruptions)/Actual running time

Quality Rate

The **quality rate**, *Q*, is an index of the equipment's ability to produce output that is nondefective or conforms to requirements:

Q = (Actual production volume – Defective output)/Actual production volume

Supposing the machine in the example produces an average of 30 defective units out of 830 units/day, its quality rate is

$$Q = 800/830 = 0.9639$$

Overall Equipment Effectiveness

The measure of equipment effectiveness that incorporates availability, performance efficiency, and quality rate is the **overall equipment effectiveness**, OEE,

$$\text{OEE} = \text{A} \times \text{PE} \times \text{Q}$$

For the example,

$$OEE = 0.7431 \times 0.7756 \times 0.9639 = 0.5555$$

OEE can have a dramatic effect on plant productivity and on choices for improving productivity. In the example, the machine produced 830 units in two shifts. Were demand to double to 1,660 units it would be necessary to add a second (theoretically identical) machine to meet demand. An alternative, however, would be to improve the OEE of the current machine from 0.5555 to 0.8333.

Improving the OEE not only increases product throughput, it also reduces variability in product quality and production schedules, and, as a result, reduces the need for inventory, overtime, rework, and other costly ways of dealing with output variability. This is why lean manufacturing plants strive for high OEE plantwide, that is, high effectiveness of all equipment.

In many traditional plants the only equipment treated as if effectiveness matters are bottleneck machines. They get regular preventive maintenance because any breakdown would have immediate and severe impact on plant throughput. However, when preventive maintenance is focused on a few high-use machines, all other machines are neglected. Eventually, these machines fall into a state of disrepair, causing innumerable bottlenecks by virtue of their low availability, efficiency, or rate of quality. High-use equipment should receive maintenance priority, but all equipment should receive preventive maintenance and be able to meet minimal standards of availability, efficiency, and quality. With small WIP inventories, every piece of equipment is a bottleneck waiting to happen.

Preventive Maintenance Program

The aim of preventive maintenance is to improve equipment performance. This is done through a number of steps that address and rectify the major causes of equipment problems.

Causes of Equipment Problems

Chronic equipment problems often stem from more than one cause. Kiyoshi Suzaki[7] gives the following five causes that act individually or in some combination to cause equipment problems:

1. Deterioration. The fact is, parts wear out. Moving parts like gears, bearings, and belts wear down or break, and electrical components burn out. Most kinds of operational equipment eventually deteriorate, but neglect or abuse hastens their deterioration.

2. Equipment ill-suited for the purpose. The equipment is utilized for purposes other than those for which it was designed. The material, size, or operation of the equipment cannot handle the expected load, which causes accelerated deterioration, breakage, and product defects.
3. Failure to maintain equipment requirements. Equipment is dirty, lubricant is not replenished, dust and grime foul the mechanism, and so on.
4. Failure to maintain correct operating conditions. The equipment is operated at speeds, temperatures, pressures, and so forth, in excess of recommended design levels.
5. Lack of skills of operators, maintenance crew, and setup people. Operators do not know standard equipment operating procedures and cannot detect or do not care about emerging equipment problems; maintenance people replace parts but do not question why breakdowns occur; setup people use the wrong tools, fixtures, or adjustment settings; operators, maintenance staff, setup people, and engineers seldom talk to each other about equipment problems, causes, and solutions.

To address these problems, PM programs emphasize the need to

- Maintain normal operating conditions
- Maintain equipment requirements
- Keep equipment and facilities clean and organized
- Monitor equipment daily
- Schedule preventive maintenance
- Manage maintenance information
- Use predictive maintenance

Maintain Normal Operating Conditions

All machines have design limitations. Often, purchased machines come with operating manuals that list recommended, normal, and maximum operating conditions. Companies often operate equipment at the maximum condition; though theoretically the equipment is designed to handle that condition, continual operation at that level forces accelerated deterioration. Preventive maintenance starts with knowing the normal operating conditions (speed, pressures, temperatures, etc.) and not running equipment in excess of those conditions. Normal operating requirements should be posted at every machine, and operators should monitor conditions to make sure these requirements are not exceeded. Derating the equipment (running it at below normal operation conditions) further reduces deterioration and extends the equipment's useful life. The life of a bearing, for example, is inversely proportional to rotational speed, so reducing the speeds by one half can double the life of the bearing.

Maintain Equipment Requirements

Equipment has physical needs. These needs are, like operating conditions, sometimes listed in manuals that accompany the equipment; sometimes, however, they must be determined by operational experience. An example of an equipment requirement is lubricant. Most equipment with moving parts requires oil, grease, silicone, and so on, which must be checked and periodically replenished or replaced.[8] In one General Motors plant, 70% of the operators felt that the main reason for machine downtime was inadequate lubrication.[9]

Another equipment requirement is tight bolts and fasteners. Bolts and fasteners that become loose during machine operation induce greater vibration, which introduces defects in products and hastens equipment deterioration. One Japanese factory estimated that it reduced equipment breakdowns by 80% just by periodically retightening tens of thousands of bolts on its equipment.[10] Some managers would say that loose bolts on equipment are a sure sign of poor PM.

Another equipment requirement is the use of proper tools and fixtures in machine setup. Machines malfunction when they are incorrectly configured or adjusted during setup. Reviewing setup procedures, training operators and setup people in proper setup techniques, and using setup checklists are ways to reduce machine problems from setup mistakes.

Keep Equipment and Facilities Clean and Organized

Cleanliness and organization are essential to preventive maintenance. Noise, oil leaks, wobbling fixtures, cracks, discolorations, and other indicators of problems that might otherwise be concealed by dirt, show up readily on a clean machine. Calibrations and settings on a clean machine are readily visible and make it easier to see when the machine needs resetting.

Dirt and disorganization are themselves major causes of equipment problems. Grime gets into working parts and causes scratches and friction, which slows down or jams machinery. Clutter in the shop leads to confusion in locating repair parts and tools, and mistakes in setting up, operating, and repairing machines. Clutter impedes efficient response to breakdowns and increases MTTR.

Responsibility for keeping machines and workplaces clean and organized is assigned to machine operators, and this assignment is an early step in increasing employee involvement in managing the shop floor. It is also important for developing employee attitudes regarding the importance of paying attention to details and caring about the workplace, which correlate with attitudes about product quality and the impact of their work on quality. Of course, managers must recognize that cleaning takes time and allow for it. Workers with tight production quotas ordinarily will not do anything that detracts from meeting their quotas.

Monitor Equipment Daily

Periodic cleaning of equipment is not enough. Equipment must be monitored daily or, ideally, in real time so that early signs of problems are promptly detected and fixed. Operational monitoring for subtle signs of problems, such as increased heating, vibration, jamming, defects, and so on, provides advance warning of emerging problems (see Figure 7.6). Often, quick detection of an emergent problem allows a simple solution like tightening a bolt, adding oil, or inserting a new part, whereas, without detection, the problem would grow into an expensive, catastrophic proposition.[11]

The most practical way to monitor and maintain equipment in real time is for the operators do so. Giving operators responsibility for basic machine upkeep helps ensure that problems will be detected early, minor problems will be quickly fixed, and major problems will be avoided. Giving them this responsibility also helps justify why they must keep machines clean and organized. Again, management must allow some time each day for workers to exercise this responsibility.

Schedule Preventive Maintenance

Even machines that are properly operated and maintained need periodic attention from the experts. Instead of waiting until machines break down, PM programs allow for periodic, scheduled downtime

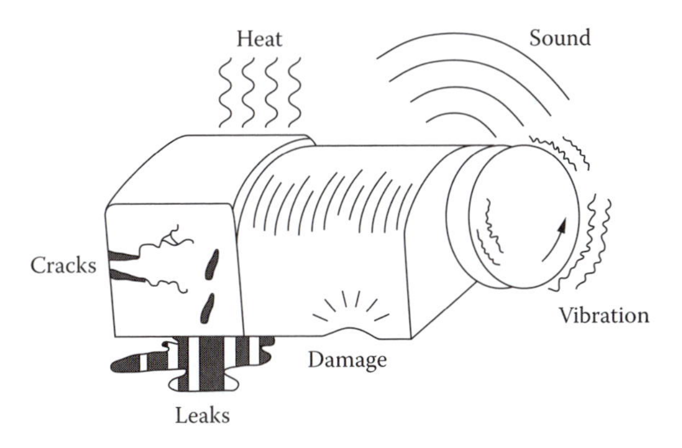

Figure 7.6 Signs of emerging equipment problems.

during which experts inspect, replace parts in, and overhaul equipment. As a rule, the relatively little downtime scheduled for PM affords big benefit in terms of fewer random and unscheduled machine breakdowns. With scheduled PM, everything is stopped at the same time and production schedules are not disrupted. Overtime and missed deliveries from breakdowns are averted.

Ways of Scheduling PM

Scheduled time for PM can be based on the following:

- Clock or calendar time intervals. Some examples follow: Run a two-shift operation, then devote the third shift to PM; run a three-shift operation, using 30 minutes at the start and finish of each shift for PM; perform PM tasks monthly, quarterly, and annually (Figure 7.7 shows an example).
- Cycles of usage. For example, schedule PM for every 1,000 hours of machine operation, every 5,000 units produced, or every 50 setups.
- Periodic inspection. Schedule PM whenever a periodic inspection indicates impending or possible malfunction or failure.

Typical scheduled PM includes opening up equipment for thorough cleaning and replacing major internal components such as motors, bearings, valves, seals, and belts. All of this is performed by skilled maintenance workers, and is in addition to the daily and weekly cleaning and inspection activities performed by operators.

With scheduled PM there must be ample time kept open every day and week for maintenance. Never, or only rarely, should production work be scheduled to consume an entire 8-hour shift and three full shifts.

Scheduled PM and Failure Pattern

Preventive maintenance often involves replacing a component that is functioning perfectly. This is because the component is believed to be approaching the end of its estimated **useful life**, the point at which the probability of the component failing greatly increases. The estimated useful life is shown in Figure 7.8 for the three failure patterns discussed earlier.

Description:

MONTHLY MACHINE PREVENTIVE MAINTENANCE

______1. Complete all weekly checks.

______2. Check all flexible lines and cables. Repair or replace any that are cracked or damaged.

______3. Remove and clean the air intake filter to the hydraulic power supply.

______4. Make a visual inspection of the hydraulic oil; either excessive darkening or milkiness indicates a need to replace the oil.

______5. Check edge locators parallel to X and Z axis.

______6. Check sliding covers and curtains. Replace if necessary.

______7. Tighten contacts at all terminals.

______8. Clean relay contacts with contact cleaning spray.

______9. Remove Z axis cover and clean chips from cavity.

______10. Remove B axis cover and inspect gears for smooth operation.

______11. Install both covers and silicone on the leading edge of the Z axis cover.

______12. Inspect tool changer arm for any wear and freeness.

Description:

QUARTERLY MACHINE PREVENTIVE MAINTENANCE

______1. Complete all monthly checks.

______2. Check hydraulic supply relief valve pressure setting 48.2 bar 700 psi.

______3. Check hydraulic pump pressure setting 41.4 bar (600 psi).

______4. Check all axes reference positions.

X Axis______________ Y Axis______________

Z Axis______________ B Axis______________

Figure 7.7 Monthly, quarterly, and annual PM tasks. (Courtesy of ITT McDonnell & Miller Corp.)

______5. Check operation of tool changer. Confirm all switch adjustments and tool changer travel motions and speeds.

______6. Check axis and spindle drive electronic adjustments.

______7. Check axis lost motion and correct as needed.

______8. Take oil samples for analysis.

______9. Replace oil filter element.

______10. Run DCS fingerprint.

______11. Clean condensor and evaporator.

Description:

ANNUAL MACHINE PREVENTIVE MAINTENANCE

______1. Complete all semiannual checks.

______2. Check the axes drive belts for looseness or wear. Adjust or replace if necessary.

______3. Replace wipers of guideway bearings.

______4. Replace tool magazine drum rollers if worn.

______5. Check spindle bearings for temperature. (Run spindle at 1,200 rpm until temperature of spindle nose casting stabilizes. Temperature should be 110 deg F to 160 deg F maximum.)

______6. Check the machine geometry. Correct as necessary. Refer to Section 9.0 of manual for instructions.

Figure 7.7 (Continued).

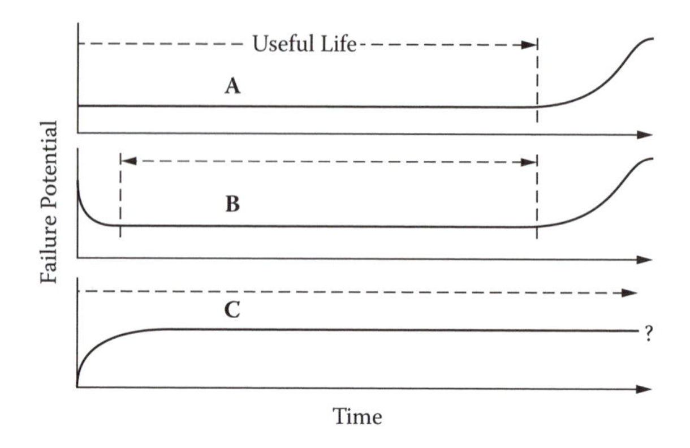

Figure 7.8 Useful lives in common failure patterns.

Components with patterns A and B in Figure 7.8 have a somewhat obvious useful life, which is used in PM to determine when they should be replaced or refurbished. For example, components on the airframes and engines of aircraft are replaced at a specified number of operational hours, which is their estimated useful life. Though they still function, replacing them at that time is clearly safer than waiting until they fail. In many cities, crews replace streetlights block by block, street by street because (assuming all lights were installed at the same time) all are near the end of their normal lives. Replacing them then, all at once, is less costly than replacing them individually as they burn out. The same kind of rationale applies to PM of all kinds of equipment in all kinds of facilities.

To extend the useful life of equipment approaching the wear-out region, equipment can be run at lower-than-normal speed (derating), or exposed to less than normal workload or stress level.

But for several reasons, however, it is not always easy, or even desirable, to schedule replacement or renovation of equipment based on the concept of useful life. First, periodic scheduled replacement of components can actually increase the risk of failure in complex machines or processes. Every replacement of a component that has a bathtub failure pattern (pattern B) at first reintroduces to the machine or process an increased risk of failure stemming from the infant-mortality risk of the component.[12] Other than that, however, component replacement in machines comprising numerous components results in pattern C for the overall system since it has a combination of new and old parts with a range of expected useful lives.

Finally, the useful life of an item is commonly estimated by statistical analysis of failure of the item and the resultant failure pattern. But for most components and equipment the useful life is unknown and will always be unknown. To derive the useful life, a component must be allowed to fail many times (alternatively, many units of the component must be allowed to fail once). But as John Moubray points out, the reason that companies do PM is because they cannot allow components to fail, ever. This poses a contradiction, he says, in that successful PM entails preventing the collection of data that we think we need to decide what PM we ought to be doing.[13]

Because of these difficulties, replacement and overhaul of PM is often predicated on periodic equipment inspections to reveal impending or possible failure or malfunction. This practice of relying on inspection to determine when and what kind of PM tasks to perform is referred to as condition-based or *predictive maintenance*, and is discussed later.

Scheduled PM cannot prevent every breakdown or malfunction, and there will always be need to perform unscheduled repair maintenance. Need of repairs is identified by workers or automatic sensors that monitor equipment performance using standards or control charts, or that detect problem signs such as oil leaks or cracks. Although some repairs can be delayed until the next scheduled maintenance period, often they must be performed immediately. Regardless of the amount of effort devoted to scheduled PM tasks, there must always be adequate expert support staff to meet on-demand (unscheduled) requests for breakdown repairs.

Manage Maintenance Information

Preventive maintenance requires keeping good records about the performance and breakdown history of equipment and the costs of operations and repairs. In many companies no one keeps track of the frequency of repairs, kinds of repairs requested, sources of problems, or costs. In a big company that has much equipment, the maintenance people might keep track of only the key machines. The operators know about the problems on their machines, but they do not report the problems or suggest solutions to them. And no one asks them. In a multishift plant, the operators on one shift are often unaware of equipment problems that happened on other shifts.

Effective PM requires a good system for tracking equipment performance, breakdowns, repairs, and related costs. The tracking system should be part of a larger computerized maintenance management system (CMMS) that processes work orders for repair maintenance; maintains PM procedures and schedules; releases work orders for all scheduled preventive, predictive, and repair maintenance; and prepares summary reports. Figure 7.9 shows an example of a summary report giving the tasks and costs for monthly PM on one machine.

Information retained by the CMMS about machine performance, breakdowns, and costs is used to compile statistics on equipment availability, MTTR, efficiency, and quality, and to assess maintenance activities. These statistics are useful for determining the maintenance requirements and schedules for components and equipment, and for improving maintenance procedures. The CMMS reports hours lost and costs for both repair and prevention activities and shows the tradeoffs. Company engineers use this information to upgrade old equipment and procure new equipment. Maintenance staff uses it to establish PM procedures and schedules, determine operator training, and know what kinds of spare parts to carry.

Creating a CMMS begins with a plant register or list of all equipment. For all equipment in the register, the following information must be compiled:

Type of machine
Serial number
Date put in service
Manufacturer
Dates of upgrade or changes
Location in plant
Location of manuals, schematics and drawings, and spare parts

For all equipment in the register the CMMS should also include information such as standards of performance, ways in which the equipment could fail to meet those standards, potential causes of failure, and consequences. The consequences determine the criticality of PM. Equipment in which malfunction would result in process shutdown or safety hazards is the most critical; equipment where the consequences are not immediate or less significant is less critical. Such information is necessary for determining PM procedures, schedules, and priorities.

When establishing an equipment database, the best sources of information are people most knowledgeable about the equipment's functioning, operation, and uses, usually the operators,

WORK HISTORY FILE REPORT

Work Category equal to: PM

Date Issued greater than or equal to: 06/01/10

Equipment No. equal to: 05010

Job # greater than or equal to: 0002

WO Number	06199504041	Parts Available? Y	
Work Category	PM	WO Status	
Failure Code		Number Printed 1	
Priority	Job # 0002	Requested By	
Date Issued	06/19/10 07:18	Account # WO1111	
Date Required	07/14/10 00:00	Schedule Date 00/00/00	
CAUSE CODES		Earliest StarFolder Name Shift 0	
Equipment No.	05010	Date completed 07/21/10 Actual 00/00/00	
Name	CNC M/C, HOR, K&T		Time 07:17
Dept/FF	FF-1	Meter Reading	0.00
Location	K-11	Previous Reading	0.00
Warranty Date	00/00/00	Downtime	0.00 % Productive 0
Cost	LTD 74144.33	Lost Operating Cost	0.00
Problem:			

Labor Cost	1,097.14
Outside Costs	0.00
Projected WO Cost	0.00
Parts Cost	0.00
Work Order Cost	1,097.14

Figure 7.9 Monthly PM summary report for a machine. (Courtesy of ITT McDonnell & Miller Corp.)

Corrective Action:
MONTHLY K&T PREVENTIVE MAINTENANCE
Text Codes:
Text Code Text Code Name
1 WKTM K&T MONTHLY PM
Tools:
Parts:
Part #/Name/Location
Labor:

	Clock#	*Craft*	*Date*	*EST Hrs*	*Reg Hrs*	*OT Hrs*	*DT Hrs*	*CI Hrs*
1	004	HELP	07/06/10	0.00	10.50	0.00	0.00	0.00
					ACTION CODE		**Craft Cost**	**103.74**
2	004	HELP	07/05/10	0.00	6.50	0.00	0.00	0.00
					ACTION CODE		**Craft Cost**	**64.22**
3	091	MECH	07/08/10	0.00	2.00	0.00	0.00	0.00
					ACTION CODE		**Craft Cost**	**31.20**
4	091	MECH	07/11/10	0.00	8.00	0.00	0.00	0.00
					ACTION CODE		**Craft Cost**	**183.30**
5	055	MECH	07/18/10	0.00	6.50	0.00	0.00	0.00
					ACTION CODE		**Craft Cost**	**101.40**
6	055	MECH	07/20/10	0.00	8.50	0.00	0.00	0.00
					ACTION CODE		**Craft Cost**	**132.60**
7	230		07/21/10	0.00	6.50	0.00	0.00	0.00
					ACTION CODE		**Craft Cost**	**89.57**
8	055		07/21/10	0.00	1.00	0.00	0.00	0.00
					ACTION CODE		**Craft Cost**	**15.60**
9	230		07/18/10	0.00	8.50	0.00	0.00	0.00
					ACTION CODE		**Craft Cost**	**117.13**
10	230		07/19/10	0.00	8.50	0.00	0.00	0.00
					ACTION CODE		**Craft Cost**	**117.13**

Figure 7.9 (Continued).

supervisors, and repair staff. A way to tap that information is to form small discussion groups focused around on the equipment in a process (a line or cell) or department. Equipment manufacturers sometimes recommend PM procedures and time intervals; often, however, PM schedules and procedures are better determined from the experience and knowledge of people who work with the equipment.

One problem with scheduling periodic PM is that scheduled PM tasks must take into account the workload of the maintenance staff. A good CMMS sets PM schedules such that not only are the PM requirements of equipment fulfilled, but the workload imposed on the maintenance staff is leveled and uniform.

Use Predictive Maintenance

For cases where it is difficult or impossible to determine the useful life of a component or machine, a common practice is to employ **predictive maintenance** to give warning about potential failures. Rather than perform periodic scheduled maintenance tasks, predictive maintenance performs periodic, scheduled inspections, the results of which are used to determine specific PM tasks and replacement of components. The operators might do the periodic inspections, though they are more often done by the maintenance staff because they require high-level skill.

Predictive maintenance is sometimes classified as a function separate from PM. However, to the extent that predictive maintenance is another way to prevent breakdowns and malfunctions, we consider it here to be an element or aspect of PM.

Most failures give some form of advance warning, and in predictive maintenance the warning indicates that a replacement or overhaul is necessary. The warning might give enough time to order replacement parts (when not stocked), transfer production to other machines, or schedule maintenance tasks. Predictive maintenance is also called **condition-based maintenance** because items remain in service on the condition that potential failure is not detected. When inspection indicates an impending problem, a closer inspection or immediate remedial action is scheduled.

Predictive maintenance often involves monitoring the vibration, speed, temperature, sound, and other physical phenomena of machinery. For example, failure of spindle bearings is a common cause of defects and breakdowns in all kinds of rotating machinery, and vibration can be an indication that a bearing fault is developing. In addition to monitoring the machine directly, measures of the output of the machine can sometimes be used to predict machine malfunction. For example, points on a control chart for product quality that show a process is moving toward an out-of-control situation might indicate a nascent machine problem.

Human sensory inspection often provides adequate results, though sometimes it does not because signs of emerging problems are too subtle. This is where detection technology comes in. One example of this technology is x-ray radiography. The human eye can easily see stress cracks on a machine, but it cannot see where metal is fatigued and about to crack. From x-raying the metal (in a manner similar to x-raying people), it is easy to see on exposed film areas of discontinuity in the metal. Two other examples of technologies for predictive maintenance are infrared thermography and ultrasound.[14] All equipment has thermal patterns that are easily detected with thermal imaging devices; some devices can detect temperature differences smaller than 1/10°F. Thermal devices are useful for finding hot spots caused by high resistance in electrical connections, and friction in bearings and couplings. With a portable infrared thermography machine, thousands of bearings in a hundred yards of conveyor can be inspected in a matter of minutes. All operating equipment also produce a broad range of sounds, but leaks in valves, pumps, gaskets, and seals in high-pressure and vacuum systems produce only high-frequency, short-wave sounds. Though these high-frequency sounds are inaudible, ultrasound instruments easily separate them out from background noise and determine their origins.

The amount of warning from inspection can vary from microseconds to weeks or more. Shorter warning time mandates more frequent inspections, and some equipment requires 24-hour surveillance using sensors linked to a computer. On-line computerized monitoring can be expensive but sometimes is a cost-effective alternative to human inspection and data logging, especially when the inspection is for components located in hidden or inaccessible places or when the inspection must be highly accurate.[15] Corporations like General Motors, Eaton, and Monsanto rely heavily on on-line monitoring. GM monitors the wear of bearings on ventilation fans on heat-treat furnaces, and tool wear and breakage on production equipment. At GM's wind tunnel, which runs 24-hours a day, 5 days a week, bearings on motors and blowers are monitored. To replace a main drive motor bearing means 28 weeks of downtime since the building must be partially disassembled to reach the motor; however, with advance warning as much as 20 weeks of this downtime can be avoided.[16]

Role of Operators

Though equipment monitoring sometimes requires sophisticated technology, the watchful eye of an operator is often sufficient to determine when equipment is performing abnormally and needs adjustment or repair. To achieve and retain high equipment effectiveness, there is no substitute for operators who believe in the importance of keeping equipment well functioning.

Most plants have hundreds or thousands of pieces of equipment—production and materials handling equipment, support machinery, fixtures, and tools. Expecting a small pool of experts from maintenance and engineering to keep up with all this equipment is illogical. On the other hand, operators deal with the same equipment daily. If they care about the equipment, they will treat it well; and if something goes wrong, they will notify someone to fix it. If they don't care about the equipment, they will abuse or misuse it; if something goes wrong, they will ignore it, try to work around it, and not report it until the problem is serious or catastrophic.

When equipment cleaning and basic upkeep are included in the responsibility of operators and frontline workers, when every eye is watchful of equipment problems, then the frequency of undetected problems will drop and the plant OEE will rise. Even if a worker sits at one machine all day, her responsibility will be increased enormously by simply transferring to her custodial care and basic upkeep of her machine. She has more control over her work world than before, and she is likely more proud of her work. In many lean companies operators proudly put their names on equipment.

Additional responsibility must be accompanied with training in additional skills. Workers should be trained to perform basic maintenance procedures (cleaning, lubricating, tightening fasteners and connections, replacing filters, etc.). Staff from the maintenance department conducts most of the training. Of course, just giving operators partial responsibility for machine upkeep will not eliminate machine problems. Management must acknowledge the time and skills associated with the responsibility, and upgrade job descriptions and pay accordingly. Experts from maintenance, engineering, and equipment manufacturers must be ready to give quick, reliable service to deal with serious and complex equipment problems.

Initially, the operators are guided by maintenance people and rely on their instructions. Eventually as the operators develop their skills, they will move on to the next step, at which time they are given greater autonomy to participate in performing basic preventive and repair maintenance, develop improved maintenance procedures, develop their own inspection check sheets, and meet in groups to discuss, diagnose, and resolve equipment problems. This next step is but one aspect of total productive maintenance.

Total Productive Maintenance[17]

Total productive maintenance (TPM) is a commitment to machine usage and upkeep that goes beyond preventive and predictive maintenance. In some ways TPM is similar to TQM (total quality management):

- All employees are involved in satisfying customer needs, where the customer is the person at the next stage of the process. For TPM this translates into providing maximum support and service to all users of equipment.
- A machine breakdown is seen as a form of defect, and TPM is committed to preventing breakdowns and malfunctions from happening in the first place.
- TPM is a further aspect of continuous improvement. It is an ongoing process of educating and involving workers, upgrading and redesigning equipment, instituting fool-proofing devices, monitoring equipment performance, and eliminating sources of equipment waste.

TPM uses a life-cycle perspective of equipment. The likelihood of equipment not performing during its useful life is a function of many factors, including the equipment's design, construction, operation, and upkeep. Equipment deteriorates with usage, and with deterioration comes diminished reliability and effectiveness. TPM seeks to reduce deterioration and increase effectiveness and useful life by focusing on all the factors that govern equipment life, which not only includes upkeep and operation, but also construction and design. In TPM, the emphasis and spending switches from procuring new or more equipment to monitoring and upgrading existing equipment.

Just as TQM seeks zero defects, TPM seeks zero machine malfunctions and breakdowns. A malfunction is the end of a series of linked causes: lubricant runs low, or dust and dirt accumulate, which causes scratches and friction, which cause loosened, cracked, or fatigued parts, and so on. As long as the initial and intermediate causes go undetected, they remain untreated, which eventually leads to malfunction. Nakajima[18] suggests the following steps for revealing and treating the hidden causes of equipment problems:

- Perform TPM preventive maintenance.
- Develop in-house capability to restore and redesign equipment.
- Eliminate human error in operation and maintenance.

Perform TPM Preventive Maintenance

TPM preventive maintenance includes all the PM features described earlier, with the addition that operators bear more responsibility for doing PM tasks and basic repairs. Operators are trained to take on many of the maintenance tasks formerly done by maintenance experts. They follow procedures listed on daily and weekly checklists. Like setup tasks, PM tasks are separated into those that can be done safely while the machine is running (inspections and cleaning of external components) and those that must be done while it is stopped (checking of moving or internal components). This operator involvement enables basic PM tasks and minor repairs to be performed more frequently and, often, more efficiently than the maintenance staff is able to do. This results in fewer equipment breakdowns and frees up the maintenance staff to take on other responsibilities, such as equipment renovation and redesign.

Case in Point: Tennessee Eastman Co.[19]

Some machines at Tennessee Eastman have a small rubber disk that ruptures and shuts down the machine if the vacuum pressure gets too high. It used be that a ruptured disk was the start of a long chain of events: the operator informed the production manager, who informed the maintenance manager, who located a maintenance expert; the expert got a replacement disk from the storage area, then went to the machine, removed four bolts, replaced the disk and aligned it, and reattached the bolts. The average downtime was 4 hours. This happened about 200 times a year, so total downtime was 800 hours.

With TPM, operators were trained to do the repair themselves, and it took them only 1 hour. However, they did not like doing the repair and because of that they started monitoring the vacuum pressure more closely so the disk would not rupture. As a result the number of ruptures dropped to just 10 a year—a 95% reduction. The machines now have 790 more hours annually to produce, and the maintenance department has 800 more hours to do other things.

In traditional union shops with rigid job classifications all equipment-related work is handled by trade specialists. Although this is supposed to result in higher quality and greater efficiency, for equipment maintenance and repair it is counterproductive. A breakdown problem is often better handled by someone with a good general understanding of a machine than by a group of specialists, each of whom understands only a limited part of it. Being well rounded, however, should not preclude maintenance staff from also being specialists in any one area. Trained electricians, technicians, and plumbers are still needed to keep a factory running.

The maintenance staff's new role influences equipment-related procedures and decisions in important ways. The new role includes teaching operators basic maintenance skills, assisting them in PM, restoring deteriorating equipment, assessing weaknesses in equipment design, upgrading equipment, and developing new operating and maintenance requirements for all equipment. Figure 7.10 summarizes the new roles of operators and maintenance staff in TPM.

Develop In-House Capability to Restore and Redesign Equipment

Another way that TPM differs from simpler PM is that TPM includes emphasis on developing in-house capability to restore, redesign, and fabricate equipment, fixtures, and tools. Many companies subcontract equipment maintenance, restoration, and design so they can better focus on equipment usage (i.e., on manufacturing). But equipment usage, design, and upkeep cannot be treated independently. Companies that rely entirely on suppliers to produce and maintain equipment and tools never truly understand their own equipment needs. These manufacturers are at the mercy of their suppliers for recommendations about new equipment and for repair or PM services, even though those recommendations or services might not best suit their needs.

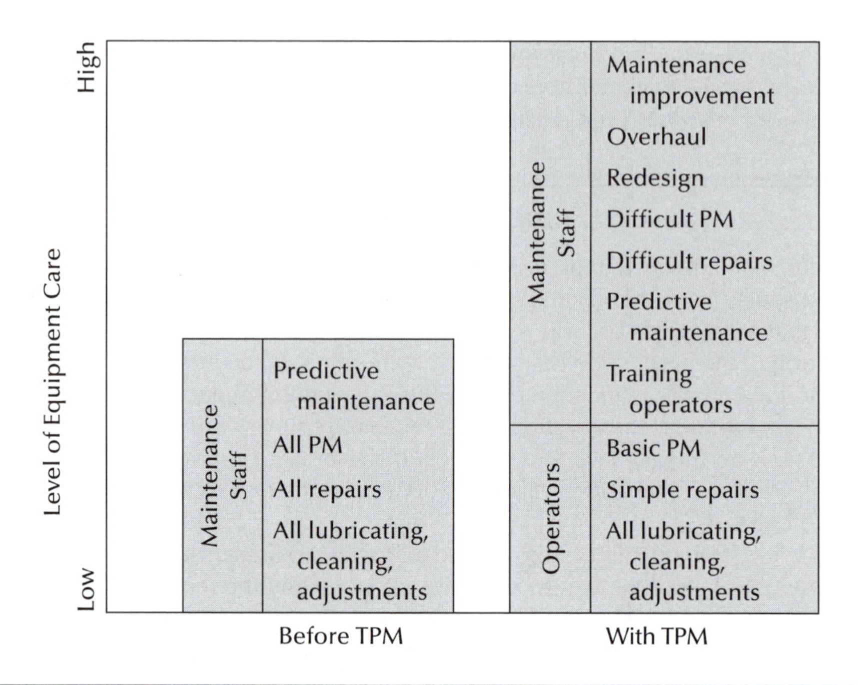

Figure 7.10 New roles of operators and maintenance staff.

A company that develops in-house technical capability is in control of the useful life of its equipment, and it often can perform most of its own PM and repair maintenance more effectively than equipment suppliers. When people in the company become involved in the upkeep and renovation of equipment, they develop an understanding of the workings and technology of that equipment, and that enables them to alter the design of the equipment to better suit the unique needs of the company. In many companies, the same equipment is used for 40 or 50 years, over which time there is ample opportunity to upgrade the equipment in terms of maintainability and operating performance. As they upgrade equipment, the company's engineers and technicians are utilizing and upgrading their own technical skills.

The ultimate outcome of in-house design capability is that the company can redesign its own machines, tools, and fixtures to improve equipment performance, problem detection, and maintainability. One example of this is that the equipment itself can perform functions that eliminate or speed up maintenance, much like the common office copier. Most copiers detect a potential problem and display the location and error code for the problem. Often the problem is small (e.g., add toner) and quickly remedied by users of the copier. When the problem is more serious, the machine shuts itself down. Even then downtime is reduced because an error code tells the service specialist where to find the problem.

Most companies do not have the resources to do sophisticated or large-scale redesign, though they do have the resources to do smaller, less sophisticated design changes such as replacing bolts with clamps, improving access to internal components, and attaching permanent fixtures or components to simplify machine operation or changeovers. Companies that are able to do this, even on a small scale, gain a competitive advantage, particularly if the in-house design is for a patentable process or a machine that outperforms machines used by competitors.

Effective equipment overhaul, design, and procurement require that all aspects of the equipment life (operational uses, setups, and maintenance) be considered together. To this end, equipment decisions in TPM are made by teams comprised of technicians from production, engineering, and maintenance, as well as setup technicians and equipment operators. Though operators might not know how to make design changes in equipment, they know about equipment usage and related problems, and can tell others about what improvements are needed.

Eliminate Human Error in Operation and Maintenance

The TPM view of equipment problems is that problems don't just happen, they happen because managers, workers, and maintenance staff are either doing things incorrectly or are not doing things that should be done. The TPM solution is threefold: education and training, foolproofing, and improved maintenance procedures.

Education and Training

Besides being trained to operate equipment properly and to perform basic PM and simple repairs, operators are trained on how to inspect for and recognize signs of deterioration and hidden causes of problems. They also learn the basic tools of data collection, analysis, and problem solving, which they apply in small-group sessions. Training is based upon equipment needs, the trainability and motivation of the operators, as well as the time and resources available. Training is provided in a variety of ways such as 30-minute minisessions on the plant floor, in-house seminars and workshops, take-home videos, and courses at junior colleges and trade schools. TPM consultants and equipment vendors might do some training, but in-house trainers and maintenance staff do most of it.

To help motivate operator participation in TPM and to ensure adequate skill development, operators should be tested for certification upon completing a course of training. The certification formally acknowledges that an operator has mastered some skill requirement. The certification can be general (e.g., lube all of a certain type of machine) or for a particular machine (see Figure 7.11, left side). In many plants, operator certifications are displayed at each machine, department, or workcell (Figure 7.11, right side).

Likewise, maintenance staff members are trained so they are able to take on an expanded role that emphasizes increased maintenance speed and efficiency, fewer maintenance errors, and greater usage of their technical skills. Increased maintenance speed and efficiency requires that all maintenance people, in addition to being skilled in some special trade, have general skills in electronics, hydraulics, and so on, as well as knowledge of basic equipment technology, equipment functioning, and new technology. The purpose of these skills is to make them more well rounded at diagnosing equipment problems, repairing equipment, and performing PM procedures, no matter what equipment they work on. It also prepares them to assist in in-house design and renovation of equipment. Some of this training happens via in-house seminars and on the job, while others come from special outside seminars, courses, and workshops.

Foolproofing

Not all equipment problems caused by human error can be resolved simply by training. Some errors are inadvertent, and the way to eliminate them is to determine their root causes and install ways to prevent them from happening. Any mechanism, device, or procedure installed to preclude an inadvertent error

Machine *LU29*
Operator *Gary Owen*

Skill set		TPM Certification
External lube	3	Level 1 Basic
Replace filter	3	
Adjust belt	3	
Adjust fairing	3	
Internal lube	3	Level 2 Intermediate
Adjust sprocket	3	
Adjust drive oil seal & bushing		
Replace sprocket		Level 3 Advanced
Replace drive seal & bushing		
Adjust magazine camber		

Machine *LU29*

	TPM Certification Level		
Operator	1	2	3
Gary Owen	3		
Marge Jones	3	3	3
Renee Lopez	3	3	
Horst Jankowski			

Dept. 24

	Machine, TPM Certification Level				
Operator	LU26	LU29	GOB1	GRB13	GRB15
Gary Owen		1	2	2	2
Marge Jones	3	3			
Renee Lopez	1	2	2		
Horst Jankowski			3	2	2
•					
•					
•					

Figure 7.11 Operator certification levels and displays.

is called a **pokayoke** device. *Pokayoke* is Japanese for "foolproofing." For example, suppose a source of equipment breakdown is the operator forgetting to check the level in the lubricating oil reservoir. When the oil level is too low, the shaft on the machine freezes. As a result, the cutting tool is damaged and the part being machined is destroyed. A simple, inexpensive foolproof device is a float in the oil reservoir connected to a light and a switch. When the oil drops to a certain level, the light goes on and the switch prevents the motor from being turned on. Most pokayokes are inexpensive and originate from worker suggestions. Pokayokes and foolproofing are the subject of Chapter 12.

Improving Maintenance Procedures

Eliminating human error in equipment problems also requires improving maintenance procedures, a concept analogous to improving setup procedures. Like setup, maintenance is often done in a somewhat disorganized and ad hoc fashion. Both can be time consuming and decrease production capacity and quality. Like setups, maintenance procedures should be analyzed, simplified, and standardized so they will (1) be done more quickly (short MTTR) and (2) result in no mistakes. The same practices to reduce setup times can be applied to improve preventive maintenance: keep tools and spare parts organized, labeled, and easily accessible; use repair carts; keep instructions, tools, and parts next to machines needing frequent maintenance; and train workers to do basic repairs themselves. A study team comprised of maintenance workers, operators, and engineers analyzes machine records, video records frequently used maintenance procedures (such as PM), and determines how to streamline or the procedures.

Case in Point: TPM at ITT McDonnell & Miller Corp[20]

McDonnell & Miller (M&M) formerly did only repair maintenance; now it does PM on every machine, both daily and weekly basic PM by operators, as well as computer-scheduled expert PM by maintenance specialists. Every machine and workcell has it own PM procedures, which in some cases the operators helped prepare, as well as daily and weekly checkoff sheets for operators to follow (Figure 7.12). In each workcell is a binder showing procedures for every machine that includes machine drawings showing places that require adjustment and lubrication. Each machine or workcell also has the tools necessary for simple repairs. Tools are kept in locked cabinets, but the operators have the keys.

Information about PM procedures, breakdown repairs, labor time, and costs is stored, processed, and reported using a CMMS. The system generates monthly, quarterly, semiannual, and annual work orders for all equipment needing PM. To allow adequate time for PM and to minimize interference with production schedules, most PM is performed on the third shift.

Lubrication is taken seriously. Prior to TPM, equipment at M&M was seldom lubricated, and some had never been lubricated since purchase. Now, posted on every machine is a list showing the required lubrication, as well as small, color-coded stickers indicating points of lubrication and type of lubricant (examples, Figure 7.13). Every machine or workcell has containers for every kind of lubricant. When containers run low, operators go to a store area in the plant and take whatever lubricant they need from 12 huge barrels.

Resistance to TPM at M&M stems largely from misconceptions about the necessity to perform rigorous PM, especially when it interferes with production.

ACTION CODE WKTD

Text Code Name DAILY K&T PM

Description:

DAILY K&T CHECKLIST

TASK	*M*	*T*	*W*	*TH*	*F*	*S*
1. CHECK AIR PRESSURE GAGE (NORMAL—30–60PSI)	___	___	___	___	___	___
2. CHECK FLOOD COOLANT LEVEL AND REFILL IF NECESSARY	___	___	___	___	___	___
3. CLEAN ALL MAGAZINE SOCKETS AND TOOLS	___	___	___	___	___	___
4. CLEAN TOOL CHANGE ARM TOOL PICKUP AREA	___	___	___	___	___	___
5. CLEAN AREA BETWEEN TABLE AND COLUMN WAY	___	___	___	___	___	___
6. CHECK FOR NORMAL AIR BLAST AT THE SPINDLE NOSE WHILE SPINDLE IS RUNNING	___	___	___	___	___	___
7. CHECK ALL LIGHTS FOR PROPER OPERATION	___	___	___	___	___	___

ACTION CODE WKTW

Text Code Name K&T WEEKLY PM

Description:

WEEKLY MACHINE PREVENTIVE MAINTENANCE:

____1. Review machine log book since last PM. Resolve problems noted.

____2. Wipe down entire Machine, Power Supply, and Control Unit. Clean all exposed limit switches and their trip dogs.

____3. Listen to hydraulic unit. Is the sound normal?

____4. Check hydraulic oil for coolant contamination (milkiness).

____5. Check system pressure—41.4 bar (600 psi). Relief valve setting—48.2 bar (700 psi).

____6. Check all exposed oil lines and repair any existing leaks.

____7. Check the tool transfer and magazine index. Did all of the tools index and change properly?

____8. Clean or replace the air intake filter for the control.

____9. Clean tape reader.

Figure 7.12 Daily and weekly maintenance checklists. (Courtesy of ITT McDonnell & Miller Corp.)

For some machines, maintenance takes longer than the third shift, and the production department balks at machines being held up for scheduled PM during scheduled production hours. The acceptance of TPM by operators has been good, in some cases too good. A few overzealous operators want to do more PM than required, including things such as electrical work for which they are not trained, which can be hazardous and result in injuries.

Has TPM paid off? Prior to TPM, the maintenance staff of M&M consisted of nine people, and they were faced with a continuous backlog of repairs.

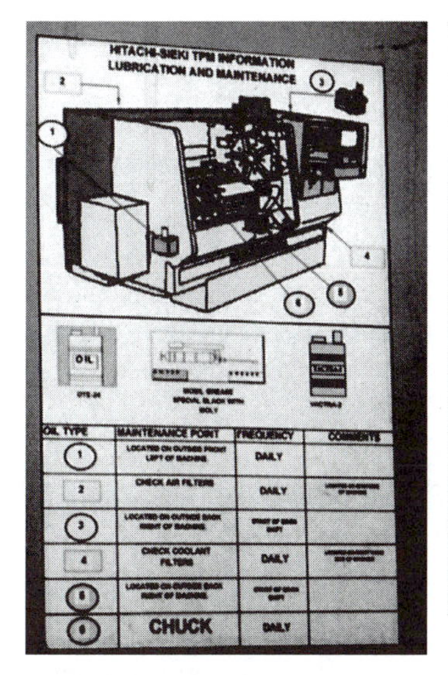

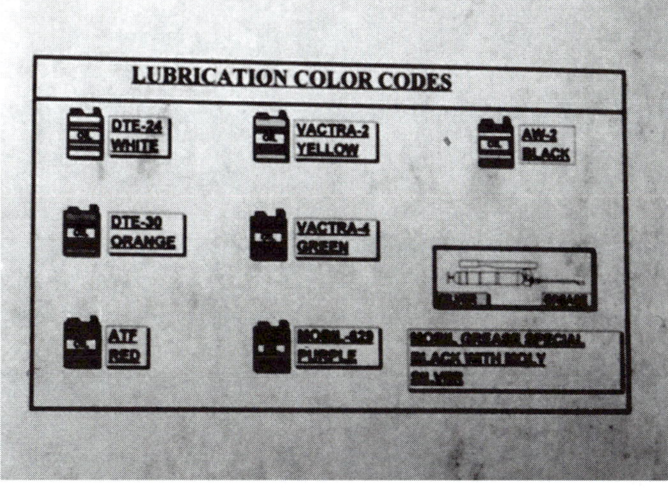

Figure 7.13 Diagrams on machines showing points for necessary lubrication and kind of lubrication. (Courtesy of ITT McDonnell & Miller Corp.)

> Less than 2 years after the start of TPM, the backlog is gone and requests for breakdown repairs readily dispatched. Even with all the extra PM being performed, the staff has had enough time to construct and equip a new central spare parts store, install a new CMMS, train operators, and develop PM procedures for every machine. They were able do all this even though, as a result of two people retiring, the current maintenance staff is smaller than before.

Implementing TPM

If TPM is eventually to be spread plantwide and impact management policy, worker responsibility, and staff roles in the functions of production, maintenance, and engineering, then TPM implementation should be guided by a **steering committee** that consists of managers from those functions. The purpose of the steering committee is to formulate TPM policies and strategies and to give advice. A top-level manager should serve on the committee to show management's endorsement of TPM and serve as champion for the TPM program.

The actual oversight and coordination of implementation activities are done by a **TPM program team**. The team includes managers, staff, and technicians from the production and maintenance departments, a TPM program manager, trainers, and sometimes a consultant. This team oversees the TPM program feasibility study, sets detailed program objectives, selects target areas, and prepares a master plan for implementation. The groups and their relationship are shown in Figure 7.14.

Program Feasibility

Prior to initiating the program, a feasibility study is performed to gather baseline data for planning the program and assessing the program's likely costs and benefits. The study is performed by a small group of operators, maintenance experts, and engineers. Through observation, interviews, and records, the study group gathers data about the current state of equipment in terms of OEE,

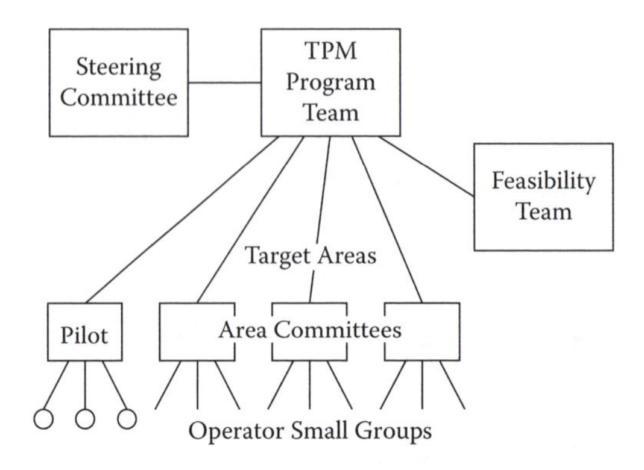

Figure 7.14 Groups involved in TPM program implementation.

production losses, downtime, MTBF, and so on. The team assesses current skills of plant personnel, other skills they will need for TPM, and their ability and motivation to learn additional skills. The study team assesses all matters relating to maintenance—current housekeeping, PM efforts, procedures for scheduling and managing PM and repairs, and so on—and the organization's culture and ability to adapt to change.

Program Objectives and Master Plan

The program team sets definitive annual objectives for the TPM program.[21] The objectives might address performance measures such as MTBF, MTTR, equipment availability, performance efficiency, rate of quality, OEE, number of accidents, percentage of employees participating in TPM, and skill level acquired by participants. An example of an objective is to reduce the amount of downtime from equipment failure by 30% within a year of TPM kickoff.

The program team also prepares a master plan for implementing TPM plantwide. Rather than dictate precise events in the program, the master plan lays out the general sequence of activities for implementing TPM in particular target areas and plantwide tasks such as training of managers and maintenance staff, setting up or upgrading the CMMS, and reorganizing the maintenance function.

Target Areas

Although ultimately every area of every plant will practice TPM, the implementation should begin in just a few target areas—particular departments, focused factories, or processes in the factory. The first target area will serve as a pilot area for trying out TPM ideas and for demonstrating TPM concepts and results to the other target areas. A pilot target area should be selected based upon high likelihood of success, that is, an area that has important but resolvable maintenance problems and where operator–manager relations are good, workers are trainable, and operators are willing to try new ideas. The program team selects the pilot and other target areas based upon recommendations from the feasibility study.

Operators in each target area are trained to take on maintenance responsibilities. Most operators somewhat accept the new responsibilities, and much of the resistance from the others tends to diminish once they see how TPM works. Implementation in each area is tracked by the program team and the target area committee (described next), and problems and mistakes are noted. As the program is spread plantwide, the lessons learned in early target areas are applied.

Target Area Committees

Implementation of TPM in each target area is planned jointly by the TPM program team and a **target area committee**. The area committee is made up of the maintenance manager and staff, and the manager, supervisors, and operators from the target area. The program team and area committee work together to determine particular TPM tasks to implement, schedules for implementation, and people who will be responsible for the tasks. The following specific tasks are assigned:

- Determine for every machine how much improvement is needed; also identify sources of problems and ways to eliminate them. (To avoid improvement for improvement's sake, the initial focus is on high-priority machines—bottlenecks, near bottlenecks, or machines that delimit the capacity or quality of the target area.)
- Decide which PM and repair tasks can be transferred to operators; consider which tasks (a) operators want to do, (b) operators are capable of doing, and (c) the maintenance department is willing to relinquish.
- Train operators in basic PM, then set daily and weekly PM procedures for the operators, and monthly, quarterly, and annual PM procedures and schedules for the maintenance staff. (The time required for PM must be estimated and compared to available labor hours; PM procedures and schedules must allow adequate time for the operators and maintenance staff to do a good job of PM maintenance.)
- Determine equipment that requires predictive maintenance; define inspection procedures, frequency of inspection, and any special inspection equipment and skill requirements needed for inspection.

As additional areas of the plant are targeted for TPM implementation, the program team coordinates tasks with the target area committees to maximize sharing of lessons learned and minimize conflicts.

Plantwide Issues

The TPM project team initiates and oversees umbrella activities that will affect every department or focused factory in the plant. The following activities are included:

- Reviewing and revising existing company policies and procedures to include maintenance issues in future decisions concerning equipment renovation, redesign, and procurement decisions.
- Educating maintenance staff about the philosophy and focus of TPM; broadening their skills to enable them to fulfill their new roles.
- Determining maintenance-related issues that need special attention and forming teams to address them, (for example, selecting particular equipment for renovation, deciding whether to redesign or buy equipment, and finding ways to reduce downtime for repairs and PM).
- Determining how best to reorganize the maintenance department. If the department is to be decentralized, the roles of newly formed decentralized units as well as the remaining centralized unit must be determined.

- Setting up a new (or enhancing the existing) CMMS system. This involves establishing system requirements, procuring and installing hardware and software, developing procedures and schedules for data collection, and determining how to improve efficiency of data collection (for example, putting bar codes or radio frequency identification (RFID) tags on all machines).
- Subdividing TPM program objectives into objectives for target areas, operators, and pieces of equipment and establishing a system for rewarding and recognizing work areas and operators that meet or exceed objectives (e.g., awarding gold, silver, and bronze decals for operators to display on their machines).[22]
- Coordinating TPM with setup reduction, layout changes, standard work studies, and employee involvement. There is often considerable overlap in these activities, in which case they should be implemented together as a coordinated package.

During implementation, the program team and target area committees assess progress as well as find ways to remove barriers. In the beginning, TPM will appear as little more than a source of expenditures (and headaches). It will be a while before benefits are seen, and until then many will argue against the program and resist the necessary changes. For the first year or two, a major task of the TPM team is to work toward minimizing resistance, maximizing support, and helping the program gain inertia. Once the benefits of TPM start to appear and the program gains momentum, the program team can be dissolved, and responsibility for planning and daily coordination of the remaining implementation tasks can be transferred to the maintenance department.

Management Support

TPM cannot succeed without the support of top management. Workers, supervisors, and staff need to know that top management is committed to the change. During a kickoff meeting involving everyone to be affected by the TPM program, top management explains the overall TPM philosophy, the mission of the TPM program, the organizational changes required as part of the program, and the time and resources it will commit in support of the program. Throughout implementation, the program champion and steering committee work together to retain that commitment and to show that management stands behind the program.

Maintenance Organization

In traditional plants, a single pool of maintenance people services the entire factory. As a result, maintenance workers never master skills for particular equipment because they are too busy maintaining all equipment; this is especially true in large plants. For the same reason, maintenance workers never develop cordial working relationships with operators or managers in particular work areas. They never understand the needs of production people. Of course, the production people never come to understand the necessity for maintenance (other than for breakdown repair). TPM requires removing the barriers between maintenance and production departments.

Decentralization

A good arrangement for removing interdepartmental barriers is to decentralize the maintenance function, that is, to assign maintenance workers to particular areas, processes, or departments in a factory. Being affiliated with a particular production area enables maintenance workers to learn about particular processes and machines, become acquainted with individual operators and supervisors, and better understand their equipment needs. They learn the idiosyncrasies of each machine, and when something needs attention, they are there on the spot. Though maintenance workers still report to the maintenance department, most days they work in the same preassigned production areas next to the same supervisors and operators. Eventually they are viewed no longer as outsiders, but as members of the production team. One drawback of decentralizing the maintenance staff is that equipment problems are not uniformly dispersed throughout a factory, so staff assigned to some areas might be overworked, while those in other areas have little to do. Thus, decentralized staffing should be based on need: high-maintenance areas have a dedicated maintenance staff, but low-maintenance areas share a common staff.

Central Maintenance

Even with decentralization, a central maintenance function is still necessary to handle maintenance issues that are plantwide or do not fall within a particular department. The central maintenance function is responsible for certain storage of expensive spare parts, while other parts are stored in areas near the points of use; inexpensive, high-use items (like fasteners, fittings, and lubricants) can be kept in areas where needed and made freely available without requisition and lubricants) can be kept in areas where needed and made freely available without requisition.

The central maintenance department also has responsibility for monitoring the overall TPM program. To keep TPM current and focused for competitive advantage, the department identifies the best maintenance practices of other organizations and applies them as benchmarks for improvement.

Summary

Lean production seeks to identify sources of equipment-related waste and eliminate them. Eliminating equipment waste begins with determining measures for equipment effectiveness, which is the degree to which a piece of equipment is available to do work, perform at the expected level of efficiency, and produce no defects. The measures are used for assessing equipment-related waste and for setting goals for equipment improvement.

Fundamental to achieving high equipment effectiveness is a program of preventive maintenance. The program must emphasize maintaining equipment requirements and operating conditions, keeping equipment clean, monitoring daily for signs of potential malfunction, performing regularly scheduled preventive and predictive maintenance tasks, and keeping good equipment records. Effective PM requires that operators take responsibility for proper equipment operation, daily cleaning, monitoring, and basic equipment upkeep.

Total productive maintenance represents a commitment to equipment performance beyond original equipment design parameters and a move to make equipment a source of competitive advantage. One goal of TPM is zero breakdowns, the virtual elimination of equipment malfunction and equipment-related sources of product defects. A second goal is equipment restoration and redesign such that equipment performs better than new and in ways that competitors' equipment cannot. TPM shifts much of the responsibility for basic maintenance to operators, which frees up the maintenance staff to perform additional scheduled PM, and to participate in equipment redesign and restoration projects.

Implementing TPM plantwide is a major program that requires support from top management. Large-scale TPM is best implemented through a series of initiatives directed at target areas such as departments or work areas. Implementation in each of these areas is managed directly by a target area committee with assistance from the TPM program team. The TPM program team oversees initial planning and coordination of all TPM efforts, though ultimately this responsibility is transferred to the maintenance department.

This and the previous chapter explained aspects of equipment that, historically, many managers have all but ignored. Though both chapters stressed the importance of plantwide efforts to improve setups and equipment effectiveness, the immediate point of focus in both was on setup and maintenance improvement of individual pieces of equipment. Now here is the rub: To improve the overall production system, improvement of equipment is not enough. In fact, systems theory relates that if you try to improve the elements of a system without looking at them in the larger context, you could end up worsening the system. Thus, despite the necessity of equipment maintenance and setup efforts, they alone are not sufficient to improve a manufacturing process enough to make it more competitive. Pieces of equipment are elements of the larger production process, and it is necessary to look at them in that larger context. The next three chapters shift focus from equipment to the processes within which it is used.

Notes

1. Adopted from S. Nakajima, *Introduction to TPM: Total Productive Maintenance.* (Cambridge, MA: Productivity Press, 1988).
2. S. Nakajima, in *Continuous Improvement in Operations,* ed. Alan Robinson. Cambridge, MA: Productivity Press, 1991), 295.
3. Ibid., 302.
4. There are actually six possible failure patterns. For discussion of these, see J. Moubray, *Reliability-Centered Maintenance* (New York: Industrial Press, 1992), Chapter 5.
5. A simpler method for calculating performance effectiveness, derived from the product of the formulae for SE and RE, is

$$\text{Performance efficiency} = (\text{Design CT} \times \text{Actual production volume})/\text{Actual running time}$$
$$= (0.5 \times 830)/535 = 0.7757$$

   (The small difference in results between this and the other formula is due to rounding error in the other.)
6. E. Hartmann, *Successfully Installing TPM in a Non-Japanese Plant* (Pittsburgh, PA: TPM Press, 1992), 61, 63.
7. Adapted from K. Suzaki, *The New Manufacturing Challenge* (New York: Free Press, 1987), 116.
8. Frequently cited in automobile literature is that the single most important thing owners can do to minimize engine problems is to change oil periodically. *Consumer Reports* recommends a change every 7500 miles, even in normal or light driving conditions.
9. R. Schonberger, *World Class Manufacturing* (New York: The Free Press, 1986), 68.
10. Suzaki, *The New Manufacturing Challenge,* 119.
11. *Catastrophic* here means the condition where failure of one component leads to failure of another, sometimes in a chain-reaction fashion. For example, when a turbine blade fails, parts of it can fly off and crash into other components, including other blades, causing much damage and possible loss of the entire turbine engine.
12. Moubray, *Reliability-Centered Maintenance,* 13.
13. Ibid., 221.
14. See J. Snell, Infrared thermography: New solutions for both maintenance and production problems, *P/PM Technology* 8, no. 5 (October 1995): 54–58; and G. Mohr, Technology overview: Ultrasonic detection, *P/PM Technology* 8, no. 4 (August 1995): 56–61.

15. See E. Page, Spurring wider use of on-line condition monitoring for predictive maintenance, *Industrial Engineering* (November 1994): 32–34.
16. Ibid., 33.
17. Adopted from Moubray, *Reliability-Centered Maintenance*, Chapter 10; J. Wright, "Unleashing a Maintenance Department's Full Potential Through Reliability-Centered Maintenance," *P/PM Technology*, no. 1 (February 1996): 50–60; Hartmann. *Successfully Installing TPM*; H. Steudle and P. Desruelle, *Manufacturing in the Nineties* (New York: Van Nostrand Reinhold, 1992), 330–334, 337–340.
18. Nakajima, *Introduction to TPM.*
19. From Hartmann, *Successfully Installing TPM*, 46.
20. Interview with Lawrence Kocen, Plant Engineer, ITT McDonnell & Miller Corp.
21. John Wright argues that besides objectives, maintenance programs should have a mission statement stating what the program is to accomplish and how it will do it. Examples: being responsive to customer needs; maintaining equipment to allow continuous operation to meet material requirements; applying predictive and preventive maintenance, and evaluating their cost effectiveness; providing resources for all crafts to keep up with expanding technologies; fostering a work environment that will enhance morale and pride in a job well done. See Wright, "Unleashing a Maintenance Department's Full Potential Through Reliability-Centered Maintenance," 52.
22. Hartmann, *Successfully Installing TPM*, 206.

Suggested Reading

J. Levitt. *Complete Guide to Predictive and Preventive Maintenance*. New York: Industrial Press Inc., 2002.

Questions

1. List the consequences of equipment malfunction and breakdown in terms of cost, quality, safety, and lead time.
2. Contrast the following kinds of maintenance: repair, preventive, total productive.
3. How does investing in maintenance help an organization?
4. What are the six sources of equipment waste?
5. Define and give examples for each of the following terms: maintainability, reliability, availability, equipment efficiency, equipment quality rate.
6. What is the difference between repair time and downtime for repair?
7. What are the five principal causes of equipment problems?
8. What is the failure pattern of a system or component?
9. What is the bathtub function? What is the burn-in period? What is the wear-out period? How does knowledge about these help in maintenance scheduling?
10. Why should a component that is well functioning be replaced?
11. Where does probability data come from for determining failure potential?
12. What is meant by predictive or condition-based maintenance?
13. A warning light on a copier indicating the toner level is low is an example of a device for condition-based maintenance. Give some other examples.
14. Discuss at least five common practices in preventive maintenance programs.
15. What role do equipment operators serve in PM?
16. Describe the meaning of in-house equipment restoration and redesign. What potential advantages does it offer over outsourcing of equipment restoration and redesign?

17. What steps are taken in TPM to eliminate human error in equipment operation and maintenance?
18. In the implementation of company-wide TPM, what roles do the following groups serve: steering committee, TPM program team, target area committees.
19. Discuss the difference between centralized and decentralized maintenance staff. What are reasons for decentralizing the maintenance function?

PROBLEMS

1. During a 3-week period, two identical machines have the following record of downtime (in minutes) for repairs:

Machine	*Week 1*	*Week 2*	*Week 3*
1	45, 18, 30	32, 55, 20, 15, 12	22, 38, 19, 15
2	136, 98	166, 124, 56	98, 107

   a. Which machine is better in terms of maintainability?
   b. Which is better in terms of other criteria? Explain "other" criteria.

2. The manufacturer of a machine component states that its MTBF is 4,000 hours. When should the maintenance staff schedule replacement of the component? What reliability did you use? Discuss the result.

3. A valve for a hydraulic press has an average failure rate of once every 6 months. The press is to be used for a critical job that will require 2 weeks operation during which everything possible must be done so the press does not break down. What is the reliability of the valve during this period of operation? What are the options to increase its reliability?

4. Bill Sorn supervises an operation that habitually requires readjustment by a skilled machinist once every 3 weeks (120 hours). It takes the machinist 2 hours to make the adjustment. In addition, the operation is usually idle for 2 hours before the machinist arrives. Bill thinks that the operator should be taught to do the adjustment. If the operation is not held up for anything other than the adjustment:
   a. What is the availability of the operation?
   b. What are the issues concerning the operator doing the adjustment?
   c. What other issues should be addressed here?

5. In the course of subjecting 100 components to 5,000 hours of operation, five of the components had failed as of 2,500 hours but the remainder survived the entire time. What is the failure rate in terms of
   a. Percent failure?
   b. Number of failures per hour per component?
   c. If each component is utilized 1,750 hours per year, what is the average number of failures per year per component?
   d. If 500 of this kind of component are to be installed in machines in a plant, how many can be expected to fail during the next year?

6. Sonya Marx operates a machine 340 minutes a day. The actual cycle time per part on the machine is measured to be 0.8 minutes, and the machine produces 360 parts a day. The machine specifications indicate that it should have a cycle time of 0.75 min/part. What is the performance efficiency of the machine?

7. Sonya's machine (described in problem 6) is scheduled to operate 390 minutes a day. About six parts per day from the machine are rejected as nonconforming. What is the machine's OEE?

8. A certain manufacturing process requires a continuous supply of water. The water comes from a tank and is maintained at a constant level with a pump. Should the pump stop, an alarm sounds, and the tank has enough water to last 2.5 hours. If the tank runs dry the process must be stopped, and the cost is $8,000 per hour. Suppose the most common failure mode of the pump is a seized bearing. A bearing has a MTBF of 3 years and requires 4 hours to replace. It is possible to anticipate when a bearing is about to fail by checking the sound of the pump once a week. Such a bearing can be replaced during a shift when the process is not running. It takes 20 minutes for a $30-an-hour inspector to perform the sound check. The process runs 50 weeks a year.
   a. Is it cost effective to perform the weekly sound check?
   b. A common way to avoid system failure is with standby equipment. Suppose a standby pump is to be installed for use when the first pump fails. Is the standby pump a good idea if the annual capital and operating cost of the pump is $800?

9. A machine has a failure pattern like curve A in Figure 7.8. The MTBF is 24 months. To prevent failure, an overhaul must be done every 18 months. Suppose the downtime and repair cost of failure is $6,000; the cost of overhaul is $2,000.
   a. Assuming it is done every 18 months, is the overhaul cost effective?
   b. Assume that daily PM (annual cost: $1,200) extends the MTBF of the machine to 3 years. If everything else remains the same, is the PM cost effective?

Chapter 8

Pull Production Systems

No sooner said than done.

—Terrance

Don't give your advice until it is called upon.

—Desiderius Erasmus

A manufacturing process typically involves numerous stages that all must be coordinated such that materials produced at the earlier stages arrive in the right quantities and at the right times at the later stages that need them. Ensuring the right quantity gets to the right place at the right time—the function of **production control**—is not trivial because equipment breakdowns, material shortages, absent workers, and other unanticipated problems keep changing the rates of supply and demand everywhere in the process. One way to keep materials moving through the process regardless of problems is to build in a cushion of inventories between stages of the process. In plants that manufacture products with numerous parts and production stages, however, that inventory cushion can be huge.

As part of his work on the Toyota Production System, Taiichi Ohno spent years looking for a way to improve production coordination without the need for large inventories. The solution he finally adopted was based on his observation of American supermarkets. He noticed that rather than stock large amounts of food at home, Americans made frequent trips to the supermarket to buy items as needed. Similarly, rather than carry large inventories, the supermarkets replenished food items according to the rate at which items were removed from the shelves. Since customers take only what they immediately need from the shelves, and since supermarkets order just what they need to replenish stock on the shelves, the customers, in effect, *pull* material through the system. This is the concept of **pull production**.

This chapter elaborates on pull production—its methods, advantages, and limitations, and how it compares to the more traditional form of mass production, called **push production**. The chapter describes various pull methods for controlling inventories and authorizing production, including **Kanban,** the system developed by Ohno at Toyota.[1]

Production Control Systems

The purpose of production control is to ensure that production output closely conforms with demand. Ideally, it ensures that products are made in the required quantities, at the right times, with the highest quality. It should do these things for the lowest cost and enable production problems to be easily identified and remedied.

Most production control systems accomplish these ends, though sometimes just barely. Take a simple example, a product that can that be produced in several possible configurations and where demand for each varies. Because of changes in the product configuration and demand, it is not suitable to manufacture the product on an assembly line. When a customer order arrives, the operations routing sequence for the particular product configuration must be determined, and a schedule prepared showing the expected date when the order will be processed at every operation in the sequence.

With a **push production system,** the schedule is based on the time when a job order is expected to arrive at an operation, plus the time when the operation is expected to complete any preexisting jobs and be available. Typically a central staff responsible for scheduling all operations for all job orders prepares the schedule. As a job order moves from operation to operation, work is performed in batches where, often, the size of the production batch is the same as the customer order. Accompanying the job through the plant from operation to operation are the route sheet, schedule, and work instructions for every operation. When there are several jobs waiting at an operation, the supervisor often has to decide which gets priority, though the priority will not necessarily be the same as that assumed by the central staff when they created the schedules. Changes in priorities and delays from parts shortages, machine breakdowns, and other unexpected events make the production schedules obsolete almost as soon as they are created so they must be constantly revised.

Of course, switching a job's priority by putting it ahead of others so it gets back on schedule (called expediting) disrupts the schedules of other jobs and operations, and makes it likely that, later, other jobs will have to be expedited, other operations will have to be rescheduled, and so on.

Many of the difficulties associated with push production come from trying to schedule every operation for every job order in advance and relying on a remote central staff to update schedules to account for current shop-floor conditions. It is a fact that the central staff is too far removed from the shop floor, physically and temporally, to ever be able to keep schedules current. It is also a fact that workers often do not need the schedules anyway and tend to ignore them. In pull production, detailed production schedules for every operation are eliminated, and workers, using a simple system that connects operations throughout the process, make immediate decisions about the quantities and timing of work.

Pull Systems and Push Systems

As an example of how a **pull system** works, consider the chain of events initiated by a child drinking milk. Whenever the child, call him Josh, wants a drink he goes to the refrigerator. He is never concerned about there being enough milk because his mom takes care of that. When the milk gets down to, say, 0.25 gallon, she goes to the store for more. The Josh–Mom process is an example of a two-stage, producer–consumer pull system with a stock point between the stages. Josh is the consumer, Mom the producer, and the stock point is the refrigerator (Figure 8.1). In pull systems, the inventory in the stock point is kept as small as possible, usually by holding it in containers of standardized size and by restricting the number of containers. In this example, the inventory at the stock point is held to a maximum of 1 gallon.

Figure 8.1 Pull system example.

The process is a pull system because the consumer initiates it at the downstream location. The consumer withdraws whatever material is needed from stock, and when the amount in stock reaches some minimum level, that signals for the producer at the upstream location to replenish it. The producer then makes or procures the material in some prespecified quantity and puts it into stock.

The charm of the pull system is its effectiveness and simplicity. With relatively little inventory and only minimal information, the system keeps material flowing to meet demand. If the fluctuation in demand (when the weather is hot, Josh drinks more; when it is cold, he drinks less) is not too great, the amount held in the stock area is sufficient to satisfy demand. It is the responsibility of the producer to keep enough in the stock area. In the simplest form of pull system, the producer knows when to make or procure more by simply *looking* at the current inventory in the stock point; that amount of inventory is one of the few bits of information needed to regulate the system. In the example, Mom does not have to anticipate Josh's thirst or schedule how much to buy. The replenishment decision is based on the level of milk remaining. When the level is high, Mom does nothing; when the level reaches some minimum (0.25 gallon), she goes to the store. Only rarely, when there is an anticipated shift in demand (Josh having friends over or the family leaving for vacation) or in supply (milk drivers go on strike) does Mom need to schedule the purchase, that is, go to the store in advance or change the replenishment amount (say, buy 2 gallons or buy none).

Pull Production Process

The previous example of a two-stage process with a stock point in between can be extended to processes with any number of stages since, conceptually, any process can be viewed as a string of successive two-stage pairs. Figure 8.2 shows a production process consisting of four operations with a buffer stock between each pair of them. The term buffer refers to a small amount of in-process material held between workstations to offset small imbalances between them in terms

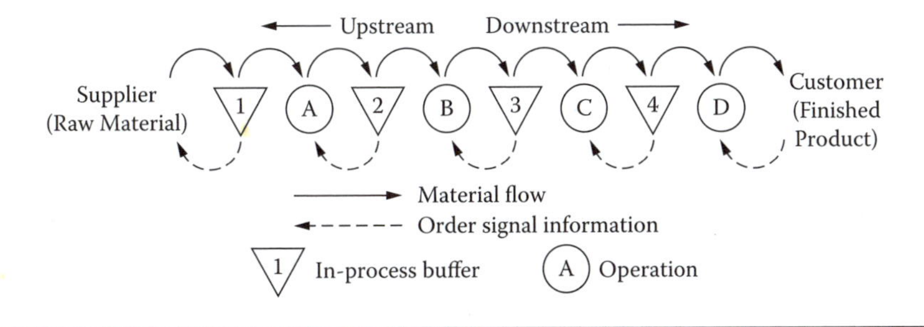

Figure 8.2 Flow of material and signals in a pull system.

of production rate and demand rate. Think of each buffer as consisting of a certain number of standard-sized containers holding materials.

The process in Figure 8.2 works as follows. The last or most downstream stage produces the finished product. When a customer order arrives at that stage, operation D, material is withdrawn from buffer 4 to begin assembly to fill the order. When the size of buffer 4 drops to a certain amount (which might be determined by the number of full containers remaining there operation C begins production to replenish the buffer. For its own production, operation C withdraws material from buffer 3, and when that buffer drops to a certain level, operation B begins production to replenish it. Operation B withdraws material from buffer 2, and when buffer 2 drops to some level, that signals production to start at operation A. Buffer 1 contains raw materials, which upon dropping to some prespecified minimum, is replenished by a supplier. Throughout the process, material is pulled from operation to operation.

The solid arrows in Figure 8.2 represent the flow of materials from left to right, from supplier operation to consumer operation, just like an assembly line. The difference between a pull system and an assembly line is that in the latter every operation continuously produces, so material flows continuously. Also, the kinds of material flowing between pairs of operations in an assembly line tend to stay the same since the final product stays the same (assuming the line makes only one kind of product).

In a pull system, production at each operation is contingent on a **signal**, or authorization, coming from a downstream operation. Thus, material flows only when authorized from a downstream stage (in effect, the authorizations move upstream, as indicated in Figure 8.2). Also, the kind of material moving between each supplier–consumer pair can vary. If the supplier operation is capable of producing different kinds of products or parts (such as in a workcell, described in Chapter 10), then it will make whatever is requested from downstream. (In the example, sometimes Mom buys milk, sometimes juice or pop, depending on when the level for each in the refrigerator gets low.) There are various ways to authorize replenishment of a buffer, the simplest being when the buffer level drops to some predetermined level.

Why Pull Production Cannot Be Stockless

Pull production is sometimes called **stockless production** because its goal is to eliminate in-process inventory, that is, to function with zero inventory in the buffers between operations. Pull production is also called **just-in-time production** because it seeks for every stage in a process to produce and deliver materials in the exact quantities and at the exact times requested.[2] Although it is possible with a pull production process to operate with very little in-process inventory, it is not possible to produce just-in-time with no in-process inventory. Some inventory must be held in the buffers.

As an example, refer back to Figure 8.2 and suppose that the production and delivery lead time for any size of order at each operation is 0.5 day. Consider first what happens if all the buffers are empty (no in-process inventory). When a customer order arrives, operation D has nothing to work on since buffer 4 is empty; thus, it must order material from operation C and wait for the order to be filled. Since buffer 3 is empty, operation C has nothing to work on either, so it must send an order back to operation B. Operation B has nothing to work on, since buffer 2 is empty, so it has to wait for operation A. Since buffer 1 is empty, an order must be sent to the supplier. Even if all of this ordering happens instantaneously, since every one of the five stages of the process (four operations plus the supplier) takes 0.5 day, then the soonest the process can complete the order is 2.5 days after the order is received. Every operation must wait for the operations upstream to provide material so it has something to work on. All this waiting is not exactly what you call just in time.

Now, if there were some stock in each buffer, then none of the operations would have to wait. Each could begin production as soon as it received an authorization. Every operation would be working, using material from its upstream buffer to produce material to replenish the downstream. Of course, the ability for each operation to keep producing without running short depends on the amount of material in the buffer and the upstream operation's ability to replenish that buffer. These topics are covered later. The paradoxical point is that to achieve stockless production you must carry stock.[3]

The following example is an illustration of the application of pull production in a plant that manufactures multiple kinds of products.

Example 1: Pull Production for Multiple Products

A product is manufactured in four different models: W, X, Y, and Z. The production process of all the products involves four workstations: Stations 1, 2, and 3 make parts and subassemblies for the models, and station 4 assembles the parts and subassemblies to produce finished products.

Figure 8.3 shows the product structures for the four models, the major subassemblies, and their parts. It also shows the following:

- Components A and C are subassemblies produced at workstation 2 from parts M, N, and O.
- Components B and D are molded parts made at workstation 3 from parts S, V, and T.
- Parts M, S, and V are produced at workstation 1 using raw material R.
- Parts N and O, and raw materials R and T are purchased from suppliers.

Figure 8.4 shows the locations of the workstations and inventory buffers between them (remember, some amount of material must be held in the buffers). When a process produces multiple kinds of products, the buffers must hold some of every kind of part needed for every product. Think of the buffer for each kind of part as consisting of a small number of standard-size containers of the part.

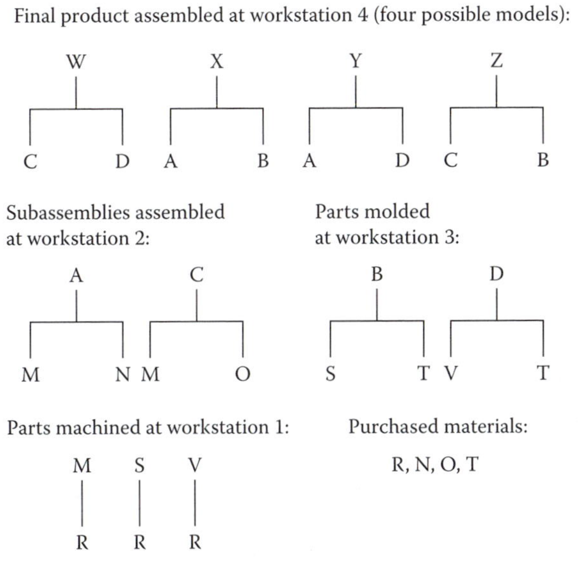

Final product assembled at workstation 4 (four possible models):

W: C, D
X: A, B
Y: A, D
Z: C, B

Subassemblies assembled at workstation 2:

A: M, N
C: M, O

Parts molded at workstation 3:

B: S, T
D: V, T

Parts machined at workstation 1:

M: R
S: R
V: R

Purchased materials:

R, N, O, T

Figure 8.3 Product and parts structure diagram.

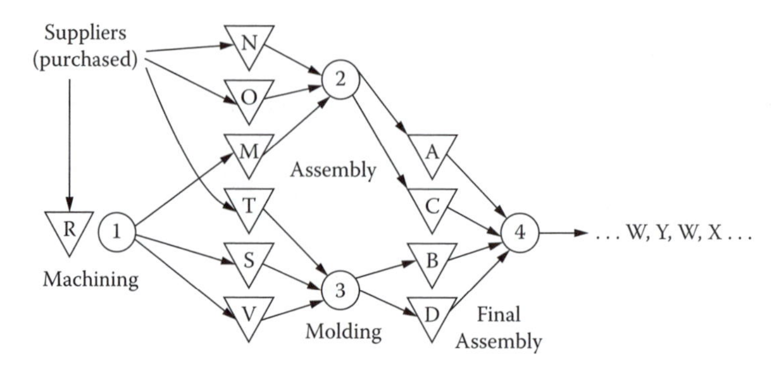

Figure 8.4 Layout of workstations and buffers.

For simplicity, presume that final workstation, station 4, has a schedule that specifies the quantity of each of the four product models required. Unlike in push production, which requires a schedule for every workstation, station 4 is the only station in the process with a schedule.

The system works like this: Suppose the final assembly schedule first specifies some quantity of product W. Workers at station 4 withdraw components from the buffers for components C and D as needed to complete assemblies of product W. If the schedule next calls for a quantity of product X, workers withdraw the appropriate quantities of A and B from the buffers. When the buffers for any of the A, B, C, or D components reach a predetermined minimum level, then workstations 2 and 3 begin to produce in prespecified quantities to replenish the buffers. To do this, of course, workstations 2 and 3 must withdraw materials from their own buffers, and when these reach predetermined levels, replenishment will begin for them, either from workstation 1 or from parts suppliers. The suppliers might be notified electronically that a buffer minimum is impending and replenishment is needed, or they might learn this on their own from periodic observation.

Smooth running of a pull system requires buffers between the stages, one for each kind of part used at each stage of the process. Given that there can be innumerable parts, you might wonder how a pull system is any less costly or wasteful than a traditional batch-oriented push system. Part of the answer is in the overall quantity held in the buffers. If the buffer consists of standard-sized containers, the number of containers and their size is determined so that only the minimum necessary amount of material is held. The size and number of containers in the buffer are based upon production rates and setup times, and are described later. As a result, although pull production might require numerous buffers, each is tightly controlled and kept at the minimal level necessary for the system to operate smoothly. As a result, even with all these buffers, the overall in-process and raw material inventory in a pull system is still but a fraction of that in comparable push systems making the same products.

Push Production Process

In push production, materials are processed at each workstation in batches according to a schedule, then moved (pushed) downstream to the next workstation (see Figure 8.5) where they are processed again according to a schedule. The materials must usually wait until the workstation completes earlier jobs, changes over, and is ready to process them. In a factory that produces many kinds of products with different routing sequences and demand rates, the wait can be unpredictable. As a result, schedules must be padded to account for waiting time as well as for material shortages, machine breakdowns, and so on. This uncertainty and consequential padding of

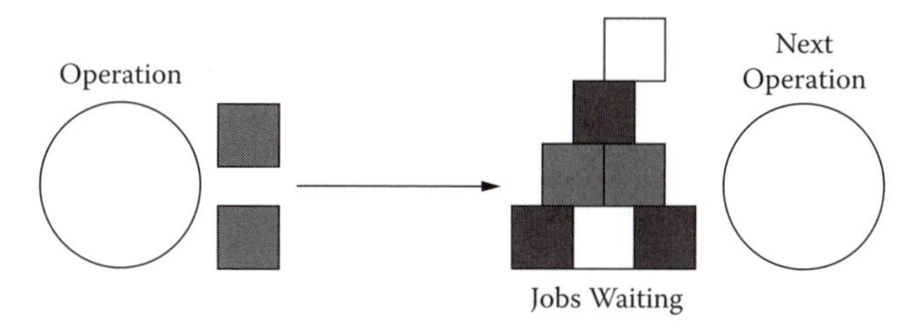

Figure 8.5 Typical push system.

schedules leads to long lead times, high variability in lead times, and large in-process inventories as the following example illustrates.

Example 2: Push Production

Consider a push production system for making the four product models described earlier. Job orders are batch processed according to the production sequence in Figure 8.4, that is, they are processed starting at workstation 1, then moved to workstation 2 or 3 for the next operation, then to workstation 4 for final assembly. Given that each station processes several types of jobs, each job requires a work order and schedule. Schedules for every job order at every workstation are centrally prepared, possibly by a **material requirements planning (MRP)** system.

The scheduling process starts with the desired completion dates for each order, then works backward using production or delivery lead times to determine when each workstation or supplier should commence production (or delivery) of a part or component. Though the processing times might be short, a few hours or less, schedules are usually prepared using daily or weekly time buckets because of the uncertainty about how long each job will wait at an operation before it is processed. The result is that jobs requiring only hours or minutes to process will be assigned, at minimum, a day or week at each operation.

For an example of push scheduling, refer to Figure 8.6, which shows the product structure for product W. The figure also shows the workstations that perform the operations.

Suppose an order is received for a quantity of 400 units of model W to be shipped at the start of day 6, and suppose the nominal lead time for every operation and purchase is 1 day. The schedule will thus indicate that final assembly (workstation 4) will process the order on day 5 (a day before the ship date). Similarly, the schedule will show workstations 2 and 3 working on the order on day 4, workstation 1 and the purchase of materials O and T happening on day 3, and purchase of raw material R occurring on day 2.

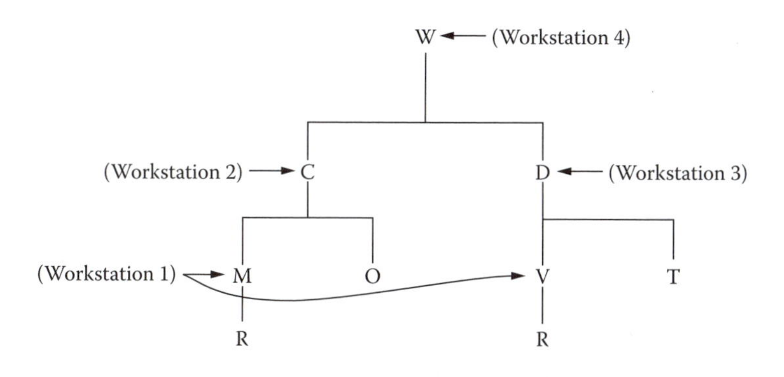

Figure 8.6 Product structure diagram and work sequence for Model W.

Suppose the orders for the other three models to be on shipped day 6 are as follows:

Model	*Quantity*
X	200
Y	200
Z	40

Work on models X, Y, and Z is scheduled by the same procedure, though with multiple products it is necessary to cross-check the schedules to ensure that every workstation has enough capacity to produce the parts, components, and final assemblies for all the products.

If the production system uses a **lot-for-lot** order policy, then the weekly gross requirement for parts and components based on the product structures in Figure 8.3 will be as shown in Table 8.1. These quantities represent roughly the maximum inventory for each item in the system in a given week. Of course, the inventories will be diminished throughout the week as materials are processed into higher-level items and, eventually, into finished products. Still, the average weekly inventory will be about half the amounts in Table 8.1.

Besides causing large in-process inventories, the schedules will require considerable updating effort. If a batch of component A scheduled to be assembled at station 2 gets delayed, then the schedules at station 2 and at station 4 (which is expecting the batch of A next) must both be revised (refer to Figure 8.4). This revision will ripple throughout the production process to cause delays and inventory pileups for other products and other stages. Without component A, job orders for models X and Y cannot be filled, and any completed batches of parts B and D will have to wait in inventory. Every schedule must be revised, otherwise workers will not be sure what jobs to do next or what to do with delayed batches of material once they arrive. Since it takes time for work-status information to get from the shop and into the schedules, by the time new schedules are released they might again be outdated.

Table 8.1 Gross Requirement Quantities for Components, Parts, and Materials

Component	*Part*	*Material*	*Quantity*
A			400
B			240
C			440
D			600
	M		840
	N		400
	O		440
	S		240
	V		600
		R	1680
		T	840

Pull Production and Push Production Contrasted

In practice, some aspects of push systems and pull systems are identical or similar. Pike and Cohen contrasted **Kanban** (pronounced conbon; is the Japanese equivalent for "card" or "visual record"), the card signaling system used at Toyota, which is considered the classic pull system, and MRP, which is considered the classic push system. They found the following.[4]

- Timing (time when production or transfer of batches is signaled). This is the only aspect that is completely different for the two systems. MRP systems specify order releases of production schedules using lead times and factory-wide information from the master schedule. In Kanban, an order release occurs when the level of inventory at the downstream buffer drops to a prespecified minimum.
- Batch size (size of production batches and order shipments). Batch sizes in MRP systems are determined in advance by a central planning staff using lot-sizing rules and master schedule requirements. Batch sizes in Kanban are determined at the shop-floor level according to the demand and replenishment requirements of downstream inventory buffers. However, the size of containers for carrying inventory and the number of containers are based on the schedule at the final operation, which is determined from product demand information—the same as in MRP systems.
- Priorities (basis for priority of orders when there are multiple orders for production and/or shipment). Schedules in MRP systems may incorporate priorities based on rules (e.g., earliest due date, shortest processing time, or first come, first serve), although, often, final priority decisions are made on the shop floor depending on the status of work at each workstation. Workstation operators in Kanban use categories on a sequence board (described later) to determine job priority. When multiple orders fall in the same category, the supervisor at the upstream operation decides which gets priority, just as in push systems.
- Interference (procedure for handling unanticipated orders that require immediate attention). In MRP systems, decisions about unanticipated orders rest with upstream workstations, though ordinarily the schedules are frozen so that no new requests will be honored. Sometimes, however, the schedules are regenerated and priorities reset, in which case the decision is based upon plantwide information. Kanban systems conform to daily quotas using somewhat-stable schedules for the final operation, although daily production at each workstation always depends on the requirements of downstream workstations. When an emergency order arrives, however, it will be honored. In both systems, some policy must be established a priori to determine how and where production capacity should be allocated when there are multiple, competing requests from downstream.

Every production system has elements of both push and pull; some, like MRP are more push-like, while others, like Kanban, are more pull-like. The remainder of this chapter focuses on systems that are more pull-like.

Containers and Cards

In a pull production process, often the same standard-sized containers are used throughout all or most of the process. In the process illustrated in Figure 8.4, all of the buffers might utilize the same-sized container to hold parts, components, and assemblies. To clearly distinguish the contents of containers, however, a card must be attached to each. The card provides information such as the kind of material in the container, the quantity of material, the origin or producer of

Figure 8.7 Standard containers of material with accompanying cards.

the material (upstream source), and the consumer of the material (downstream destination). An example of containers and accompanying cards is in Figure 8.7.

Besides describing the contents of the containers, the cards serve another purpose: to delimit the amount of inventory in the system. In pull systems that use cards, every container of material must have an attached card. Once the minimal amount of material needed in a buffer to keep the system functioning has been determined, then the number of containers needed is also determined. One card is created for each container. Therefore, no container in the system can be filled or moved unless it has a card. By restricting the number of containers in the system to a prespecified number of cards, the total quantity of in-process inventory never exceeds a predetermined maximum.

Another, related, purpose of the cards is authorization. When materials in a container are first withdrawn at a downstream operation, the operation posts the card as a signal to replenish the container. The posted card authorizes the upstream operation to produce (or procure) a container of material and move it downstream to replenish the buffer.

The best-known card signaling system is Kanban. In the Kanban system, a card signals the need for and subsequent authorization to produce and move a container of materials, parts, or subassemblies. The number of kanban cards corresponds to the number of containers in each buffer. In the following discussion, Kanban (with an uppercased K) refers the Kanban system of pull production, while kanban (with a lowercased k) refers to the card.

Rules for Pull Production

Successful functioning of the card-based pull system requires understanding of and conforming to certain rules. Table 8.2 lists the rules of Kanban pull production systems.[5] The first rule specifies how, when, and in what order materials are moved between stages of the process. The second rule specifies the amount and priority in which items are produced. The quantity and

Table 8.2 Rules of Pull Production

Rules
1. Downstream operations withdraw only the quantity of items they need from upstream operations. This quantity is controlled by the number of cards.
2. Each operation produces items in the quantity and sequence indicated by the cards
3. A card must always be attached to a container. No withdrawal or production is permitted without a kanban.
4. Only nondefective items are sent downstream. Defective items are withheld and the process stopped until the source of defectives is remedied.
5. The production process is smoothed to achieve level production. Small demand variations are accommodated in the system by adjusting the number of cards.
6. The number of cards is gradually reduced to decrease work-in-process and expose areas that are wasteful and in need of improvement.

Note: The term *card* assumes cards are used in conjunction with containers; in cases where cards are not used, the rules apply to the containers themselves.

sequence of production are determined by the number of cards. When multiple kinds of material are requested, they are produced in the order in which demand occurs (i.e., they order in which containers are emptied or cards are sent upstream). The third rule prevents the overproduction or movement of items not required. The fourth rule encourages cooperation and defect prevention. A defective item discovered downstream is never used but is returned upstream, where it is analyzed and the defect source is identified. Defects are corrected and replaced on the day they occur. The fifth rule mandates a uniform rate of production. The final stage of the process is the only one that has a daily schedule, and all operations elsewhere in the process respond to that schedule using only cards. When demand level changes, the daily schedule is adjusted, but adjustments are kept small so the impact on upstream processes will be manageable. The sixth rule is based on the premise that inventory hides problems, and that progressively reducing the number of kanbans reduces inventory and enables more problems to be exposed and resolved. Reducing the number of cards is a mechanism for continuous improvement.

These are rules, not recommendations. For successful pull production they must be observed. According to Taiichi Ohno, attempting pull production without obeying the rules "will bring neither the control expected ... nor the cost reduction ... [A] half-hearted introduction of Kanban brings a hundred harms and not a single gain."[6] He should know; he devoted over two decades to perfecting Kanban at Toyota.

How to Achieve Pull Production

Several questions need to be answered in setting up a pull production system: (1) When and how should authorization signals for replenishment of buffers be sent upstream? (2) What size should the buffers be? (3) How should operators keep track of what they are supposed to do? The last question arises whenever an operation produces multiple kinds of items, or when it is located in a place where the operators cannot see the buffers they are supposed to replenish (analogous to Mom having to decide when to replenish milk without being able to look in the refrigerator first). These questions can be answered in different ways; we will consider several, taking them in order of increasing complexity.

Pull System as a Fixed-Quantity/Reorder-Point System

The pull system is, in effect, a variant of the simple reorder-point system where a replenishment order is placed whenever inventory falls to a critical level. This level, the **reorder point** (ROP), is based upon the estimated amount of material used between the time when the order is placed and when the replenishment batch is received. The formula for ROP is

$$\text{ROP} = \text{D(LT)} + \text{SS}$$

where

D = Demand (consumption rate)
LT = Lead time (elapsed time between order and replenishment)
SS = Safety stock

Safety stock, discussed in Chapter 5, is an amount of stock to buffer the variability in D and LT. The greater the variability in D or LT, the larger SS. When D and LT are relatively constant, SS can be zero. We will start with the assumption that LT and D between any two stages of a production process are relatively constant.

An application of the reorder-point system is the **two-bin** system. Material is held in two bins. Material needed to satisfy demand is withdrawn from one of the bins; when the bin is emptied an authorization is sent for another full bin. Meantime, material needed to satisfy demand is withdrawn from the second bin. The amount held in each bin is the specified ROP quantity—enough to meet demand until a full bin arrives. The process repeats.

In a pull production system, the LT is the total time required to replenish a buffer. Usually LT is separated into two categories, **production time** (P) and **conveyance time** (C), that is, LT = P + C. P is the total time to produce the quantity ordered, including the setup time, processing time, and any planned waiting time. C is the time to convey the order to the upstream operation that will fill it, plus the time to move the materials to the downstream operation that initiated the order.

Suppose demand for an item is 105 units per week; given a 5-day week, then demand is 21 units per day. If production time is 0.1 day and conveyance time is 0.4 day, then

$$\text{ROP} = 21(0.1 + 0.4) = 10.5 \text{ units; rounding up, ROP} = 11 \text{ units}$$

This says that whenever the inventory level in the buffer reaches 11 units, an order should be placed. In a two-bin system, each bin would hold 11 units (plus a small safety quantity, described later).

Case in Point: Two-Bin Kanban for Replenishing Hospital Supplies[7]

Stocks of supplies are found everywhere in hospitals, though commonly the amount of each item held in stock is either too much or too little. Oftentimes little- or never-used items take up valuable space, while frequently used items run short. Fearful of shortages, medical personnel hoard and hide their own supplies. In one laboratory, tubes of a particular blood type were hidden in drawers and shelves at eight locations; this took up valuable storage space and heightened the risk that sterile tubes would expire before use.

To control inventories, hospitals go to costly extremes, employing sophisticated computerized systems and elaborate manual procedures. At one

Figure 8.8 Medical supply room with two-bin stock.

hospital the operating room nurses devoted 600 hours a year to counting and replenishing supplies.

A simple alternative to these schemes is a Kanban system that consists of two bins (containers) of each item. Attached to each bin is a card showing the item name, number, source, and other information. When a bin is emptied it is placed in a designated area. This is the signal to replenish it. Empty bins are picked up each shift and are returned full during the pick up next shift. Each bin holds the average demand for one shift plus a small buffer amount. Figure 8.8 shows a supply room with two bins for most items stocked.

The same Kanban system can be used to connect all stages of the hospital supply chain, for example, to connect exam rooms to departmental supply rooms, departmental supply rooms to hospital central stores, and hospital central stores to outside medical vendors and distributors (Figure 8.9). The system assures adequate supply of materials everywhere and eliminates shortages and the need for counting.

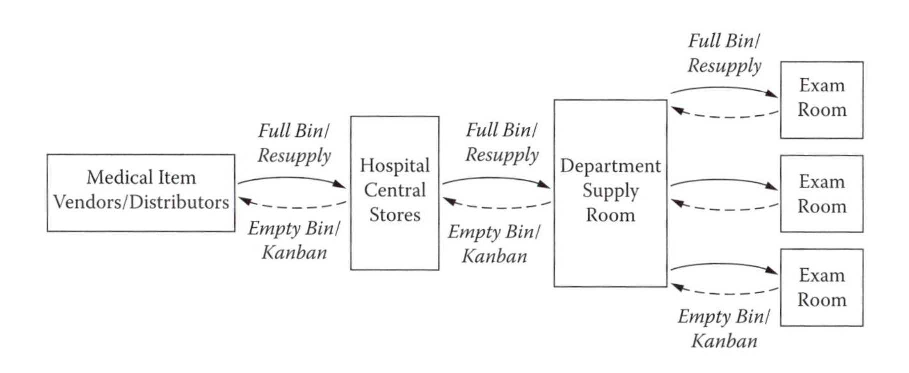

Figure 8.9 Kanban system connecting multiple stages of a hospital supply chain.

Containers in a Buffer

Common practice in pull production is to use **standard-sized containers** for holding and moving parts. The answer for when to order still uses ROP, but the ROP is expressed as the number of containers instead of number of units. Suppose *Q* represents the capacity of a standard container; thus, the ROP as expressed in terms of the number of containers, K, is

$$K = D(P+C)/Q$$

The term *K* represents the maximum number of completely full containers in a buffer. As soon as a container is emptied and another container is accessed, a replenishment order is sent upstream. A simple way to place the order is to send the empty container and its card to the upstream operation that will refill it.

As mentioned, pull systems often use cards, and in that case K refers to the number of cards. (If you like, think of K as representing the number of kanban cards.) Given that usually there is one card for each full container, it does not matter whether *K* represents the number of cards or containers. The manner in which these cards are used to authorize the production and movement of materials is discussed later.

Container Size

In pull production the size of buffers is kept small by using containers that are small. Another reason for small containers is that they are easy to move and materials inside are easy to access.

A rule of thumb is that a container should have the capacity to hold about 10% of the daily demand for the material it holds. Using the previous example, if D = 21 units/day, then Q = 0.10(21) = 2.1 units. Allowing for upward fluctuations in demand, we would specify the container as holding a quantity of three units.

Example 3: Simple Pull System

To put all this together, let P = 0.1 day, D = 21 units, Q = 3 units, and C = 0 (assume the buffer rests snugly between two workstations and move time is zero), then

$$K = 21(0.1)/3 = 0.7 \text{ kanban; rounding up, } K = 1 \text{ kanban}$$

Therefore, the buffer should hold one full container. In practice, we need at least two containers between workstations: one initially empty, the second initially full. The upstream station puts new stock into the one while the downstream station withdraws old stock from the other. In this simple two-container system, one container is ideally fully replenished just as the other is fully depleted. Because of the necessity to round up in the computation of K (in this example, but in most cases as well) to get an integer (here, from 0.7 to 1), the one container will be fully replenished at some time before the other is fully depleted. In the meantime, before the other is depleted, there will be two containers having some material in them.

Material Handling

Next consider the case where containers in the buffer must be moved between operations for replenishment (i.e., C > 0). In that case, workers at either the downstream or upstream operations can be given responsibility to move the container. In the example in Figure 8.4, for instance, whenever one container is emptied in the buffer for material D, an operator from workstation

4 would take it to workstation 3. As soon as workstation 3 fills the containers, an operator from there takes it back to the D buffer at workstation 4. Another way is to employ material handlers, workers whose sole responsibility is to monitor the buffers and transfer containers back and forth throughout the plant.

Suppose the producer and the consumer workstations cannot be located next to each other or that the inventory buffer cannot be located right next to workstations. For example, a buffer that serves several workstations might be located in a place central to all of them, yet not be close to any of them. The time it takes to transmit replenishment orders between workstations, or to move the materials between workstations and buffers, then becomes an important matter. Additionally, suppose each buffer holds several kinds of materials, in which case a method is needed to keep track of which materials require replenishment and which orders get priority. Such situations are common, as described next.

Outbound and Inbound Buffers

Workstation 1 shown in Figure 8.4 makes parts M, S, and V, which are used by workstations 2 and 3. Suppose the three workstations are situated some distance apart and it is not possible to put the buffers between them in a location that is close to all of them. In such a case it is necessary to have two kinds of buffers for each part, an **outbound buffer** and an **inbound buffer**. This is shown in Figure 8.8, where the outbound buffers holds the output material of workstation 1 (the producer station), and the inbound buffers holds the materials for workstations 2 and 3 (the customer stations).

Both outbound and inbound buffers consist of one or more containers for each kind of part. In Figure 8.10, workstations 2 and 3 withdraw parts from containers in their respective inbound buffers, while workstation 1 deposits parts into containers in its outbound buffers. Whenever workstations 2 and 3 need more parts, full containers are withdrawn from the outbound buffers at workstation 1.

Now, to ensure there are enough containers of the different parts in workstation 1's outbound buffers to meet demands at workstations 2 and 3, and to ensure the production process proceeds smoothly, some form of communication mechanism is needed to **signal** to workstation 1 when to (1) move a full container downstream, and (2) produce additional items of a material. Also, in case multiple move or produce signals are sent upstream at once, a scheme is needed to tell operators and material handlers at workstation 1 which to attend to first. A simple way to do all this is to use kanban cards. The following sections discuss three different kinds of kanban cards.

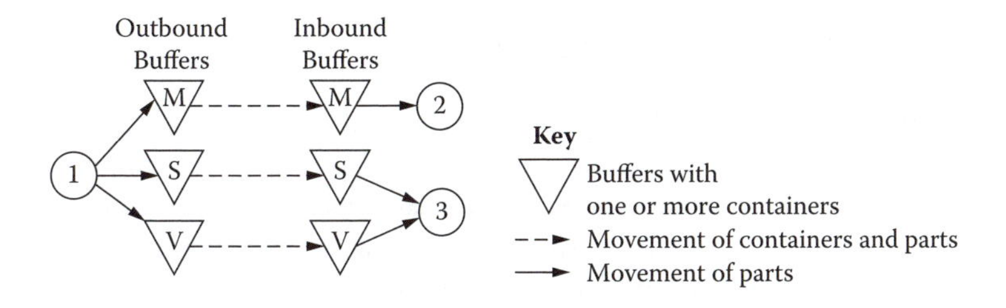

Figure 8.10 Outbound and inbound buffers.

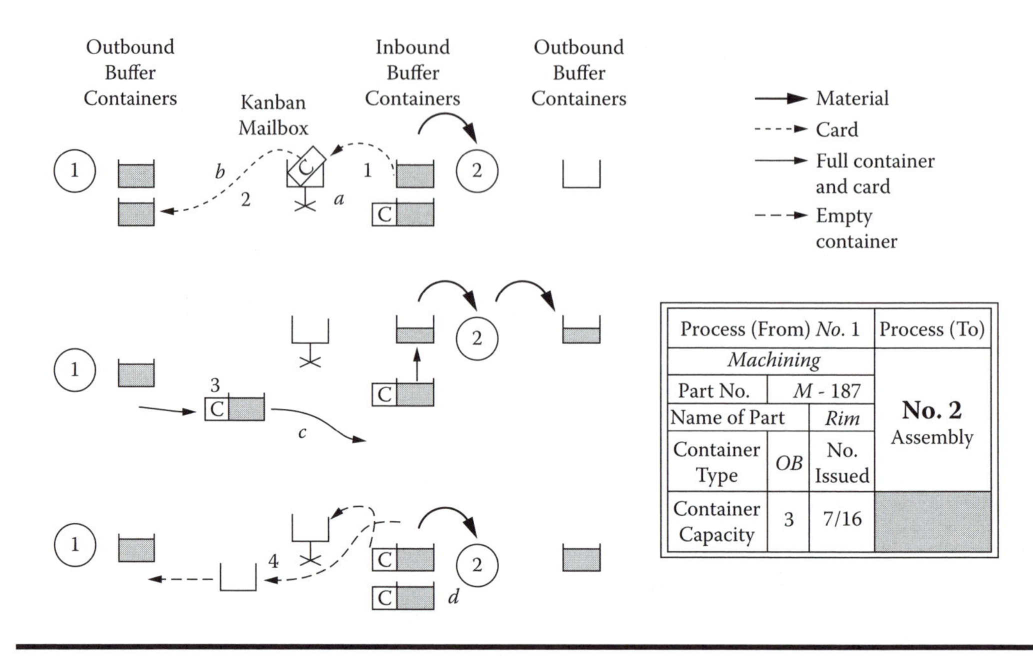

Figure 8.11 Single-card Kanban system using only C-kanban (example of card shown on right).

Conveyance Kanbans[8]

A **conveyance kanban** or **C-kanban** (also termed a move or withdrawal kanban) is an authorization to move a container from an upstream, outbound buffer to a downstream, inbound buffer. No container can be withdrawn from an outbound buffer unless a C-kanban has been issued. The conveyance process illustrated in Figure 8.11 works as follows:

Step 1: When operators at station 2 first access a full container, they take the C-kanban from it and place it in a kanban mailbox. (The mailbox is a place where cards are kept or posted, as shown in Figure 8.12.) The C-kanban specifies the kind of material needed and the upstream station from which to get it.

Step 2: A material handler reads the C-kanban and takes it to the specified upstream station (here, station 1).

Step 3: The material handler affixes the C-kanban to a full container (located at the outbound buffer of station 1), then takes the container back to station 2.

Step 4: Whenever station 2 empties a container, the material handler takes the container upstream to station 1. (Often steps 2 and 4 are combined so that the material handler takes the card and empty container in a single trip.) The process then repeats.

The example in Figure 8.11 is called a **single-card Kanban system** because it uses only one card, the C-kanban. Upstream workstations (e.g., station 1) produce according to an order list or a daily schedule, possibly one generated by an MRP system using an expected daily demand rate. To prevent buildup of outbound stock in the event of downstream interruptions, production is limited to a prespecified number of full outbound containers.

A downstream workstation (station 2) can accumulate no more material in its inbound buffer than it has immediate need for because containers of material cannot be moved to it without a C-kanban.

Figure 8.12 Kanban mailbox and posted cards.

Therefore, the maximum number of full containers at the inbound buffer corresponds exactly to the number of C-kanbans. Recall the formula for the number of containers between two processes,

$$K = D(P + C)/Q$$

To determine the number of C-kanbans, K_C, focus only on C and ignore P. Thus,

$$K_C = D(C)/Q$$

where C is the time from when workers at the inbound buffer remove a C-kanban from a full container, to when they remove a C-kanban from the next full container. That time, called the **conveyance cycle time**, is the sum of the times shown by the lowercase letters in Figure 8.11. Specifically, this is the time that the C-kanban

a = Waits in the mailbox
b = Moves to the upstream station
c = Moves back to the downstream station (in Figure 8.9, with a full container)
d = Waits in the downstream buffer until the container is accessed and the kanban is put back in the mailbox

Example 4: C-Kanban Computation

Suppose D = 21 units/day, Q = 3 units, and, in minutes, a = 32, b = 52, c = 60, d = 48, and 1 day = 480 minutes. Then,

$$C = (32 + 52 + 60 + 48)/480 = 0.4 \text{ day}$$

so

$$K_C = 21(0.4)/3 = 2.8 \text{ C-kanbans; rounding up, } K_C = 3 \text{ C-kanbans}$$

Thus, the inbound buffer of workstation 2 will at most consist of three full standard containers.

Production Kanbans

Besides the C-kanban, another principal kind of kanban is the **production kanban**, or **P-kanban**, which is used to authorize production of parts or assemblies. In a system that utilizes this kind of card, no production is allowed without it. Except for at the final operation in the process, there are no production schedules, just P-kanban authorizations. A system that uses both C-kanbans and P-kanbans is called a **two-card** pull system for obvious reasons. As shown in Figure 8.13 the system works as follows:

Step 1: When operators at station 2 access a full container, they remove the C-kanban and place it in the C-kanban mailbox. The C-kanban specifies the material needed and the station upstream that produces it.
Step 2: A material handler takes the C-kanban and an empty container to the specified upstream location (in the figure, station 1).
Step 3: The material handler removes the P-kanban from a full container at station 1, puts it in the P-kanban mailbox, then affixes the C-kanban to that container.
Step 4: The material handler leaves the empty container at station 1 and takes the full container to station 2.

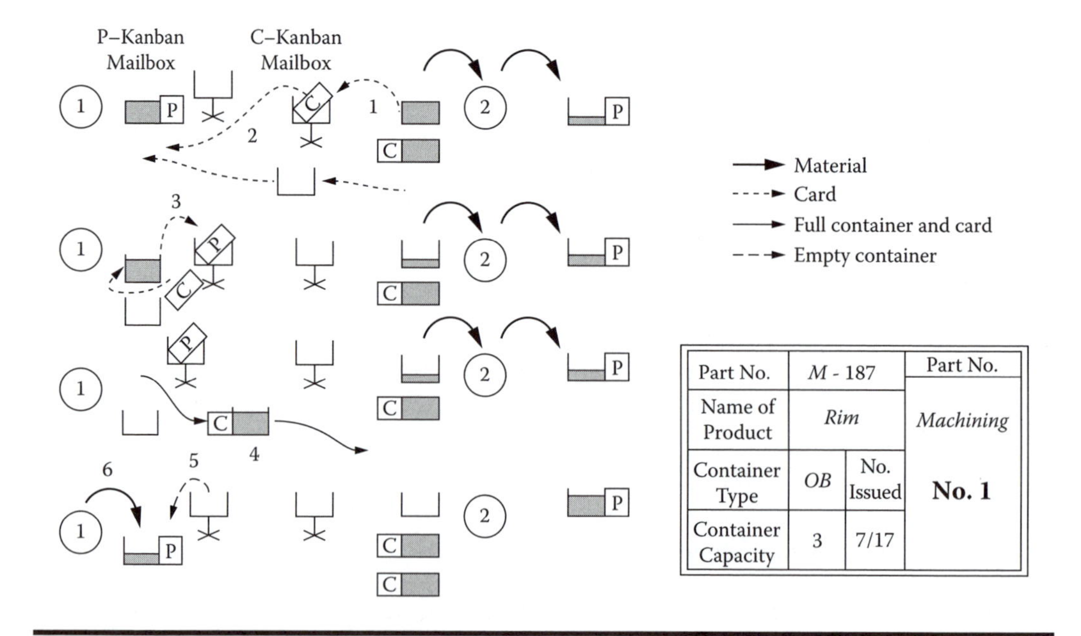

Figure 8.13 Two-card kanban system (P-kanban example shown on right).

Step 5: The P-kanban in the mailbox authorizes Station 1 to produce enough material to fill the empty container. An operator removes the P-kanban from the mailbox and affixes it to an empty container.

Step 6: Station 1 produces just enough material to fill the empty container.

(Meanwhile, the same procedure is happening at station 2, that is, production does not begin until a P-card is posted in its own P-kanban mailbox. When a card is posted, station 2 then begins producing, which requires that it access a full inbound container. The C-kanban from the container is posted, and the process repeats.)

The two-card system gives tight control over buffer inventories. No container can be moved or filled unless there is a C-kanban or P-kanban, respectively, authorizing it. Since standard-sized containers are used everywhere in the buffer, the size of both the transfer and production batch size is the container capacity, Q.

Just as the number of C-kanbans specifies the maximum number of full containers at an inbound buffer, the number of P-kanbans specifies the maximum number of full containers at an outbound buffer. The number of P-kanbans is

$$K_P = D(P)/Q$$

where P is the total time elapsed from when workers or material handlers remove the P-kanban from a full container and post it at the outbound buffer until the time they remove the P-kanban from the next full container. P is the **production cycle time**. To demonstrate what it means, suppose station 1 is a multistation workcell like that shown in Figure 8.14. The time P is the sum of the times

a = The P-kanban waits in the P-kanban mailbox
b = For the P-kanban to be moved to the order post at the first operation
c = The P-kanban waits at the order post
d = To process the quantity to fill the container (= setup time + run time + in-process waiting time)
e = To move the full container to the outbound buffer, and
f = The container waits in the buffer.

Often there is no separate order post, so times b and c are zero.

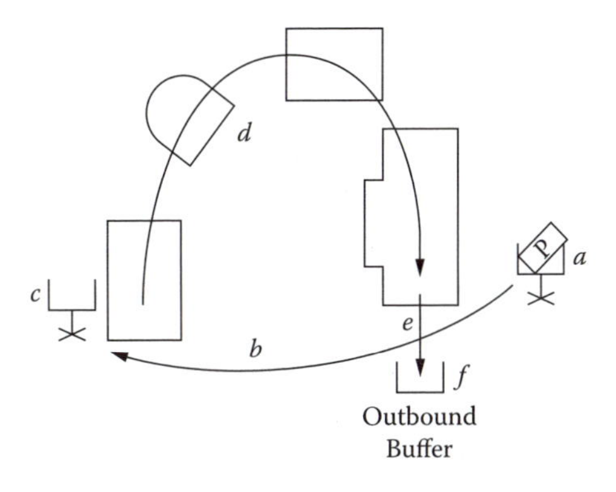

Figure 8.14 Time elements in computing *P*.

Example 5: P-Kanban Computation

Suppose D = 21 units/day, Q = 3 units, and, in minutes, a = 15, b = 0.5, c = 0.5, d = (setup = 6, run = 3/unit, in-process wait = 0), e = 0, f = 17, and one day = 480 minutes.

Thus, P = [15 + 0.5 + 0.5 + (6 + 3(3) + 0) + 0 + 17]/480 = 0.1 day, so

$K_P = 21(0.1)/3 = 0.7$ P-kanban; rounding up, $K_P = 1$ P-kanban

Thus, there will be at most one full standard container of this material at the outbound buffer.

Combining this result with the result of the previous example yields $K = K_P + K_C = 4$ kanbans for the outbound and inbound buffers separating stations 1 and 2. Technically, this stage of the process should be able to operate with four full containers of materials (and one empty container).

Safety Factor

It is common in determining K to include a **safety factor**, X, to account for minor fluctuations in demand. The more leveled the demand, the less the need for a safety factor, though some margin should be included to cover minor glitches and breakdowns in the system. The safety factor results in the earlier-mentioned safety stock being added to the buffer. The more problems contributing to fluctuations in supply and demand between workstations, the larger X has to be to offset them. To explicitly account for X, the K formula is modified:

$$K = D(LT)(1 + X)/Q$$

where LT can be P, C, or the sum, P + C.

Given that the value for K is frequently rounded up to arrive at an integer value, the safety factor is often accounted for even when X is not included in the original computation. In a previous numerical example, K_C was computed as 2.8. When the number was rounded up to 3, a factor of (3 – 2.8)/2.8, or 7.1% is built in. If managers think this margin is too small, the number of containers can be increased to 4. For K_P the computed value was 0.7. When rounded up to 1.0, this gives a factor of (1 – 0.7)/0.7 = 43%. The rule of thumb is to start with a safety factor of about 10%, then try to decrease it to whatever practical experience shows is most workable.

Another Single-Card System

If workstations are located closely adjacent with the buffer stock in between them, then only one card, a P-card, is needed. Since the move time is almost zero, C-kanbans are not necessary. When the station downstream accesses a container in the stock area, it posts the P-kanban card from the container, which authorizes the upstream station to fill another container with the item specified on the card.

Even when workstations in a process are not adjacent, it is still possible to use a single, P-card system. Imagine a rack of shelves that holds a large number of different components used by two workstations that each assembles a different product. Suppose all the components on the shelves are produced by four upstream operations. Each of the two assembly areas withdraws the type and quantity of components it needs to make products according to a daily production schedule. The components are boxed, and inside each box is a card. Whenever an assembler accesses a new box of components, he posts the card in a nearby mailbox. Someone stops by frequently, removes all the cards from the mailbox, sorts them by upstream operation, then puts them into the mailbox by each of the four upstream operations. Each card is a P-kanban, an authorization for the operation to produce another box of the specified component.

Signal Kanban[9]

A signal kanban is a special kind of production kanban. There are actually two kinds of signal kanbans. The first, an **SP-kanban** or **production-signal kanban**, is for ordering production of large batches or quantities in excess of one container. Large batch production is necessary when the changeover time between products is time consuming enough so as to restrict the number of daily production runs to a relative few. This is the case for processes such as punch pressing, die casting, forging, and injection molding.

Whereas a P-kanban restricts the batch size to the capacity of a container (Q), with an SP-kanban the batch size can be any number of containers. Using an SP-kanban is like collecting several P-kanbans before starting production. If a process involves a batch size of 12 units and the container size is 3 units, then four P-kanbans would have to be accumulated before starting; for the same case, only one SP-kanban is needed.

The other kind of signal kanban is an **SM-kanban** or **material signal kanban**. It is like a C-kanban in that it is used to authorize movement of materials, except the SM-kanban is used in conjunction with an SP-kanban to authorize the transfer of materials needed to produce the batch authorized by the SP-kanban.

Use of signal kanbans is predicated on the requirement that batch sizes exceed one container. In general, the minimal batch size, *B,* for an item is determined by the maximum number of times a process can be switched over; that, in turn, is a function of **total (Internal + External) setup time**[10] and the number of different items to be produced.

Suppose workstation 3 is a molding machine that requires a total time of 2 hours for each changeover. Also, suppose the machine must produce two different kinds of parts each day. Assuming an 8-hour workday,

$$\text{Maximum number of setups} = 8 \text{ hrs/day} \div 2 \text{ hrs/setup} = 4 \text{ setups/day}$$

If two kinds of parts must be produced each day, then for each part the maximum number of setups is

$$S = 4 \text{ setups/day} \div 2 \text{ parts} = 2 \text{ setups/day/part}$$

Assume that demand for one of the parts is 30 units per day. Thus, the minimum batch size for this part must be

$$B = D \div S = 30 \text{ units/day} \div 2 \text{ setups/day} = 15 \text{ units}$$

To allow for a 10% minimum safety factor, increase this to 17. Now, the batch size must be expressed in terms of the container size, Q, and if Q = 3 units, then B, rounded up, must be 18 units, which is 6 containers. Thus a batch size of 6 containers should be produced whenever production is authorized using an SP-kanban.

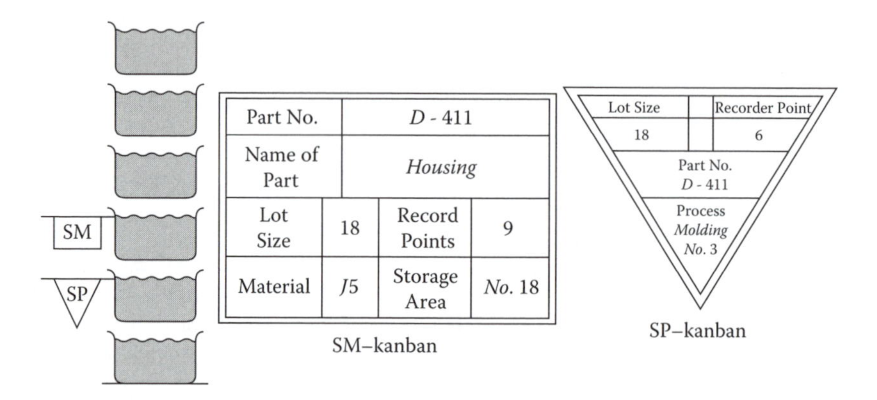

Figure 8.15 Signal kanbans and their locations.

To show how SP- and SM-kanbans are used, imagine a buffer where full containers are stacked atop each other. For each kind of item there is a stack, and for each stack there is one SM- and SP-kanban each attached to the stack as shown in Figure 8.15. As containers are withdrawn from the buffer, the stack gets smaller. When the level of the stack reaches the container with the SM-kanban, that kanban is sent to the indicated upstream workstation (operation or parts storage) as authorization to withdraw and move materials. When the level of the stack is further depleted to the container with the SP-kanban, that kanban is posted to authorize an upstream station to begin production. The materials authorized by the SM-kanban will arrive at just about the time they will be needed at the upstream station to begin production.

The whole process must be synchronized, and for that to happen the SM- and SP-kanbans must be located at the right locations in the stack. These locations are determined using variants of the familiar formula

$$K = D(LT)/Q$$

except that here K represents the level of full containers in the stack at which the SM- and SP-kanbans are to be located.

Consider first the SP-kanban. Its position in the stack is that level where enough containers remain to satisfy demand for the item while a batch of additional items is being produced and transferred to the buffer. This, the number of full containers in the stack where the card is located, is

$$K_{SP} = D(P)/Q$$

where D is the consumption rate, Q the container size, and P the time between when the batch is ordered and when it is filled and received at the buffer. The time P, here being the **SP cycle time** (refer to Figure 8.16), is the sum of the times

- a = The SP-kanban waits in the mailbox,
- b = For the SP-kanban to be moved to the order post prior to the first operation
- c = For the SP-kanban to wait at the order post
- d = To process the specified batch quantity (Setup time + Run time + In-process waiting time)
- e = To move the batch of full containers, along with the SP- and SM-kanbans, to the outbound buffer

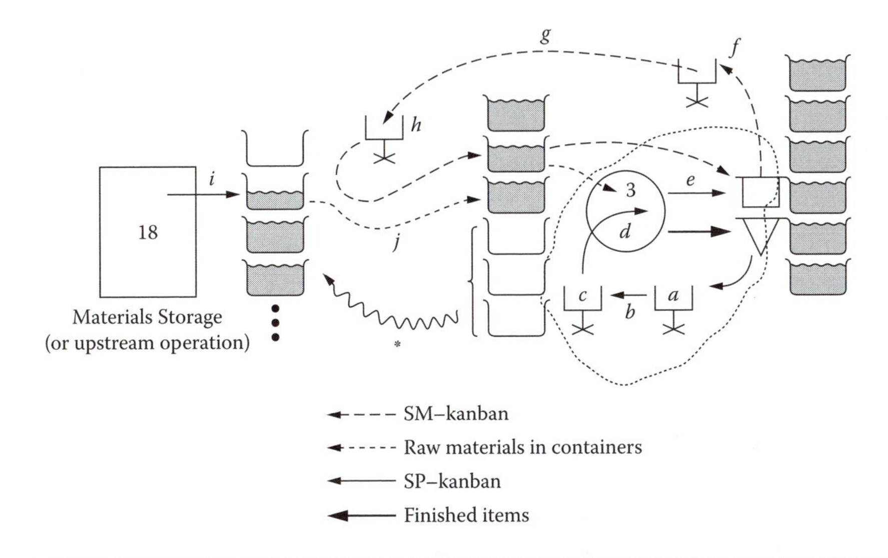

Figure 8.16 Example of SP-kanbans and SM-kanbans.

(Note: Times for *f* through *j* are explained in the next section.) When there is no separate order post, times *b* and *c* are zero.

Example 6: Computation of SP-Kanbans

Suppose D = 30 units/day, Q = 3 units, and, in minutes, *a* = 15, *b* = 0.5, *c* = 0.5, *d* = (setup = 16, run = 2 per unit, in-process wait = 2), *e* = 5, and 1 day = 480.

From earlier, assume the batch size is 18 units, then the SP-kanban cycle time is

$$P = 15 + 0.5 + 0.5 + (16 + 18(2) + 2) + 5 = 75 \text{ min}/480 = 0.15625 \text{ day}$$

Therefore,

$$K_{SP} = 30(0.15625)/3 = 1.5625 \text{ containers}$$

Since the position of the SP-kanban must be an integer, round up to 2 containers; this rounding affords a safety margin of 0.4375/1.5625, or 28%.

In summary, when the buffer level reaches two full containers, a production order for a new batch (six containers) will be placed in the kanban mailbox.

Next, determine the location of the SM-kanban in the stack using

$$K_{SM} = D(C - P')/Q + K_{SP}$$

where D and Q are the same as before, C is the total time between when materials are first ordered and when they arrive for usage, and P′ is the time between when production is first ordered and when setup begins. Setup here refers to internal setup tasks only (that portion of the changeover procedure that must be performed when the machine or operation is stopped). If the internal setup procedure requires adjustments through trial and error, then material must be delivered in time for the adjustment procedure.

The time P′ is the sum of the first three time elements in the **SP cycle time**, described earlier, and the internal setup time. In Figure 8.16 it is the sum of the time the SP-kanban

a = Waits in the mailbox
b = Moves to the order post prior to the first operation
c = Waits at the order post
d′ = Internal setup time

If there is no separate order post, times *b* and *c* are zero.

The time C is called the SM cycle time. From Figure 8.16, it is the sum of the times

f = The SM-kanban waits in the mailbox
g = For the SM-kanban to be moved to the upstream (to parts storage or previous operation)
h = For the SM-kanban to wait at the order
i = To fill the order (withdraw or produce parts)
j = To convey filled containers to place of usage
* = Empty containers are taken back upstream on the return trip; the time for this is irrelevant

If (C – P′) has a negative value, round it to the next lowest negative integer (e.g., –0.4 become –1).

Example 7: Computation of SM-Kanbans

Suppose D = 30 units/day, Q = 3 units, and, in minutes, *a* = 15, *b* = 0.5, *c* = 0.5, *d′* = 16, *f* = 6, *g* = 10, *h* = 1, *i* = 15, *j* = 10, and 1 day = 480. Thus,

$$C - P' = (6 + 10 + 1 + 15 + 10) - (15 + 16 + 0.5 + 0.5) = 10 \text{ min}/480 = 0.0208 \text{ day}$$

From before, K_{SP} = 2 containers, so

$$K_{SM} = [30(0.0208)/3] + 2 = 2.208 \text{ containers; rounding up, } K_{SM} = 3 \text{ containers}$$

Summarizing, the signal system in this and the previous examples works like this: authorize materials transfer when 3 full containers remain; then authorize production when only 2 full containers remain; each time, authorize enough material transfer and production to make 18 units (6 containers).

These are the basics of Kanban pull production. To illustrate how different cards and containers move through the system, all the kinds of cards discussed in this section using the original example in Figure 8.4 are illustrated in Figure 8.17. With the exception of the last operation, station 4, which operates according to a schedule that specifies the products to be assembled and their sequence, cards dictate the production and movement of material everywhere in the system. Creating the final assembly schedule is covered in Chapter 13.

Another kind of kanban card that has not been discussed is the **supplier kanban**. If suppliers of the raw materials and parts in Figure 8.17 are just-in-time suppliers, then the card system can be extended to them using supplier kanbans, or **S-kanbans**, cards that link buffers of raw materials and parts to external suppliers (the *S* loops in Figure 8.17). S-kanbans are discussed in Chapter 16.

One other place where the pull concept can be applied is to link the organization to its customers. If the organization represented in Figure 8.17 is itself a just-in-time supplier, then the finished products storage (on the lower right) can be considered an outbound buffer, to be replenished by authorizations (kanbans) coming from customers. This concept is also discussed in Chapter 16.

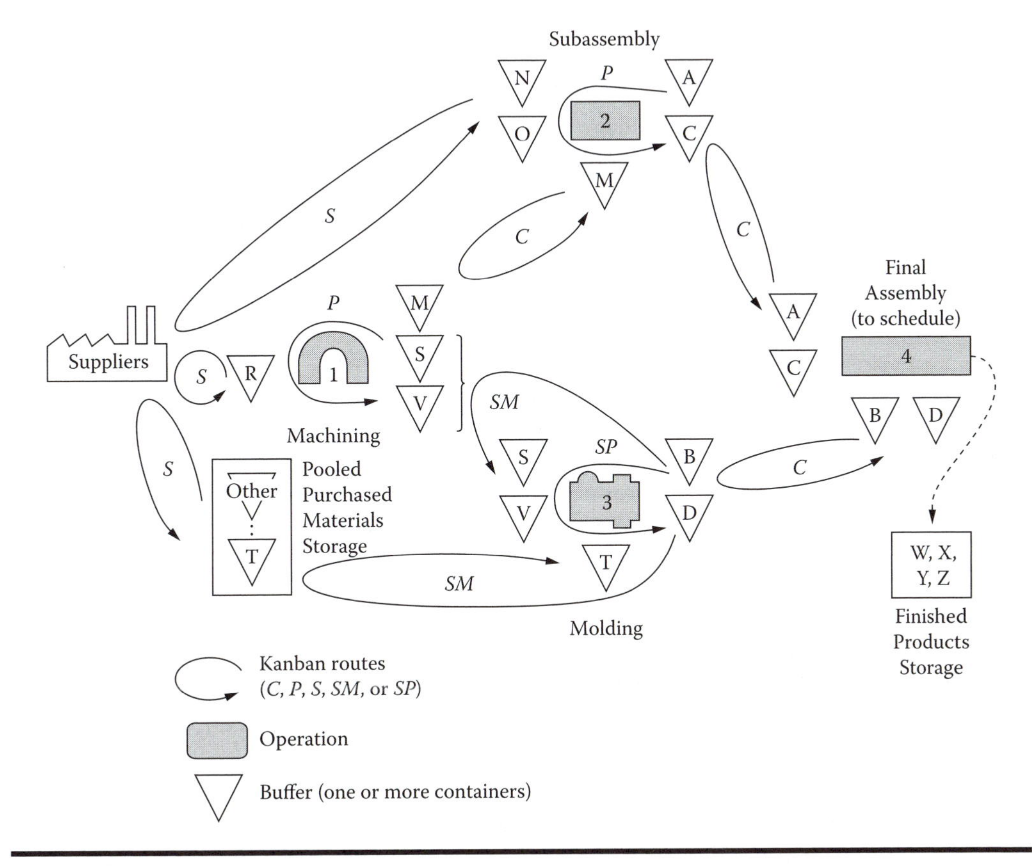

Figure 8.17 Kanban cards, routes, and buffers.

Whether or not a pull producer is a just-in-time supplier, its final assembly operation must operate to keep pace with demand. Pull systems work well when demand is fairly uniform; they are somewhat limited in ability to adjust to large, short-term fluctuations in demand. As a result, pull system producers must carry some amount of finished goods inventory to absorb demand fluctuations. These and other matters about accommodating pull systems to demand are discussed later in this chapter and in Chapter 13.

What, More Cards?

Move, production, and signal kanbans cover most but not all of the shop-floor authorizations in pull production. Others are needed for producing or moving items not included in the normal production sequence or for compensating for unanticipated situations. Pull production is not immune (and is perhaps prone) to unanticipated events, and there must be cards to cover every eventuality.

Express Card

This card is used whenever problems result in a shortage of items (e.g., defective parts) or threaten to interrupt production (e.g., equipment breakdowns). It is issued to expedite production or to move "emergency" items.

Temporary Card

This card is issued whenever production must temporarily deviate from the normal pull pattern. It is used to authorize a temporary increase in production necessary, for example, to provide more material to downstream operations to fill special orders or to build up outbound buffers to cover temporary downtime for machine maintenance. A temporary card is also issued to authorize production of trial parts for new products or special parts for test, engineering, or purposes besides customer demand.

Odd-Number Card

Defective items discovered in a container are not used and result in the number of usable items in the container being less than the standard quantity. When that happens an odd-number card is issued to authorize production of just enough to fill the container. That container is then given priority for the next withdrawal.

The different kinds of cards have different colors for quick identification: green for normal work, orange for rework, blue for temporary work, yellow for higher-than-normal production, red for express, and so on.

Other Mechanisms for Signal and Control

Kanban cards are just one of many possible ways for signaling and controlling pull production, and often they are not even necessary. If an operation produces only one kind of part, the mere presence of an empty container can be the signal to authorize production to replenish it. Even when multiple kinds of items are being produced or moved, information from the cards (the type and quantity of the items inside it, and the workstations between which it is moved) can be written on the containers themselves, and the cards eliminated. Operators often become so familiar with the procedure that, as a matter of course, they do not need the information on the cards or containers to know what to do. Still, it is a good idea to place that information somewhere clearly visible, if not on a card at least on a container, to prevent workers from getting careless. Since the number of cards and containers controls production and buffer stock levels, putting this information in an easily seen place helps ensure the correct levels will be maintained.

Though the concept of a container might conjure an image of a handheld shopping-basket thing, actual containers can be of any size and configuration—whatever effectively contains the quantity of the item produced and moved. A container can be a small box, lifted and moved manually, or a larger pallet or skid transferred using a forklift truck. One manufacturer of farm equipment uses large, wheeled iron frames, called dollies, as containers. Each dolly serves as both a support structure for assembling the equipment chassis and as the vehicle for moving the chassis from operation to operation.

In addition to cards and containers, there are many other ways to authorize and control production in pull systems. Some of these are explained next; the first three are illustrated in Figure 8.18.

Wheeled Carts

Wheeled handcarts or pallet carts are lined up in lanes painted on the floor. Each kind of item has a separate lane, clearly marked, just long enough to hold a certain number (K) of carts. Typically, each cart has a kanban card attached showing its contents and the necessary production/move information. Each lane might also have a line painted across it to show when the number of

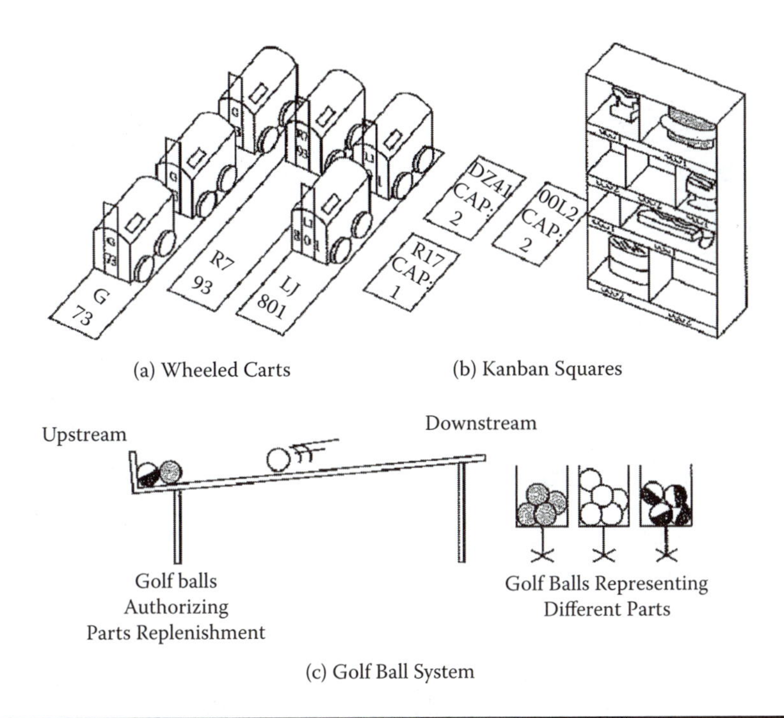

Figure 8.18 Different kinds of pull signal mechanisms.

remaining full carts is low or when replenishment is needed (analogous to a signal kanban). Figure 8.19 shows wheeled kanban carts on a factory floor.

Kanban Squares

Sometimes no container is needed, just a space to place things in. The space might be a square painted or taped on the floor or on a shelf, just large enough to hold a certain number of units (with tape, the size and location of the square can be easily changed). An empty square authorizes production or conveyance for the number of units that will fill the square. When a buffer holds different kinds of items, then a different square is provided for each item. Kanban squares are especially useful for items that are readily movable without a container.

Golf Balls

Another signal mechanism is golf balls. Whenever a downstream station needs more of an item, it rolls a golf ball on a gravity chute upstream. An arriving golf ball is authorization for a station to produce or move something. Golf balls have different colors and markings to represent different types and quantities of items. The sequence of golf balls arriving at a station specifies the production and move sequence.

Electronic Kanban

The golf ball system can work wherever it is practical to connect operations with a network of golf ball chutes. A chute network that would be too unwieldy can be replaced with an electronic kanban that uses monitors and keyboards. In the latter system, workers at downstream stations

Figure 8.19 Kanban carts.

enter on a keyboard the items they need, which then appear on monitors at upstream stations that supply them. The electronic kanban can also be used in conjunction with, or as a replacement for, conventional kinds of signals such as cards.

Clothespin Clips

Imagine a rack of shelves between two workstations, say a circuit-board assembly area and a circuit-board test area. The rack has 50 pigeonholes, and each hole holds up to four of one kind of circuit board.

On the corner of each circuit board is a detachable clothespin-like clip with a one-inch tag that shows the stock-keeping number of the circuit board. Whenever a worker from the test area withdraws a board from a pigeonhole, she detaches the clip and puts it on a small rod at the top of the hole. As boards are withdrawn, clips accumulate on the rods. Looking at the number of clips on the rods next to each pigeonhole, a worker in the assembly area can see which boards need to be assembled. A rod with four clips gets highest priority, one with three has next priority, and so on. The worker removes the clips from the rod, assembles the type of board specified, and attaches the clips to the completed boards and puts them into the pigeonhole.

Milk Run

A basic requirement of Kanban production is that there must be somewhat continual demand for a product—not necessarily large demand, just continual demand. To fulfill that requirement, the final stage of production must produce a continual stream of final product. For the final stage to produce continually, however, it is not necessary that all the materials feeding into it also be produced continually. As long as upstream operations produce materials in sufficient quantity and in time to satisfy the requirements at the final stage, they can produce those materials in intermittent batches.

When materials are produced intermittently, a way is needed to ensure the regular transfer of these materials to the final stage. One way is for a material handler to make a periodic "milk run,"

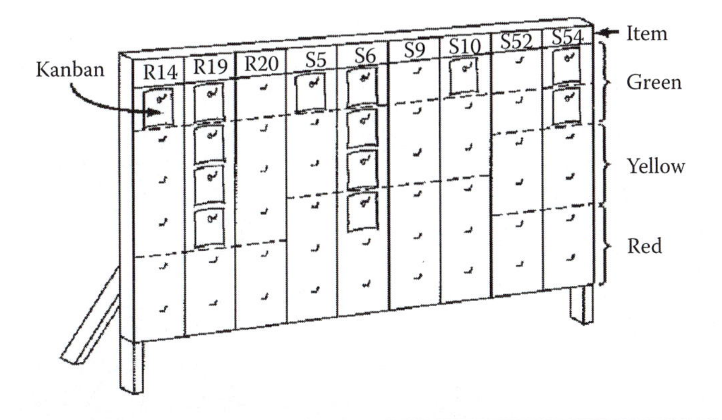

Figure 8.20 Kanban sequence board.

say once every 2 hours. The handler starts at the final stage, picks up the empty containers for all parts, and goes to the operations that produce these parts. After visiting every operation and exchanging the empty containers with full ones, the handler delivers the full containers to the final stage and repeats the cycle. Since pickups and deliveries happen on a regularly scheduled basis, and the quantities and types of materials required tend to be somewhat fixed, the upstream operations know what they need to produce (items and batch sizes) in the time available to fill the containers dropped off earlier.

Kanban Sequence Board

Ordinarily, P-kanbans are processed in the sequence in which they arrive or are posted at an operation. When several cards arrive at once or a backlog of cards develops, there is a risk that some items will not be produced in time and that downstream operations will be idled. To maintain a balanced workload and prevent habitual shortages of some items, production of items that require different cycle times must be carefully sequenced (short cycle times interspersed with long cycle items). One way to enable operators to sequence jobs when a backlog of cards accumulates is with a **kanban sequence board**. As cards arrive at a workstation, they are sorted and hung on the board according to type of item. As shown in Figure 8.20, the board is demarcated into regions of importance (green, yellow, and red) to represent increasing priority. The cards are hung starting from the top; cards first reaching the red region have priority over those in yellow, which have priority over those in green.

Clearly, the variety of ways to authorize and control inventory in pull production is limited only by the imagination of managers and workers. One further example is the following case in point.

Case in Point: Using Kanban for Batch Production at Strombecker[11]

Strombecker Corp. uses Kanban in the manufacture of toy cap pistols. Parts used in the assembly of the pistols are held in standardized bins, each with an assembly kanban card inside (Figure 8.21).

Figure 8.21 Assembly area showing parts bins.

When a bin of the die-cast parts is depleted, a material handler takes it and the kanban card to the die-cast store area where full bins are kept, each with a die-cast kanban inside (Figure 8.22). At the store area the handler takes the assembly kanban from the empty bin and places in a full bin (Figure 8.23, step a). At the same time, he takes the die-cast kanban out of the full bin

Figure 8.22 Store area: Full bins, foreground; empty bins, background.

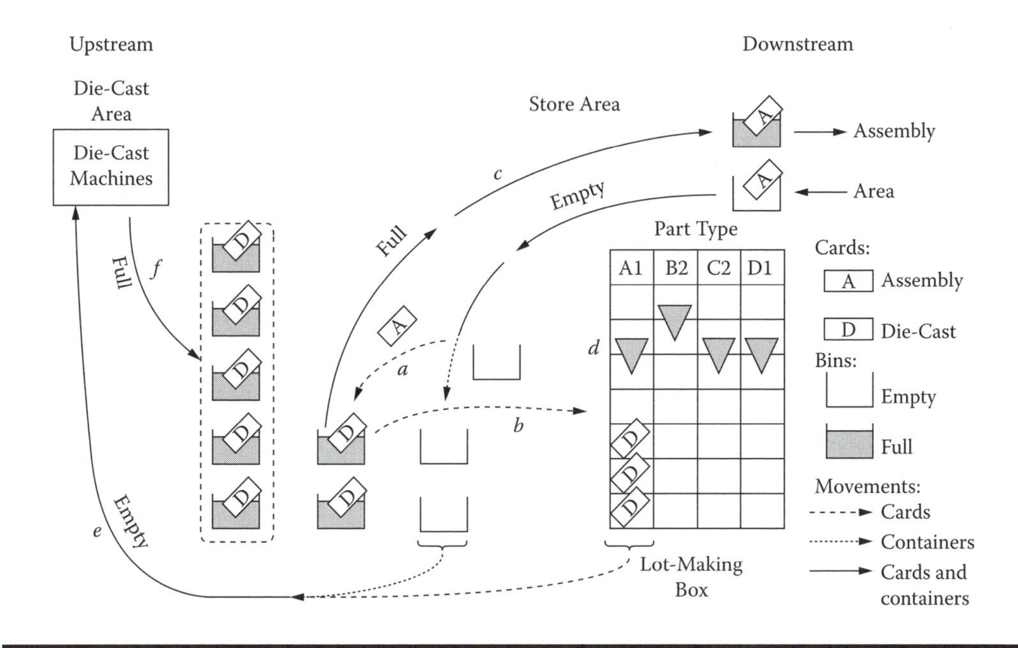

Figure 8.23 Kanban procedure.

and places it in a pigeonhole in the lot-making box (Figure 8.23, step b). The handler then takes the full bin to the assembly area (step c) and leaves the empty bin behind.

The lot-making box, Figure 8.24, is divided into columns, one for each kind of part. Cards are added to pigeonholes starting at the bottom. When the number of die-cast kanban cards for a part reaches a particular level (Figure 8.23, step d), that indicates that a batch for that kind of part should be die-cast. All the cards in the column are removed and taken with an

Figure 8.24 Lot-making box.

equal number of empty bins to the die-cast area (step e). A die-cast machine produces the batch of parts in the quantity as specified by the sum of the die-cast kanbans; this quantity has been determined in advance as a function of the daily production requirement for each part, the processing and setup time, and the number of different parts to be produced. When die-casting of the batch of parts is completed, the parts are distributed among the bins, a die-cast kanban card put into each bin, and the bins taken back to the store area (step f).

Process Improvement

In pull production, continuous improvement is enacted by slowly reducing the number of kanbans—the point of rule number 6 in Table 8.2. By reducing the number of kanbans, buffers are reduced and problems in the production system are exposed. To enable the system to function with ever-fewer kanbans, management has to face up to and resolve the problems exposed by less inventory.

Suppose the buffer between two operations consists of 8 containers with 10 units each, or a total of 80 units. Now, slowly reduce the number of containers until shortages begin to occur. Suppose with 7 containers the process works fine, but with 6 the buffer occasionally runs out and downstream operations have to wait. The buffer level would then be set at 7 containers, and an analysis done to determine reasons why shortages occurred when there were only 6 containers. Suppose the main reason is that setup time at the upstream operation takes too long. Suppose the setup time is shortened, and with that it is then possible to reduce the size of the production batch from 10 units to 9. Further, suppose that when the standard quantity per container is also reduced to 9 units, no shortages occur. Thus, the new buffer size will be 63 units (7 containers, 9 units each). The next steps are to try to further reduce the buffer size, look for more problems, solve them, and so on.

Some organizations establish a pull production system but then seldom try to reduce the size of the buffers. Managers are ecstatic because they got the pull system to "work," and because it resulted in, say, a 50% reduction in inventory. Of course, 50% of a large quantity is still a large quantity, so the organization still retains much more inventory than necessary, not to mention the problems that inventory covers up and which never get resolved. A pull production system wherein the buffers are never reduced misses one of the major benefits offered by pull production: the opportunity to expose sources of waste and eliminate them.

Practical Matters

Many managers do not rely too much on formulae to determine the number of kanbans, either because they do not trust the results or they know the production system has many problems, and that the formulae are not reliable. When they begin to install a pull system, they are skeptical that a process can operate with so little in-process inventory. As a result, they start with a number of kanbans that they are sure will work, which is usually around three or four times the number the formulae will give. (Even with that, the overall inventory in the system is much less than managers are accustomed to.) As they gain confidence and tinker with the process, they gradually reduce the number of kanbans. It is common, however, that the number of kanbans settled on is about twice the number suggested by the formulae (just to be safe), though in many cases that is not a bad idea since the plant must deal with somewhat unpredictable demand, long setup times, equipment unreliability, and so on, problems for which the formulae do not account. But as mentioned, the fact that large inventory buffers are necessary is a symptom that the system has many unresolved problems.[12] It is a practical matter that pull systems often start out with large buffers, but it is bad management when no continuing attempt is made to reduce the size of those buffers by resolving the problems that make them necessary.

Necessary Conditions for Pull Production

Pull production systems are demanding in terms of prerequisites. Among the conditions that should exist before a pull production process is instituted are the following:

1. More *planning and control responsibility* must reside in the hands of frontline supervisors and worker teams.
2. Production emphasis must be on *producing to meet demand* (not overproducing).
3. Throughout the process, the motivation everywhere must be to *reduce in-process inventories* and remove unnecessary stock.
4. Equipment *preventive maintenance* efforts must be ongoing and geared to eliminate breakdowns. Since pull systems carry minimal buffers, breakdowns will halt production flow.
5. *Quality assurance* efforts must be aimed at *preventing defects* from happening. Since pull systems carry minimal buffers and allow no defectives to proceed through the process, defects and quality problems will halt the process.
6. *Setup times must be small*, otherwise operations that produce a variety of outputs will require too much time for changeovers and have not enough time for production.
7. The *plant layout must facilitate linking* of all operations into the process. For smooth material flow between operations, every operation must have roughly the same capacity and be capable of producing at a rate dictated by the final stage of the process.
8. *Production schedules must be somewhat uniform*. Large variations and seasonal patterns that can be anticipated are leveled in advance to minimize the impact on upstream operations. This is necessary since a Kanban system with a given number of cards can effectively deal with demand variations in the range of only ±10%. The leveling process is described in Chapter 13.
9. Training, job descriptions, and compensation of workers must be geared toward developing *cooperative work attitudes and teamwork* since much of the planning and control decision making in pull production is done by worker teams. Managers must provide workers opportunity to develop and expand their skills and utilize their capabilities, and they must show workers respect and recognition for their ability as problem solvers.

From this list it should be obvious that pull production will be a long time in coming for some organizations, and will never come for others. Some of the conditions can be achieved incrementally and require changes in policy rather than big capital investments (e.g., reducing inventories and moving from repair to preventive maintenance). Others, however, are more difficult to meet. Engineering studies and projects for reducing setup times, preventing breakdowns, and rearranging facilities, as well as training programs to give workers skills in equipment setup, basic maintenance, inspection and quality, and scheduling and control are all costly and time consuming. Management must understand, however, that these costs are short-term and justifiable given the long-term benefits of reduced inventory and improved flexibility, responsiveness, and quality of pull production.

It should be noted that none of the above conditions pertains exclusively to pull production. Efforts toward achieving these conditions in any kind of production system will have the same result: improvement. Even in traditional job shops and push-production systems, efforts to reduce setup times, enhance preventive maintenance, and improve worker skills can only make the system better.

Pull Production and Repetitive Production

Pull systems work best for products where the demand is stable and continuous. In most pull systems, every kind of part or material must be held in a buffer, and to justify all those buffers requires that the materials be in somewhat continuous demand. Also, to keep down the proliferation of buffers, pull production is often confined to a relatively few standardized products. Such production of standardized items on a continuous basis is called **repetitive** production.

For repetitive production, demand does not have to be large, just stable enough so that the final assembly schedule can be smoothed (i.e., have relatively level daily production output). To achieve stability and enable leveled production scheduling, some organizations combine production of different versions of products that were formerly produced separately. Combining production of different products is feasible as long as product differences are add-on features or options and not differences in fundamental design, major components, or production processes. Grouping different products by **product family** is a way to produce multiple products in a single process. This topic is discussed in the next chapter.

Without stable product demand, the only way to achieve a leveled production schedule is to stockpile the finished product. For example, seasonal variation can be accommodated with leveled, repetitive production, but excess output from periods of slack demand must be stockpiled to cover production shortfall during periods of peak demand.

When Pull Does Not Work

Even if demand is stable enough so that, conceptually, a leveled production schedule can be developed (discussed in Chapter 13), and if products can be grouped to get around the limited product variety requirement, there are other factors that can make pull production impossible or impractical. Hall lists the following as typical:[13]

1. Despite stable demand, the assembly of the final product cannot be executed in a level-enough fashion to provide steady demand for upstream operations. This happens when the assembly involves trials, matching, adjustments, testing, or anything that causes the time for assembly to vary with every unit.
2. Some operations must be started in advance of pull signals. This happens when operations require special, lengthy, or difficult setups that cannot be simplified or significantly shortened, and must be scheduled in advance.
3. The product is made in so many options, and the demand for each option is so small or unstable, that it is impractical to carry buffer stocks for all parts everywhere in the process.
4. The high defect rate causes too many interruptions to permit continuous flow, and technology is such that the defect level cannot be reduced significantly.
5. Products must be produced as integrated batches throughout the process for reasons of quality control or certification. An example is pharmaceuticals.

Now, although any of these factors can make implementation of pull production impossible or very difficult, none of them pose a particular problem for a push system like MRP. Besides, MRP readily accommodates products that are nonstandard and have fluctuating or erratic demand. The only requirement for MRP is that the demand can be forecast far enough in advance to be input into a master schedule.

Pull and Push Systems, Both at Once

In any situation where a leveled final assembly schedule is feasible, a pull system can be implemented by starting at the final assembly stage and working backward to link upstream stages into the process. The upstream operations are linked together using pull signal methods, one stage at a time. Those operations that cannot be linked into the system (because of factors like those listed above) can remain outside of the pull process and be treated separately. These outside operations will be treated as job shops that serve the repetitive pull process, but which use traditional push methods for scheduling and control. In other words, two kinds of systems will be used: a pull system for the repetitive portions of the process and a push system for everything else.

When products can be **modularized** (built as a collection of standardized options and subassemblies, described in Chapter 13), often then the modules can be produced repetitively in a pull system, even though the final assembled product cannot and must be produced with a traditional push system schedule.

Even when a company is able to implement a complete pull production process, say, for products it manufactures throughout the year, it still might need a push-oriented system for other products it manufacturers only seasonally or on demand. Again, two systems, or a mixed system, are needed depending on the products. Many large manufacturers utilize multiple kinds of production systems (pull lines, as well as batch-oriented push lines); whatever best fits the requirements of the product and process. As described in Chapter 15, every pull production system needs some form of system for forecasting demand, production and material requirements, and for capacity requirements planning, all of which is handily done with MRP-II and ERP (enterprise resource planning) systems.

Getting Started

In many organizations the initiative to implement pull production starts with midlevel management. To test its feasibility and demonstrate its benefits, pull production is introduced in stages that are limited to the segments of the process that best meet the necessary conditions. Trial and error at each stage enables the application to be fine-tuned and provides the knowledge and experience necessary to expand pull production to larger portions of the process. Many firms first experiment with pull production by using it to link only two or three operations in a much larger process.

If the factory experiment is to eventually take hold plantwide, top management must get involved. The changes and investments required to implement pull production wide-scale are too great for the implementation to succeed without commitment and backing from top management. A top manager must serve as the pull production champion.

Often the greatest challenge in implementation is in overcoming cultural barriers. Implementing pull production requires teamwork and moving the locus of control away from centralized staff and toward the shop floor. It requires significant changes in organizational roles, worker responsibilities, job descriptions, and pay systems. These issues are covered in some depth in the discussion on workcell implementation in Chapter 10.

Summary

Pull production is a way of controlling a production process and reacting quickly to changes without relying on inventory. In a pull system, each stage of a process produces exactly what the immediate downstream stage requests; in effect, this results in material being pulled through the process since each stage produces only what is demanded of it from the next stage. This contrasts to push production wherein every stage produces according to a preplanned schedule, then pushes material to the next stage, whether or not the next stage is ready for it.

Production and movement of material through a pull process are contingent on signals coming from downstream. No material is moved, and nothing is produced unless a signal comes from downstream. The most familiar kind of signal system uses cards and is called Kanban. A kanban is a card authorizing production or movement of a standard quantity of material and showing the kind of material needed, its source, and destination.

A pull system reacts immediately to changes or problems anywhere in the process. If a stage faces a problem and ceases to send signals upstream, the upstream stages also cease, at least until the signals resume. A small amount of inventory called a buffer is held between stages to allow demand from downstream to be filled immediately while upstream operations work to meet requests for more. The amount held in the buffer is controlled by using a prespecified number of standard containers. The number of containers corresponds to the number of cards (kanbans) deemed adequate for the system to meet demand. Because the number of cards (hence, containers) is regulated, the amount of in-process inventory required for a process to function smoothly can be held to a small quantity.

Various kinds of containers and signals are used in pull production. Depending on the item and the need, containers can be small boxes, carts, pallets, or large dollies, and signals to produce or move items can be produced using special P-cards, C-cards, or other cards, by using computer monitors linking stages of the process, or deduced visually from empty squares or spaces on a shelf or floor—whatever fits the need.

Pull production is not universally better than push production. First, pull production requires repetitive manufacturing—fairly smooth, continuous production of somewhat standardized items. It also requires that the production system be free of the usual sources of interruptions (e.g., product defects, long setups, equipment malfunction). Products where demand is highly variable or unstable or that requires lengthy setups, trials, adjustments, or testing; or that are uniquely made to customer requirements cannot be produced using pull production. On the other hand, push systems and conventional planning and control systems using MRP can readily accommodate such situations.

Pull production is often used in combination with push production: some processes in the same plant use push, others use pull, depending on the product/process conditions. For products made from modular components, often the demand for the modules is stable enough that they can be produced using pull, even though demand for the assembled product is unstable and requires push procedures for scheduling and control.

Pull production can be introduced in stages, starting with the last stage of the process and working upstream. Pull production requires commitment and resources from top management, and a shift of most responsibility for production control to frontline worker teams.

Notes

1. Kanban and the Toyota system are described by Taiichi Ohno in *Toyota Production System—Beyond Management of Large-Scale Production* (Cambridge, MA: Productivity Press, 1988). Much the same topical material is also in Yasuhro Monden, *Toyota Production System,* 2nd ed. (Norcross, GA: Institute of Industrial Engineers, 1993).
2. Use of the term *just-in-time* as being synonymous with pull production is declining. As discussed earlier in this book, JIT today more generally implies lean production, a management philosophy that emphasizes continuous improvement, value added, and elimination of waste. Pull production is but a technique in JIT and lean production.
3. There is a method called "broadcast" that permits portions of a production system to operate on a pull basis without inventory buffers for every kind of material. That system, however, utilizes production schedules and is thus not a pure pull system. Another method, called CONWIP, also does not require buffers for every kind of material. Though not a pure pull system, it has most advantages of pull systems; see W. Hopp and M. Spearman, *Factory Physics: Foundations of Manufacturing Management* (Chicago: Irwin, 1994), Chapters 10 and 14.
4. See D. Pyke and M. Cohen, Push and Pull in Manufacturing and Distribution Systems, *Journal of Operations Management* 9, no. 1 (1990): 24–42.
5. Adopted from K. Suzaki, *The New Manufacturing Challenge* (New York: Free Press, 1987), 155–157; Japan Management Association, *Canon Production System* (Cambridge, MA: Productivity Press, 1987), 133–137; R. Hall, *Zero Inventories* (Homewood, IL: Dow Jones-Irwin, 1983), 53.
6. T. Ohno, Evolution of the Toyota Production System, in *Continuous Improvement in Operations*, ed. A. Robinson, (Cambridge, MA: Productivity Press, 1991), Chapter 11.
7. Adopted from M. Graban, *Lean Hospitals* (Boca Raton, FL: CRC Press/Productivity Press, 2009), 112–113.
8. For further discussion of conveyance and production kanbans, see Monden, *Toyota Production System*; and H. Steudel and P. Desruelle, *Manufacturing in the Nineties* (New York: Van Nostrand Reinhold, 1992).
9. Ibid.
10. Steudel and Desruelle, *Manufacturing in the Nineties.* Refer to Chapter 5 in this book for minimum process lot size and the ramifications of internal versus external setup times.
11. Conversation with Vern Shadowen, Operations Manager at Strombecker Corp.
12. R. Inman, Inventory is the *flower* of all evil, *Production and Inventory Management Journal* 34, no. 4 (1993): 41–45.
13. R. Hall, *Zero Inventories* (Homewood, IL: Dow Jones-Irwin, 1983), 308–309.

Suggested Reading

Productivity Press Development Team. *Kanban for the Shopfloor* (Shopfloor Series). New York: Productivity Press, 2002.

Questions

1. Describe the fundamental differences between push production systems and pull production systems. How are they the same?
2. What are the necessary conditions for pull production?
3. Discuss the relationship between pull production and demand stability and variation.
4. Under what situations is pull production inappropriate?
5. Describe the reorder-point system. How is it different from a Kanban system? How is it the same?
6. How is pull production more than just a way of controlling inventory?

7. List various ways for authorizing material movement or production in a pull system. How does the proximity of operations determine which of the ways are most appropriate?
8. Explain how some pull systems operate without the use of cards.
9. For the following, indicate whether MRP or Kanban would be more effective for production control.
   a. An automobile assembly plant that produces 10 car models.
   b. A shop that produces a large variety of custom-ordered products.
   c. An assembly operation that involves 20,000 part numbers.
   d. An assembly operation that involves 50 part numbers.
   e. An assembly line plant where all parts are outsourced.

10. What is the role of Kanban in process improvement?
11. Four kinds of signal and control mechanisms are (1) kanban cards and general-purpose containers, (2) special purpose containers that had only one kind of item, (3) kanban squares, and (4) an automated conveyor system with no containers. Explain which is best for each of the following situations.
    a. Final assembly of MP3s.
    b. Production and finishing of metal castings of common parts that involve casting, tumbling, heat treatment, and machining.
    c. Repetitive production of kitchen and dining room furniture.
    d. Replenishing stock of parts supplying an assembly line.
    e. Producing small batches of replacement parts in a five-operation process.

12. Explain how each of the following works: (a) a two-card pull system and (b) a single-card pull system. Explain when one or the other is necessary. When are the two systems equivalent?
13. Explain the purpose of signal kanbans for production and material movement. When are signal kanbans used?
14. The assembly shop of a manufacturer of large air conditioners for commercial and industrial buildings has numerous units at various stages of completion, most sitting unattended for lack of parts. As parts and subassemblies arrive from other departments, assembly workers install them. Is this a push or pull system? Suggest how the system might be improved.

PROBLEMS

1. Calculate the required number of kanbans for items X, Y, and Z.

	X	Y	Z
Demand	110 per day	90 per week	25 per hour
Cycle time per kanban	5 hours	1 day	20 minutes
Container size	40 units	10 units	5 units
Safety margin	10%	0%	30%

2. Superseven Corp. uses a two-card kanban production system. Inventory levels are roughly proportional to the number of kanbans. For one part, the average usage rate is 1,400 parts per day, container size is 50 parts, and cycle time per batch is 2 hours. Superseven is working to reduce the lead time to 100 minutes. Assuming no safety stock is carried, what effect would this have on average inventory?

3. An assembly line pulls containers of parts from machining centers at a rate of 600 per day. Each container holds 20 parts and typically waits 2 hours at a machining center before it is processed. The setup and processing time for 20 parts is 6 hours. How many containers are needed if the safety margin is 10%?

4. The daily demand for parts from machining workcell JMB to an assembly workcell is 1,600 units. The average processing time is 25 seconds per unit. A container spends, on average, 6 hours waiting at JMB before it is processed. Each container holds 250 parts. Currently 10 containers are being used for the part.
   a. What percent safety margin of stock is being carried?
   b. If you wanted to retain this safety margin, but remove one container, to what value must the waiting time be reduced?
   c. What would happen if the demand for parts increased to 1,900 with K = 10?

5. The BOM for an assembly is shown in the following (numbers in parentheses indicate the quantity of part for each assembly).

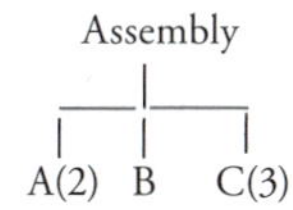

Each 8-hour day the assembly line produces 400 units. Because the assembly line is located in a different area of the plant than the areas where the parts A, B, and C are produced, a two-card pull system is used. Information about producing and moving the three parts is shown in the following (all times in hours):

	A	*B*	*C*
Container size	100	50	100
Conveyance:			
Total wait time	1.0	1.5	1.0
Total move time	0.5	0.5	0.5
Production:			
Total wait time	1.0	1.0	0.5
Setup time/container	0.2	0.5	0.5
Processing time/unit	0.008	0.02	0.006

If a safety margin of 20% is to be maintained, how many of each of conveyance and production cards are needed for each of the three kinds of parts? Draw a schematic diagram showing the assembly and parts production areas, and the location of containers and cards.

6. A worker sorts parts at a rate of 400 per hour. A material handler arrives every half hour, drops off empty bins and departs with bins of 100 parts. How many kanbans are needed for a safety margin of 20%?

7. For the previous problem, which of the following has the greatest influence on reducing the number of kanbans? (Treat each independently.)
   a. Cutting the material-handler delivery cycle in half
   b. Doubling the capacity of the bins to 200 parts
   c. Eliminating the safety margin

8. A fabrication workcell uses two components, QR1 and QR2, each produced at the rate of 300 per hour. Because the two components are different sizes, they have different-sized containers: the container for QR1 holds 20 units, the container for QR2 holds 30. The operation supplying QR1 takes 15 minutes to fill the container; the operation for QR2 takes 8 minutes to fill the container. Total move and wait time between each operation and the assembly workcell is 10 minutes. How many kanbans are needed between each operation and the workcell? Assume a 10% safety margin.

9. At Fabfour Products Inc., daily demand forecast for the top 10 products is:

Product	*Daily Demand*
A	1,600
B	1,000
C	800
D	600
E	400
F	200
G	100
H	50
I	25
J	25
Total	4,800

   The final assembly area produces 600 units per 8-hour workday.
   a. Products are produced in batches of 50 units, which is also the containers' size. In terms of number of batches, what is the daily production schedule for the 10 products?
   b. The cycle time for all products is 2 hours per container. Using a 20% safety margin, how many kanbans are needed for the products?
   c. If inventory were proportional to the number of kanbans, by what percentage would inventory decrease if the cycle time per 50-unit container were reduced by one hour?
   d. Assume case (c): What is the benefit of reducing the batch and container sizes to 25 units?

10. The average wait time before starting an order at the machining workcell is 2 hours. It then takes an average of 25 minutes to set up the cell for the order. The machine cycle time on each part is 120 seconds. If a container holds 20 parts, management allows a 20% safety margin, and 8 containers have been allocated for the part, how much average daily demand can the machining cell handle?

11. Fabfour Products Inc. receives kits of parts from a supplier. Each kit contains all of the small parts required for an assembly. The kits are boxed in quantities of 50, and boxes are delivered directly to point-of-use racks in the assembly area. Each time a box is opened, a replenishment order is sent to the supplier. The average daily demand is 45. The lead time on replenishment from the supplier (total order and delivery time) is 10 days. The company likes to hold enough stock for a 20% safety margin. Assuming one kanban card per box, how many cards should there be?

12. Colored wooden blocks used in small-toy assemblies are painted using a tumbling vat and batch sizes of 2,000. Six different color blocks are produced using three vats. A request for additional colored blocks from the assembly area usually waits 3 hours before the paint area begins to process it; it then takes 1.25 hours to clean out a vat and prepare it for a different color, 0.25 hours to paint a batch, and 3 hours for the batch to dry and be moved to the assembly area. No blocks are needed at the paint vat while it is being cleaned and readied for a new color. The time between when additional colored blocks are requested from the assembly area and when material handlers move the unpainted blocks from storage to the paint room is 3 hours. Each color of block is used by the assembly area at the rate of 200 per hour. Standard size containers holding 250 blocks are used for both painted and unpainted blocks. Containers of finished painted blocks are stacked for use in the assembly area.
    a. Assuming 10% safety margin, at what position in the stack of containers should the paint production signal-kanban be placed?
    b. At what position in the stack should the material signal-kanban (signal to authorize movement of unpainted blocks to the paint area) be placed?

13. In the previous problem, what is the effect on the kanban positions of reducing the batch size to 1000? Interpret the results. In general, what is the implication when the number of containers in the batch size is fewer than the kanban signal position?

14. Refer to the process in Figure 8.25. The product assembled at operation OpD is made from two subassemblies, each produced in a process that has three operations. The operations within each process are located next to each other, and operation OpD is located next to operations OpC and OpG. The hourly production capacities (fixed hourly production rates) are shown in Figure 8.25. The current demand requires that the output for operation OpD be 75 units per hour. Assume negligible setup and move times, and that suppliers are able to fulfill any likely demand rate. The same containers are to be used throughout the process.
    a. Discuss the size of the container for the overall process.
    b. Discuss the location of stock areas and number of kanbans at each stock area, assuming kanbans are used to link successive operations.

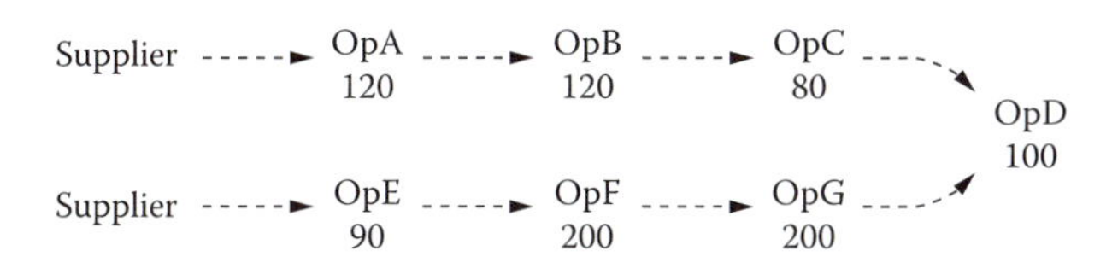

Figure 8.25 Product assembly procedure.

Chapter 9

Focused Factories and Group Technology

> Movement and rest have their definite laws: according to these, firm and yielding lines are differentiated …
>
> As the firm and yielding lines displace one another, change and transformation arise.
>
> **—I Ching**

A production process is implemented within a physical facility. The facility consists of machines, workers, and material-handling equipment assigned to workstations, work centers, and functional departments to which they are assigned. The kind of process and the spatial arrangement of the operations, workers, and machines have a tremendous effect on production lead time, cost, quality, and flexibility.

Part of the difficultly in designing production processes and facility layouts stems from proliferation of products. As will be discussed, the greater the number of different kinds of products manufactured in a single facility, the more difficult it is to produce each one efficiently. Nonetheless, competition drives product proliferation and diversity. A company that can modify or customize its products to meet the needs of existing and future customers has a potential advantage over competitors that cannot. Many manufacturers are able to prosper by producing a large variety of products for many different customers.

Diversity in products is often not just a matter of colors, add-ons, and gimmicks; it often represents major, fundamental differences between products. Regardless, as a way of holding on to its competitive advantage, a company must be able to quickly switch over and make each product in a diversified line in small batches, or make every one of them at virtually the same time.

At some point, however, proliferation and diversity in products can strangle a company. Simply, no company can do everything well. The diversity itself must be delimited and managed, and the waste associated with producing multiple kinds of products must be kept to a minimum. One solution to the problem of managing diversity is to outsource some products to suppliers—a topic addressed in Chapter 16 on supply chain management. Another is to use **group technology** to identify similarities among different products and group them accordingly, and then produce each group in a single place with the same workers and equipment, that is, within a **focused factory**.

Both group technology and focused factories are considered in this chapter, starting with their place among various ways of doing work. The concept of a **workcell**, which is a particular kind of focused factory commonly found in pull production systems, is also introduced; the next chapter covers workcells and related issues in depth.

Ways of Doing Work

The common ways of doing work are as projects, jobs, batches, and repetitive and continuous operations. As shown in Figure 9.1, the suitableness of each of these ways depends on the volume of the end item (product or part) to be made and the amount of resources needed to make each unit of the end item.

A **project** is a unique, large-scale work effort directed at one or a few end items. Each item is tailored to fit unique requirements. Large one-of-a-kind systems, structures, or machines such as buildings, dams, highways, ships, and automated manufacturing systems are examples. The work involves diverse, often cross-functional activities, the nature of which varies with each project.

A **job** is a small-volume, somewhat small-scale work effort where the output is one or a few identical items, custom made to fit an order. A custom-ordered machine fitting, a special-purpose casting, and a hand-crafted dining room set are examples of jobs. A job can be considered a small-scale project done in a small plant or shop, called a **job shop**, by a group of professionals, craftsmen, or skilled tradespeople. When a job involves producing several or many identical end items, the items can be produced in a **batch**.

Repetitive and **continuous** operations produce similar or identical items in high volume. Both kinds of operations are called **flow shops** because material moves through them somewhat smoothly and with few interruptions.

Products made in *repetitive operations* are discrete units such as cars, refrigerators, computers, and pens. Often, the equipment used in repetitive operations is designed for single-purpose, high-efficiency operation and the workers are narrowly trained to perform one or a few tasks each.

Products made in *continuous operations* actually flow through the process; examples are fluids, minerals, and mixtures such as foodstuffs, paint, steel, paper, coal, and petroleum. Such products are produced in large volume and, often, are considered commodities in that they are

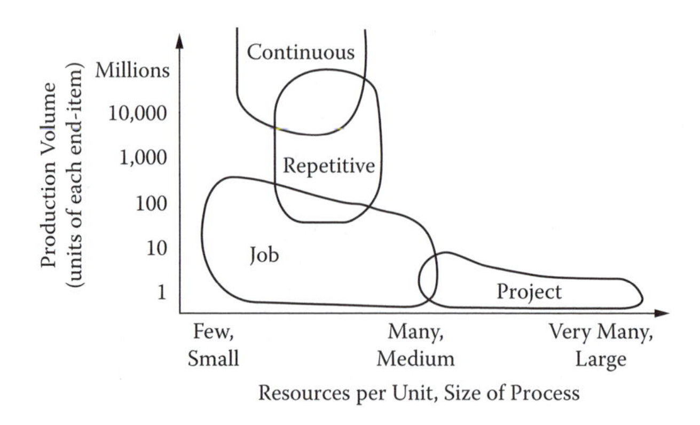

Figure 9.1 Ways of doing work by quantity of output and size of process or operation.

indistinguishable from products made by other producers. In a continuous production process the product seldom changes. Production is scheduled and controlled by volume output, not by discrete units. Often, not until the product is packaged is it identifiable as discrete units; steel, dairy products, soft drinks, and paper are examples.

Hybrid ways of doing work are also common, such as batch production within continuous and repetitive processes. A variation of the hybrid theme is to repetitively produce a batch of one kind of product, then modify the process to repetitively produce a batch of a slightly different product. The batches are usually large and changeovers are infrequent. An example is a repetitive manufacturer of garden tools that in midsummer switches over to produce rakes, then switches in fall to produce snow shovels, then in midwinter to again produce garden tools.

Variety–Efficiency Tradeoff

When a manufacturer changes its usual method of work from job production to batch production and then to repetitive and continuous production, several things happen: material flow through the plant becomes more steady, less intermittent, and more continuous; work task efficiency increases; and unit time and cost of production decrease. Smoother, steadier flow happens because the number of stops, starts, and changeovers in daily production decreases. In a job shop virtually every end item requires a setup, whereas in repetitive manufacturing setups are relatively infrequent, as seldom as a few times a year or every several thousand units. Production efficiency increases because more attention, knowledge, and skill are directed at fewer kinds of things. As a result, less time per unit is wasted, work-in-process (WIP) inventory is smaller, throughput rates are higher, and unit costs are lower.

This does not mean that continuous or repetitive work is better than job or batch production on all counts. As a company shifts from job production to repetitive production, it also shifts emphasis from product variety to output efficiency. Historically, at least, it seems that as a company increases its ability to produce things efficiently and in large quantity, it decreases its ability to produce things in large variety. Henry Ford was able to achieve high efficiency with his Model T, but he only produced one kind of car that had few options (the saying was, "Any color you want, as long as it's black").

Solutions to the problem of achieving efficiency and variety lie partly in the way products are designed, partly in the way jobs are planned and scheduled, partly in the way operations and equipment are physically arranged in the factory, and partly in the way products are grouped for production. The first way, the subject of design for manufacture and assembly (DFMA), is beyond our scope.[1] The second way, planning and scheduling, is discussed in Chapters 13 through 15. The other two ways, physical arrangement of factory facilities and grouping different products into families for production purposes, are discussed next.

Facilities Layout

The three common types of facilities layout are **fixed-position**, **process,** and **product**.

Fixed-Position Layout

In a fixed-position layout, the end item remains in a stationary position while it is being produced. Fixed-position layouts are common in project work where the end item is large and difficult to move (e.g., ships, hydroelectric turbines, electrical transformers, aircraft), and in job work where items are custom produced by craftspeople (racing cars and machine tools). This kind of layout was used by Henry Ford and other early mass producers. At Ford, each car was assembled at one location, materials were brought to it, and workers walked around it to perform tasks.

As Ford discovered, however, it is often easier to keep the workers and equipment stationary and to move the end items to them, especially when the end items are readily movable and the production volume is large. Actually, this is true even for large end items such as aircraft. Bombers mass produced during WWII were moved to staging areas throughout cavernous assembly plants. At each staging area (in effect, a fixed-position shop) a major portion of work was completed—such as attaching wings to the fuselage—then the airplane was moved to the next staging area for additional work, like attaching engines to the wings.

Process Layout

In a **process layout**, similar types of operations (similar equipment and tools, workers with similar skills and expertise) are clustered into functional work areas or **departments**, and each job is routed through the areas according to its routing sequence of operations. Take, for example, four products with the following characteristics:

Product	*Operations Sequence*	*Annual Volume*
X	J-B-L-T-P	100,000
Y	T-L-P-T	8,000
Z	P-L-B-J	20,000
W	J-L-P	11,000

In a process layout, all the J-type operations would be done in one area, say, department J, all the B types in department B, and so on. Figure 9.2 shows a process layout with departments J, L, P, T, and B, and the routings for the four products. Each department has one or more machines and workers that perform one particular kind of operation such as milling, drilling, painting, printing, casting, or assembly. The production machines within each area are similar or identical, and are considered general purpose in that they can perform work on a variety of different products. Most equipment (including hand tools, drills, mills, presses, and PCs) in machine, wood, and printing shops is general purpose. Job shops, which manufacture items individually or in small batches, use process layouts.

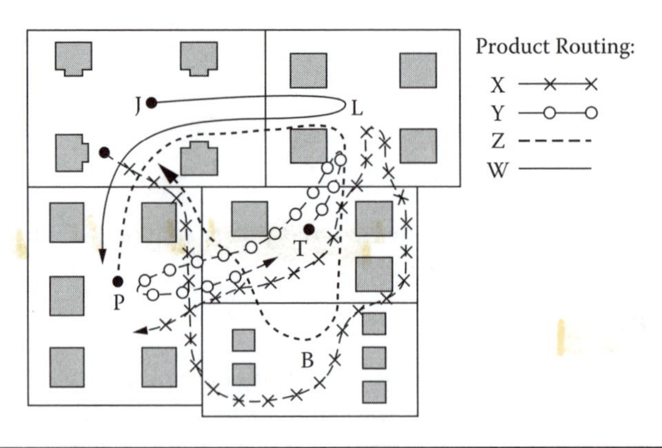

Figure 9.2 Process (functional) layout and product routings.

The advantage of process layouts is that they can produce any product that requires work in any of the departments, no matter the production volume or routing sequence of the product. The fact that annual volume among the aforementioned five products varies has no bearing on the layout. Production output can be increased or decreased, new products added, old ones dropped, all without changing the basic departmental layout of the facility. Custom, made-to-order products are simply routed through areas of the plant in the appropriate sequence. Capacity in the plant is increased by adding overtime or machines to each area, as needed.

Beware, however, that beneath the simple appearance of a process layout lies potential complexity and waste. Consider job routing. If, for example, five products can be produced by any combination of five operations, there are 5! = 120 possible routings. Of course, in an actual factory many of these combinations would be eliminated as illogical routing sequences. Nonetheless, the fact remains that the number of possible routings is large and many of them require moving material over long distances (as is obvious in Figure 9.2). Moving and handling of material is, of course, nonvalue added.

Process layouts also require considerable effort for job scheduling, routing, sequencing, and tracking of jobs. When a job arrives at a work area, it must wait until other jobs are completed. Jobs waiting throughout a typical plant represent sizable WIP inventory.

Further, jobs for different end items require machine setups in each department. If setups are time consuming, that encourages production of large-size batches, with the attendant quality, cost, and time drawbacks. If each takes a long time to process, that causes long waiting times for all other jobs. Because of long waiting times, a job routed through several departments can take weeks or months to complete, even though actual value-added processing might take only hours or minutes.

To summarize, process layouts are flexible and can accommodate a variety of products, regardless of differences in demand or processing steps, but they are also wasteful in terms of time, material handling, defects, and inventory. These wastes can be mitigated by the lean methods already described such as small batch production, small-size transfer batches, reduced setup times, and by methods described later such as standard operations and mistake-proofing. Nonetheless, waste cannot be eliminated from a process layout. Unless the product mix consists primarily of small-quantity, custom-designed, or one-of-a-kind products, the goal should be to move away from process-type layouts toward product layouts.

Product Layout

Facilities with repetitive or continuous processes are usually arranged in a **product layout**. The layout consists of all the necessary operations for producing a product arranged in a sequence on a line—a production line, assembly line, or flow line. Product X, for example, requires the sequence of operations J-B-L-T-P, so the layout for it would consist of machines and workers at workstations as laid out in Figure 9.3. Plants that produce repetitively or continuously using a product layout are called **flow shops**.

Continuous/repetitive production and product layouts go together. Only one or a few kinds of end items are produced, and they all follow the same routing sequence through the line. Work scheduling in a product layout consists of determining the flow rate or cycle time necessary to satisfy demand, then designing the line so it produces at that rate. Since material moves continuously from operation to operation and neighboring operations are adjacent, there is little or no inventory waiting

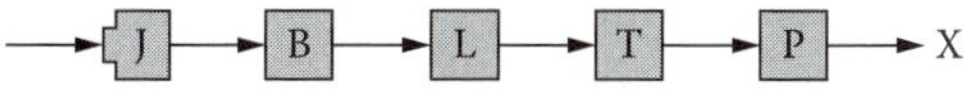

→ J → B → L → T → P → X

Figure 9.3 Product layout for product X.

between operations. Throughput time per unit is not much more than the processing time in the line. Material handling consists of transferring items from one operation to the next. The transfer can be automated, or for small, light items, manual. On a fully mechanized **transfer line**, material at each machine is automatically loaded, machined, unloaded, and moved to the next machine.

The big drawback of a product layout is, of course, that it is dedicated to producing just the one kind of thing. Once the line has been created, that is, once the operations sequence and transfer system have been installed, it can produce only products that follow that one sequence and require all the operations built into the line. It is termed *product* layout because it is designed specifically around one kind of product. Though minor variations in the product might be introduced at places in the line, they are usually cosmetic and constrained to whatever the sequence of operations and production rate allow.

A product layout is also constrained in its output capacity since the production rate is designed into the layout. To alter the rate, the line must be rebalanced, which means the operations must be redistributed among the workstations. The more mechanized the line, the more difficult and costly it is to rebalance it. The process must be stopped and changed over, which can take hours or days, even after setup-reduction efforts. The easiest way to increase the line's output is to simply run the line for more hours or shifts. To decrease capacity, the line is periodically stopped.

The capital investment in a product layout can be large. To achieve high efficiency and rapid flow rates, lines often use special-purpose equipment and transfer systems. The equipment is often custom made and one of kind, so it is expensive to design, fabricate, install, and maintain. Since resequencing and rebalancing the line to accommodate changes in product options and production rates is costly, producers try to keep variety to a minimum. Workers do not see much variety either. Their tasks are often narrowly defined, and their minute-to-minute tasks stay the same and can become boring.

Product layouts and process layouts represent two pure types of layouts at opposite ends of a continuum. Just as many plants use different work processes, so they use mixed or hybrid layouts. Some areas of the plant are arranged as process layouts to produce parts and components for other areas of the plant, which are arranged as product layouts that assemble the parts and components into finished products.

Variety–Volume Tradeoff

There is a perplexing dynamic in manufacturing, a variant of the variety–efficiency tradeoff discussed earlier: to increase its product appeal and broaden its market scope, a producer should strive to offer many kinds of products; at the same time, to reduce production waste and increase profits, it should focus on relatively few products and produce them in high volume. The question is how can a manufacturer achieve production efficiency without sacrificing product options and variety? Referring to the aforementioned taxonomies of types of work and facility layouts, it seems that a choice must be made between variety and volume, and that efficiency will simply be the by-product of that choice: high volume, high efficiency; low volume, low efficiency. The problem is demand for many products is small, so not every product should be produced in high volume. Also, many companies choose to fill their production capacity with a wide variety of small-volume end items. So, must companies that produce many kinds of things, each in small volume, be content with low efficiency for all of them?

The answer lies in the way production resources are allocated to individual products. We know that for efficient production of discrete units, repetitive production is much better than job-shop production. Therefore, if items can be produced over and over again, and if, at the same time, the production process can accommodate some variety in those items, then efficient production of a variety of products will result.

To achieve repetitive production, it is necessary to focus on a few kinds of products at a time. Now, the way we incorporate a variety of things into repetitive production starts with the way we define a *kind* of product; that is, we define each kind of product so it can actually represent many products, each that is somewhat different. The point is that we start with a large number of different products, then collapse that number for purposes of production into a much smaller, more manageable number. This is the purpose of group technology.

Group Technology

Most of the waste in a job-shop environment comes from trying to use a single, all-purpose process layout to make items that each have unique components, operations, and routing sequences. In most job shops the number and variety of products and parts is large and ever expanding, and in some shops the proliferation of products and parts is epidemic.

Group technology (GT) is grounded on the premise that, given multiple means to achieve the same end, simpler is better. *Ceteris paribus*, if products can be reduced in number and made more standardized, and if the processes to make them can also be reduced in number, then production waste will be reduced and efficiency will be increased.

An important feature about GT, however, is that it does not seek to reduce variety in the kinds of products offered to customers; that kind of variety is appreciated and retained. Rather, GT seeks to reduce variety in the kinds of products produced by the manufacturer. This is achieved by identifying and exploiting similarities between different products, similarities such as physical characteristics, dimensions, geometrical shapes, and materials. Beyond appearance and composition, GT also identifies similarities in the processes and operations employed to make products. Products that look quite different might require the same or similar operations and, thus, from a GT perspective might all be the same. Product similarities are identified using coding and classification schemes.

Product Coding and Classification Schemes

Product coding refers to assigning a multidigit, alphanumeric code to a product. For example, a product might be coded as "6932AQ," where each of the digits and letters represents a feature or attribute about the product like its dimensions, materials, and machining requirements. A code like this can be used in several ways, such as to specify the design categories into which different products belong or to classify the products into groups that all use a similar manufacturing process. Three basic schemes are used to code products: hierarchical, chain, and hybrid.[2]

Hierarchical (Monocode) Structure

This code structure is interpreted as an inverted-tree hierarchy. The code for each product is a sequence of digits created by starting at the top or trunk of the tree, then moving downward through whatever branches fit the product. For example, the code for the part shown on the lower right in Figure 9.4 is 0011. It was created by moving through the hierarchy, top to bottom, and assigning digits 0 or 1 depending on the product's dimensional features: the product is cylindrical (0), has L/D ratio < 1 (0), has I/D ratio > 0.5 (1), and has inside tolerance > 0.0001 (1).

The advantage of this structure is that it is compact and assigns only the number of digits necessary to define each part. That is, the number of branches on the inverted tree differs depending on the part being coded. As a result, some parts have seven digits, others only three. However,

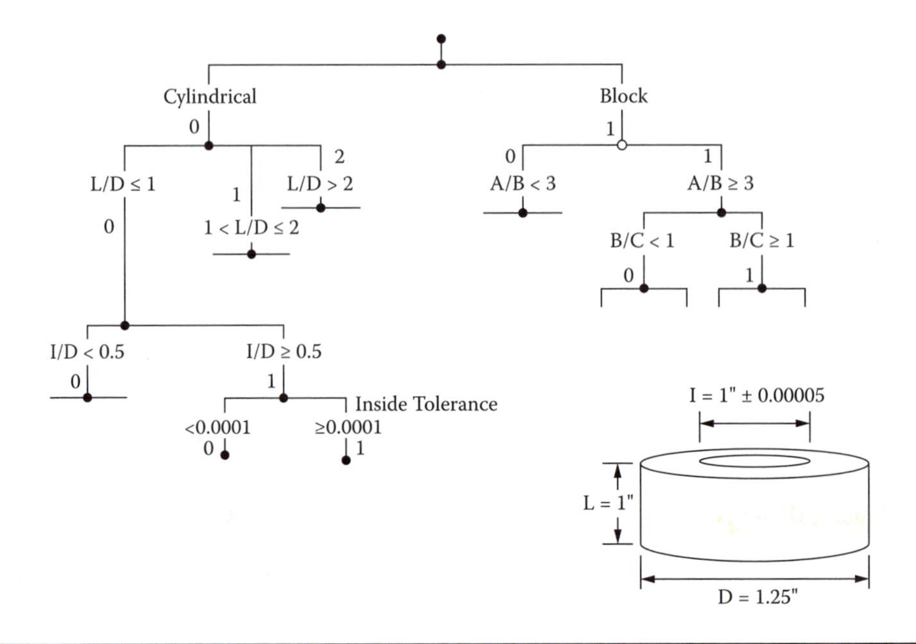

Figure 9.4 Hierarchical coding structure example. For the part shown (cylindrical, length 1″, diameter 1.25″, inside bore 1″ with 0.0001″ tolerance), code = 0011.

because of the different possible meanings for each of the digits in a sequence, the code is more difficult to interpret than the next coding scheme.

Chain (Polycode) Structure

In this structure, each digit's position in the sequence always represents the same attribute or feature of a part. Given a five-digit code, the first digit would always refer to, for example, the class of the part, the second digit would refer to the part's external shape, the third would refer to its internal shape, and so on. Each digit has values 0 to 9 or A to Z and has a particular meaning, which is maintained in a reference table. For example, part code 37357 might be interpreted from the reference table as the following:

Feature/Attribute	
Digit Position	*Interpretation from Reference Table**
1: class of part	3 = turning part with L/D > 2
2: external shape	7 = cone
3: internal machining	3 = functional groves
4: surface machining	5 = external planed surface
5: gear teeth forming	7 = with bevel gear teeth
* *Note:* Reference table not shown.	

A part code is thus created by stringing together appropriate digits as established from the reference table. Because each digit position always refers to the same part characteristic, chain codes are easy to interpret. However, chain codes are less efficient than hierarchical codes because the number of digit positions is fixed (in our earlier example, at five), even though features might be absent or meaningless for a particular kind of part. A simple cubical part that involves only one or two operations will thus have a code with the same number of digits as a geometrically complex part that requires numerous turning, planning, and drilling operations.

Hybrid Structure

This structure consists of subgroupings of digits that each follows either a monocode or polycode structure. For example, in the code 396-56098-2722 the first three digits and last four digits might use a monocode structure, while the middle five use a polycode. Hybrid codes are versatile because they can be tailored to represent both general attributes of parts as well as process-specific or company-specific features.

Product Families and Focused Factories

Products that follow similar manufacturing processes as identified by GT coding can be grouped together and classified as a **family** of products, parts, or components. All of the items in a product family are generally made from the same material, have similar overall dimensions, and require similar machines, tooling, or routing sequences. Suppose, for example, the seven-digit chain code for two products, U and V, are 3020011 and 3120001, respectively. Aside from digits in the second and sixth positions, the codes for the two products are identical. Suppose the first digit represents the kind of material; the second digit special storage conditions; the third digit the kind of machining process; the fourth, fifth, and last digits dimensional aspects of the products; and the sixth digit a final plating process. Given that both products are made of the same material, use the same machining process, and are almost dimensionally identical, they might be classified as members of a family of products with common materials and production processes.

If sufficient combined demand exists for the products in a family, then a **focused factory** can be set up to exclusively produce that family of products. In one sense, a focused factory is like a product layout: it clusters together the required equipment, tools, and workers to enable production of the product family on a repetitive basis. Unlike product layouts, however, the focused facility is capable of producing every item in the family, not just one. Making things in focused factories thus affords efficiencies and economies associated with high-volume production. High volume, however, is achieved by processing different, but similar kinds of items, even though the volume for any one product might be small. The focused factory combines the variety and flexibility of a job shop with the simplicity and efficiency of a flow line.

GT and Product Design

Though our interest here is on GT coding for the purpose of classifying products into families, we briefly note GT's role in new product design. Whenever a part is called for in a new product, the designer can access the GT database to see if a part already exists for current products that is similar to the one needed in terms of function and physical features. If so, that part can be used either as-is or with modification. With GT codes, new parts can be designed for interchangeable application in more than one product or to make use of existing processes and operations. The results are fewer unnecessary new and unique parts, more applications for new and existing parts, simpler bills of materials, and simpler production and inventory planning and control.

To Code or Not to Code

The decision about whether to code products, for whatever purpose, is not trivial since the coding process can itself be difficult and expensive. Although coding new products in a new facility is relatively easy because everything is done from scratch, collecting data, assigning codes, and forming families for existing products and facilities can be complex, frustrating, and costly. Just selecting, purchasing, and installing the coding software can be time consuming and expensive.

As an alternative to commercial software, the coding and classification procedure can be developed internally. In that case a company starts by reviewing the process route sheets of all parts it produces, revises the sheets for errors, then assigns in-house codes. Small companies can store part numbers and codes in a personal computer using database software and keep any additional information about the parts in a filing cabinet for reference. Although unsophisticated, this kind of system is often sufficient to eliminate redundancies in product design and to identify part families around which focused factories can be formed.

Focused Factory[3]

A focused factory is also called a plant within a plant or a *subplant*. Each focused factory is a portion of a plant devoted to making a product family of somewhat-similar products. Walls, temporary partitions, or lines on the floor serve as boundaries to separate subplants. Each subplant is like an autonomous factory with its own equipment and workers, which can number between 10 and several hundred (around 30 is most common). Each subplant might also have its own maintenance, purchasing, and engineering support staff nearby on the plant floor. The staff is small because much of the preventive maintenance, housekeeping, and simple setups are done by frontline workers. Oversight of a focused factory is handled in different ways: a big focused factory might have a manager, a small one a supervisor, and a very small one (three or four workers) a supervisor that also oversees other small focused factories. A focused-factory manager ordinarily reports directly to the plant manager.

Most focused factories achieve efficiency through repetitive production of a product family. However, even focused factories that produce in small- and medium-sized batches are somewhat efficient, and are certainly better in terms cost and quality than job shops. One reason is because the workers and manager in a focused factory are more focused; they have greater control over planning, scheduling, and controlling output, and over costs and quality. Small-sized focused factories (small physical size, few workers) tend to be more efficient then large ones because employees are more directly involved, know the overall process better, are more motivated, and a single supervisor or manager can oversee everything. In small subplants the physical distances are also small, which reduces waste from material handling, communication, and work coordination. It is common that within a subplant, the manager, employees, customers, and vendors work with each other.

On What to Focus

In first writing about the focused factory concept in 1974,[4] Wickham Skinner suggested organizing focused factories around individual products. Today, the organization of a focused factory depends at least as much on manufacturing processes and operations as on products. Although

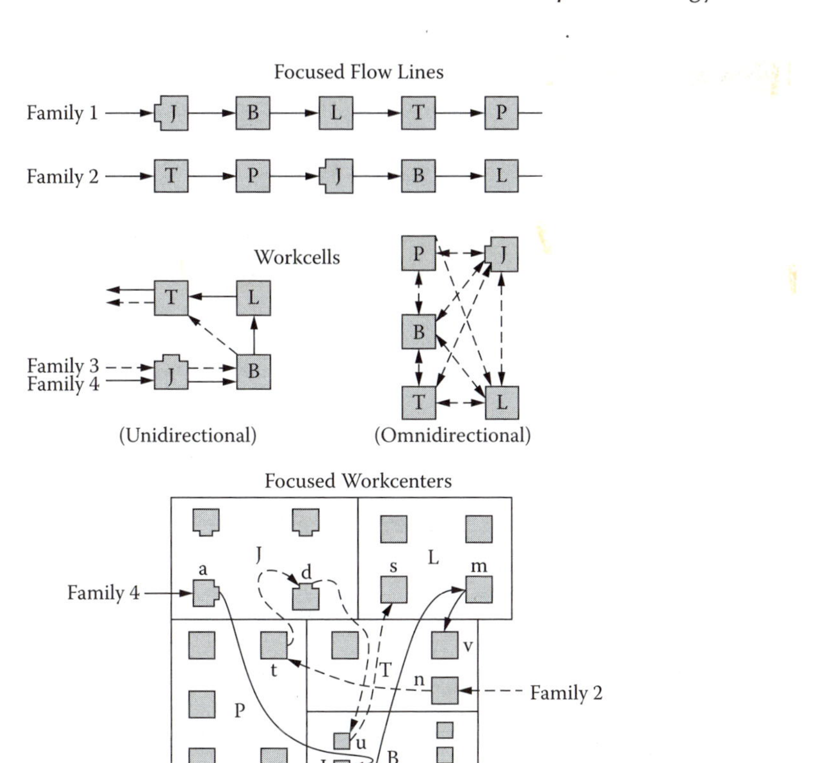

Figure 9.5 Some examples of focused facility layouts.

some modern focused factories are centered on the requirements of special customers (such as military, aerospace, or construction contractors), others are devoted to product families and commonality of production processes. The design of each focused factory includes consideration of the individual products or part families to be produced, the routing sequences they follow, and the operations and machines they require.

Focused factories take different forms, three of which are shown in Figure 9.5. A **focused flow line** is similar to a product layout, except it can produce all the parts in a part family. It is appropriate when every item in the part family follows the same sequence of operations and requires about the same processing time on each. Variety within product families is achieved by using different but similar parts and components in assemblies, quick changeover machines, and multiskilled workers at each operation. Typically a conveyor, gravity slides, or an automated transfer system connects the operations.

A **workcell** has more general application. The workcell on the left in Figure 9.5 can make products that use any of the four operations in the same general sequence. Operations not needed for a product are skipped. The workcell on the right can make products that require any of the five operations in any combination or sequence.

A workcell is actually a dedicated job shop with features of a product layout, except it is capable of economically producing items in small batches, even of size one. Ideally the routings within a workcell are unidirectional to allow parts production on a flow-line basis. If the flow is unidirectional, a conveyor or simple gravity-chute system connects the operations. For omnidirectional flow, materials moving between stations are hand or cart carried, or placed on tables in the center of the workcell or between successive operations. Typically, the machines and workstations in a workcell are arranged closely in a U-pattern so workers can quickly move among them.

A **focused workcenter** is similar to a process layout in that machines are clustered into areas by functional type, except that in a focused workcenter certain machines in each functional area are dedicated to producing only certain part families. For example, whenever a job for any product in family 4 is processed, it always goes to machines a, j, m, and v, whereas any job for a product in family 2 always goes to machines n, t, d, u, and s. This simplifies machine scheduling and setups, and improves product quality.[5] A focused workcenter arrangement makes sense when each department has multiple machines that are difficult or impossible to move, or where product volumes and product mix change so fast that it is impractical to move machines into focused flow lines or workcells.

All of these kinds of focused factories enable better production control and are less wasteful than job shops. The following examples illustrate uses of different kinds of focused factories in manufacturing and hospital settings.

Case in Point: Focused Factories at Iekian Industries

Iekian Industries manufactures two families of compressors, large and small. The small family has three products, S1, S2, and S3; the large family has four products, L1, L2, L3, and L4. Demand for the different products varies greatly between them, though for each of them it is somewhat steady. The difference in physical size between the smallest and the largest compressors (S1 and L4) is substantial. The small compressors can be moved by hand or on small carts, but the large ones require forklift trucks or overhead cranes. The manufacturing and assembly equipment also vary with the size of the product; the small compressors require relatively small equipment and the large compressors require large equipment. Aside from some electrical components that are outsourced, the two families have little parts commonality.

Given the differences between the two product families, Iekian reorganized its plant from a process job shop into two focused factories, one for each product family. Additionally, it subdivided the focused factories into focused flow lines based upon commonality of equipment, material handling, and production volume. The principal tasks on each line are main component subassembly, product final assembly, and product checkout. In the focused factory for small compressors are two lines, one for S1 and another for S2 and S3. Each is a dedicated product line. While demand for S2 and S3 was too low to justify a line for each, demand for the combination of the two justified dedicating a line to both. The new, focused layout is shown in Figure 9.6.

Using similar rationale, the focused factory for the large compressors is divided into three focused flow lines, one for L1 and L2, one for L3, and one for L4. Demand for L4 is small, so the line is frequently stopped. Since, however, the equipment to make L4 is too large for use on the other compressors, dedicating it to the L4 line does not affect its utilization. Aside from L4, all the flow lines produce repetitively.

In addition to the compressor assemblies, Iekian also produces many of the components for the compressors. Component production is done on a batch-repetitive basis. Many of the machines and processes to make components for all of the compressors are the same, regardless of compressor size. Thus, in addition to the two focused factories formed around products

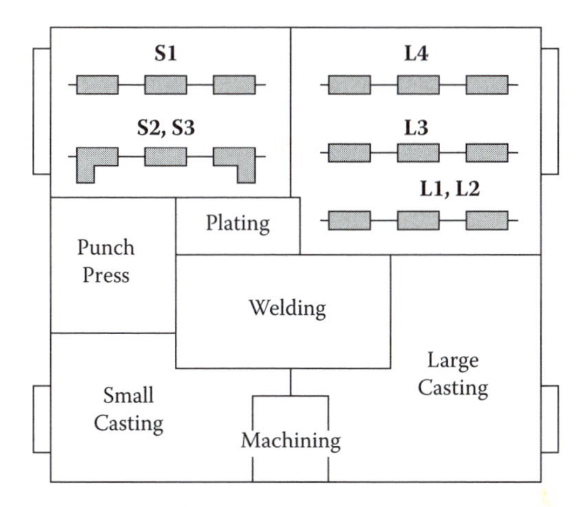

Figure 9.6 Plant layout after reorganization into focused facilities.

(one each for the S products and L lines), four other focused factories are formed around processes—punch press, welding, plating, and casting. These focused factories serve as parts suppliers to the product focused factories. Because the metal castings vary in size from small to very large, and because they require a variety of machining operations, the casting focused factory was split into three focused factories, one for small castings, one for large, and one for casting machining.

Case in Point: Hospital Focused Factories

An alternative to a mammoth all-purpose hospital is a specialized, niche-focused facility—a hospital focused factory.[6] A hospital focused factory is a facility that is dedicated to a group of patients with similar needs.

An example is Shouldice Hospital in Ontario, Canada.[7] The hospital specializes in doing one thing —abdominal hernia repair. As a consequence of specialization, it is able to perform this service for lower cost, lower relapse rates, and higher patient satisfaction than traditional hospitals. The hospital performs a much higher volume of hernia repair cases than other hospitals, which results in a higher rate of learning among devoted staff and a higher utilization of facilities.

Another example is a dedicated 31-bed trauma unit (TU) located in a large university medical center.[8] The TU has all the resources to perform immediate medical procedures, x-rays, and most lab tests. The unit is managed by a head nurse and nurse managers, and is staffed with nurses, therapists, social workers, case managers, a pharmacist, security personnel, and a cleaning crew. The unit also has many of its own physicians and receives assistance from orthopedic surgeons and medical specialists from other areas of the hospital as needed.

The TU is located in a separate wing of the hospital and includes the subareas of receiving, septic beds, intensive care unit (ICU), and step down. Prior to the TU, these areas were located in different parts of the hospital, and the staff in the different areas seldom talked to or even saw each other. Patients would have to be routed from one area to another, and the care provided among the different areas was minimally coordinated. Now, patients in the TU stay in the same general location and are treated by the same staff, even as they "move" between subareas within the TU.

A research study comparing patient care before and after the TU concluded that the focused unit somewhat reduced the patient hospital length of stay and contributed to improved hospital revenues.[9]

Decisions about which operations to include in a focused factory depend not only on demand and commonality of products and processes, but also on the required or available speed and capacity of the operations. If, for example, a machine can produce parts at a rate of 50 units per hour, but the focused factory it feeds requires parts at a rate of 10 per hour, then the machine produces at a rate 5 times the requirement. If dedicated to that focused factory, the machine would be utilized 20% of the time. Now, if no other processes make use of that machine, then low utilization is irrelevant and the machine can be assigned to the focused factory. If, however, the machine is one of a kind and other processes use it, then it should be available for use by other processes and should remain outside the focused factory.

A one-of-a-kind machine can be put in a central area, and the focused factories that need it can be located nearby. However, routing parts from multiple processes through one machine always poses potential scheduling problems. If the machine becomes a bottleneck, it will be impossible for the focused factory to autonomously schedule and control its work since the bottleneck is outside its control. Such a situation erases much of the advantage of using focused factories. If the capital expense is low relative to the benefits of dedicating machines to focused factories, more machines should be purchased so each focused factory has its own and does not have to route work outside.

Microdesign Issues

The guiding principle behind focused factories is form follows function, where function is defined by product-mix and product-volume requirements. Each focused factory should be thought of as a flexible, temporary arrangement that, if necessary, can be changed to meet new requirements. Flexibility in a factory results from flexible workers and flexible, mobile machines. Other things being equal, multiskilled workers who can be assigned to different tasks and processes, as needed, offer much more flexibility than a greater number of workers each able do but one task. Additionally, multiple machines that each perform fewer operations, but are small and can easily be moved (on casters or rollers) offer greater flexibility than a few machines that each can perform many operations, but are large and immobile. These matters are discussed in the next chapter.

Flexible Flow Lines

A traditional product layout is a highly automated transfer line or an assembly line of workers at stations connected by a conveyor system. In either case, the layout can be difficult to modify to meet changes in product mix and product volume. A flexible alternative is a series of work benches.

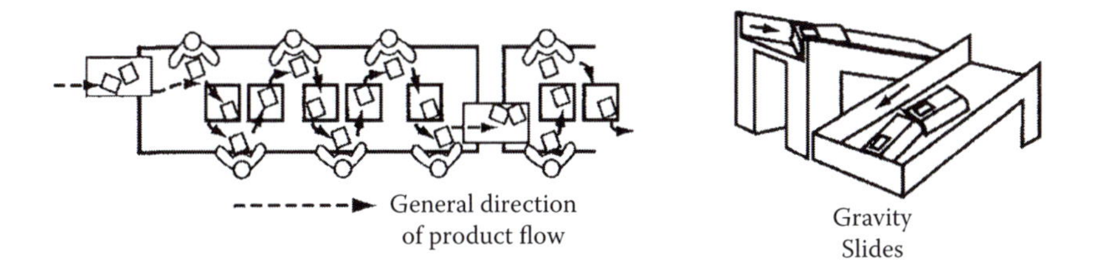

Figure 9.7 Assembly benches and gravity slides.

Schonberger suggests using benches of roughly 5 feet by 30 feet, and putting them end to end to form a product line.[10] The benches can be easily moved around, added, or deleted to suit current product requirements. Further, each bench should be a focused factory, that is, the first bench should be for making parts and components, the next bench for subassemblies, the next for final assembly, and the last benches for testing and packaging.

As shown in Figure 9.7, workers sit on both sides of the benches and are in close contact. Parts are passed between workers and between benches on simple gravity slides. Forming a focused factory at each bench encourages focused problem solving and continuous improvement, and simplifies overall line balancing. The amount of WIP is tightly controlled either by kanbans or by an amount as specified in standard operating procedures, as discussed in Chapter 11.

Flexible U-Lines and S-Lines

Product lines should be curved. For a small assembly or machining process with, say, three to six workstations, the line can be U-shaped, with workers sitting or walking around inside the U. The U-shape minimizes distances between operations and allows a few workers to quickly move between and operate several machines. The U-shape is common for workcells, as discussed in the next chapter.

For a process that has many workstations the line should be serpentine or S-shaped. This shape minimizes space requirements and the average distance between workers. Putting people into a smaller space encourages face-to-face interaction and teamwork, and putting materials in a smaller space puts attention on raw-material and in-process inventory as well on finding ways to minimize it. The relative location of workers, and the actual distances between them, depends on ergonomics and such things as ease of workers reaching for parts and passing them to the next station.[11] Although, in fact, each worker might not be any closer to adjacent workers than on a straight line, on a curve line each worker is closer to all other workers on average.

Flexibility in a manual assembly process is easily achieved by changing the length of the line (adding or deleting benches) and by flexing the line (changing the arc radius of the curves). This is shown in Figure 9.8.[12]

Working Out the Final Layout

The actual position of machines and workstations in a focused factory is often determined by trial and error. Workers should be able to contribute ideas to new layout proposals and should have final say in new layout proposals. One way is to put the floor plan on a table in the shop cafeteria. Pieces of paper cut to the shape (footprint) of different machines are located on the plan according to the proposed arrangement. Workers are encouraged to scrutinize the plan, move the pieces of paper around, and try better arrangements. Because workers are often more knowledgeable about

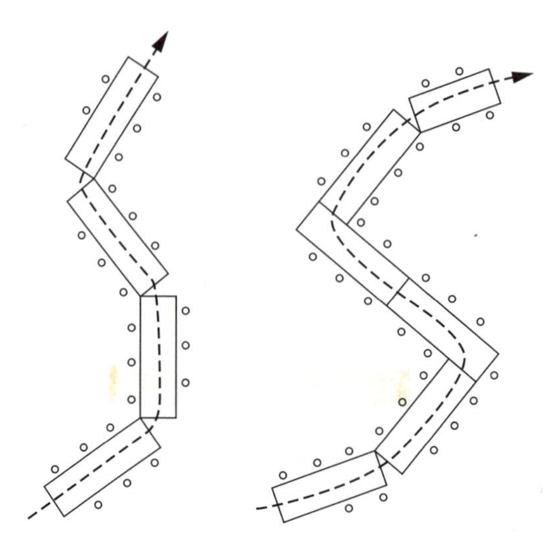

Figure 9.8 Flexing an assembly line.

the size, weight, shape, and other features of machines than are managers and consultants, they have good arguments about why their plan is better than the one proposed.

Drawings, computer layout analysis, and planning on paper are a good start, but they can never give the full picture (you can never tell about roominess or traffic patterns in a plant by looking at blueprints; you have to be *in* the plant, with machines, and walk around). Thus, the next step is to actually move machines and workstations to the proposed location, and then move them again if necessary until everyone is satisfied. That is sometimes easier said than done, especially when the machines are big and bulky.

Case in Point: Changing Layout at Hamilton Standard[13]

Hamilton Standard's Windsor Locks plant used a paper doll layout approach for planning facility changes. Since many of the machines are enormous, moving them in trial-and-error fashion is impossible, and workflow analysis on paper is inaccurate. So to analyze different layouts, cardboard pieces were cut out with the same floor dimensions as the machines. The pieces were then arranged and rearranged in an empty area of the parking lot until people were satisfied with the layout and walk patterns. The machines in the plant were then relocated according to the new pattern. Consequently, required floor space was reduced by 75% and required travel distance between machines was reduced by 85%.

Product-Quantity Analysis

As with other manufacturing decisions, the product variety–volume tradeoff is a key element in determining which products to produce in a focused factory.

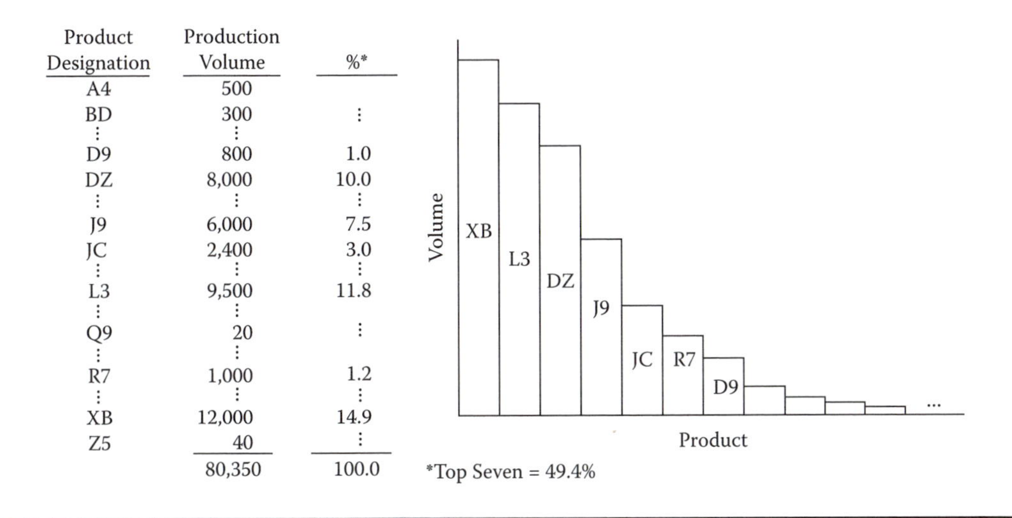

Product Designation	Production Volume	%*
A4	500	
BD	300	⋮
⋮	⋮	
D9	800	1.0
DZ	8,000	10.0
⋮	⋮	⋮
J9	6,000	7.5
JC	2,400	3.0
⋮	⋮	⋮
L3	9,500	11.8
⋮	⋮	
Q9	20	⋮
⋮	⋮	
R7	1,000	1.2
⋮	⋮	⋮
XB	12,000	14.9
Z5	40	⋮
	80,350	100.0

Figure 9.9 Pareto diagram of product volume.

One principle for establishing focused factories is that the volume of the production output of each focused factory must justify the cost of dedicating workers and equipment to producing that output. The decision about what to produce in a focused factory, then, must consider product volume.

As an example, suppose a company makes 800 products, and Pareto analysis indicates that of those, 7 account for 49.4% of factory output (Figure 9.9). If one or more of these seven products could be produced in one or more focused factories, not only would they be produced more efficiently, but the reduced plantwide workload and inventory would also improve scheduling and production of the other 793 products.

While the emphasis is first on the highest-volume products, the volume of one or a few products alone might not be sufficient to justify a dedicated focused factory. In that case emphasis switches to a focused factory that can produce multiple products and, in particular, products that constitute a product family. This is where the concept of grouping products according to commonality of operations comes into play in forming focused factories.

Establishing Product and Machine Groups

Forming a focused factory involves two things: (1) forming a cluster or group of products or parts that are similar in terms of processing requirements (the product family) and (2) forming a cluster or group of operations, machines, workers, and tools (the focused factory to produce that product family). From here on the term *group* will refer to the combination of products and the machines/operations that produce them. The ideal situation is where groups are completely independent, which means that each focused factory contains all the operations required to produce a complete product family. In forming focused factories the goal is to maximize the degree of independence of groups, although complete independence between groups is often difficult to achieve. Two kinds of approaches are used for configuring groups: (1) parts coding and classification and (2) cluster analysis.[14]

Coding and Classification

When GT product codes include information about the production process, they can be used to classify products into groups. Classifying products based upon production information results in the formation of part families by common processes, though classification is only the first step since it does not assign machines, much less give the final, physical layout of the factory. While a preexisting GT product code database can be used to create product groups, an organization that does not have such a database should consider using simpler and less costly methods to form focused factories, such as those described next.

Cluster Analysis

Like coding and classification, **cluster analysis** looks for similarities in product features and production requirements to form homogeneous groups of products. Unlike coding and classification, cluster analysis does not require codes and uses information that is readily available. Since cluster analysis methods often rely on shop personnel's knowledge about how different products are made, information needed for the analysis is relatively simple and inexpensive to obtain. Based on their experience in more than 400 factories, Harmon and Peterson state that coding and classification is impractical and unnecessary for establishing focused factories,[15] and they suggest instead starting with information about product routings through operations. One plant where they used this approach was able to determine the product–machine groups to form workcells and specific layouts for the workcells in only a few days. The same plant had earlier employed a consultant who spent 6 months coding and analyzing parts, only to conclude that the parts were too dissimilar for focused factories.

As illustrated next, when the number of products and machines is not too large, cluster analysis can be performed by just looking at the process routings.

Example 1: Group Formation by Visual Process-Routing Analysis

The group formation procedure should begin by focusing on products with the highest volume. Suppose a product-quantity analysis gives the results as shown in Figure 9.9, and the resulting seven largest-volume products are selected for process-routing analysis. Table 9.1 shows the process routings (the sequences of machine operations) required for each product.

Table 9.1 Product Routing Sequences through Machines

Product	*Volume*	*Process Routing (Machine Sequence)*
XB	12,000	A→B→C
L3	9,500	C→B→A
DZ	8,000	E→D
J9	6,000	D→E
JC	2,400	A→B→C
R7	1,000	C→A
D9	800	C→D→E

Table 9.2 Routing Sequences, Reorganized to Show Product-Machine Clusters

Product	*Volume*	*Process Routing (Machine Sequence)*	
XB	12,000	A→B→C	
JC	2,400	A→B→C	
L3	9,500	A←B←C	
R7	1,000	A←C	
D9	800		C→D→E
J9	6,000		D→E
DZ	8,000		D←E
		Workcell 1	Workcell 2

Products with similar operations are clustered together. The example includes only five operations, so it is easy to cluster products according to similar operations and similar sequences. The result, Table 9.2, shows two natural groups: the first includes machines A, B, and C, and produces products XB, JC, L3, and R7; the second includes machines D and E, and makes products D9, J9, and DZ. Each group is a candidate to be a workcell. Except for product D9, each of the products falls into either of two groups and could be wholly produced by a single cell. The situation with D9 is common since products often involve operations that do not entirely fit into only one workcell. Perhaps for D9, operation C could be performed in cell 1 on a batch basis, then the batch moved to cell 2 for machining on D and E.

This example considers only the seven largest-volume products, though ordinarily the routing analysis would include as many products as practical. Although products might each have small volume, when grouped into part families their collective volume might justify a focused factory.

Visual methods are practical if the number of products and machines under consideration is somewhat small. When the volume of the few highest-volume products is insufficient to justify a focused factory, then products with successively lower volume in the Pareto ordering are included. Focused factories commonly produce a large number of similar products where the demand for each product is somewhat small.

Production Flow Analysis

Production flow analysis (PFA)[16] is another methodology for forming groups of part families and machines. The term PFA has also come to refer to many subtechniques for simplifying material flow systems at any level of an organization. Like the visual method, PFA techniques use information about process operations to simultaneously form product and machine groupings; the techniques, however, are more rigorous than the visual method.

One group of such techniques performs *factory flow analysis* that divides a factory into major, largely independent focused factories, where each factory completes all the parts it makes. These techniques also connect the focused factories such that products flow in a simple, unidirectional fashion throughout the entire plant. Another group of PFA techniques performs *group analysis,* which divides a department into groups or workcells that are largely independent and responsible for performing most or all the operations to make a product family. Other PFA techniques address

	Product						
Machine	*XB*	*JC*	*L3*	*R7*	*D9*	*J9*	*DZ*
A	1	1	1	1			
B	1	1	1				
C	1	1	1	1	1		
D					1	1	1
E					1	1	1

Figure 9.10 Process matrix for example in Table 9.2.

other levels of the production system. From here on we will focus on techniques primarily for group analysis and forming independent workcells.

PFA techniques usually start with product–machine information as represented on a product–process matrix, hereafter termed the **process matrix**. Figure 9.10 is the process matrix corresponding to Table 9.2; it shows the machines required for each product, though not necessarily the sequence. Each column in the process matrix represents a product or part; each row a machine or operation. A "1" means that the part visits (is processed by) the machine. In Figure 9.10, product XB visits machines A, B, and C; product D9 visits C, D, and E; product DZ visits D and E; and so on.

Notice in Figure 9.10 that the 1s fall into two dense blocks (clusters) along the main diagonal. In general, dense blocks of 1s suggest natural groupings of products and machines. The dense blocks in Figure 9.10 suggest the obvious: form one workcell with machines A, B, and C to make products XB, JC, L3, and R7; form the other with the remaining machines to make the remaining products. Only product D9 is considered an exceptional product because it requires operations in more than one group. (Machine C could be considered an exceptional machine because it is needed by products in more than one group.) Many of the PFA techniques use the process matrix to form workcells. Following is an example.

Binary Ordering Algorithm

In terms of the process matrix, assignment of products and machines to groups happens by rearranging 1s so they lie in diagonal, dense-block form. The **binary ordering algorithm** (BOA) is a procedure for transforming any N-column, M-row, binary (0–1) matrix into this form. BOA rearranges rows so that similar rows are adjacent, and then does the same for the columns. Consider as an example the process matrix in Figure 9.11 (the term *part* is used instead of *product*, though the terms are interchangeable, depending on whether the end item is a component part or completed product).

Example 2: Binary Ordering Algorithm Procedure

The procedure works as follows:

1. Assign a value to each column k, where the value is 2^{N-k}. In Figure 9.11, N = 9, k = 1, 2, 3, ..., 9. These values are shown at the bottom of the matrix in Figure 9.12.

Initial Machine	*Binary Ordering Algorithm* *Part* 1	2	3	4	5	6	7	8	9
A					1	1		1	
B					1	1		1	1
C		1		1	1		1		
D	1		1						
E					1	1			
F	1		1	1					
G				1	1	1		1	1
H		1		1			1		

Figure 9.11 Initial process matrix.

2. For each row, obtain the sum by adding the 2^{N-k} values wherever a 1 appears in that row. These sums are shown at the far right in Figure 9.12. For example, for machine A, a 1 appears in columns 5, 6, and 8 (circled in Figure 9.12), so the row Sum = 16 + 8 + 2 = 26.
3. Rearrange rows in decreasing order of row Sum—F, D, C, H, etc. Figure 9.13 shows the new row ordering.
4. Assign a value to each row k, where the value is 2^{M-k}. These values are shown on the far right in Figure 9.13; there, M = 8 and k = 1, 2, 3, ..., 8.
5. For each column obtain a sum by adding the 2^{M-k} values wherever a 1 appears in that column. These column Sums are shown on the bottom of the matrix in Figure 9.13. For part 1, for example, a 1 appears in rows F and D (circled in Figure 9.13), so the column Sum = 128 + 64 = 192.
6. Reorder the columns in decreasing order of column sums. The new column ordering is 1, 3, 4, 2, etc. This is shown in Figure 9.14, the final matrix.

Machine	Part 1	2	3	4	5	6	7	8	9	Sum
A					①	①		①		㉖
B					1	1		1	1	27
C		1		1	1		1			180
D	1		1							320
E					1	1				24
F	1		1	1						352
G				1	1	1		1	1	59
H		1		1			1			164
2^{N-k}	256	128	64	32	⑥	⑧	4	②	1	

Figure 9.12 Process matrix, steps 1 and 2.

Machine	Part 1	2	3	4	5	6	7	8	9	$_2$M-k
F	(1)		1	1						(128) +
D	(1)		1							(64) =
C		1		1	1		1			32
H		1		1			1			16
G				1	1	1		1	1	8
B					1	1		1	1	4
A					1	1		1		2
E					1	1				1
Sum	(192)	48	192	184	47	15	48	14	12	

Figure 9.13 Process matrix, steps 3, 4, and 5.

PFA techniques are more efficient than GT coding for forming workcells because they identify product families solely by processing requirements, not by appearance or physical features. PFA techniques are more versatile than the visual method because they permit analysis of larger numbers of products and machines, and allow for other considerations such as constraints on group size and machine availability. PFA methods rely on existing information about processes and equipment, and so they can be used to analyze alternative facility layouts with minimal expense.

A drawback of PFA techniques is that they do not guarantee answers that can be implemented. Because the PFA algorithm ignores aspects of manufacturing such as product demand, operations sequence, number of each kind of machine, availability of machines, machine utilization, transport costs of groupings, problems of exceptional machines and products, and many others, the groups are at best provisional, and final decisions about grouping and layout require considerable additional analysis. Many algorithms have been developed that address additional, important features of manufacturing, although they too have so many other variables that their results require additional analysis and fudging to arrive at final product–machine arrangements.[17]

Machine	*Part* *1*	*3*	*4*	*2*	*7*	*5*	*6*	*8*	*9*
F	1	1	1						
D	1	1							
C			1	1	1	1			
H			1	1	1				
G			1			1	1	1	1
B						1	1	1	1
A						1	1	1	
E						1	1		

Figure 9.14 Final matrix; diagonal, dense block form.

Dense Blocks, Then What?

Three dense blocks are very apparent in Figure 9.14. These indicate three possible (obvious) workcells to which parts and machines could be assigned:

Cell 1: Machines F and D; parts 1 and 3
Cell 2: Machines C and H; parts 2 and 7
Cell 3: Machines G, B, A, E; parts 6, 8, 9

Parts 4 and 5 are exceptions: Part 4 straddles all three blocks, and Part 5 straddles the second and third. There are several possible ways to handle these: (1) redesign Part 4 so it does not need machines F and G, and redesign Part 5 so it does not need a machine C; (2) add machines F and G to the second cell so Part 4 can be entirely processed there, and machine C to the third cell so Part 5 can be entirely processed there; (3) combine all three groups into one group; (4) combine any two groups into one group, and add whatever machines are still needed to the remaining group; or (5) route part 4 through all three groups, and part 5 through machine groups 2 and 3. Of course, the viability of any of these alternatives is open to question and depends on the availability of machines from elsewhere or capital funds to procure machines.

Demand, sequence of operations, transportation distance and cost, and setups are also factors that the initial assignment ignores. Consider the assignment shown in Figure 9.15 and the question where to put machine C. If demand for part 3 is greater than for part 4, put it in workcell 1. If the cost of transporting part 4 is much greater than part 2, then put C in workcell 2. Suppose the parts routing is A-C-B for part 3, and C-D-E for part 4; then put C in workcell 1 so part 3 does not have to backtrack (go from cell 1 to cell 2, then back to cell 1). A 1 in the process matrix indicates only that a machine is used to make the product, not how many times it is needed. Suppose the machine sequence for part 3 is A-B-C, and for part 4 it is D-C-E-C. In that case machine C should probably be put into workcell 2, otherwise part 4 would have to backtrack (cell 2 to cell 1, then back to cell 2).

Machine setup time can also influence the product–machine assignment. If the setup procedure is lengthy, then it is desirable to cluster products with machines that require similar or identical setups. The result might be several workcells that contain identical machines, but where, within each cell, the machines are all set up to process a particular product group.

		Part				
	Machine	*1*	*2*	*3*	*4*	*5*
Workcell 1	A	1	1	1		
	B	1	1	1		
	C			1		
Workcell 2	D				1	1
	E				1	1
					1	

Figure 9.15 A part–machine grouping.

	Part								
Machine	*1*	*3*	*4*	*2*	*7*	*5*	*6*	*8*	*9*
F	1	1	1						
D	1	1							
C			1	1	1	1			
H			1	1	1				
G			1			1	1	1	1
B						1	1	1	1
A						1	1	1	
E						1	1		

Figure 9.16 Process matrix; dense block form.

Machines arranged for unidirectional flow, either as focused flow lines or unidirectional workcells (shown in Figure 9.5), give the most efficient processing. Referring to Figure 9.16, a workcell with four sequential operations G-B-A-E could process part 6 efficiently if the part had that sequence, though it could not process part 5 efficiently if that part had the sequence A-B-C-E-G. However, by skipping operation E that same cell could efficiently process part 8 if the part had the sequence G-B-A. In general, the more a focused factory is designed for unidirectional, repetitive production, the more efficiently the items that follow the flow can be processed. As with product layouts, however, the routing sequence restricts the number of products the focused factory will be capable of processing. Even within a focused factory, the tradeoff between production efficiency and product variety remains.

The factory arrangement finally chosen might consist of a variety of focused-factory configurations. Referring again to Figure 9.16, suppose analysis of demand, routing sequences, and so forth leads to the following layout (where FF refers to focused factory):

FF1, Focused workcenter: F, D, E
FF2, Unidirectional workcell: C-H
FF3, Focused flow line: G-B-A

With this assignment, some parts will be processed entirely within a single focused factory (parts 8 and 9 in FF3, parts 1 and 3 in FF1, and parts 2 and 7 in FF2), some by being routed through two focused factories (part 4 in FF3 and FF2; part 6 in FF1 and FF3), and part 5 by being routed thorough FF2, FF3, then FF1.

It might appear from the discussion that a large degree of subjectivity is involved in establishing final product–machine groupings. The degree of subjectivity is lessened with more sophisticated PFA techniques (beyond the scope of our presentation), however all PFA techniques are limited in the scope of the things they consider. Design of focused factories in plants that have many tens of operations and similar numbers of products is complex and time consuming, and there are literally millions of possible clustering combinations. For these, computer simulation and what-if analysis should be used to narrow the possibilities and evaluate the tradeoffs.

Throughout this chapter the terms *operation* and *machine* have been used interchangeably. In many cases, the operation will not involve a machine, either fully or partially, but will be a manual procedure such as a filing, inspection, or assembly. The following chapter distinguishes between manual and machine operations in focused factories, and the differences between an assembly workcell (wherein all operations might be manual) versus a machining workcell (all operations might be automated). Focused factory design and implementation includes organizational and human behavior issues, addressed in the next chapter.

Advantages and Disadvantages of Focused Factories

Focused factories offer advantages over job shops and flow shops. Compared to product layouts, focused factories require less capital investment and are significantly more flexible. Compared to the process layouts, focused factories are significantly less wasteful. According to Steudle and Desruelle, the following results were achieved:[18]

- 70%–90% reductions in lead times and WIP inventories
- 75%–90% reductions in material handling
- 20%–45% reductions in required factory floor space
- 65%–80% reductions of overall machine setup time
- 50%–85% decreases in quality-related problems
- Simpler shop-floor control procedures and reduced paperwork
- Greater worker involvement, productivity, flexibility, satisfaction
- Reduced overhead costs (cross-trained workers perform tasks formerly done by support staff).

Similar improvements were reported by Ewaldz:[19]

- 95% reduction in WIP
- Reduced response time to customer demand from months to hours
- Increased ownership of tasks and functions by operators
- Work imbalances and bottlenecks better handled by workers

Reductions in lead times, WIP inventory, material handling, and quality problems happen because production in focused facilities is more similar to continuous/repetitive production than to process, job-shop production. Any well-researched move from a job-shop to a product-focused approach will bring such benefits. Required space is reduced because machines are relocated closely together into groups, and where possible, the groups themselves are located closely together. Overall time devoted to setups is reduced because parts in product groups require similar kinds of setups, so both the frequency of setups and the extent of each setup are reduced. Special fixtures and tools can be developed to speed the setup process, and dies and anything else needed can be stored at or near the focused factory. As will be discussed more in the next chapter, workers take on greater responsibility; at minimum, each worker is capable of operating several or all of the machines in the focused factory. Each worker might also be responsible for machine setups, routine machine PM, quality control, and job scheduling. This requires considerable training, but the result is less boredom and, possibly, greater satisfaction.

Case in Point: Lockheed Aeronautical Systems

Lockheed Aeronautical Systems' plant in Marietta, Georgia, is structured into 13 focused factories, each a manufacturing cell devoted to similarly processed parts such as extrusions, machinings, sheet metal parts, tubing, and composite parts. The focused factories were implemented and refined over a 3-year period. Since completion of the implementation, throughput time for parts has improved by a minimum factor of 2.5. For one part in particular, small extrusions, throughput time fell from 65 days to 11 days, and in-process inventory from 6,000 units to 500 units.

Case in Point: Sun Microsystems

Sun Microsystems' plant in Milpitas, California, manufactures desktop computer workstations.[20] When the production became incapable of meeting swings in demand, Sun set up a pilot line in a vacant area of the plant. On the pilot line, called SimplePlant, one worker can assemble a complete workstation in less than 15 minutes. Unlike the original facility, which was a highly automated assembly line, most of the tasks in SimplePlant are manual, including transfer of components and products with inexpensive handcarts.

Morale improved because workers in SimplePlant control the pace of the process and are involved in all or most of the steps, not just one like before. As a result, idle time and work stoppages were virtually eliminated and productivity increased by 12%. After 3 months, Sun decided to push the limit and require SimplePlant to double its output, which it did with no problem. To everyone's amazement, SimplePlant was producing state-of-the-art workstations at high productivity and quality levels, yet with only simple hand-me-down workbenches and carts to transfer parts and products.

Based on the results of the pilot, the factory was restructured into three discrete, identical cells. Whereas the former automated line had to be run at between 20% and 100% of capacity to meet changes in demand, the three cells can be run in any combination, each at 100% capacity. Also the assembly and package areas of each cell have mirror-image sides that can be turned on or off, which allows for six possible levels of production.

One disadvantage of focused factories is the potential for low machine utilization since any machine dedicated to a focused factory is delimited in its use. The greater the constraint on how a machine is to be used, the greater the potential that it will be used less. Thus, product–machine grouping may require the purchase of more machines, even though some of the machines might be utilized very little. But, again, this might be of small consequence if the aforementioned advantages are realized.

Plants that shift from a process layout to focused factories always lose flexibility. Sure, they are more flexible than flow-line layouts and less wasteful than job shops, but each focused factory's effectiveness depends on its having adequate customer demand for the product group it produces. As that demand decreases or shifts, the focused factory might no longer have sufficient demand to justify its existence. This is an argument for employing multiskilled workers and equipment and machinery that is mobile and can readily be relocated to form new groups.

Despite reports about workers being happier in workcells, most of them are anecdotal, and one of the few comprehensive studies to research the subject came up with mixed results.[21] In that study, attitudes of cell workers and functional workers were compared in two plants that had been using workcells for over 2 years. While the cell workers showed greater satisfaction about some aspects of their jobs, they showed less satisfaction about others and expressed feelings of greater role conflict and role ambiguity, and less commitment to the company than workers in traditional functional jobs.

From a management perspective, a big disadvantage of focused factories is that they require considerable effort and expense associated with designing and implementing them. Simply moving equipment into groups is not enough. Workers must be cross-trained and taught to work in teams, individual machine setup times must be reduced, and equipment reliability increased. Also, reassigning workers and equipment is disruptive to ongoing operations. Suffice it to say, implementing focused factories is a long-term commitment and a big change to the way many organizations are accustomed to operating.

Summary

Process design and facility layout are the major determinants in a manufacturer's ability to produce a wide variety of products in any volume—high, medium, or low. Job shops accommodate large variety but are inherently wasteful. Repetitive operations are significantly better in terms of manufacturing speed and cost, though their use is traditionally restricted to relatively high-volume production of very similar or identical products. To achieve repetitiveness, however, it is not necessary to repeatedly produce the same product over and over; it is only necessary to repeat the same process. This is true even with products that are physically different or that follow different routing sequences.

Group technology is a way to identify physical, functional, and processing similarities between different products. GT coding is a tool for describing properties of parts and products, and for classifying parts and products by commonality of function, appearance, or production. Knowing the physical or functional properties of existing parts enables better utilization of those parts in new products, and reduces the proliferation of new parts designs. Group technology is also a tool for grouping products according to common production processes.

A focused factory is a portion of a larger factory devoted to a particular kind of product or group of products. The focused factory achieves efficiency by exploiting commonalities among the end items it produces. In the focused factory, setups and changeovers are reduced because of similarities in products, and workers operate with high proficiency because they direct attention to fewer things. Handling and transportation wastes are reduced because workers and machines are clustered closely together. Many focused factories are able to operate on a repetitive basis by producing items that share the same process or operations, even though the items are physically different and each have small demand. For these reasons, focused factories are less wasteful then job shops.

The next chapter focuses on one kind of focused factory, the workcell. Workcells have been growing in popularity and are especially common among lean manufacturers.

Notes

1. See G. Boothroyd, W. Knight, and P. Dewhurst. 2002. *Product Design for Manufacture & Assembly Revised & Expanded*, 2nd ed. (Boca Raton, FL: CRC Press, 2002); D. Anderson, *Design for Manufacturability & Concurrent Engineering; How to Design for Low Cost, Design in High Quality, Design for Lean Manufacture, and Design Quickly for Fast Production* (Cambria, CA: CIM Press, 2004); C. Poli, *Design for Manufacturing: A Structured Approach* (Amsterdam: Butterworth-Heinemann, 2001).
2. See, for example, H. Opitz, *A Coding System to Describe Workpieces* (New York: Pergamon Press, 1970); and U. Rembold, C. Blume, and R. Dilman, *Computer Integrated Manufacturing Technology Systems* (New York: Marcel Dekker, 1985).
3. Perhaps the best reference on the subject is R. Harmon and L. Peterson, *Reinventing the Factory* (New York: The Free Press, 1990), 12–35.
4. W. Skinner, The focused factory. *The Harvard Business Review* (May–June 1974): 113–121.
5. Because every machine is unique, using different machines to perform a given operation causes machine-specific variations in the product. When a product family is consistently assigned to the same machines, each machine can be set up and adjusted to make allowances for its uniqueness, which results in reduced variation.
6. G. Leung, Hospitals must become "focused factories," *BMJ* 320 (April 2000): 942.
7. R. Herzlinger, *Market-Driven Health Care: Who Wins, Who Loses in the Transformation of America's Largest Service Industry.* (New York: Basic Books, 1999).
8. N. Hyer, U. Wemmerlov, and J. Morris, Performance analysis of a focused hospital unit: the case of an integrated trauma center, *Journal of Operations Management* 27(2009): 203–219.
9. Ibid., 210–217.
10. R. Schonberger, *World Class Manufacturing: The Next Decade* (New York: The Free Press, 1996), 158.
11. D. Alexander, *The Practice and Management of Industrial Ergonomics* (Englewood Cliffs, NJ: Prentice-Hall, 1986); M. Sanders and E. McCormick, *Human Factors in Engineering and Design* (New York: McGraw-Hill, 1987).
12. Schonberger, *World Class Manufacturing: The Next Decade,* 159–160.
13. G. Galsworth and L. Tonkin, Invasion of the Kaizen Blitzers, *Target* 11, no. 2 (March/April 1995): 30–36.
14. One study of workcell design identified six categories of ways to form workcells: GT coding and classification, clustering methods based on machine or product similarity, visual examination, key machine, production flow analysis, and naturally-apparent product lines. The discussion in this chapter is of the most commonly used of these methods. See F. Olorunniwo and G. Udo, Cell design practices in U.S. manufacturing firms, *Production and Inventory Management* (Third Quarter, 1996): 27–33.
15. Harmon and Peterson, *Reinventing the Factory*, 124.
16. J. Burbidge, *The Introduction of Group Technology* (New York: John Wiley & Sons, 1975).
17. For a review of clustering algorithms, see C. Cheng, A. Kumar, and J. Motwani, A comparative examination of selected cellular manufacturing clustering algorithms, *International Journal of Operations Management* 15, no. 12 (1995): 86–97; A. McAuley, Machine grouping for efficient production, *Production Engineering* 2 (1972): 53–57; P. Waghodekar and S. Sahu, Machine-component cell formation in group technology: MACE, *International Journal of Production Research* 22, no. 6 (1984): 937–948; A. Ballakur and H Steudel, A within-cell utilization heuristic for designing cellular manufacturing systems. *International Journal of Production Research* 25, no. 4 (1987): 639–665.
18. H. Steudel and P. Desruelle, *Manufacturing in the Nineties* (New York: Van Nostrand Reinhold, 1992), 119–123.
19. D. Ewaldz, Caveats for cellular manufacturing, *Tooling and Production* 61, no. 7 (October 1995): 9.
20. K. Laughlin, Increasing competitiveness with a cellular process, *Industrial Engineering* (April 1995): 30–33.
21. S. Shafer, B. Tepper, J. Meredith, and R. Marsh, Comparing the effects of cellular and functional manufacturing on employees' perceptions and attitudes, *Journal of Operations Management* 12 (1995): 63–74.

Suggested Reading

J. Burbidge. Production Flow Analysis for Planning Group Technology. New York: Oxford University Press, 1997.

Questions

1. Define, compare, and contrast job operations and repetitive operations.
2. Compare the features, advantages, and disadvantages of product layouts and process layouts.
3. For what manufacturing situations are process layouts appropriate? For what situations is a product layout appropriate?
4. What is a product family?
5. What is a focused factory? Describe each of the following kinds of focused factories: focused flow line, workcell, focused workcenter.
6. How does a focused factory combine the features of a product layout and a process layout?
7. What is group technology? What is its purpose? How it is used in product design? How is it used to form product families?
8. Compare and contrast GT polycode, monocode, and hybrid code structures.
9. What are the drawbacks of using GT codes for forming focused factories?
10. What is production flow analysis (PFA)? Why use PFA instead of simple visual techniques?
11. What are the limitations of results from PFA analysis?
12. Discuss the potential major advantages and disadvantages of focused factories.

PROBLEMS

1. Four components—W, X, Y, and Z—are classified using an 11-digit code. Each digit of the code refers to a particular production operation, material, and general physical feature of the component. The coding numbers are as follows:

	Code Number of Operations				*Code Number of Materials*				*Code Number of Features*		
Digit Component	*1*	*2*	*3*	*4*	*5*	*6*	*7*	*8*	*9*	*10*	*11*
W	2	2	2	2	1	1	3	1	1	2	2
X	1	1	2	1	1	2	3	1	1	3	4
Y	2	1	2	2	1	1	3	1	1	1	2
Z	1	2	2	1	2	3	3	2	1	3	2

Cluster components:

a. According to most-shared operations.
b. According to most-shared materials.
c. According to most-shared features.
d. For focused-factory production. Explain your rationale.

2. Shown in the following table are seven manufactured components and their corresponding production volumes and machine sequences.

Component Part	*Volume*	*Machine Sequence*
L7	50,000	A, B, C, D
R6	30,000	S, M
P1	9,000	C, D
O3	10,000	S, M, R
F7	1,000	S, R, A
J4	15,000	B, A, D
HS	1,000	B, D, M

Assume the plant has one of each kind of machine.

a. Form workcell groupings using the visual method (i.e., determine the machine–component clusters). Discuss the result and how you would handle exceptional parts or machines.
b. Create the process matrix. Form workcell groupings using the binary ordering algorithm. Discuss the result.

3. Shown in the following table are six manufactured components and their corresponding production volumes and machine sequences.

Component Part	*Volume*	*Machine Sequence*
PAC	10,000	Q, S, T
QBB	100	X, Q, S
RCC	100	X, M, Q, S
PDL	1,000	X, M
JGR	5,000	Q, T
CHB	5,000	T

Assume the plant has one of each kind of machine.

a. Form workcell groupings using the visual method. Discuss the result.
b. Create the process matrix. Form workcell groupings using the binary ordering algorithm. Discuss the result.
c. Assume two Q machines, two S machines, and one of every other machine. Discuss the resulting workcell groupings.
d. Besides the information given, what other factors would you need to know to determine the appropriate workcell groupings.

4. Ten manufactured products and the corresponding weekly production volumes and machine sequence are shown next.

Products	*Volume*	*Machine Sequence*
SA	9,000	A, B, C
SB	8,000	B, C, E
SC	80	E, F
SD	6,600	C, D, E
SE	40	A, B
SF	600	A, E, F
SG	10,000	A, B, D
SH	6,000	D, E, F
SK	1,000	B, C, D
SL	2,100	A, B, D

Assume one of each kind of machine.

a. Form workcell groupings using the visual method (i.e., determine the machine–component groupings). Discuss the result.
b. Create the process matrix. Form workcell groupings using the binary ordering algorithm. Discuss the result.
c. Assume all products require about the same production time per unit on all machines. Suppose each machine has a maximum weekly output capacity of 30,000 units. Does this influence the workcell grouping?
d. Suppose each machine has a maximum weekly output capacity of 20,000 units, and you have a capital budget to acquire at most three additional machines. Which machines would you acquire, and how would you allocate them among workcell groupings?

Chapter 10

Workcells and Cellular Manufacturing

One hand washeth the other.

—Seneca

When love and skill work together, expect a masterpiece.

—John Ruskin

Focused factories are a common sight in lean production because they are more efficient than job shops and more flexible than flow shops. They are also common in pull production processes because pull production requires repetitiveness, and, often, focused factories allow for repetitive production, even of low-volume products. Probably the most common kind of focused factory is a *workcell*. The concept of performing all of the operations necessary to make a part, component, or finished product in a workcell is called **cellular manufacturing**.

Cellular manufacturing and pull production go hand in hand. Often, stringing together many workcells can create an entire large-scale pull production process. The workcells produce parts, components, and subassemblies that are assembled into the final product at the last stage of the process. Because each workcell can produce a variety of parts and components, the overall production system is capable of producing a variety of products.

This chapter begins where the last chapter left off—with the assumption that product or part families exist, and that the combined volume of products in the family justifies dedicating machines and workers to focused workcells. It looks in depth at cellular manufacturing applications, the design of workcells, and how workcells function individually and as elements of pull production systems. Worker staffing, equipment availability, and organizational and behavioral issues of workcell implementation are also addressed.[1]

Workcell Concepts

Workstations, Workers, and Machines

The basic building blocks of a workcell are workstations, workers, machines, and means for holding and transferring items between workstations. Workstations (or **stations**) are the places where operations are performed, and in a workcell they are located close together, ideally in the routing sequence for a product or product family. The tasks in a workcell might be performed entirely by workers. Manual assembly of components is an example: the role of workers in such a cell is to perform assembly tasks, inspect items, and transfer them to the next station.

Machines at some or all stations might also perform the tasks in a workcell. The role of workers in that case is to set up and monitor machines, turn machines on and off, load and unload parts, inspect parts, and transfer parts between the machines. The functions of loading, unloading, inspecting, and transferring parts might also be automated so that the workcell operates with little human intervention.

Some workcells can be run with as few as one worker. Figure 10.1 shows a workcell with six machines and one worker who walks between them. The machines are arranged in the process routing sequence so that parts can be produced one at a time, continually. Parts are transferred between stations by hand or using gravity chutes or small carts. The stations are arranged in a U-shaped layout so the worker can move quickly around the cell. Obviously, the worker must know how to operate all the machines to keep everything in the cell running smoothly.

Workcell Output and Number of Workers

The output rate of a workcell can often be manipulated by changing the number of workers. For example, Figure 10.2 shows two possibilities of the six-machine cell in Figure 10.1 but with two workers. In Figure 10.2a each worker moves among three machines and has responsibility for half of the cell. If the output rate of the cell is a function of the time it takes the workers to do their tasks, then adding a second worker like this will roughly double the production rate of the workcell. This will be explained later. Figure 10.2b shows another way of adding a second worker, called **rabbit chase** (named after a popular children's game in Japan) where both workers move around the entire cell, one leading (or chasing) the other. This, too, will be discussed later.

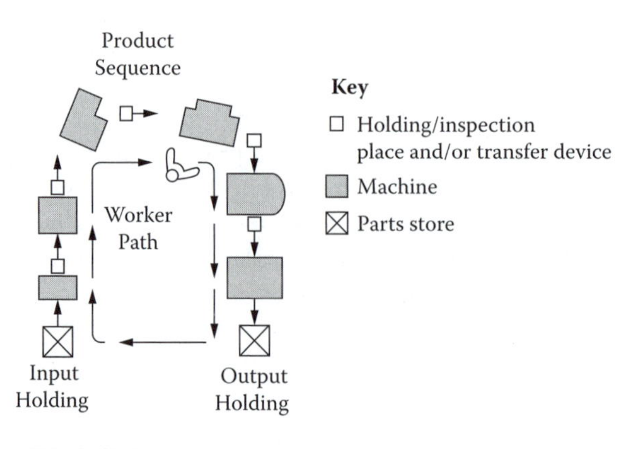

Figure 10.1 One-worker, six-machine cell.

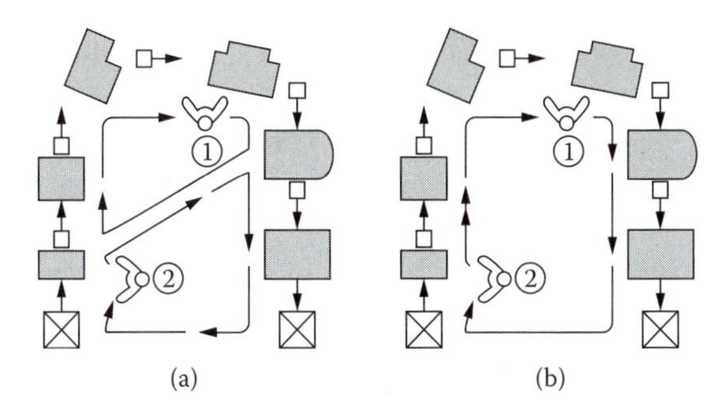

Figure 10.2 Cell with two workers: (a) divided sub-cells, (b) rabbit chase.

Workcell Applications

Typical Workcell End Items

Often an entire product can be produced in a single workcell, though such a product is usually somewhat simple in terms of the number of components and operations it requires. It is typically a one- or few-pieced item such as a metal casting on which a series of drilling, boring, and finishing operations is performed, or a simple assembly of components, such as a DVD drive. In the service sector, workcells process end items such as mortgage applications and insurance claims.

Workcells can also be used to manufacture products that are more complex and involve numerous operations such as small electronic devices and communications equipment. Producing somewhat complex products in a cell is practical as long as the skills and abilities of workers fit the range of tasks required. Many workers welcome the opportunity to learn and apply a broadened range of skills, and larger cells that produce more complex items offer that opportunity.

The physical size of the cell necessary to encompass all of the operations involved is also a determining factor when considering complex end items. The number of workstations in the cell and the distances between them must be small enough so workers do not become overwhelmed by the number of tasks or waste excessive time walking among stations. Further, a larger workcell might require many machines, which takes more machines away from non-cell-produced parts and products. Also, in larger cells with more workers, teamwork suffers.[2] For problem solving, work coordination, and cohesiveness, a group size of five to seven people is optimal. Workcells with two to six workstations are very common; those with ten or more are less common. For all these reasons, workcell production of complex components and assemblies in their entirety are less common.

Though a workcell usually focuses on producing one product family, it can be designed to produce multiple product families when the families require similar operations and sequences. Figure 10.3 shows a workcell for three product families that all follow the same general routing sequence, although some families skip some operations. To help workers avoid confusion about which machines to use for each family, lights are mounted on the machines. To make a product in the X family, for example, the worker presses a button at the master panel for X, which illuminates lights on machines A, C, E, and F.

Linked Workcells and Subcells

Products that involve numerous machining and assembly steps and that are made from parts that also require numerous steps can also be made in workcells, but by first dividing all the

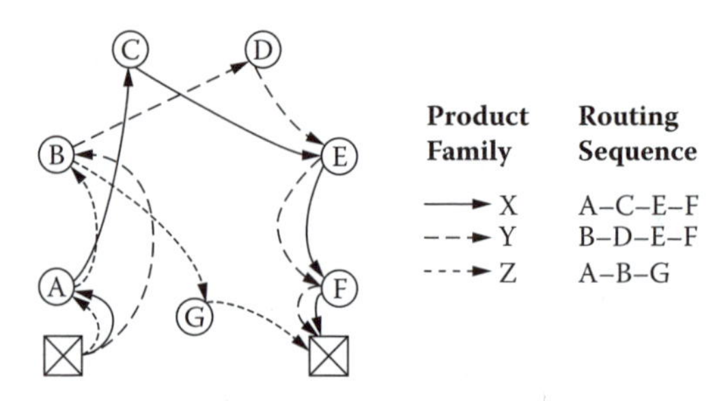

Figure 10.3 Cell for producing multiple product families.

operations into several workcells, then linking the workcells so parts flow from one cell to the next in a coordinated manner.[3] Figure 10.4 shows an example of several workcells (the small U's) linked in sequence to produce families of parts that are produced on two subassembly lines and a final, main assembly line.

Workcells can be linked piece by piece with conveyors or mechanical feeders, in which case the material flow between workcells is somewhat continuous, or they can be linked by material handlers, in which case the flow is intermittent. In the latter case, the transfer of material is authorized by kanbans (in a pull system) or schedules (in a push system).

When workcells are located immediately adjacent to one another, they can perform as if they are subunits of a larger workcell. For example, Figure 10.5 shows two kinds of linked workcell processes: (a) a cluster of four clearly distinguishable workcells and (b) four subcells that function together as one.

Parts in cluster Figure 10.5a move between cells in small lots using pull signals. As cell D produces items, it withdraws parts from containers at its inbound stock area. When the parts in the stock area reach a certain level, a replenishment order is sent to the feeding cell, cell C. Some stock of finished parts is held in containers at each cell's outbound stock area, which enables orders from downstream cells to be filled immediately. The stock also serves as a safety stock for minor delays and breakdowns in the process. As containers are withdrawn to replenish the inbound stock at cell D, the outbound buffer stock of cell C is reduced. When it reaches some minimum level, that is a signal for cell C to replenish those items in its outbound stock up to some maximum level. Cell C must then withdraw items from its own inbound stock area, and when that stock reaches some

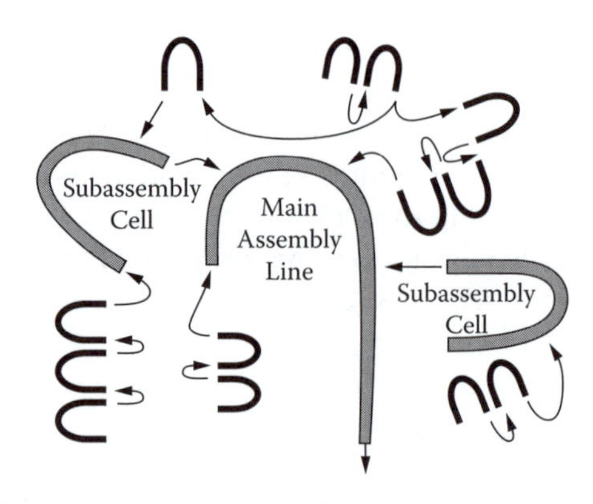

Figure 10.4 Facility of linked workcells.

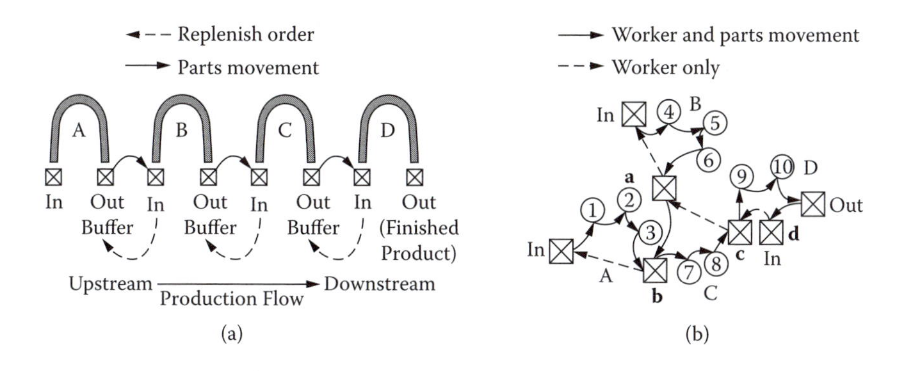

Figure 10.5 Linked cell and subcell examples.

level, a replenishment order for additional parts is sent to its feeding cell, cell B. This procedure repeats at cell B and then at cell A.

This kind of linked-cell system is a classic pull production system (but using container stock levels instead of cards as the reorder function). Each cell only produces enough to bring its outbound stock quantity up to the maximum allowable number of containers.

As in all sequential processes, the output rate of the entire chain of cells depends on the output rate of the slowest, or bottleneck, workcell. Thus, whenever the output of the process must be increased, adjustments to the process begin with the slowest workcell. Ideally, the output rates of all the cells are balanced.

In the process in Figure 10.5a, each workcell has only one upstream cell feeding it. A more complex process is where each workcell has multiple upstream workcells feeding it, in which case the entire process looks like a tree diagram, with final assembly at the main trunk. Moving upstream, each cell branches outward into multiple feeding cells, and they, in turn, branch outward into still more cells. Although the process might appear more complicated than the single-sequence case, the dynamics are the same: the workcells are linked with a pull system and the production rate of the overall system is dictated by the slowest workcell. (In actuality, the production rate of the system is set in advance, and that rate determines the production rates of all of the workcells.) The rates of the workcells are adjusted by altering the number of workers or the length of the workday, as described later. Details of this are addressed in Chapter 14.

Now look at the process in Figure 10.5b, which can be viewed alternatively as four separate but linked workcells (denoted A, B, C, D), or as one big workcell with four subcells. The process works almost the same as the one in Figure 10.5a except the subgroups are located closer to one another, and instead of containers in the stock areas between the subgroups are kanban squares or racks (denoted as **a, b, c, d**) at which workers withdraw or deposit parts. Rather than sending replenishment orders, workers simply look at the number of items in each square, and when the quantity gets low enough, the upstream station produces a quantity to replenish it. Each subgroup can receive inputs from multiple feeding subgroups (subgroup C receives items from subgroups A and B, picked up at squares a and b, respectively; subgroup D receives items from subgroup C and elsewhere, picked up at squares c and d). The production rates of each subgroup are balanced so that supply rates and demand rates between subcells are roughly equal.

Aside from the way replenishment orders are placed, and the way materials are held and transported, this arrangement of subcells operates in the same manner as the linked workcells in Figure 10.5a. Conceptually and analytically, the two can be treated as identical, except that in the linked-workcell case, Figure 10.5a, allowance must be made for time delays in communicating orders and transporting materials between workcells. In the subcell case, communication and transportation times are negligible.

All of the workcells in a linked process must produce at a rate that enables the overall process to satisfy demand for the final end item. If the production schedule for the final end item specifies one unit every minute, then that is the minimum allowable rate for every workcell. When the required rate is greater than the maximum rate capable for a given workcell, then additional workcells for making the same part must be added to the process. Suppose a workcell is already producing parts at its maximum rate, say 1 unit per minute, and the process requires that part at a rate of 1.5 units per minute. It will be necessary to add another workcell to the process with an average rate of 0.5 unit per minute (or 1 unit every 2 minutes) such that the two workcells combined can meet the demand rate of 1.5.

For pull production, material flowing to and between workcells everywhere in the process must be synchronized and coordinated. Everything discussed in Chapters 5 to 8 now comes into play. Since delays or shutdowns anywhere will affect the entire process, setup time must be short, equipment cannot break down, and materials ordered from suppliers must arrive on time. The methods for placing orders and transporting materials, the size of containers, and the frequency of transport all require special attention.

Balancing the system so every cell is capable of producing at the required rate is a dynamic procedure. For every change in the final assembly schedule, corresponding changes are required in the production rates of the workcells.

Workcell Design

There are two fundamental kinds of workcells, assembly cells and machining cells. In **assembly cells** the work tasks are entirely or mostly manual. Usually, the tasks performed in these cells are difficult or costly to automate, for example, hand assembly, welding, and testing of multiple components. Assembly workcells produce components or completed products such as electronics (DVDs, keyboards, circuit boards, phones), furniture (wooden and metal chairs, desks, tables, cabinets), toys (wagons, bikes, remote-controlled models), electric motors, hand power tools, small appliances for home and industry, and many other products. In contrast to assembly cells, in **machining cells** the work tasks are usually simpler, more easily automated, and largely or entirely performed by machines. Machining cells usually produce single-piece items that require no (or little) manual assembly. The workcell process involves a series of machining operations on a piece of metal, wood, plastic, or other material. Because both kinds of cells produce items piece by piece, cycle time, discussed next, is a crucial design parameter.

Brief Digression: Cycle Time Concept

The central concept in workcell facility design is that of **cycle time.** Cycle time is the time between when units are completed in a process.[4] If, for example, the cycle time for a workcell is 10 minutes per unit, that implies that the cell produces a completed unit once every 10 minutes. Expressing production in terms like this—time per unit—is important because it automatically gets people to think in terms of piece-by-piece product flow.[5] The cycle time concept is also central to pull production because it implies a repetitive, smooth, and steady material flow throughout a process.

Cycle time can also be thought of as the inverse of **production rate**, since, for example, saying that a process has a cycle time of 10 minutes is almost the same as saying it has a production rate of six products per hour. The difference is that, again, cycle time implies smooth, steady flow whereas production rate does not.

In the design of a production process, we distinguish between the required cycle time and the actual cycle time: **Required cycle time**, or **takt time**,[6] is the production target rate of a process or operation. It is based on the demand for the item being produced. If the demand is 80 units per day, that translates into a required cycle time of

$$CT_r = \text{Time Available} \div \text{Demand} = 480 \text{ min/day} \div 80 \text{ units/day} = 6 \text{ min/unit}$$

To satisfy demand, the workcell, process, or operation must be designed so that its actual cycle time does not exceed 6 minutes per unit.

Actual cycle time represents the actual production capability of a process or operation. In a workcell, it is determined by physical conditions in the cell such as actual time to perform manual or automatic operations, to walk around the cell, to fetch materials, and so on, as will be discussed shortly.

Establishing and standardizing work in a process such that the actual cycle time is as close as possible to the required cycle time—an important aspect of lean production—is discussed in Chapter 11.

Assembly Workcells

Figure 10.6 is an example of an assembly workcell. Depending on the required cycle time, it can be operated by as few as one worker or as many as eight workers.

Figure 10.6 shows work-in-process (WIP) holding spaces between stations. When a workcell has more than one worker, each workstation has a container or holding space in which to put items just completed.[7] As a way to control WIP within the cell, each holding space has room for only one or a few items. When the space is full, no more items can be added and work stops at the preceding workstation. In effect, items are pulled through the workcell since each station produces only enough items to replenish those withdrawn by the succeeding workstation.

The large X boxes in Figure 10.6 represent inbound and outbound stock areas for incoming parts and outgoing end items. Also shown are four small x boxes that represent areas for stocking additional parts and components at their points of use. Each of these stock areas is replenished using kanbans or schedules, depending on whether the process is part of a pull system or push system.

The actual cycle time of an assembly cell is entirely a function of the cell **manual time**, which is the time required for workers to perform their tasks and move between workstations. If only

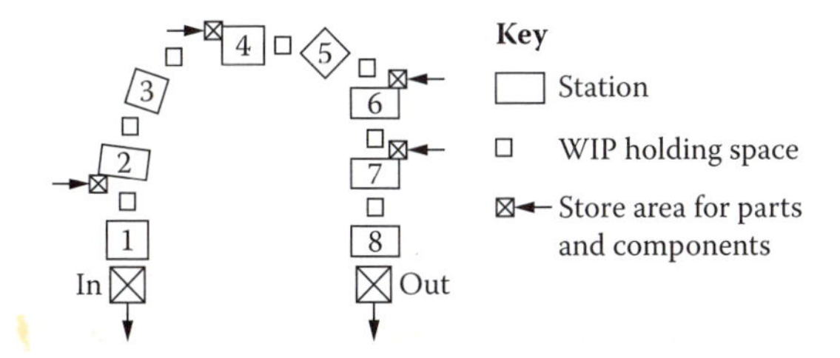

Figure 10.6 Assembly cell example.

one person operates the cell, the actual cycle time, CT_a, is the sum of the operation times at every station and the walk times between them:

$$Cell\ CT_a = \Sigma \text{ operation times} + \Sigma \text{ walk times}$$

Given the actual cycle time, the capacity of the workcell is then

$$\text{Cell capacity} = \text{Time available/Cell } CT_a$$

This is illustrated next.

Example 1: Workcell CT and Capacity

Suppose the cell in Figure 10.6 is operated by one worker who walks from station to station. Figure 10.7 shows the worker's route around the cell and the relevant times: the number next to each station is the time required for the worker to perform the operation; the number by each arrow is the time to walk between locations (including time to pick and place items at the locations).

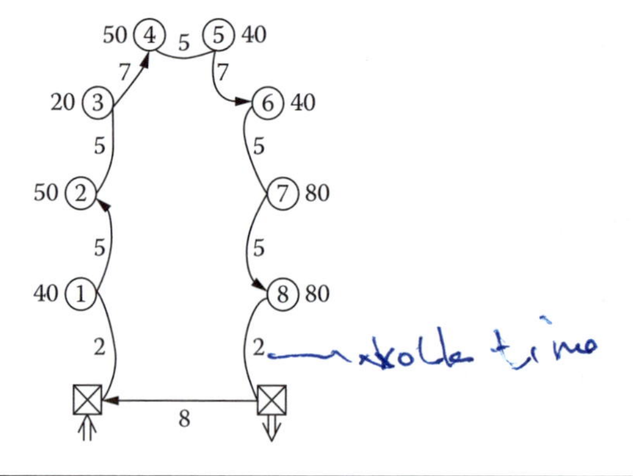

Figure 10.7 One-worker assembly cell.

From earlier, the actual cycle time is

$$\text{Cell } CT_a = 400 \text{ sec} + 51 \text{ sec} = 451 \text{ sec/unit}$$

Then, assuming an 8-hour workday,

$$\text{Cell capacity} = (8 \text{ hr} \times 60 \text{ min} \times 60 \text{ sec}) \div 451 \text{ sec/unit} = 63.9 \text{ units/day}$$

With only one worker, the cell's production capacity is at its minimum. To increase the capacity of the cell, subdivide the cell into subcells and add more workers. Each worker walks around and tends exclusively to his own subcell. If each subcell were independent of the others, then its CT would be determined solely by the equation to compute cell CT_a. The subcells, however, are not independent since each gives to or picks up material from others. Given that the CT of the subcells are not all equal, subcells with shorter CT must wait on the subcells with longer CT. As a result, the cycle time of the entire workcell is determined by CT of the subcell that takes the longest, or

$$\text{Cell } CT_a = \max \text{ (CT of each subcell)}$$

Each time another worker is added to the cell, the cell CT gets smaller. The smallest CT happens when there is a worker at every station. At that point, workers no longer walk between stations, and the cell capacity is at its maximum. The following example illustrates these concepts.

Example 2: Reducing Assembly Workcell CT by Adding Workers

Suppose the workcell in Figure 10.7 is divided between two workers as shown in Figure 10.8: the first worker is assigned to stations 1, 7, and 8 (and the inbound and outbound holding spaces); the second worker to stations 2 through 6. Note in Figure 10.8 the addition of spaces **a** and **b** as buffers to hold items from each worker until the other worker can get to them.

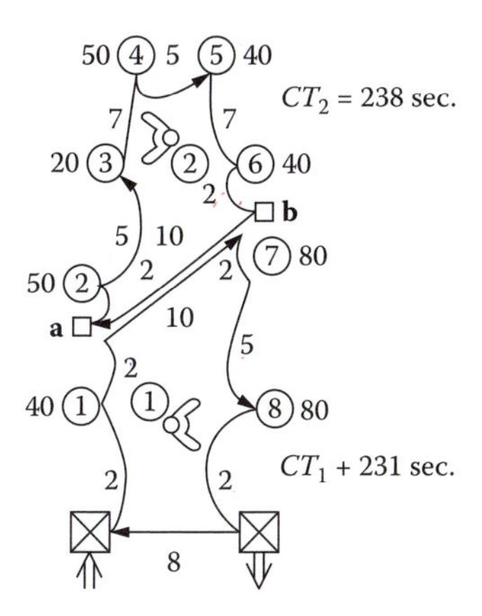

Figure 10.8 Two-worker assembly cell.

The sum of operation and walk times for the first worker is 231 sec/unit; for the second worker it is 238 sec/unit. Thus the CT for the workcell is 238 sec/unit.

The workcell performs like this: After completing operation 1, the first worker drops off an item at holding space **a,** then proceeds to holding space **b** to pick up an item for processing at stations 7 and 8. Although the first worker has a shorter CT than the second worker, she cannot proceed to stations 7 and 8 until an item is dropped off at **b**. Since an item is dropped off at **b** every 238 seconds, by default, 238 seconds becomes the CT for the first worker, and, thus, the CT_a for the workcell.

As another example, the cell is subdivided among five workers as shown in Figure 10.9. The subdivision was determined by trial and error while attempting to balance the CT of workers.

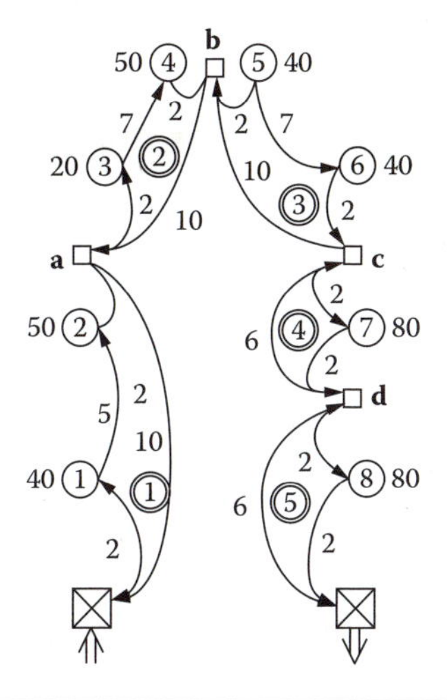

Figure 10.9 Five-worker cell.

Notice the addition of holding spaces at four locations (**a–d**) where the workers interface. Summing the operations and walk times for each worker, the CT are as follows:

Worker	*Subcell Stations*	*CT (sec)*
1	1 and 2	109
2	3 and 4	91
3	5 and 6	101
4	7	90
5	8	90

The cell CT_a is 109 sec/unit; the cell capacity is

$$(8 \text{ hr} \times 60 \text{ min} \times 60 \text{ sec}) \div 109 \text{ sec/unit} = 264 \text{ units/day}$$

Notice that because of the imbalance of CT among the workers, all workers except the slowest have to wait. Since worker 1 is the slowest, the total idle time for the other four workers is the difference between their CT and 109 seconds, or 18 + 8 + 19 + 19 = 64 seconds per unit.

As a final example, suppose every workstation is assigned a worker. Walk times are eliminated, but suppose picking up and placing items at each station takes 4 seconds. The cell CT_a will be the longest workstation time, which is at stations 7 and 8:

$$CT_a = 80 \text{ sec} + 4 \text{ sec} = 84 \text{ sec/unit}$$

The resulting cell capacity is 342 units/day.

Notice again the amount of worker idle time in this arrangement. Only workers at stations 7 and 8 require the full 84 seconds for each unit, while all the others require considerably less. Station 3, for example, needs only 20 + 4 = 24 sec per unit; the remaining 60 sec per unit is idle time.

An alternative to adding workers and subdividing them among subcells is to add workers using the **rabbit chase** scheme shown in Figure 10.2b. Every worker carries, slides, and carts the work piece or item from station to station with him. There is no need to balance subcell CT (since there are no subcells, just one big cell) and no need for special WIP buffer spaces. With rabbit chase, the cycle time of the workcell is thus

$$\text{Cell } CT_a = \text{Cell CT, one worker} \div \text{Number of workers}$$

where *Cell CT, one worker* is the CT for the slowest of the workers. (Rarely do workers perform at exactly the same pace, and in rabbit chase the workers end up having to follow the slowest worker around the cell.) For example, using the five workers in rabbit chase as an alternative to the subcell scheme in Figure 10.9, and assuming 451 seconds per unit is the CT for the slowest worker,

$$\text{Cell } CT_a = 451 \text{ sec/unit/worker} \div 5 \text{ workers} = 90.2 \text{ sec/unit}$$

With rabbit chase there are no inequities since everyone works the same amount of time for each unit produced. One might think that continuously walking around the cell would fatigue workers, and fatigue is a problem when the CTs are somewhat short (less than 5 or 10 minutes) and tasks are simple. When CTs are longer and tasks more challenging, rabbit chase can actually increase worker alertness and productivity above stationary workers doing only one task. The drawback of rabbit chase in such cases is that every worker must be skilled at performing tasks at every station in the workcell.

Machining Workcells

Machining workcells differ from assembly workcells in several ways. First, machines do virtually all operations, with one or a few machines located at every workstation. These machines are often **automatic, single-cycle** machines that stop after the machining operation has been completed. Additionally, stations and machines are connected to one another using a variety of devices called **decouplers** because they allow machines in a sequence to operate somewhat independently. Like a holding space or container between workstations in an assembly cell, each decoupler holds one or a few parts to enable the workcell to continue operations even though subcells or workstations are not perfectly balanced.

Decouplers between operations can serve additional functions:[8]

- WIP control: The preceding machine is automatically stopped when the number of units in the decoupler reaches the maximum (one or a few parts).
- Transportation: Decouplers automatically transfer parts from operation to operation; examples include gravity chutes, slides, or mechanical conveyors.
- Worker freedom of movement: Because of WIP control and transportation, workers (or robots in an automated cell) can move in any direction around the cell, even counter to the product-flow direction.
- Automatic inspection: Mechanical or electronic sensors on decouplers perform inspection of critical dimensions as parts move from one operation to the next; in cells that produce multiple kinds of parts, sensors check features of a part to determine to which of several possible downstream machines the part should be routed and to what parameters the next machining operation should be set.
- Part manipulation: A decoupler reorients the part so it is ready for insertion into the next machine.
- Leap-frog or skip operations: A decoupler identifies different kinds of parts and allows downstream operations to be selectively bypassed.
- Converging or branching: A decoupler enables multiple machines to feed into a single machine or a single machine to branch into multiple machines.

Because little or no assembly work is done in a machining cell, all material used in the cell arrives at the inbound stock area. Upon completion of the machining sequence, the finished parts or products enter the outbound stock area.

The actual CT of a machining cell is based upon the CT of the machines in the cell and the CT of workers. Assume that each station in the cell has a single-cycle automatic machine, that is, a machine that automatically stops after its operation has been completed. In that case, the **machine CT** is the time per unit to set up the machine (unload, change over, and load machine) and for the machine to perform its operation. The **worker CT** is the time for the worker to complete a trip around the cell. Specifically, it is the time for the worker to unload, change over, load, and start every machine (the task times) plus the times for the worker to walk between all the stations:

$$\text{Worker CT} = \Sigma \text{ Task times} + \Sigma \text{ Walk times}$$

Now, in a one-person machining cell, one of two things will happen as the worker walks around the cell: either the worker arrives at a machine before the machine has finished its operation, or the machine finishes its operation before the worker arrives. Thus, the cell CT for a machining cell, CT_m, depends on whichever takes longer—the worker CT or the CT of the machine in the cell that takes the longest:

$$\text{Cell CT, one worker} = \max(\text{Worker CT, longest machine CT})$$

Just as with an assembly cell, to decrease the CT of a machining cell, add more workers and divide the cell into subcells. Subcells with shorter CTs must wait on subcells with longer CTs; thus, the cycle time of the entire cell, CT_m, depends on the subcell with the longest CT:

$$CT_m = \max(\text{Subcell } CT_m\text{'s})$$

The following example illustrates this concept.

Example 3: Reducing Machining Cell CT by Adding Workers

Start with the one-worker machining cell in Figure 10.10. The number between adjacent stations is the worker walk time. Assume the setup task time (manually load, unload, start machine) is 10 seconds per machine per unit. Thus,

$$\text{Worker CT} = 8(10 \text{ sec}) + 51 \text{ sec} = 131 \text{ sec/unit}$$

In Figure 10.10 the number next to each station is the automatic CT—the time for the machine to automatically perform one cycle, then stop. Assume the machine CT is the automatic CT plus the 10 second setup task time per unit. The longest machine CT is at station 7 or 8,

$$\text{Longest machine CT} = 70 \text{ sec/unit} + 10 \text{ sec/unit} = 80 \text{ sec/unit.}$$

Therefore,

$$CT_m = \text{Max}(131, 80) = 131 \text{ sec/unit}$$

Assuming an 8-hour workday, the cell capacity is

$$8 \text{ hr} \times 60 \text{ min} \times 60 \text{ sec} \div 131 \text{ sec/unit} = 219.8 \text{ units/day}$$

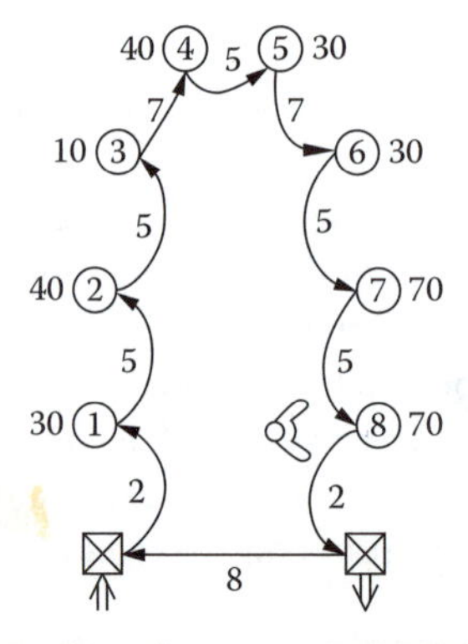

Figure 10.10 One-worker machining cell.

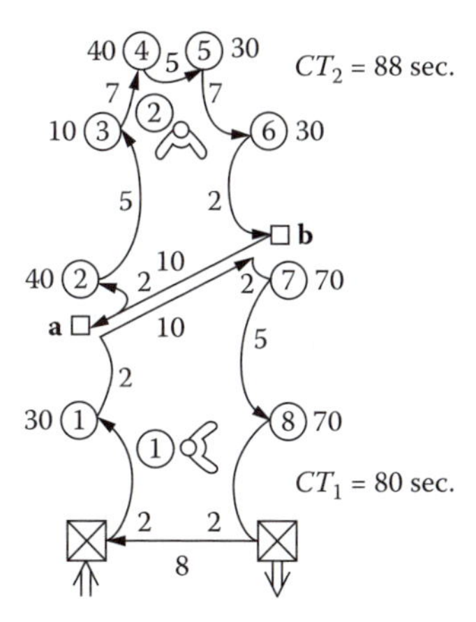

Figure 10.11 Two-worker machining cell.

To decrease the actual CT, suppose we add one more worker and divide the cell as shown in Figure 10.11. Thus,

Subcell 1: Worker CT = 3(10 sec) + 31 sec = 61 sec/unit

Subcell 2: Worker CT = 5(10 sec) + 38 sec = 88 sec/unit

Since the longest machine CT is still 80 sec/unit,

CT_m = max (Longest worker CT, longest machine CT)

= max (88, 80) = 88 sec/unit.

The workcell capacity is now (8 × 60 × 60)/88 = 327 units/day.

Keep adding workers and eventually the longest worker CT will be less than the longest machine CT. When that happens, the cell CT becomes 80 sec/unit, the longest machine CT. Suppose that the required cell CT (takt) must be less than 80 sec/unit. If the existing workcell grouping must be retained (i.e., existing product family and machine combination) then the only way to reduce the cell CT below 80 seconds is to add another machine at the bottleneck station.[9]

In the example, an additional machine must be put at stations 7 and 8; the CT at those stations would then effectively be reduced to 40 sec per unit (80/2). Worker 1 operates machines 8A and 8B on an alternating basis, and worker 2 does the same on machines 7A and 7B. So now, the theoretical longest machine CT is 50 sec/unit at stations 2 and 4. Suppose we have three workers and subdivide the cell as shown in Figure 10.12. In that case,

Subcell 1: Worker CT = 2(10 sec) + 30 sec = 50 sec/unit

Subcell 2: Worker CT = 3(10 sec) + 32 sec = 62 sec/unit

Subcell 3: Worker CT = 3(10 sec) + 26 sec = 56 sec/unit

Since the longest worker CT, 62 sec/unit, exceeds the theoretical longest machine CT, 40 seconds, the cell CT becomes 62 sec/unit. The cell capacity becomes 464.5 units/day.

Besides adding workers, cell CT can be reduced by lowering the manual task times. In the examples the task setup time for each machine was assumed to be 10 seconds. Shaving a second or two off the setup at each machine would reduce the cell CT by many seconds.

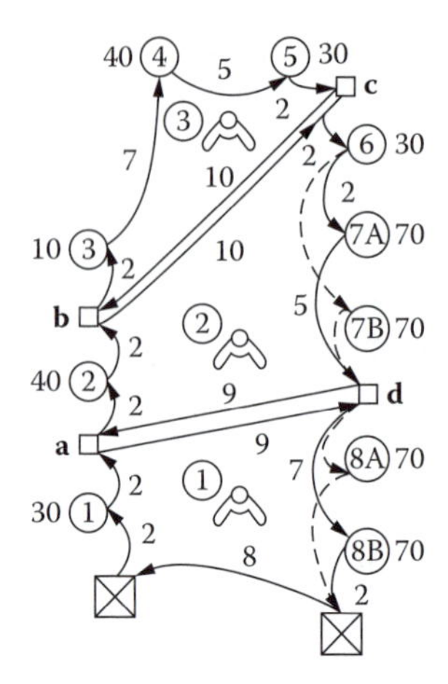

Figure 10.12 Three-worker machining cell.

One benefit of machining workcells is that as long as the machine CTs are less than worker CTs, the workcell output depends not on machine speed, but on the number of workers, which can be easily altered. In the example, the big disparity in machine CT (20 sec/unit at machine 3 versus 80 sec/unit at machines 7 and 8) would have no effect on cell capacity as long as the cell required CT was less than 80 sec/unit.

Workcell Capacity

One objective in the design of workcells is to determine the number of workers necessary to meet the required CT and, hence, achieve the desired output capacity. To avoid overproduction, desired output capacity should be based upon demand. However, since demand varies, design alternatives should be considered. Design issues include not only the number of workers but also the number of machines in the workcell, procurement of new machines, cross-training workers, preparation of backup cells, and expansion of existing cells.

To account for variable demand, at least two scenarios should be considered: most likely and maximum likely. This requires producing two cell designs, one to handle the expected or most likely demand, the other the largest realistic demand, then determining whether there is a practical and efficient way to alternate between the designs using different numbers of workers and machines. Besides machines and workers, requirements for resources such as tools, decouplers and conveyors, and floor space must also be assessed for different demand scenarios. The same approach is used to plan for demand changes regardless of source, including seasonal variation.

In designing workcells for adequate capacity it is impossible to cover all eventualities since events such as machine breakdowns, material shortages, worker absenteeism, and erratic demand spikes cannot be predicted. However, two ways to be prepared for most eventualities are (1) backup workcells, and (2) planned-in excess utilization. For every product family there should ideally be a backup cell to which work can be offloaded should the original cell become temporarily overloaded or incapacitated. The backup cell is ordinarily used for other purposes, but can with slight changes readily accommodate another product family. When it is impractical to establish a

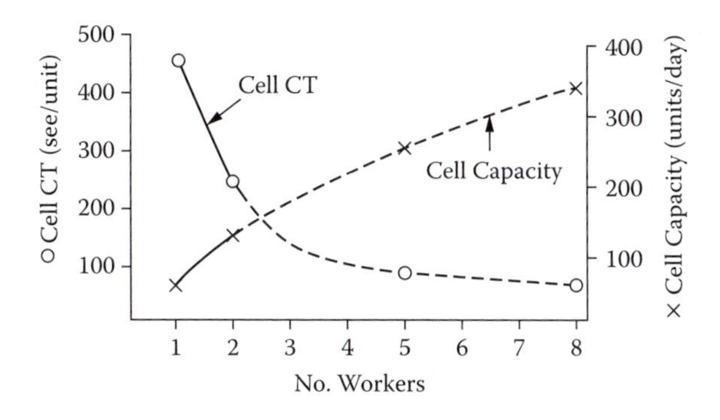

Figure 10.13 Effect of adding workers on cell output.

backup for every cell, then priority should go to creating backups for cells that produce the longest running or bread-and-butter products.

The second way to prepare for unanticipated events is to schedule work such that normally it falls well within a cell's capacity; in other words, schedule work such that the cell can accommodate current demand and then some. A cell normally scheduled to run, say, 36 hours a week will have 4 hours excess capacity to accommodate demand spikes or work interruptions. Scheduling for excess capacity is necessary for backup cells too because unless a backup has excess capacity, it cannot take on work from cells it is supposed to be backing up.

Cost–Capacity Tradeoff Analysis

Changing the number of workers alters not only workcell capacity but also the unit manufacturing cost of the cell. Particularly if a cell is intended to be long standing, and if workers must be newly hired and trained, the unit cost might be a key factor in determining the feasibility of assigning multiple workers to the cell.

Adding workers to a cell increases the cell output rate, but the marginal increase in output rate gets smaller with each additional worker. Figure 10.13 illustrates this for the eight-station assembly cell from Example 1 and 2. Also, with each additional worker, the direct labor operating cost of the cell increases by the worker's wage. Now, given that with each additional worker the marginal increase in the output gets smaller as the direct operating cost gets larger, at some point the costs will overtake the benefits. The following simple cost analysis shows this.

Example 4: Cost–Capacity Tradeoff Analysis

Consider again the eight-station assembly cell example and the four staffing levels investigated earlier. Reviewing the results,

No. Workers	*Cell CT (sec)*	*Output = 60 sec × 60 min/cell CT (units/hr)*
1	451	7.98
2	238	15.13
5	109	33.03
8	84	43.86

Suppose the machine operating cost of the cell is $80/hr, and the labor rate per worker is $10/hr. The total direct labor cost is then $10/hr *times the number of workers,* and the total unit manufacturing cost is

No. Workers	*a: Direct Labor (Cost $/hr)*	*b: Machine Operating (Cost $/hr)*	*Unit Cost = (a + b)/ Output ($/unit)*
1	10	80	90/7.98 = $11.28
2	20	80	100/15.13 = 6.61
5	50	80	130/33.03 = 3.94
8	80	80	160/42.86 = 3.73

In this case, it is clear that the increasing rate of output from additional workers more than offsets the increasing direct labor costs. In general, if the labor rate is low relative to other cell costs, then it is cost effective to operate the cells with many workers, which is what happened here ($10/hr labor rate versus $80/hr machine operating cost). With increases in the labor rate, however, the results can change. Suppose the labor rate is $20/hr:

No. Workers	*a: Direct Labor (Cost $/hr)*	*b: Machine Operating (Cost $/hr)*	*Unit Cost = (a + b)/ Output ($/unit)*
1	20	80	100/7.98 = $12.53
2	40	80	120/15.13 = 7.93
5	100	80	180/33.03 = 5.45
8	160	80	240/42.86 = 5.60

The lowest-cost staffing level is now five workers. (Further analysis might reveal a still lower cost at four or six workers had those options been considered.)

In situations where the cells process many small jobs of different products with frequent changeovers, the number of workers is determined primarily by the required cell CT (that is, by demand) and less by unit manufacturing costs. This is because the same workers are regularly shifted among different cells in the plant on an as-needed basis. Cross-trained, the workers are assigned to whatever cells need them to meet the CT requirements of immediate jobs. Shifting workers like this among cells takes advantage of worker competencies and gives the company maximum flexibility. The point is that adding or deleting workers at a particular cell represents no net change in the number of workers to the company, so the true manufacturing cost does not increase or decrease. Workers are transferred from one place to another, and in that case, questions about the effect of labor cost on manufacturing cost are moot. The same is true about the effect of machine cost when the same machines are regularly moved among different cells as needed.

Cells for Batch Size = 1

One advantage of workcells is that they can produce any batch quantity, even a quantity of one, assuming the setup times are sufficiently small. Suppose that a cell forms parts A, B, C, and D

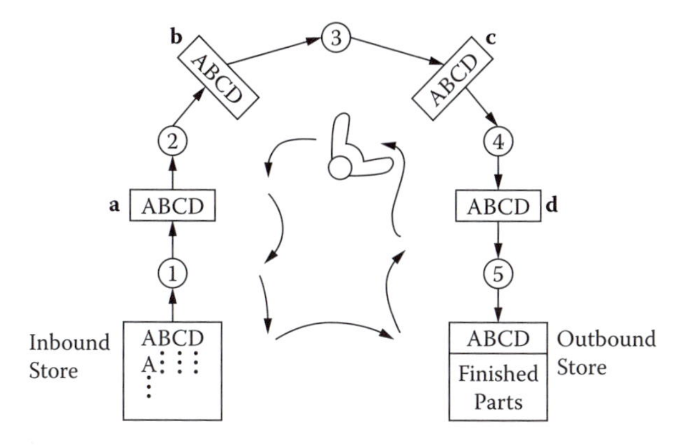

Figure 10.14 Example of cell for making one-unit batches.

in a five-machine process and that demand for the parts from a downstream cell is in one-unit quantities. To conform to the demand the cell must be capable of producing parts in batches of size one. Figure 10.14 shows the cell for making these parts. Between each pair of workstations is a rack large enough to hold one unit each of every part A, B, C, and D. When the downstream cell withdraws a part from the outbound store area, that is a signal to replenish the part. Suppose a part B is withdrawn from the outbound store. Starting at machine 5, the cell operator takes a part B from holding area d, loads the machine and turns it on. She then goes to machine 4, loads a part B from holding area c, and turns the machine on. She also goes to machine 3, then machine 2, then machine 1, loading the B parts and turning on the machines. Upon returning to machine 5 she unloads the part and puts it on the outbound store rack. This process is repeated every time a part is withdrawn from the outbound store rack.

The machines are all single-cycle, and they might automatically eject the parts to the next holding area at completion of machining. If not, the operator simply unloads the part before loading the next one. Except for the last machine, parts can also be left in the machines until the next time the operator cycles around the cell (the machine becomes the holding area for one part). Perhaps obvious is that the task of changing over between making different parts must be quite simple, ideally a one-touch operation that takes only seconds.

Sequential Changeover Tasks

Regardless of batch size, whenever a workcell must produce a different part or product, machines and fixtures at each station must be changed over. One way to do the changeover is to stop the cell and change everything at once. A less disruptive way is to integrate the changeover into the cell's sequence of operations. Instead of shutting down the cell, the changeover is done on only one station each time the operator walks around the cell. Suppose a cell has five machines, is currently making part X, and is to be changed over to make part Y. As shown in Figure 10.15, the operator completes the changeover after five trips around the cell. On the first trip the operator changes over machine 1 from X to Y, then walks around the remainder of the cell to process the last four part X's on the remaining four machines. On the second trip the operator processes the first part Y on machine 1, walks to machine 2 and changes it from X to Y, then walks to the other three machines to process the last three part X's. With three more trips around the cell, the changeover from part X to part Y is complete.

Cycle	Machine				
	1	2	3	4	5
1	Changeover	X	X	X	X
2	Y	Changeover	X	X	X
3	Y	Y	Changeover	X	X
4	Y	Y	Y	Changeover	X
5	Y	Y	Y	Y	Changeover

Figure 10.15 Sequential changeover of machines.

Setup reduction efforts are especially important in workcell operations. So that the entire changeover procedure causes minimal interruptions to the process, changeover procedures must involve minimal steps, no trial-and-error adjustments, and result in parts that are defect-free and meet specifications, starting with the first try.

Productivity Improvement

Workcell productivity improvement is aimed at achieving the required CT with the minimum number of workers. Recognize that the emphasis in improvement is decidedly *not* on reducing the cell CT, since reducing the CT below the required CT and without any anticipated increase in demand only results in overproduction. Workcell improvement efforts start with the required CT and then seek ways to achieve it with the fewest workers. Suppose, for example, that to meet a required cell CT of 123 seconds per unit, the five-worker assignment shown in Figure 10.9 is initially adopted. With this assignment, the actual worker CT and idle times (assuming workers conform to the 123 seconds requirement) are as follows:

Worker	*Subcell Stations*	*Cycle Time (sec)*	*Idle Time (sec)*
1	1 and 2	109	14
2	3 and 4	91	32
3	5 and 6	101	22
4	7	90	33
5	8	90	33
			Total: 134

The total idle time for the assignment, 134 seconds, is rather high. In fact, given that it exceeds the required CT of 123 seconds, the theoretical implication is that one less worker is needed than the five currently assigned. To reduce the number of workers, a way must be found to reduce task times and reassign the workers so that most of the idle time falls on just one worker. Suppose through productivity improvement efforts the task times are reduced and workers are reassigned such that the following results:

Worker	*Subcell Stations*	*Cycle Time (sec)*	*Idle Time (sec)*
1	1 and 8	120	3
2	2 and 6	120	3
3	3	30	93
4	7	94	29
5	5 and 6	101	12
			Total: 140

This assignment might at first seem worse than the initial assignment because total idle time is now 140 seconds. But most of the idle time, however, is with worker 3, so if ways can be found to reduce the task time by only 1 second at station 3 or 7 (which total 124 seconds), then both stations could be combined and handled by worker 4. Worker 3 would be removed from the cell, leaving just four workers. The assumption is that worker 4 has the skills necessary to take over tasks for the removed worker.

In practice, any worker selected to be removed from a cell owing to productivity improvement should be the one recognized as being, overall, the most highly skilled worker. There are two reasons for this. First, workers removed from one cell are reassigned to another cell, and the more skilled the worker, the easier it is to reassign her to another cell. Second, the message to be conveyed to all workers is that reassignment is a reward for good performance, not a penalty for poor performance. Also, this practice identifies to everyone the top-performing workers, though it does not expose or handicap average or below-average workers. An example of this is at Allen-Bradley in Milwaukee, where a group of highly experienced and able workers in a volunteer group called the SWAT team accept assignments anywhere in the plant, based on demand.[10]

Quality Control

Workers monitor product quality as pieces move through the cell. No item identified as defective is allowed to proceed to the next station. Sometimes colored lights, called **andons**, are located above each station to indicate work status. The lights are green as long as work is proceeding normally, but are switched to yellow when a worker needs assistance and a possible delay is expected from the station. If the problem is severe and the process must be stopped, the worker switches the andon to red, and the entire cell stops until the problem has been resolved. Lights are used particularly where noise or other factors make verbal communication difficult, or where workcell status must be quickly communicated to areas of the plant that provide inputs to the cell or that rely on the cell's outputs—as in a pull production system formed by linked workcells.

Workcells Beyond Manufacturing[11]

Workcell concepts can be applied to any situation that repetitively processes similar end items. Beyond manufactured products, workcell applications extend to the service sector and include processing of end items such as mortgage applications and insurance claims. McKinsey &

Company reports that financial institutions employing workcells claim benefits of a 25% to 40% reduction in cycle times, a 15% to 25% boost in productivity, lowered labor costs, freed up space, and faster site consolidations. One Asian mortgage lender that replaced its traditional processing system with cells saw a 50% jump in productivity. A U.S. wholesale lender was able to consolidate its operations from more that 20 locations to 10 centers, each with three to five workcells.

As in manufacturing, a workcell in the service industry would include all the steps to process a family of end items, for example, home mortgage applications. A workcell that handles mortgage applications would include the steps of application receipt, clearing, and closing the loan. Such a workcell is commonly staffed with about 14 people that include the cell manager and three kinds of specialists: processors, underwriters, and account managers. The following example illustrates how the exact size of the workcell staff is determined.

Example 5: Staffing a Mortgage Application Workcell[12]

The actual number of each kind of specialist in a workcell depends on the task time for the specialist and the required CT of the cell. Assume that each mortgage application requires about 45 minutes for a processor, 75 minutes for an underwriter, and 40 minutes for an account manager. Assume also that each day the cell must process 20 applications; thus, for an 8-hour (480 min) workday the required CT is 24 minutes. To meet the required CT, the cell will need the following number of specialists:

$$45 \text{ min}/24 \text{ min} = 1.875 = 2 \text{ processors}$$

$$75 \text{ min}/24 \text{ min} = 3.125 = 4 \text{ underwriters}$$

$$165 \text{ min}/24 \text{ min} = 6.857 = 7 \text{ account managers}$$

The cell is divided into separate areas for the three kinds of specialists; each application moves from area to area, and is processed in an area by the first available specialist. With this arrangement, most applications can be processed and decisions communicated to the borrowers within a day.

Note that this cell requires 13 specialists, although it would need only 12 specialists if the underwriters could reduce their task time from 75 minutes to 72 minutes.

Additional issues in workcell staffing are discussed next.

Workers in Cells

Staffing a Workcell

Workcells are responsibility centers wherein workers have considerable autonomy and perform functions ordinarily done by staff experts, including machine setup, maintenance and basic repair, job scheduling, and quality control. Workers sometimes also perform cell-related planning and problem solving, and order parts from vendors. They even meet with suppliers to resolve issues about incoming parts, and with customers to better understand the requirements and how the cell's outputs are used. Though each cell usually has a stable core group of workers and a supervisor or group leader, as described, workers are rotated among different workcells depending on demand requirements.

Though a workcell's actual CT can be adjusted by changing the number of workers, such changes should occur relatively infrequently and as dictated by changes in the required CT (takt time). In assembly plants, required CT is set by the requirements of the final-assembly stage.

Cell supervisors on the shop floor make decisions about workcell assignments. Determining the number of workers and their assignments to workstations to meet a required CT is difficult at first but gets easier as supervisors learn which staffing combinations yield the desired cell CT. Analytical tools such as computer simulation can help with staffing decisions. In some plants, planners run computer simulation models to generate recommendations for supervisors; in other plants, the supervisors themselves run the simulation models.

Simultaneous Staffing of Multiple Cells

Workers cross-trained to perform operations in more than one cell can be assigned to operate more than one cell at a time. Strictly speaking, what was stated earlier about a cell being operated by as few as one worker is incorrect since a single worker can operate two or more cells located together. To see how this happens look at two possible staffing assignments for cells A to E in Figure 10.16. Assignment (a) has seven workers, (b) has four. In both (a) and (b) assignments, the cells are linked by each worker relying on another worker for parts or materials. Looking at (b), worker 1 cycles through cells A and B; worker 2, through cells B and C; worker 3, through cells C and D (with an additional stop at the inbound stock area for station E); and worker 4 through only cell E. Because they are linked, the cells all have the same CT, which is the CT of the worker who takes the longest. Adding workers reduces the overall CT, so in Figure 10.16 the CT for seven workers (a) will be less than than for four workers (b).

When the required output of the different cells is different, the worker assignment gets trickier. Cells needing less output (i.e., having longer required CT) can be visited by a worker on every second, third, or more cycles, but that increases the time available for them to perform operations

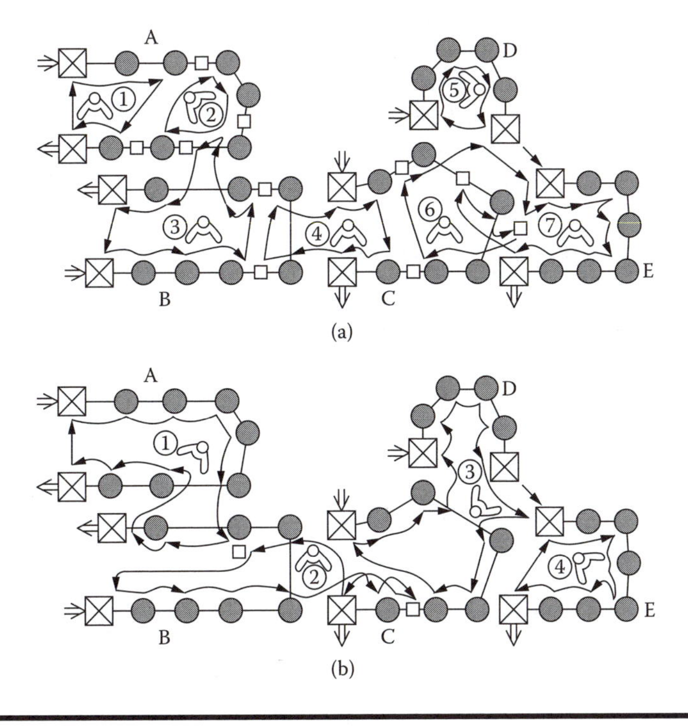

Figure 10.16 Workers with overlapping cell assignments.

in the other cells they visit. As an example, suppose in assignment (b) that the required output of cell A is roughly half that of the other cells. Worker 1 can visit cell A only on alternate trips, which then will roughly double his available time for doing tasks in cell B. More operations in cell B could be assigned to worker 1 but that would change the assignments for the other worker in cell B, worker 2.

Adjusting worker 2's assignments in cell B affects his assignment in cell C, which affects worker 3, and so on. Any attempt to change the CT in one cell has a ripple affect on the assignments of all the workers. Learning curve and skill issues further complicates the staffing decision as well. What-if analysis with computer simulation models is an almost mandatory tool for staffing decisions with multiple cells.

Equipment Issues

When not all products are assigned to a product family or production in workcells, a portion of the plant must be set aside to produce them. These odd products might outnumber or account for greater aggregate volume than products in families, and the portion of the plant set aside to produce them might considerably exceed the workcell portion. This nonworkcell part of the plant usually continues to operate as a job shop. An issue, then, is the availability of equipment for both workcell and job-shop applications.

Machine Sharing

When multiple areas (cell and noncell) of the plant require the same kind of machines, when there are not enough machines to go around, and when procurement of additional machines is cost prohibitive, then the machines will have to be shared. Figure 10.17 shows two possibilities: on the left, a shared machine located between two adjacent cells so that, theoretically, it resides in them both; on the right, a shared machine is located elsewhere and treated as a special operation. The first case is convenient for both cells, but it also interrupts the operations in both. Specifically, a shared machine must be switched over to process jobs from the cells sharing it, and while one job is being processed, jobs from other cells must wait; this further leads to WIP accumulation and increased lead times—just like in traditional batch processing. The second case has the same

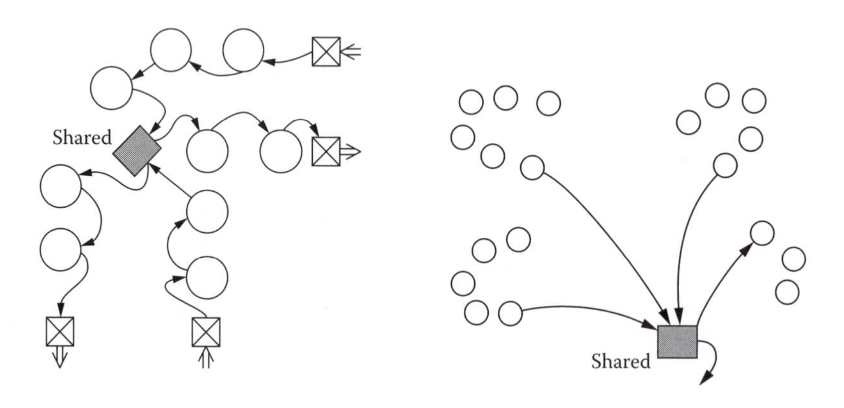

Figure 10.17 Shared machines.

drawback in addition to requiring greater material handling effort; however, it is the only solution when a machine is immobile and must be shared by many processes.

Sometimes the shared-machine approach is adopted to ensure that a machine will get high usage. If the machine is expensive and has multiple capabilities, shared usage is a way to exploit its capabilities and amortize the cost. These are not good reasons, but they are common reasons. In summary, when cells must share machines, the advantages of cellular manufacturing are diminished.

Machine Acquisition

Criteria for machine acquisition traditionally consider the tradeoff between a machine's production rate or multipurpose capability and its cost. For plants with focused factories, traditional acquisition criteria must be rethought since they can lead to results opposite as intended, such as a multipurpose machine decreasing a plant's flexibility and a special, high-speed machine increasing production lead times.

Multipurpose machines can quickly be changed over to do many things or even do several things at once. Such machines have the capability to perform a variety of machining operations on products with a wide range of sizes, shapes, and configurations. Such machines, however, usually cost much more than fewer-purpose machines, so purchasing them leaves less funds to purchase other machines, and that increases the possibility that focused factories will have to share machines. Even when a multipurpose machine can be dedicated to a single workcell, managers might try to expand the variety of the cell's output and route products from other cells or autonomous operations through the machine to take full advantage of its capabilities. The same situation can result from procuring costly high-speed machines: fewer of them can be acquired, so focused factories must share them, jobs have to wait longer, and so on.

The alternative to costly multipurpose or special-order machines is inexpensive, slower, and fewer-purpose machines, but many of them. Every workcell that needs one gets one and can then function autonomously. Although a multipurpose machine offers high flexibility through quick changeover and rapid production rate, all else equal, conventional fewer-purpose machines might provide even greater flexibility when employed in a number of manufacturing cells. Conventional machines are also simpler to operate and less costly to maintain.

Every workcell is eventually changed, either in response to changes in demand or product-family mix, or as a result of cell productivity improvements. Plantwide, workcells are in flux; some are being newly created or reconfigured, while others are being disbanded, depending on changing requirements. To minimize the time and cost of reconfiguring workcells, machines and workstations should be small, mobile, and have simple utility hookups.

Special Operations

Machine and equipment considerations in workcell design extend beyond cost and flexibility. An operation might involve large, heavy machines that need massive foundation supports and special utility hookups; it might require machines or work areas that are unique and must serve the entire plant (welding areas, paint rooms, heat-treatment rooms); or it might require uncommon procedures and isolation from the rest of the plant (for venting toxic fumes or shielding against heat or radiation). Although equipment for operations such as these might not be relocateable to workcells, there are ways to include them in a workcell process. One way is to move products back

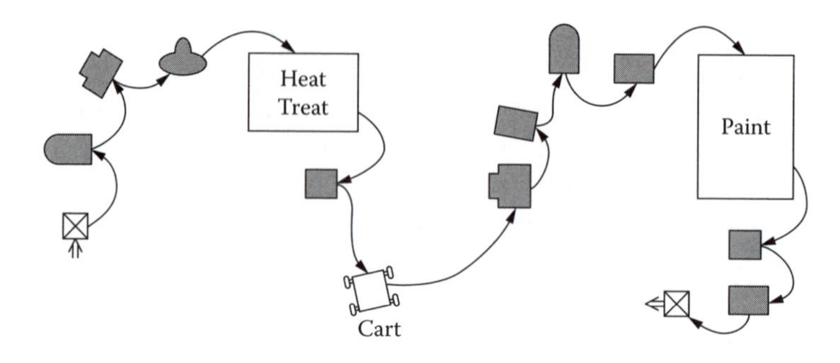

Figure 10.18 Cells built around special operations.

and forth between the special operation and workcells on a batch basis, though the batches are kept small to minimize wait times and WIP inventory.

When most of the items in a product group require the same special operation, then a workcell can be built around the special operation (Figure 10.18). If the special operation requires batching, then inventory is allowed to accumulate just ahead of it until the right quantity is built up. If the operation is continuous and not shared by other cells, then items can be routed through it like any other operation in the cell.

Cell Automation

Automation of a cell is often an evolutionary transformation, starting with installation of single-cycle, automatic machines where workers manually load, unload, and transfer parts between machines. When all the manual functions have been automated, the transformation is complete. Figure 10.19 shows a workcell with a centrally located robot that pivots to transfer, reorient, load, and unload parts at any of the stations and in any sequence. The robot and machine operations are automatic and computer controlled.

Benefits of automation include low setup time, high output rate, and low process-induced variability in the output. A manufacturing cell or system of linked cells that is completely automated by virtue of robots and computer control is called a **flexible manufacturing system (FMS)**. Like other, less automated workcells, an FMS can be an economical way to produce a variety of different products, even if each has only small or moderate demand. Economy is attained by high machine

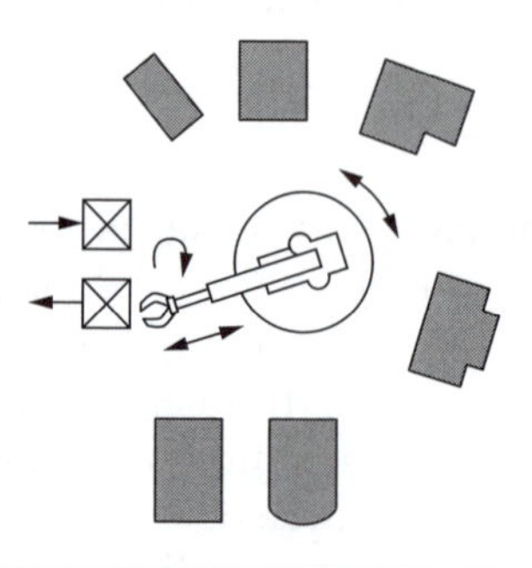

Figure 10.19 Robotic cell.

utilization from processing many products that require a similar process. With automation it is possible to achieve very high tolerance standards and low variability. Like manual workcells, FMSs can be linked together. Flow of material into a cell's inbound stock area and out of its outbound stock area (the X boxes in previous figures) can be linked by conveyors or automated guided vehicles. The linked FMSs make up the larger FMS, complete with automated machine tools, automated handling equipment, and computers that coordinate work within and between the FMS subsystems. Effective operation of FMSs still requires human operators, who provide important services such as equipment monitoring, quick intervention when operational problems arise (stuck tools, incorrectly positioned parts, an automated guided vehicle going to the wrong place), problem solving and cell improvement, and assisting engineers in the design of the parts to be manufactured.[13]

Full automation is expensive, and because "unmanned" cells still require human support, the labor savings is not always as great as expected. FMSs are also somewhat less flexible than cells having greater manual labor content, and machines and transfer systems that can be easily relocated and reconfigured. Unlike cross-trained workers that can be reassigned virtually anywhere to perform a broad range of tasks, robots are limited in the range of tasks they can be programmed to perform. The capital cost for automated systems is much higher than for simpler, manual systems, and so is the maintenance cost. Still, there are many cases where the right combination of product variety and production volume makes automation the right choice. Many cellular manufacturing plants employ a range of cellular processes, from fully manual to fully automated.

Implementing Cellular Manufacturing[14]

Focused factories and cellular manufacturing represent more than changes in production technology; they represent a different philosophy about managing and performing work. The following sections put issues associated with implementing cellular manufacturing into three broad categories: planning and control, organization, and attitudes. Many of these issues are interrelated and logically span more than one category.

Planning and Control[15]

Every production process has a planning and control system. The system anticipates production requirements for materials, labor, and facilities, and checks that they are within capability of the production process. The system performs job routing, workstation loading, detailed scheduling, job sequencing, and job dispatching, and tracks jobs as they move through the system. Although a workcell is a shop-floor phenomenon, its implementation must be integrated with planning and control mechanisms beyond the shop floor. Organizations implementing workcells usually already have centralized planning and control systems in place, so ways must be found to adapt them to the control procedures of workcells. The common material requirements planning (MRP)-type planning and control system is an example of a centralized system. With MRP the focus of attention is on order releases and completion dates for all individual components and higher-level assemblies in a product. In contrast, cellular manufacturing focuses attention on completed products, that is, on what goes out of the cell, not on what is happening within it. Unless modified, all of this order-release information from an MRP system about intermediate (within-cell) operations is of no use in cellular manufacturing.

At issue is the primary unit of focus in planning and control. In traditional noncell shops, the machine or workstation is the unit of focus. In workcell plants, the whole cell is the unit of focus.

All that matters is what goes into the cell and what comes out of it. This change of focus greatly condenses the amount of scheduling and control information generated since material is tracked at only two places—the points where it enters and leaves the cell. Only in workcells in which the throughput times are very long or the operating and overhead expenses of equipment vary greatly are the individual machines or workstations retained as the unit of focus.

This is not to say that MRP-type planning and control systems are useless for workcell operations. In a survey of 57 companies that implemented cellular manufacturing, 75% had used MRP planning and control systems before implementation. After implementation, 71% still used MRP systems (though of these, 33% had also adopted Kanban systems). Only 13% of the companies used Kanban-only systems, and only 31% used MRP-only systems.[16] Because workcells themselves have virtually no planning or forward-looking capability, they depend on a system such as MRP for future planning and order entry. Though cell operators might be responsible for daily capacity planning and job scheduling, the size and frequency of jobs arriving at the cell is determined by the centralized system. The centralized system forecasts demand, accumulates job orders, performs rough-cut capacity planning, and prepares and coordinates master production schedules. To adopt MRP systems for releasing orders to workcells, the product bill of materials must be restructured. The MRP system is then used to send jobs to workcells. Once a job arrives at a workcell, all detailed job sequencing, scheduling, and control of the jobs are performed by supervisors and workcell operators. These matters are further discussed in Part III.

Organizational Issues

Implementing workcells requires that job roles, responsibilities, and relationships be redefined. The roles of workers are expanded, teamwork is emphasized, and production output and compensation measures are group oriented. Workcell implementation usually involves considerable worker training as well as modification of performance appraisal and pay systems. As a result, many or most of the human resource policies and practices traditionally applied in shop settings require some attention.[17]

Roles and Responsibilities

The most evident change that comes with cellular manufacturing is to the role of the frontline shop worker. To take full advantage of workcell production methods, shop personnel must be empowered. The locus of production-control decisions must be moved from functional support departments to the shop floor. Workers must be made responsible not only for assembly and machine operations, but also for performing tasks previously done by staff workers such as inspection, basic maintenance and repair, job prioritizing, and dispatching.

As workers take on more of these responsibilities, the primary function of staff professionals shifts to supporting the workers. Support happens in the form of training cell operators, providing on-demand technical guidance, and performing tasks that require special expertise. For instance, instead of routine repairs, maintenance staff only do difficult repairs, and spend the rest of the time training workers, analyzing equipment failures, overhauling equipment, and improving maintenance and operation procedures; instead of routine inspection, the quality control staff does company quality audits, certifies suppliers, trains workers in inspection procedures, and does tests and inspections that require specialized skills.

Cellular manufacturing also affects people not directly involved in production planning, control, or execution. Design engineers, for example, are accustomed to being judged for the number and creativity of the designs they create. Cellular manufacturing and group technology encourage design engineers to find ways to improve product quality while taking maximum advantage of existing parts and manufacturing capabilities.

Incentive Plans

Traditional incentive plans reward workers based on individual performance. The plans are typically based on piece rates, and earnings are a function of the number of units produced.[18] Thus, each worker can maximize her pay by maximizing her output. In workcell operations that require teamwork, however, such plans are usually ineffective because workers who help others have less time to devote to maximizing their own output. Further, in workcell operations the goal is to produce output to meet demand, not to maximize output. For these reasons, it will probably be necessary to replace the traditional incentive plan with some form of a group-based incentive system to maintain productivity and to minimize worker resistance to teamwork.

Two group-based plans that promote teamwork are **pay for skill** and **gain sharing**. Both tend to promote behaviors and attitudes important to teamwork. Pay-for-skill plans increase the wages of workers for each new skill they learn.[19] The available skill options depend on worker responsibilities and include machine operation, setup, maintenance, quality inspection, and so on.

Gain-sharing plans provide team members bonuses for improvements in team performance as measured against a baseline standard.[20] Whenever the team performs better than the baseline, it accumulates bonus points in a pool. Some plans apportion the bonus pool equally to everyone in the group, others to workers as a percentage of their base salary or number of hours worked. Measures of team performance include cell output rate, cell defect rate, setup times, equipment effectiveness, cell inventory level, and average worker skill level. The team members must agree that the baseline standards are reasonable, and if they do not, the baseline can become a sore point with them and ineffective as an incentive.

Whatever compensation plan is adopted, it must not result in workers receiving less pay when they move to workcells. Implementing a pay plan that hurts workers financially will almost guarantee workcell failure. In fact, it can be argued that workers should justifiably be paid more because workcells require higher skill level and more responsibility from them.

Time and Rate Standards[21]

A **time standard**, or standard time, is the amount of time required to perform a particular task. Time standards are used for many purposes:

- Computing the quantity of a product that can be produced in a given time period
- Determining the number of workers or machines needed to meet production demand
- Distributing work among employees
- Assessing alternative production approaches
- Setting production schedules
- Estimating production costs
- Assessing employee performance in wage incentive plans

For a team-oriented process, the fundamental unit of measure must be the group. Since existing time standards are usually based upon one person performing a single task, they are relatively useless in cells where workers move among operations or perform tasks collectively. Thus, workcell implementation usually involves developing workcell-based standards, most of the information for which must be newly collected.

As examples, look back at the machining cell Example 3 (Figures 10.10 to 10.12) where the CTs for three different staffing levels for the cell were determined. Among the reasons that cell CT is so important is that it serves as the cell standard time, which is used for determining other cell standards, such as standard output rates. For instance, from the cell CTs in the cell in Example 3, the derived standard output rates are as follows:

Number of Workers	*Cell CT (sec)*	*Standard Output: Rate = (60 sec × 60 min)/Cell CT (units/hr)*
1	131	27.48
2	88	40.91
3	62	58.06

Establishing the cell CT is not necessarily easy. Recall that the cell CT is determined as follows by taking the greater of the longest worker time and the longest machine time:

$$\text{Cell CT} = \max\,(\text{Longest worker CT, longest machine CT})$$

Both the worker CT and machine CT are themselves standard times. The machine CT is the time for the machine to perform a particular single operation, and this standard time is usually already well established, and, if not, is easy to measure.

The worker CT is another matter. Worker CT is the time for a worker to walk between machines and perform tasks, but since the time can vary considerably from worker to worker, it can be difficult to ascertain as a standard. One way to get standard worker CT is to use time-motion analysis for each walk and machine-task segment, and sum the times over all segments along the route of the worker. A better way is use work sampling: observe the actual times of different workers performing a particular walk–machine-task sequence and take the average. The problem with relying on work sampling is that the data to develop standard times cannot be obtained until after the cell is formed, yet standard times are needed to plan the cell. In such a chicken-and-egg case, the initial cell standard times might be derived from times collected elsewhere at operations considered similar to those in the cell. These initial standards are used to design a pilot cell, then are revised as soon as data from that actual cell can be obtained. A pilot cell, discussed later, is useful for establishing the initial cell standards.

Another tool for developing cell standards is computer simulation. Since workcell standards vary depending on factors such as the product, workstation configuration, and number and skill level of workers, computer simulation is a good way to establish initial cell standards that fit a given combination of all these factors. Simulation also can take into account variability in machine and worker times, and show the likely range of expected performance outcomes.

Time and output standards, once determined, are used to set the labor standard. A labor standard is the average labor time required to produce a unit of product. The labor standard for a cell is computed by dividing the standard cell output rate by the number of cell workers. For example, using the aforementioned standard output rates,

Number of Cell Workers	*Divided by Standard Output Rate (units/hr)*	*Equals Labor Standard (hr/unit)*
1	27.48	.0364
2	40.91	.0489
3	58.06	.0517

Labor standards are used for capacity planning, product costing, work scheduling, and deriving gain-sharing baseline measures.

Team Education and Training

Obviously, for workcell teams to be able to perform an expected wide range of tasks, training is essential. In addition to the basics of workcell operation, workers should learn to do machine setup, basic machine upkeep and PM, simple machine repair, product inspection and quality control, and to work as members of a team. They are also trained to operate many or all of the machines in the workcell and possibly in other workcells, too. The best training for machine operation, maintenance, and setup is hands-on experience with assistance from veteran operators and support staff. As workers become proficient, the assistance is phased out.

Like the workers, supervisors should be capable of working in a variety of areas, depending on demand requirements. Rotating supervisors among different work areas improves not only plant staffing flexibility, but human relations too. One plant, for example, had a history of supervisors in different areas constantly arguing over schedules and parts shortages. After being required to exchange jobs, the supervisors stopped bickering because they better appreciated the kinds of problems facing other supervisors.

A record of employee skills (as shown in Figure 10.20) is maintained for workcell staffing assignments and pay-for-knowledge incentives. Once workers have gained new skills, they should

	Machine and Task Skills							
Name	M1	M2	M3	M4	M5	Insp'n	SPC	Comments
Bill	1	2	2			2	1	
Jane	2	2		1	2	2	2	Goal: master M4
Wonda				1	1	1		Trainee
Igor	2	2	2	2	1	2	1	Goal: master M5 and SPC
Talia			1	2	2	1		Goal: master M3 and insp'n, basic SPC

Notes: Competency levels: 1 = basic (operation); 2 = master; 3 = 2 + preventive maintenance; 4 = 3 + repair maintenance.

Figure 10.20 Worker skill chart.

be rotated daily or weekly among the stations within a cell, or in different cells in a focused factory, so they can utilize and improve those skills.

In addition to classroom and shop-floor training, workers learn about workcell concepts and operations by visiting other plants that have successfully adopted workcells. They discuss implementation issues with workers at the site and learn about procedures, mistakes, and how to do things better.

Attitudinal Issues

Adopting workcells requires more than a shift to new layouts and procedures; it requires a basic shift to worker empowerment. Many staff people and supervisors will perceive that transfer as a reduction in their own power. Power, by whatever means achieved, is not something most people willingly relinquish, and those with the perceived loss will try to resist the change.

Shop-Floor Workers

Education and training alone will not modify people's attitudes. Many shop workers who are accustomed to performing narrow tasks will not eagerly welcome the chance to learn and utilize new skills. They will have to be convinced that the changes are an improvement, that their work will be more challenging and less boring, and that their ideas and opinions, not just their physical labor, will be valued. Although there is much anecdotal evidence that workers in cells are more satisfied with their jobs, the evidence is not unequivocal. Union contracts outlining job descriptions and pay rates will have to be rewritten to account for the cross-functional, team-orientation of workcell jobs. Union spokespeople might be skeptical and view workcells as a ploy to squeeze more productivity out of workers.

Supervisors

Supervisors will also be skeptical.[22] Job security is one of their concerns: they wonder whether their role will become unnecessary once workers are empowered. While management often makes guarantees to hourly workers about their jobs, no comparable guarantees are made to supervisors. Supervisors are also anxious about their job function. They wonder what they are supposed to do. A third concern is the amount of added work. Implementing workcells requires additional work for training, coordinating, and problem solving, all for which supervisors might be responsible but for which there may be no additional pay.

Management

Management too may oppose many of the necessary changes in organizational structure and responsibility and have to be convinced about the benefits of focused factories and workcells. The employee training, equipment relocation and retooling, and new setup and maintenance procedures associated with workcell implementation often require considerable time and some capital outlay. For managers to understand the benefits of workcells, they must first understand the importance of reduced defects, shortened lead times, reduced inventories, and increased reliability and flexibility of operations. Analysis to justify workcells requires multicriteria methods that besides financial costs consider the intangible benefits of quality, flexibility, employee satisfaction, and improved management, as well as the intangible risks (market and technology) of not adopting them.[23]

One reason cellular manufacturing programs fail is because management tries to cut corners.[24] For example, by cutting training costs, workers end up incapable of performing well in workcells. By avoiding costly relocation of equipment, cells end up being formed around pseudo products instead of actual product families. Like all things in lean production, cellular manufacturing requires long-term commitment. It cannot be done quickly or cheaply, and it requires the support of management.

Resistance to workcells is certainly not universal, and in many organizations people welcome (even get excited about) implementing workcells and being assigned to work in them. At the McDonnell & Miller plant, a union shop that years ago began using machining and assembly workcells, enthusiasm about being assigned to cells is so high that workers are on a waiting list. Cell workers say they enjoy the added responsibility and variety of the tasks, and the absence of close scrutiny by supervisors.

Getting Started

Initial planning for workcells should be conducted by a multifunctional steering committee. This committee, comprised of representatives from the functional departments and the union, frontline workers and machine operators, management, and possibly a consultant with workcell experience, addresses the issues described above. The committee oversees high-level planning and coordination of all activities, including product-family/machine grouping, layout and equipment changes, space allocation, training, site visits, contract negotiations, changes in jobs, modifications to planning and control procedures, and major equipment procurement.

Besides the steering committee, a core team should be formed for each workcell to handle detailed workcell design and final implementation. The team should include people from industrial engineering and production management, and the supervisor and workers who will eventually staff the cell. The team is provided financial and technical support from the steering committee and specialists in functional departments. Members of the core team should be selected for demonstrated willingness and ability to work in a team, skill qualifications, adaptability to change, and acceptance of new roles and ideas.

Many companies start workcell implementation using one or two pilot cells and a product family that accounts for a sizable proportion of the firm's volume.[25] A pilot cell should be stacked for success, meaning it should be implemented in an area where it is very likely to succeed. A pilot cell serves many purposes: engineers and workers tinker with it to fine-tune the cell design; managers collect data to assess its costs and benefits; workers train in it, learn to operate machines, do setups, maintenance, and quality-checking; and staff members use it to discover ways to reduce setup times and eliminate quality problems. A single pilot cell can usually be implemented in 6 months, though it can take longer. Assuming the cell performs well, the benefits it demonstrates will open the way to additional cells. When the intent is to establish a series of linked cells in a pull production process, the conversion to cells usually begins downstream and works backward, starting at the final assembly area and successively converting upstream areas into cells.

Summary

Cellular manufacturing is manufacturing done in a workcell. A workcell is a group of dissimilar operations formed to produce a product family. Benefits of cellular manufacturing, including high quality and efficiency, results from the workcell's focused nature: workers and equipment are clustered together and dedicated to producing a family of outputs on a repetitive basis. The output rate of the cell is easily modified by changing the number of workers and machines, and a workcell can efficiently produce output for virtually any batch size, even one. By linking workcells together, it is possible to produce complex end-items.

Implementing cellular manufacturing requires commitment to the principles of lean production. Setup procedures must be simplified, and machine reliability and commitment to quality must be high. Cellular manufacturing requires empowering workers. Besides performing assembly and machining operations, the workcell team assumes responsibility for most machine changeovers, basic PM and repair, and quality inspection. Sometimes the team also does its own workcell planning, priority decision making, and material purchasing.

Cellular manufacturing greatly simplifies production scheduling and control. Whereas jobs in traditional shops are scheduled and tracked at each operation, in cellular manufacturing they are tracked only at the start and the finish of a cell. Although this simplification saves on nonvalue-added paperwork and processing, it requires modification of traditional MRP-based planning and control systems for scheduling and tracking at the workcell level.

Cellular manufacturing is best implemented by starting slowly and with a pilot cell. The pilot cell should be formed such that it has the greatest chance of success. A product family that has sufficient volume to justify devoting people and machines solely to it should be chosen. A steering committee should be formed to handle high-level implementation issues and coordinate cross-functional activities. A shop-floor core team should also be formed to handle details about the pilot workcell design and implementation. The purpose of the pilot is to teach everyone about the way a cell works, to experiment with setup reduction, preventive maintenance, extended worker involvement, as well as to demonstrate to everyone the potential benefits of workcells.

Important to add is that despite the benefits, workcells are not the best way to produce everything. The pilot thus serves another purpose: if cellular manufacturing cannot be made to succeed in the pilot, then it will be clear that there is no way it can be made to succeed anywhere else in the plant.

Notes

1. Instead of *workcell* the simpler term *cell* is often used. John Burbidge, a major proponent and writer on the topic, avoids using the word *cell* because, he says, telling shop-floor workers they will be working in cells (as in prison) is likely to be counterproductive. J. Burbrige, Production flow analysis for planning group technology, *Journal of Operations Management* 10, no. 1 (January 1991): 5–27.
2. This number is based on many small-group research studies; R. Napier and M. Gershenfeld, *Groups: Theory and Experience* (Boston: Houghton Mifflin, 1973).
3. The system of coordinated cells may itself be considered a focused factory for the production of a part or product family.
4. The term *cycle time* is also commonly used to refer to the total throughput time for a product, that is, the sum of the operation times required to make the product. The definition of cycle time as used in this book is common among engineers and in the literature on pull production processes.
5. G. LaPerle, Letter to the editor, *Target* (July/August 1995): 51.
6. Ibid.; Y. Monden, *Toyota Production System*, 2nd ed. (Norcross, GA: Industrial Engineering and Management Press, 1993) 303–304.
7. With only one worker, there is no need for holding places between stations. The worker simply moves the items being worked on from station to station with him.
8. J. Black, *The Design of the Factory with a Future* (New York: McGraw-Hill, 1991), 190–194.
9. The feasibility or practicality of adding machines depends on whether additional machines can be relocated from elsewhere in the plant, the impact of relocation on other products, or whether new machines must be purchased. Although it is simple to determine where machines should be placed in a cell to reduce the CT, decisions to procure the machines require economic cost–benefit analysis.

10. R. Schonberger, *World Class Manufacturing: The Next Decade,* (New York: Free Press, 1996) 168.
11. A. Eichfeld, B. Ledbetter, and R. Thomas, *The Lean Work Cell: A Mortgage Solution*, Working Paper, McKinsey & Company (February 2007).
12. Ibid.
13. M. J. Maffei and J. Meredith, Infrastructure and flexible manufacturing technology: theory development, *Journal of Operations Management* 13, no. 4 (December 1995): 273–298.
14. Portions of this section are derived from Black, *The Design of the Factory with a Future*, 88–91; N. Hyer and U. Wemmerlov, Group Technology and Productivity. *Harvard Business Review* (July–August 1984): 140–149; R. Schonberger, *World Class Manufacturing* (New York: Free Press, 1986), 112–114; H. Steudel and P. Desruelle, *Manufacturing in the Nineties* (New York: Van Nostrand Reinhold, 1992), 133–137, 155–162; U. Wemmerlov, *Production Planning and Control Procedures for Cellular Manufacturing Systems* (Falls Church, VA: American Production and Inventory Control Society, 1988).
15. Wemmerlov, *Production Planning and Control Procedures*, discusses in detail planning and control issues and considerations in workcell manufacturing.
16. Other kinds of planning and control mechanisms, including reorder points and bottleneck scheduling were among the systems studied. See F. Olorunniwo, Changes in planning and control systems with implementation of cellular manufacturing, *Production and Inventory Management*, 37, no. 1 (1996): 65–70.
17. V. Huber and K. Brown, Human resource issues in cellular manufacturing: A socio-technical analysis, *Journal of Operations Management*, 10, no. 1 (January 1991), 138–159.
18. For an introduction to traditional, piece-rate incentive plans, see D. Miller and J. Schmidt, *Industrial Engineering and Operations Research* (New York: John Wiley, 1984).
19. A good discussion of issues and implementation guidelines for pay-for-skill plans is provided in J. Orsburn, L. Moran, E. Musselwhite, and J. Zenger, *Self-Directed Work Teams* (Homewood, IL: Business One Irwin, 1990), 182–194.
20. Ibid., for discussion of gain-sharing plans.
21. For detailed coverage of time standards and work measurement, see "standard" texts on industrial engineering or motion and time study. See, e.g., E. Polk, Methods Analysis and Work Measurement (New York: McGraw-Hill, 1984).
22. J. Klein, Why supervisors resist employee participation, *Harvard Business Review,* (September–October 1984): 87–95.
23. D. Dhavale, A new book from FAR, *Management Accounting* 77, no. 7 (January 1996): 63–64.
24. I. Winfield and M. Kerrin, Toyota Motor manufacturing in Europe: Lessons for management development, *Journal of Management Development* 15, no. 4 (April 1996): 49–57.
25. A good discussion of pilot cell implementation is in the 1994 article, Empowerment pumps Duriron up, *Tooling and Production* (October): 13–15.

Suggested Readings

J. T. Black, and S. Hunter. *Lean Manufacturing Systems and Cell Design.* Dearborn, MI: Society of Manufacturing Engineers, 2003.

C. Heckscher. *The New Unionism: Employee Involvement in the Changing Corporation.* Ithaca, NY: Cornell University Press, 1996.

N. Hyer, and U. Wemmerlov. *Reorganizing the Factory: Competing Through Cellular Manufacturing.* Portland, OR: Productivity Press, 2002.

Questions

1. In what situations would it make sense for workers to walk around a workcell in a direction opposite the parts flow?
2. Explain the difference between required CT (takt time) and actual CT?

3. How might increasing the number of workers in a cell result in a longer cell CT?
4. Why is the cost (wage rate and associated costs) of workers in a workcell sometimes irrelevant in determining the number of workers to assign to a workcell?
5. Discuss the drawbacks of not being able to include every machine needed for processing of an item within one workcell.
6. Why should workcell productivity improvement efforts be centered on the required cell CT? In what situations would the improvement effort focus on reducing the actual CT?
7. What are the advantages and difficulties in assigning workers to operate multiple cells simultaneously?
8. For what reasons might a workcell have to share machines or operations with other workcells or operations?
9. Why must schedules that are generated by a traditional MRP system be modified for use by workcells?
10. How is production planning and control simplified by cellular manufacturing?
11. Discuss how cellular manufacturing impacts the role of workers. Why can't workers' traditional roles of performing one function be maintained in workcell operations?
12. Why is performance of workers in workcells measured on a group basis?
13. How are time standards set for workcell planning?
14. Discuss the concept of a pilot cell, including what it is, what purposes it serves, and how to set one up.
15. A 12-station assembly workcell is being formed for a new process. The work at each station requires several complex assembly and inspection subtasks. The difference between the longest and shortest workstation times is 15%. Analysis indicates that to satisfy the required CT, three workers must be assigned to the workcell. Discuss the pros and cons of subdividing the workcell into three subcells versus using rabbit chase.

PROBLEMS

1. A workcell is being planned to produce a part at rate of 420 units/day. Suppose for planning purposes the company uses 7 hours as the normal workday. What is the required CT (takt time) of the cell? Should the workcell be designed so that it has an actual CT longer or shorter than the required CT? Explain.
2. A process that produces part A operates for 480 minutes a day. The required daily output for part A is 320 units.
   a. What is the required CT for part A?
   b. If the same process must also produce 130 units per day of part B, what is the average required CT for both products? What is the minimal required daily production capacity of the process?
3. A workcell with two workers is divided into two subcells; one subcell has an actual CT of 323 seconds, the other a CT of 392 seconds. The workcell must produce a part with a required CT of 410 seconds. What is the required production capacity of the workcell? Does the workcell have adequate capacity?
4. Referring to the cell in the previous question:
   a. If it produces at the required CT, what is the resulting amount of daily idle time of the two workers?
   b. If it produces at the current maximum possible rate, how much will its daily output differ from the required output?

5. Three workers cycle around a cell in rabbit-chase fashion. If they were each working alone, one worker has the capability to cycle around the cell (task time + walk time) in 13 minutes, another in 14 minutes, and the other in 15 minutes.
   a. What is the CT and capacity if workers cannot pass one another?
   b. What is the cell CT and daily cell capacity if faster workers can pass slower workers?
   c. In a rabbit-chase workcell, is it possible for faster workers to pass slower workers? Discuss.

6. This chapter distinguished between assembly cells and machining cells. What about cells with both assembly operations and automatic machines? Refer to Figure 10.7:
   a. Suppose the operation at workstation 5 is performed by a single-cycle automatic machine that takes 10 seconds to load and start, but then runs for 30 seconds and stops after completing the operation. What effect does that have on the actual cell CT? What is the new cell CT?
   b. Suppose all the operations are manual assembly with the exception of operations at workstations 7 and 8, which are each performed by single-cycle machines that take 10 seconds to load and start, then run for 70 seconds and stop when finished. What is the new cell CT?
   c. What can you conclude about the actual CT of workcells where some operations are manual and some are automatic?

7. Use the eight-station assembly cell in Figure 10.7 to answer the next several questions. Also, assume the following:
   - Task times are as shown in Figure 10.7.
   - Holding areas are placed between subgroups (for example, shown as a, b, c, and so on in Figure 10.9).
   - Workers assigned to only one station have zero walk time but need 4 seconds to pick and place items.
   - Walk time between any holding area and the nearest workstation is always 2 seconds.
   - Station-to-station walk times are as shown in the following table.
   - Walk time between any two holding areas is the median of the walk times of the four stations involved (ignore any 0's). For example, as shown in the boxed area in the table, if one holding area is between stations 2 and 3, and the other is between station 6 and 7, the walk time between them is 10.

	To Station									
From Station	***In***	***1***	***2***	***3***	***4***	***5***	***6***	***7***	***8***	***Out***
In	0	2	6	7	10	12	11	11	9	8
1	2	0	5	6	7	8	9	10	9	9
2	6	5	0	5	10	10	10	10	9	9
3	7	6	5	0	7	10	10	11	11	12
4	10	7	10	7	0	5	8	10	11	12
5	12	8	10	10	5	0	7	8	9	9
6	11	9	10	10	8	7	0	5	6	7
7	11	10	10	11	10	8	5	0	5	5
8	9	9	9	11	11	9	6	5	0	2
Out	8	9	9	12	12	9	7	5	2	0

a. Divide the cell among four workers so times for them are as similar as possible. What is the resulting cell CT and daily capacity?
b. Repeat (a) for six workers. What is the cell CT and capacity?
c. What is the unit manufacturing cost for four workers. Assume machine operating cost is $80 and a labor rate of $20/hr. Given this result and results in the chapter for other numbers of workers, what number of workers gives the lowest unit manufacturing cost?

8. Use the eight-station machining cell in Figure 10.10 to answer the following questions. Assume all machines are single-cycle, automatic, and require a 10-second setup time.
   a. Divide the cell among four workers so the times among them are as similar as possible. What is the resulting cell CT and daily capacity?
   b. Suggest a way to reduce cell CT down to at least 40 sec/unit by subdividing the cell among five workers and adding machines. Indicate the number of workers and their machine assignments; also indicate the number of each kind of machine needed at each workstation. Add 2 seconds of walk time for every machine a worker walks past (i.e., bypassing without stopping).

9. Refer again to the assembly cell in Figure 10.7. Suppose the product made in the cell is small, and that instead of walking around the cell with just one unit of product, the worker carries a rack that holds six units of the product. As a result, the worker does tasks on six units at a time at each workstation. Assume the walk times are as shown in Figure 10.7. The operation times shown in Figure 10.7 are for assembling one unit. Assume handling of each unit at each workstation takes 3 seconds. What is the average CT per unit in seconds?

10. A four-workstation cell has single-cycle machines to perform all operations. The walk time around the cell is 60 seconds. The times (in seconds) for the machine operating cycles and setups (unload, changeover, load, and start machine) are listed next.

	Machine			
	A	B	C	D
Operating cycle (sec)	152	173	175	190
Setup (sec)	23	31	52	28

    The cell produces different kinds of parts continuously, one unit at a time.
    a. What is the actual CT?
    b. Assume the cell CT must be reduced to 215 seconds. Discuss where in the cell you would have to make changes to achieve this CT. Discuss alternatives or possible actions for making the changes.

11. For the workcell in the previous problem, assume that parts are produced in 20-unit batches (i.e., 20 of one part are produced in the cell, the machines are changed over, then 20 of the next part are run, etc.). The operating times are the same as shown in the table in problem 10, except the setup times apply only when the machines are changed over between batches of parts. Between identical parts, the automatic load–unload time is 2 seconds. What is the actual CT of the workcell?

Chapter 11

Standard Operations

There's a method to this madness.

—**William Shakespeare,**
Hamlet

Effective application of focused facilities, workcells, quick setup, machine preventive maintenance, and pull production is predicated in large measure on work discipline on the shop floor, discipline established by considering the best way to do work, then monitoring to ensure the work is actually done that way. Such is the purpose of **standard operations** and **work standards**.

Standard operations are also fundamental to the process of continuous improvement. What many organizations fail to realize is that without good standards it is difficult or impossible to achieve high production efficiency, to match production output with demand, to keep work in process (WIP) small, and to improve quality.

The principal focus of this chapter is on the elements of standard operations, including standard times, standard operations routine, and standard in-process inventory. The chapter addresses the issues of how standards are developed, why they are important, and who uses them. Necessary conditions for successful development and application of standard operations, and the place of standard operation in continuous improvement are also considered.

Standard Operations

Standard operations are a group of standards that completely define all aspects of a task, operation, or process. They are also called **standard operating procedures**, though here the more abbreviated term *standard operations* is used to imply the broadest possible definition of work standards, which goes beyond standards of procedures.

Shop-Floor Relevancy

Standard operations define best practice, although what constitutes best practice is a relative thing, depending on the situation. Since no one standard of work can fit all situations, no standard should be considered rigid and unchangeable. Standard operations must be adapted to reflect, for example, changes in product demand, current worker skill levels and proficiency levels, current equipment, as well as how work is actually performed. As things change on the shop floor, so should the standard operations.

Standard operations are not the same as standards for product requirements or process performance, which are typically developed by engineers and used in product design and quality assurance. Rather, they are the work procedures, tasks, and times prescribed for the shop floor to produce a unit of output.

In lean plants, production flexibility is achieved through continuous, small adjustments to production capacity, schedules, and priorities based on decisions made on the shop floor. Ideally the facilities are arranged in focused factories and workcells where multiskilled operators are each able to perform multiple functions. Given that, standard operations provide supervisors and workers the information necessary to determine the appropriate work procedures, routing sequences, and required number of workers to meet given product demand and product-mix requirements.

Shop-Floor Involvement

Initially, the key people in development of standard operations are industrial engineers, but with a large measure of participation from shop-floor supervisors and frontworkers. Eventually, the goal should be for supervisors and workers themselves to perform the work measurement and standards development, and for engineers to provide only assistance. Supervisors usually have the most current information about the status of everything on the shop floor and often they are the most knowledgeable about the best way to schedule and use facilities to achieve production goals. When supervisors help determine the standards, they have a better grasp of the standards and can explain them better to workers.

Putting responsibility for standards development and implementation at the shop level can result in standards that are more accurate, up to date, and more accepted by workers than those developed by staff personnel (who are sometimes viewed as outsiders). It can also result in process improvements that are overlooked by staff specialists. Maffei and Meredith describe such a case, a flexible manufacturing cell where operators suggested process changes to improve the quality of production output.[1] The engineer in charge of process quality felt that the changes were unnecessary, but when the operators took the initiative and implemented the changes anyway, the scrap rate dropped from over 50% for hard-to-make products to a very small percentage.

Of course, so that supervisors and frontline workers will be capable of developing good standards and applying them to planning, scheduling, and controlling, first they must be trained in the methods and procedures of operations analysis, time–motion study, and related tools.

Benefits

A benefit of putting standards development at the shop-floor level is that whenever changes in operations occur, the standards can be revised immediately. Otherwise the revision must wait until a staff person gets around to it. Ability to revise standards quickly is especially important in plants where cell configuration and worker assignments are in constant flux. Further, creating up-to-date standards permits staff planners to prepare schedules that are feasible and that reflect shop-floor status and current capabilities. Standard operations also make it easier to identify safety hazards and procedures that lead to or permit defects and then to modify them.

In a flexible factory, workers are rotated among jobs and responsibilities, depending on demand. Posted standard operations, located where workers can easily reference them, enable workers to quickly become familiar with the standards and procedures of their new work assignments. They also make it easy for workers and supervisors to periodically check actual work against the standards, which corrects backsliding and prevents falling into bad habits.

We will consider four aspects of all standard operations: takt time, completion time per unit, standard operations routine, and standard WIP inventory.

Takt Time

The starting point for setting standard operations is the **takt time** (the **required cycle time,** discussed elsewhere in this book).[2] That's because the quantity of something to be produced should be based upon a production output goal, and that goal is best stated in terms of the frequency an item should be produced in the allotted time—the takt time. Once the takt time is set, operations are planned so that the timing of production output is as close as possible to that time. The takt time is

$$\text{Takt time} = \text{Daily time available/Required daily quantity} = T/Q$$

If, for example, 192 units must be produced and the available time is 480 minutes, then

$$\text{Takt time} = 480 \text{ min/day} \div 192 \text{ units/day} = 2.5 \text{ min/unit, or } 150 \text{ sec/unit}$$

Assuming the process runs continuously throughout the day, it must produce one unit every 150 seconds.

In determining the takt time, no allowance is made for waste in the process. The daily time available should not be reduced to allow for equipment breakdowns, idle time, or rework, and the required daily quantity should not be increased to allow for defective items. When the takt time includes allowances for waste, as it often does, the sources of waste are never addressed or eliminated. When no allowance is made for the sources of waste, attention is drawn to the wastes and they are remedied.

Completion Time Per Unit

Another aspect of standard operations is **completion time per unit**, or the average (actual) time required to process one unit. Completion time per unit is determined for every task and operation, even for nonvalue-added tasks like handling, picking and placing, inspection, and so on.[3]

To avoid confusion, the terms *task* and *operation* are distinguished as follows: A **task** refers to an elemental unit of work—a simple step or motion; an **operation** refers to a group of tasks, such as those usually performed at a workstation. For example, suppose a worker performs six tasks. If the worker does the six tasks in sequence, without interruption or doing any other tasks, then those six tasks combined form an operation.

Time to Complete a Task or an Operation

To determine the completion time per unit, start by determining the standard **task time**, which is the expected time for an average worker to perform a task at a satisfactory level. The task time must include all manual and machine time involved in the task. Suppose a part is machined at a workstation using a single-cycle, automatic machine that runs for several seconds and then shuts off. As defined in the last chapter, the automatic run time is the **machine time**. Suppose the worker at the workstation also performs several tasks by hand such as unloading and loading the machine, setting the machine, turning it on, and so on. The sum of the times for these tasks is the **manual** or **handling time**.

Determination of the manual time is based upon well-established work measurement techniques.[4] Briefly, it involves measuring with a stopwatch the actual time required for a worker to perform a task. Each task must be measured many times to get an accurate average. In addition, the standard involves a judgment call about the worker's speed, which should include consideration of the worker's skill level, the amount of effort she is expending, and particular conditions under which she is operating. Supervisors are often familiar with individual workers' skills, motivation, and work habits, which is a reason for involving them in standards development.

Calculation of the standard time is based upon the actual, observed time for the task, the performance rating of the worker, and allowance for any unavoidable delays:

$$\text{Standard task time} = \text{Actual time} \times \text{Performance rating} \times \text{Allowance factor}$$

If a worker is thought to be of average skill, then her **performance rating** is 100%. If her skill is higher or lower, then the percent is raised or lowered accordingly. The **allowance factor** takes into account delays that are unavoidable but that are expected as a normal part of the operation and cannot be eliminated. It is important that the allowance factor not account for delays that can be eliminated (sources of waste) and not be used as a fudge factor to allow for potential (unexpected) delays. Ordinarily, the worker being measured for the standard should be selected for being average in terms of speed and performance, and only when such a worker is not available will it be necessary to adjust the time with a performance rating. The worker performance rating is assessed by the worker's supervisor prior to time study.

For every newly created task, there will be a **learning curve** effect, even for an average worker. Not until a worker builds proficiency in a task will she be able to perform at the normal level. For a newly designed task (one for which no worker has reached normal ability), the standard task time will have to be periodically remeasured to account for the worker's increasing proficiency.

Suppose a time study reveals a worker takes on average 23 seconds to perform a task. The supervisor, who feels that the worker in the time study is slightly less proficient (less capable or less experienced) than other workers, sets the performance rating of the worker at 90%. The supervisor estimates that unavoidable interruptions from upstream stations will cause the station of the task being studied to be idle about 20% of the time. As a result, the standard time of the task is

$$\text{Standard task time} = 23 \text{ sec} \times 0.90 \times 1.20 = 24.84, \text{ or } 25 \text{ sec}$$

Standard task times are used to determine production capacity and work schedules. Because not every worker will perform according to the standard, workers will sometimes end up producing too much or too little in the scheduled time. A Kanban system will prevent overproduction, even if workers perform better than the standard, but shortages can occur if workers consistently produce below the standard. If the latter happens, the standard time should be revised upward so that production schedules will be realistic, and a short-term improvement plan should be implemented to improve workers' performance. An alternative is to retain the existing standard and bring in workers who are able to meet it. So, as much as possible, the standard times must accurately account for the realities of the shop floor, including workers' abilities. When several workers are rotated to do one task, then the task time must be adjusted to represent the weighted average time of all the workers.

Completion Time per Unit

The **completion time per unit** is the time to process one unit. It might be the same as the standard task time, though not necessarily. For example, referring to Figure 11.1, suppose that for the third task the inspect-part portion of the task is done only on every other part. If a part is inspected, the task time is 7 seconds; if a part is not inspected, the time is 3 seconds. Thus, the completion time per unit is the average time for two units:

$$(3 \text{ sec} + 7 \text{ sec})/2 = 5 \text{ sec/unit}$$

Part R2D2-block		*Part (description or drawing)*			
Station 4, BZ					
Operation 1 *A: 225/insp*					
Task	*Description*	*Procedure*	*Act*	*CTU*	*Notes*
1	Press eject	Check mill 225 autoselect switch: rotation stopped; machine is off.	1	1	
2	Turn autoselect to 0	Check part has ejected.	1	1	
3	Pick up part; or	Pick up part from rack. Drop part into right chute or	3	5	
	Pick up part, inspect part	slip part through sleeve (every second part only), then drop part into right chute.	7		
4	Insert part	Pick up part from left chute. Place part into mill slot. Ensure part is notch-side down. Ensure part rests snugly in slot.	2	2	
5	Press start button	Check that autoselect switch is rotating.	1	1	
6	Autocycle	Mill 225 autorun for one cycle, then stop.	30	30	
		CTU Total		40	

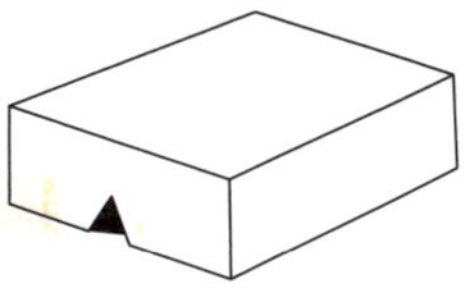

Figure 11.1 Tasks and completion times per unit for a machined-part operation.

The six tasks in Figure 11.1 comprise a simple operation on the machined part shown. In the figure, standard task times are listed in the Act column, and completion times per unit in the CTU column. Assuming the full operation comprises only the six tasks, and those tasks are performed in the sequence listed, then the completion time per unit for the operation is the sum of the completion time per unit for all the tasks, 40 seconds.

Production Capacity

Accurate completion time per unit is important because it is the basis for estimating **production capacity**, the number of units that can be produced in the available production time. Production capacity can be computed for every level of work—an individual task, an operation or workstation, a production process, or even an entire plant. In the case of batch production, accurate information about the process **lot size** and **setup time** is also necessary.

One way to compute production capacity, N, for an operation is

$$N = T/(C + m)$$

where

C = Completion time per unit
m = Setup time per unit
T = Total operation time

Using the operation in Figure 11.1 as an example, if parts at this operation are produced in batches of 300, if each batch requires a setup time of 3 minutes, and if the total operating time per day is 480 minutes, then

$$m = 180 \text{ sec}/300 \text{ units} = 0.6 \text{ seconds/unit}$$

$$N = 480 \text{ min/day} \div [(40 + 0.6)/60] \text{ min} = 709.4 \text{ units/day}$$

Standard Operations Routine

Once the completion time per unit has been determined for every task (or operation, depending on the level of analysis), the next step is to determine the sequence in which the tasks will be performed. This sequence is called the **standard operations routine**, or **SOR**. At the task level, the SOR gives the tasks and the required sequence in which a worker performs them in a given operation. At the operation level, the SOR includes the operations and the sequence in which a worker performs them.

Kinds of SORs

There are three kinds of SORs, one for each generic kind of work routine:[5]

1. SOR for a single repeated operation: This SOR gives the prescribed sequence in which a group of tasks is to be performed over and over. If, for instance, the worker in the previous example were to perform the operations in Figure 11.1 in the sequence listed, then that sequence would be the SOR for that operation. If the tasks listed in Figure 11.1 could be performed in some other sequence and still produce the same results, then these other sequences might also be considered before a final, prescribed SOR sequence is selected.

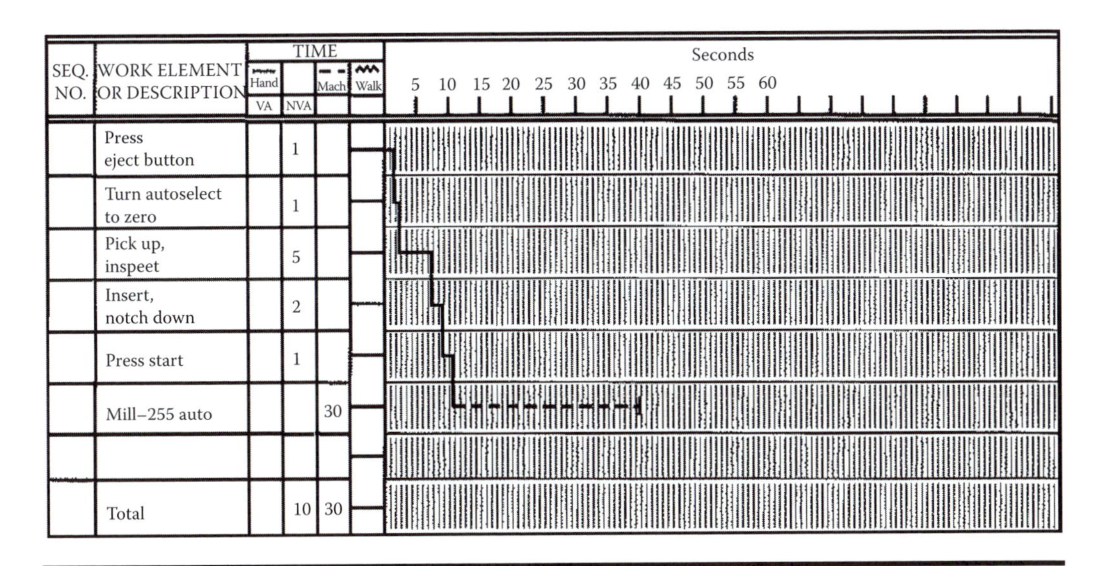

SEQ. NO.	WORK ELEMENT OR DESCRIPTION	TIME Hand VA	NVA	Mach	Walk
	Press eject button		1		
	Turn autoselect to zero		1		
	Pick up, inspeet		5		
	Insert, notch down		2		
	Press start		1		
	Mill–255 auto			30	
	Total		10	30	

Figure 11.2 SOR sheet for a single, repeated operation.

2. SOR for multiple, repeated operations: This SOR gives the prescribed sequence in which several operations are to be performed on a repeating basis. If a worker operates several machines or performs several operations at multiple workstations, as in a workcell, then the SOR will show the sequence in which the worker is to visit the workstations. An example is shown later.
3. SOR for multiple nonrepeated operations: This SOR gives a sequence of operations or tasks that vary throughout the day, but that, combined, are the same every day. For example, suppose a worker operates a machine and is responsible for setting up and starting the machine, periodically loading batches of parts into it, periodically sampling and inspecting the parts, periodically recalibrating the machine, and then stopping the machine, cleaning it, and doing simple PM. The SOR would give the sequence and approximate times throughout the day when each of these tasks and operations is to be performed.

SOR Sheet

One way to determine the SOR is with the **SOR sheet**. Figure 11.2 shows the general appearance of an SOR sheet for a single, repeated operation based on the information from Figure 11.1. The times in Figure 11.2 correspond to the completion times in Figure 11.1. In general, a solid line meandering downward through the time grid on an SOR sheet represents the time a worker spends doing the manual tasks; a dashed line represents the run time of the machine after the worker turns it on.

Example 1 illustrates an SOR sheet for multiple, repeated operations.

Example 1: SOR for Multiple, Repeated Operations

Eight operations are performed in a workcell that uses single-cycle, automatic machines. Each machine automatically stops after it has completed processing a part. Assume the eight operations are arranged in the workcell shown in Figure 11.3. All the manual (hand) times, machine times, and completion times per unit for the operations are shown in Table 11.1. For each operation the

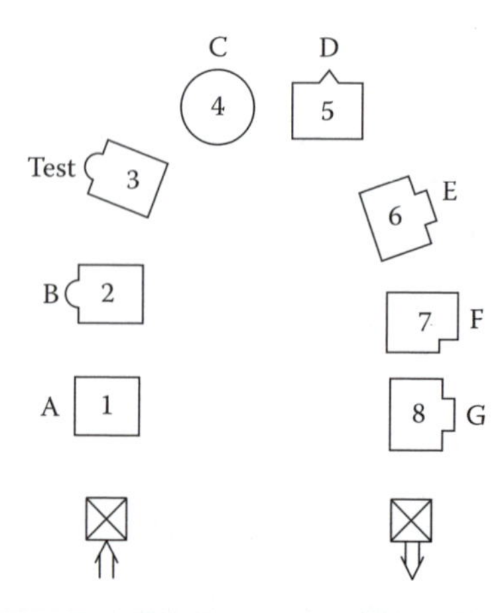

Figure 11.3 Location of operations, one per workstation.

times are standard times as developed from a time study of tasks in each operation. As a case in point, the times for operation 1 in Table 11.1 are based on the time figures shown in Figure 11.2.

The procedure for creating the SOR now follows.

1. The SOR for any group of operations must be designed such that each work cycle will be completed within the takt time (required CT). First, then, determine the takt time and draw it on the SOR sheet. Assume the takt time is 150 sec. This is shown at the vertical line marked (1) on the SOR sheet in Figure 11.4.
2. Determine the approximate number of workers and the range of operations that each worker will be able to perform within the takt time. The range (the number and kinds of operations that a worker is assigned) is determined by the worker CT, which is the time required for the worker to walk among the operations and perform the manual tasks. The worker CT cannot exceed the takt time. If the worker CT exceeds the takt time, then the worker has too many operations to perform in the available time, and some of them must be reassigned to other workers.

Table 11.1 Operations and Times for Workcell in Figure 11.3

Order of Operation	*Description of Operation*	*Handling Time (sec)*	*Machine Time (sec)*	*Completion Time/Unit (sec)*
1	A: 225/insp	10	30	40
2	B: 26/909	10	40	50
3	Test 1: A/B	10	10	20
4	C: R9/223/ZZ6/422	10	40	50
5	D: L28/899	10	30	40
6	E: 225/R76	10	30	40
7	F: R9/890/44; 67/LPT; Test 2: F	10	70	80
8	G: HU356/3; 46P/567; Test 3:G	10	70	80

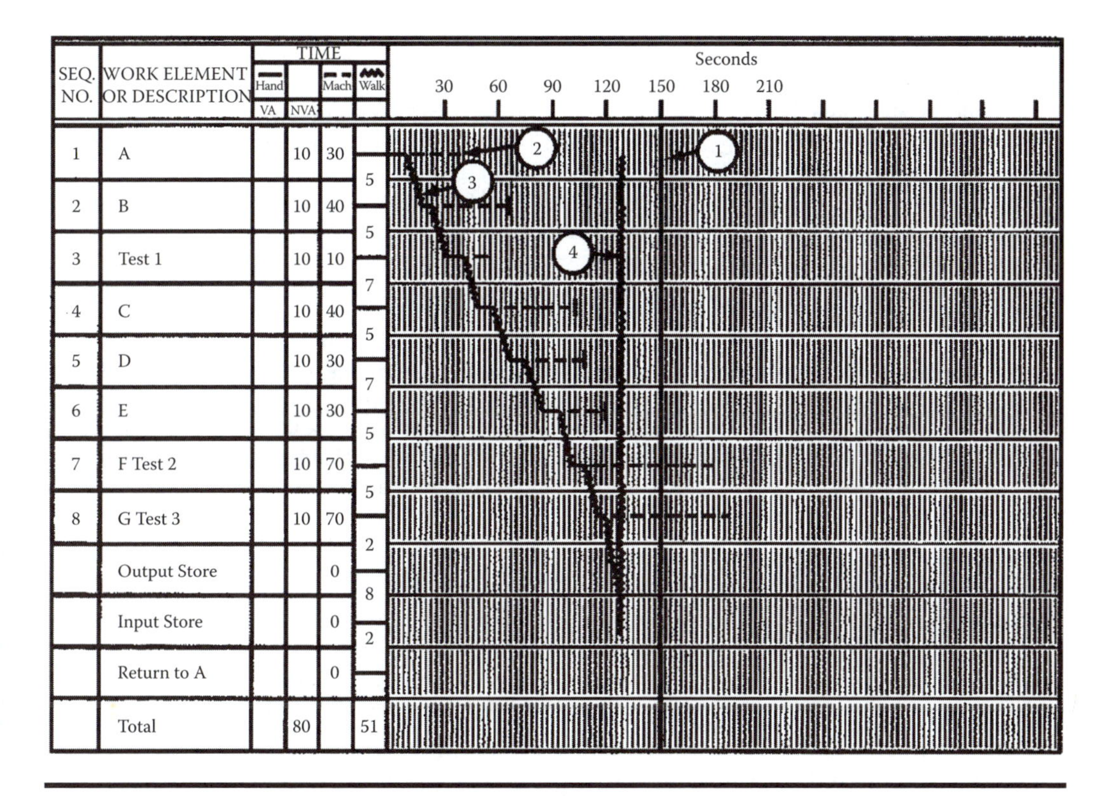

Figure 11.4 SOR sheet for multiple, repeated operations.

Likewise, the completion times per unit at none of the operations can exceed the takt time. If it does, some way must be found to reduce it. Perhaps the manual time can be shortened, some of the manual tasks can be automated or eliminated, or the entire operation replicated (thereby effectively cutting the completion time per unit in half). Simply speeding up a machine is usually not an option because machines operate at prescribed speeds with limits, and increasing the speed accelerates machine deterioration and contributes to defects.

As Table 11.1 shows, the completion times for all eight operations are well below the takt time of 150 sec. As the table also shows, the sum of the manual times for the eight operations is 80 sec, which indicates that as long as the total walk time between the eight operations in succession is less than 150 – 80 = 70 sec, the worker will be able to perform all eight operations. The theoretical way of verifying this is to divide the sum of the manual times for the operations by the takt time. In this case the resulting ratio is 80/150 = 0.533. The ratio, being less than 1, suggests that one worker is needed.

3. Draw on the SOR sheet the completion time per unit for the first operation the worker performs. On Figure 11.4 this is shown as the dashed line marked (2) on the grid part of the sheet. The solid portion of the line represents the manual time of the first operation, while the dashed portion represents the machine time. The combined length of the solid and dashed lines is the completion time per unit at that operation.
4. Whenever a worker must walk between operations, the walk times are also represented on the SOR sheet. To show the walk time, draw a wavy line from the ending of the manual portion of the first operation to the time when the next operation is to begin. Suppose in the example the walk time is 5 seconds. The wavy line marked (3) on the SOR sheet, going from the end of operation 1 to the beginning of operation 2 represents 5 seconds walk time. (Note in contrast the SOR sheet in Figure 11.2 has no wavy lines. That is because the tasks in the operation can all be performed by an operator standing or sitting at one workstation.)

5. Steps 3 and 4 are repeated for all operations the worker must perform. The result is shown in Figure 11.4 as the combined manual task and walk times, represented by the line running diagonally down the page, and the machine processing times, represented by dashed lines. In the actual workcell, the procedure would be this: The worker would walk around the cell and stop at every machine where he unloads a part, loads another part, sets the machine, and then turns it on; he would then walk to the next machine, carrying the part from the last machine with him, and repeat the procedure.
6. If the worker must walk from the last operation back to the initial operation to resume the routine (that is, walk around the cell again), then this is shown as a wavy line. In Figure 11.4 this is shown as the wavy line marked (4).
7. If the worker arrives back at the initial operation at or before the takt time (line (1)), then the operations routine sequence is verified as feasible. In the example, the worker arrives at the initial operation at 131 sec. Thus, the actual CT is 131 seconds.
8. After the SOR has been developed the supervisor performs the SOR sequence to ensure that the in-practice CT conforms to the theoretical CT on the SOR sheet. If it does, and if the supervisor feels comfortable with the operations sequence and procedure, then the SOR is retained and taught to workers.

Operations Routine and Process Routing Sequence

Part of establishing the SOR is establishing the sequence in which workers will perform tasks and operations. In the previous example the SOR followed the same sequence as the process routing sequence; in general, however, the two do not have to be the same. To avoid confusion, we distinguish the SOR as the order in which the worker performs a sequence of operations and the **process routing sequence** as the order in which operations must happen to make a part or product, that is, the route a *product* must follow from workstation to workstation.

A worker in a cell might walk from station to station in a route that is opposite the route followed by the product. This is shown in Figure 11.5. Alternatively, the cell might be divided among several workers, each performing a subset of cell operations, and in that case the SOR for each worker would be different from (and only a portion of) the process routing sequence of the product. For example, the cell shown in Figure 11.3 might be split between two workers, one at

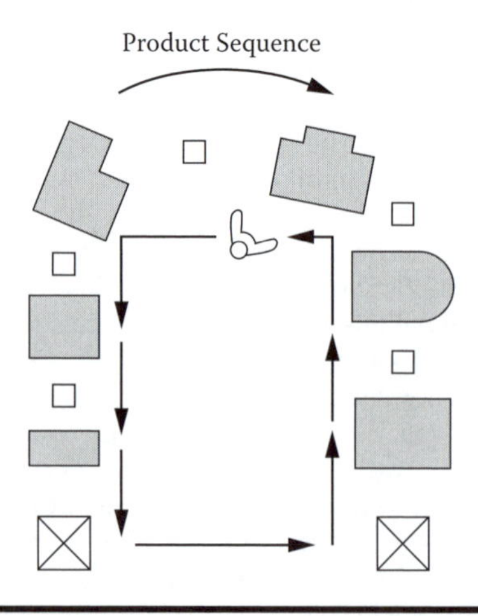

Figure 11.5 Operation routing sequence counter to process routing sequence.

workstations 1, 2, 7, 8, the other at 3, 4, 5, 6. In neither case would the worker SOR be identical to the processing sequence of 1, 2, 3, …, 8. While routing workers through operations in a sequence other than the processing sequence might be necessary, it is more difficult to determine whether the resulting worker CT conforms to the takt time.

Idle Time

In Example 1, Figure 11.4, the worker has 150 – 131 = 19 seconds of time remaining at the end of each cycle. The question arises as to what should be done with this time. In general, considerable difference between actual CT and takt time means one thing: an opportunity or mandate for improvement. If actual CT is less than takt time, that suggests a potential for reducing the number of workers (assuming more than one worker is assigned to the process) or for giving existing workers additional tasks (possibly for other products and even in other cells, as discussed in the previous chapter). Any operations can be added to the SOR sequence as long as the additional handling and walk times do not exceed the time remaining, in the example, 19 seconds.

Now, if the actual CT exceeds the takt time, that means an improvement is mandatory if the output is to meet demand. The initial improvement might be to increase the number of workers to sufficiently decrease the actual CT.

Standard Quantity of WIP

The standard quantity of WIP is the minimum in-process inventory necessary for the process to function. It consists solely of the items being processed at each operation as well as any items held between operations. Ordinarily it is the minimum quantity of material necessary to achieve a smooth flow of work, although that quantity will vary depending on the type of machine or workstation layout, and the SOR. For example, if the SOR sequence is the same as the process routing sequence, then it might not be necessary to hold any items between operations. In that case, the standard WIP quantity will be the same as the number of operations. In the machining cell example in Figure 11.3, one part will be held at each machine; since the SOR sequence is the same as the routing sequence, the worker can transfer a part from one operation to the next as he walks between machines. No additional parts are necessary, so the standard WIP would be eight parts.

If, however, the SOR sequence is the opposite of the process routing sequence, then to avoid backtracking at each machine at least one piece must be held between every pair of machines. (That way, each time the worker unloads a part from a machine, he can put it into the holding area on the right of the machine, then take a part from the holding area on the left and load it into the machine.) In the machining cell in Figure 11.3, this would make the standard WIP 15 units, not counting stock at the input and output storage areas.

Usually there are additional considerations that factor into the size of the standard WIP. When multiple workers cycle through different parts of a process, even when they follow the same general direction as the process routing sequence, some units must be held at the places where workers hand off items to one another (an example is given later). It might also be necessary to accumulate some quantity of items to, say, perform quality checks, allow time for machine temperature to cool down or heat up, or perform an operation in which batching of items is necessary.

The reason for specifying the standard WIP quantity is to keep WIP to the minimum required for effective production. An observer can easily refer to the standard WIP quantity as posted, look at the quantity actually on hand, and determine if there is a discrepancy. If there is, questions arise at to what is wrong and what needs to be done to correct it. This is another example of visual management.

Standard Operations Sheet

Information about the completion time per unit, the SOR, and standard WIP are combined and displayed in one place—on the **standard operations sheet (SOS)**, or standard worksheet. The SOS as originated at Toyota contains the takt time, the SOR, standard quantity WIP, and the actual CT, as well as diagrams indicating locations in the process to check for quality and to pay attention to safety.[6] Figure 11.6 shows the SOS sheet for the one-worker operation in Figures 11.1 and 11.2. Another example of an SOS, representing a two-worker operation, is shown in Figure 11.10. Details about this figure are explained later.

Part *R2D2-block* Employee *Finney* New Revised ✓ | Document No.

Station 4,*BZ* Supervisor *Goebel* M/D/Y 8/14/10 | *C*–33704 *TBZ*–089

Department: *Rollon*

Std Time: 39.5 Std Quant: 699 Std WIP: 2 Act Time: 40 *sec.*

Task	Description	Procedure	Std	Act	CTU
1	Press eject	Check mill 225 autoselect switch: rotation stopped; machine is off.	1	1	1
2	Turn autoselect to 0	Check part has ejected.	1	1	1
3	Pick up part or Pick up part, Inspect part	Pick up part from rack. Drop part into right chute or slip part through sleeve (every second part only), then drop part into right chute.	3 6	5 7	5
4	Inspect part	Pick up part from left chute. Place part into mill slot. Ensure part in notch-side down. Ensure part rests snugly in slot.	2	2	2
5	Press start button	Check that autoselect switch in rotating.	1	1	1
6	Autocycle	Mill 225 autorun for one cycle, then stop.	30	30	30

Work Area Layout

Left chute, Mill 225, Rack, Right chute, Slot, On, Eject, Sleeve, Autoselect

● WIP Inventory ✓ Quality Check

Part (description or drawing)

SEQ. NO.	WORK ELEMENT OR DESCRIPTION	Hand VA	Hand NVA	Mach	Walk	Seconds 5 10 15 20 25 30 35 40 45 50 55 60
	Press eject button		1			
	Turn autoselect to zero		1			
	Pick up, Inspect		5			
	Insert, notch down		2			
	Press start		1			
	Mill 225 auto	30				
	Total	30	10			

Figure 11.6 SOS sheet for one operation.

SOSs should be displayed at each operation and at each sequence of operations (such as a workcell) so workers can readily refer to them. The SOS is an important tool for areas of visual management:

1. It serves as a guide to inform workers about the SOR, completion times, and other important aspects of the operation.
2. It helps the supervisor assess whether the operations are being done according to standards.
3. It serves as a tool to evaluate performance and improvement.

The SOS is always dated to indicate when the last revision occurred.

Improvement Tool

The SOS should be revised regularly since, says Monden, standard operations "are always imperfect and operations improvements are always required in a process."[7] Standard operations should never be considered as fixed. They must be adapted to reflect changing demand, worker skills, and so on, and ongoing improvements to the process. Even after standards have been adapted to the situation, they should be continually scrutinized for places where the process can be tightened up to improve efficiency, reduce cost, and improve quality. At Toyota, the takt time is occasionally changed for the sole purpose of requiring frontline workers to review processes and create new standards that conform to the new takt time. This is one reason why Toyota is much better at adapting to changes in product demand and product mix than its competitors.

Example 2: Adjusting SOR to Meet Increased Demand

Suppose increased demand for the item produced in our eight-operation machining cell mandates that the 150 seconds takt time be reduced to 95 seconds. Clearly, the standard operations must be changed. Assuming that the completion times per unit cannot be reduced, then it will be necessary to add more workers. Suppose we add one more worker. Either we can cycle both workers through all the operations in a rabbit-chase fashion or we can split the cell between the two workers. Suppose because of the required skills, it is more practical to split the cell between workers. Figure 11.7 shows the range of operations for each worker and the sequence the workers

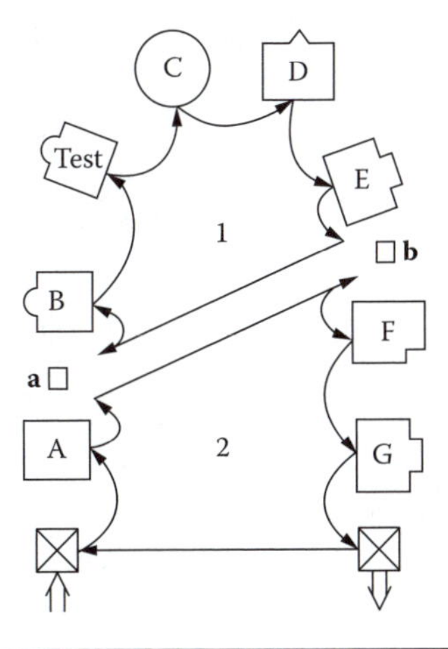

Figure 11.7 Range of operations assignments for two workers.

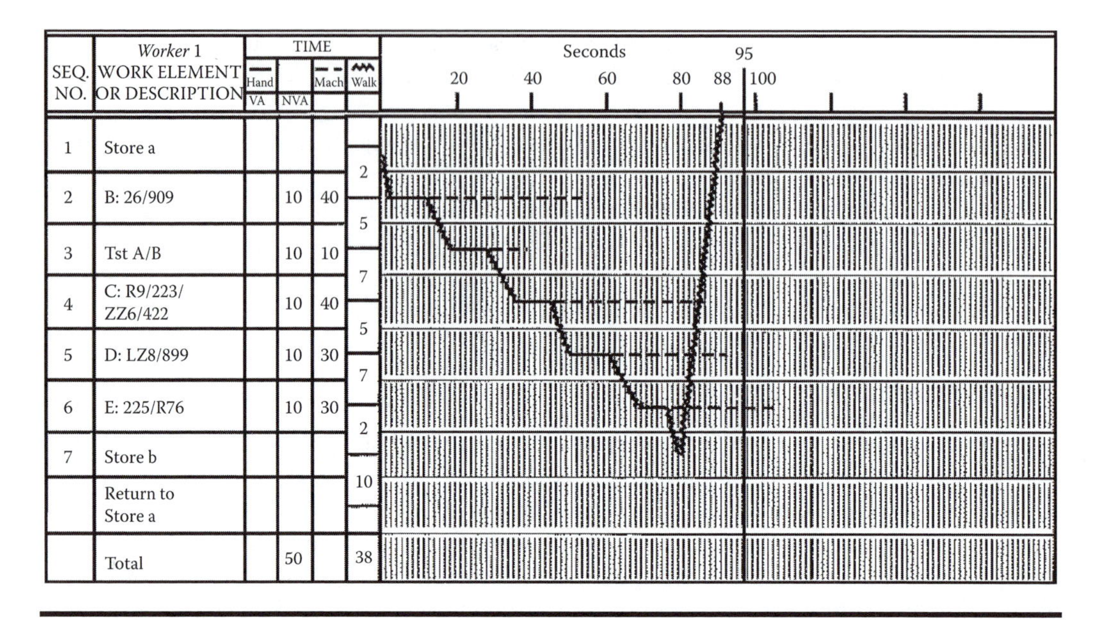

Figure 11.8 SOR sheet for worker 1. Total CT = 88 seconds.

will follow. It also shows the locations of two in-process holding areas, **a** and **b**, places where the workers hand off parts to one another. As mentioned, such holding areas are necessary whenever there is more than one worker and when their CTs differ.

Figure 11.8 shows the SOR sheet for worker 1. The worker CT is 88 seconds, which falls within the takt time of 95 seconds. Figure 11.9 shows the SOR sheet for worker 2. Two cycles of work are shown to illustrate the point that, although the worker might be capable of completing a work

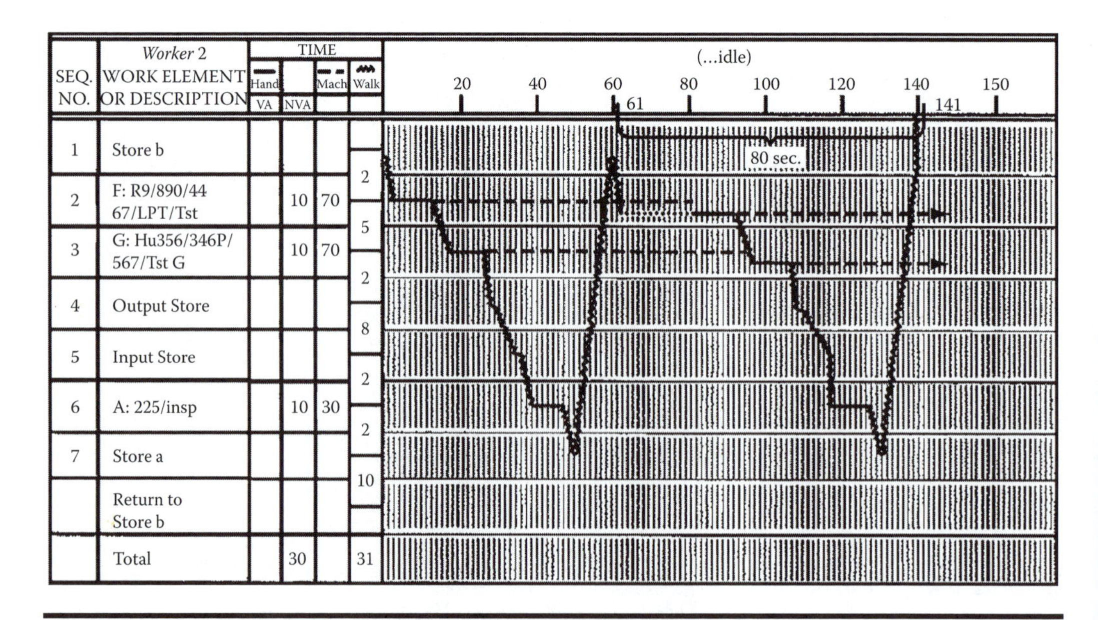

Figure 11.9 SOS sheet for two-worker cell. Total CT = 61 seconds, which is less than the 80 seconds CT at operations 2 and 3; therefore, CT = 80 seconds.

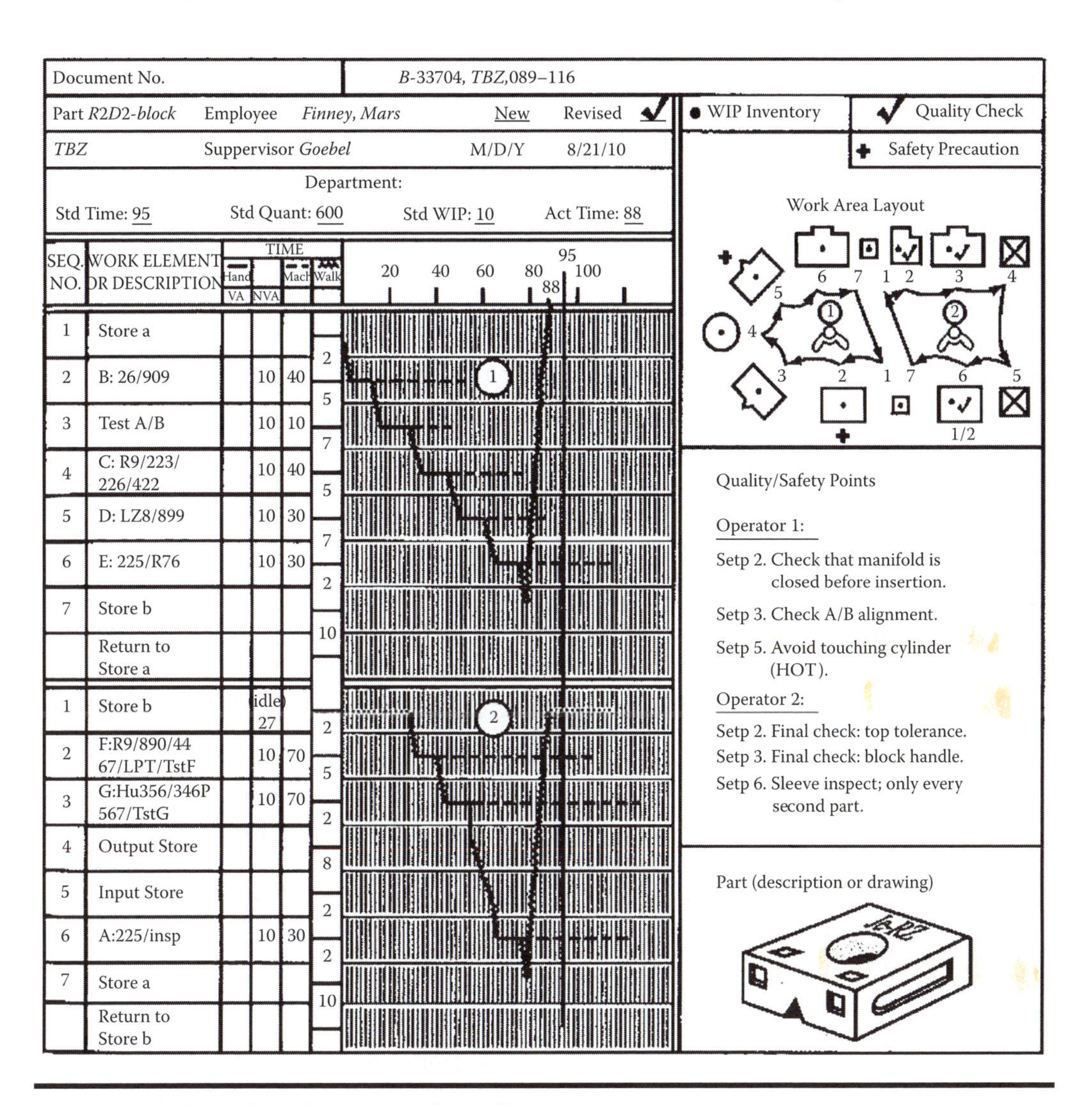

Document No.	*B*-33704, *TBZ*,089–116		
Part *R2D2-block*	Employee *Finney, Mars*	New	Revised ✓
TBZ	Suppervisor *Goebel*	M/D/Y	8/21/10

Department:

Std Time: 95 Std Quant: 600 Std WIP: 10 Act Time: 88

SEQ. NO.	WORK ELEMENT OR DESCRIPTION	Hand VA	Hand NVA	Mach	Walk
1	Store a				2
2	B: 26/909		10	40	5
3	Test A/B		10	10	7
4	C: R9/223/ 226/422		10	40	5
5	D: LZ8/899		10	30	7
6	E: 225/R76		10	30	2
7	Store b				10
	Return to Store a				
1	Store b		idle 27		2
2	F:R9/890/44 67/LPT/TstF		10	70	5
3	G:Hu356/346P 567/TstG		10	70	2
4	Output Store				8
5	Input Store				2
6	A:225/insp		10	30	2
7	Store a				10
	Return to Store b				

Quality/Safety Points

Operator 1:

Setp 2. Check that manifold is closed before insertion.

Setp 3. Check A/B alignment.

Setp 5. Avoid touching cylinder (HOT).

Operator 2:

Setp 2. Final check: top tolerance.

Setp 3. Final check: block handle.

Setp 6. Sleeve inspect; only every second part.

Part (description or drawing)

Figure 11.10 SOS sheet for two-worker cell.

cycle in only 61 seconds (shown for cycle 1), because some operations take as long as 80 seconds, the worker is forced to accept a CT of 80 seconds (shown for cycle 2). Since the completion times per unit at operations 2 and 3 are each 80 seconds, the worker has to wait at operation 2 for 19 seconds until the machine has finished.

Finally, the SOS sheet for the eight-operation, two-worker cell is shown in Figure 11.10. To create this sheet the SOR sheets for the two workers must be made compatible. The CT for worker 1 is 88 seconds, so the CT for worker 2 must also be 88 seconds. Because of the discrepancy in CTs for the two workers, worker 2 will have to wait about 8 seconds each cycle at holding area **b** to receive a part from worker 1.

The SOS sheet in Figure 11.10 prescribes procedures that result in an 88-second worker cycle time. Since the takt time is 95 seconds, each worker will be idle for 7 seconds every cycle. (For worker 2, the 7 seconds is in addition to the 8 seconds she is already idle from waiting on worker 1.) Although it is unlikely that this is enough time to introduce other operations into the cell, the idle time might be used to make other forms of improvements such as inserting tasks that would reduce defects, improve safety, or increase the reliability of the machines. Additionally, if the cell is not linked into a pull system, the two workers could run the cell at the 88 seconds cycle time, in which case they would fill the daily requirement well before the day ends. They could then use the remainder of the day for problem solving, equipment PM, and so on.

An objective of standard operations is to prescribe the operations and number of workers needed to satisfy demand. In any workcell or sequential process, the SOR should be set up such that, ideally, the operation time of at least one worker is roughly the same as the takt time. That automatically sets the actual CT of the entire process to the right level. When the CTs of all workers cannot be balanced to the takt time (the usual case), then operations and tasks should be reassigned so that as much idle time as possible falls on just one worker. The worker team then concentrates on how to improve and reassign tasks so that that position can be eliminated. This is discussed more in Chapter 14.

Conditions for Successful Standard Operations[8]

Successful development and use of standard operations is predicated on six conditions:

1. Focus on the worker. The central concern in standard operations is with the worker, not machinery. Emphasis is on labor-time improvement and reduction.
2. Job security. As a result of labor-time improvement, fewer workers are needed to perform a job. No team of workers, however, can be expected to suggest improvements that will cause someone to be out of work. Successful implementation of standard operations thus requires guarantees that no jobs will be lost as a result of workers' efforts to improve operations. In expanding markets, workers are transferred to or rotated among other jobs. To retain jobs in stable or shrinking markets, outsourced work must be returned to the factory, workers must be trained to take on responsibilities traditionally done by staff and managers, and new business opportunities must be aggressively pursued.
3. Repetitive work. Although standard operations can be applied to tasks that are unique for every job (as in the case of a pure job shop), they work best in operations and work sequences that are repeated. This is not as constraining as it might seem. For all the products in a product family, the sequence of operations and completion times are often very similar, so the SOR for one product is quite like that of the others. Even in mixed-model production where different kinds of products are produced in mixed sequence rather than in a string, all of one kind at once, and where several SORs are followed depending on the product, every SOR will be repeated numerous times throughout the day.
4. Level production. Revising the number of operators, sequences of work, and groupings of operations every day is not practical, so the production schedule must be held level for some period of time.
5. Multiskilled operators. Workers must be able to operate multiple, different machines within a workcell and, ideally, in other workcells and areas of the plant so they can be rotated weekly or even daily, as needed, to meet demand requirements for all products.
6. Team effort. Standard operations are established by a team consisting of a supervisor and workers and operators. The team is trained in analysis for standard operations. Collectively, the team agrees on the tasks, sequences, number of workers, and worker assignments. Shops where jobs are rigidly classified and based on seniority will not be able to implement standard operations. Where union contracts are involved, the contract must be renegotiated. Most unions, however, are now aware of the importance of standard operations (as well as

the importance of expanded work roles in pull production and cellular manufacturing) and are willing to work through the implementation issues with management.

Standard Operations in the Service Sector

As with most of the concepts and tools in lean production, the principles and benefits of standard operations apply to all kinds of organizations, not just manufacturing. The following examples illustrate standard operations applications in nonmanufacturing settings.

Case in Point: Standardization in Hotels and Hospitals

Marriott International Inc. relies on rigorous standardized procedures, called standard operating procedures (SOPs) to ensure consistent and predictable efficiency, quality, and safety in all aspects of its business, ranging from room cleanliness to foodservice sanitation. Many years ago founder J.W. "Bill" Marriott recognized the importance of standardization and established SOPs that included room cleaning checklists, food recipes, and guest satisfaction scorecards. SOPs are necessary to ensure consistency in results, Marriott felt, because some aspects of business operations can be boring.[9] Procedures are necessary to show employees what is expected of the end result and the steps necessary to accomplish it. Says one Marriott proponent, "You know what you're going to get … even down to the shampoo, conditioner, and various lotions in your bathroom. By following a 66-point checklist, a maid cleans a room in 30 minutes. The beauty of the system is that not only does the maid clean the room efficiently but also she knows what the end result is supposed to look like, and she won't leave the room until it does. If the maid walks into a room where there is flooding, the lights don't work, and the wall is a mess, she is authorized to call in a plumber, electrician, or painter to fix it. The maid knows what the end result is supposed to look like, and she is empowered to do whatever it takes to accomplish it."[10]

As another example, consider hospitals and clinics, and the serious problem of patients developing bedsores and pressure ulcers. Though these problems are easily prevented by properly repositioning the patient every 2 hours, the perplexing issue is how to ensure that every patient is properly repositioned every 2 hours. No amount of scolding, encouraging, or posting of signs as reminders seems to work.

Several hospitals have virtually eliminated the problem of bedsores with a simple solution—an SOR that includes a checklist posted in the room of each at-risk patient on which staff members are required to mark the time and their initials each time they reposition the patient.[11] Because checklists (like all SORs) can be fudged, the procedure also requires that the nurse supervisor in charge regularly review the checklists and verify that patients are being properly repositioned.

If the supervisor suspects the checklists are being fudged, she must determine why they are being fudged—whether because an employee does not have the time or because the employee doesn't believe a patient always needs to be repositioned. For whatever reason, the issue then becomes a matter of engaging the employee so she understands that the procedure is in the best interest of patients and of ensuring that the employee has adequate time to conform to the procedure. Issues like this regarding conformance to SORs have to be addressed in all SORs in all organizations, which is one reason for involving workers in creating the SORs. Once the SORs are implemented, workers need to know they are being monitored to conform to the SORs, and supervisors need to know that they too will be monitored, but with regard to their monitoring and enforcement of the SORs.

Summary

Standard operations are the work procedures, sequence of tasks, and times prescribed for production of a unit of output. They are developed in large measure by suitably trained shop-floor supervisors and workers, with assistance from planners and engineers. Putting responsibility for standards development at the shop-floor level results in standards that are more accurate, current, and accepted by workers.

The main elements of standard operations are the takt time (required CT), the standard completion time per unit, standard operations routine (SOR), and standard WIP. The completion time per unit is the average time required to complete a task or operation (group of tasks). A group of operations combined in a particular, prescribed sequence is the SOR. The standard WIP is the minimum in-process inventory necessary for a process to function effectively. All information about completion times per unit, SOR, and standard WIP is summarized on the standard operations sheet (SOS). The SOS is prominently displayed so workers at each operation and workcell can readily refer to it.

Standard operations serve vital functions in competitive manufacturing. They are essential for communicating and training standard times and procedures, and they provide planners and schedulers with accurate, up-to-date information about cycle times and operations capacity. With standard operations, production plans and master schedules are more realistic and likely to meet production goals. For the most part, however, standard operations are a shop-floor tool. Workers and supervisors use them for their own capacity planning, work scheduling, production control, and work method improvement. They provide an important contribution in efforts to place responsibility for production planning and control on the shop floor. Along with setup-time reduction and equipment PM, standard operations are a prerequisite for pull production, cellular manufacturing, and, more generally, for agile, low cost, high-quality manufacturing.

Notes

1. M. Maffei and J. Meredith, Infrastructure and Flexible Manufacturing Technology: Theory Development, *Journal of Operations Management* 13, no. 4 (December 1995): 273–298.
2. Takt time has become the de facto term for required CT.
3. See, for example, E. Polk, *Methods Analysis and Work Measurement* (New York: McGraw Hill, 1984).
4. Work measurement techniques are described in, for example, R. Barnes, *Motion and Time Study*, 7th ed. (New York: John Wiley, 1980); B. Niebel, *Motion and Time Study*, 6th ed. (Homewood, IL: Richard D. Irwin, 1988); and Polk, *Methods Analysis and Work Measurement.*

5. These three types are used by Canon Corp. as described by Japan Management Association, *Canon Production System* (Cambridge, MA: Productivity Press, 1987), 153–160. Other variations are possible, depending on the nature of the work.
6. For other examples, see Y. Monden, *Toyota Production System*, 2nd ed. (Norcross, VA: Industrial Engineering and Management Press, 1993), 157; and K. Suzacki, *The New Manufacturing Challenge* (New York: Free Press, 1987), 139.
7. Monden, *Toyota Production System*, 158.
8. See D. Edwards, R. Edgell, and C. Richa, Standard Operations—The Key to Continuous Improvement in a Just-in-Time Manufacturing System, *Production and Inventory Management* (Third Quarter, 1993): 7–13.
9. J. W. Marriott and K. Brown, *The Spirit to Serve: Marriott's Way* (New York: HarperCollins, 1997).
10. W. Eaton, 2004, Reframing the Conversation on Management, GovLeaders.org, http://govleaders.org/reframing_the_conversation.htm.
11. M. Graban, M. *Lean Hospitals* (Boca Raton, FL: CRC Press/Productivity Press, 2009), 65–79.

Suggested Reading

Productivity Press Development Team. *Standard Work for the Shopfloor* (Shopfloor Series). New York: Productivity Press 2002.

Questions

1. Define standard operations. What are the four key components or aspects of standard operations? Explain each.
2. Why should supervisors be involved in developing standard operations? What are the reasons for locating development of standard operations at the shop-floor level?
3. Discuss the use of standard operations for each of the following: planning, control, worker training, and improvement.
4. Explain the difference between standard task time per unit and completion time per unit.
5. How is workcell production capacity described in the last chapter consistent with production capacity as defined in this chapter?
6. What does the standard operations routine (SOR) show?
7. Why are standard operations never considered permanent?
8. Discuss the contents of the standard operations sheet (SOS).
9. Discuss how SOR could be used for setup analysis and reduction.

PROBLEMS

1. A time study indicates that a worker takes 430 seconds to perform an operation. The worker is known to be quite skilled and the supervisor estimates before the timing that her performance will be about 10% above average. During a 7-hour workday, various anticipated delays cause an estimated 20 minutes of downtime on the operation.
   a. What is the standard task time for the operation? What is the average completion time per unit?
   b. If this operation is performed once every third item, and the standard task time for each of the other two items is 380 seconds, what is the average completion time per unit?

SEQ. NO.	WORK ELEMENT OR DESCRIPTION	TIME: Hand VA	NVA	Mach	Walk	TIME SCALE 10 30 50 70 90 110 130 150 170
1	Face		17	50	3	
2	Drill & Tap	29			3	
3	Drill 1/4" NPT		12	13	2	
4	Tap 1/4" NPT		12	13	2	
5	Drill 2 Holes 3/4" NPT		10	23	2	
6	Tap 2 Holes"		10	23	2	
7	Drill 2 Holes 3/4" NPT		13	23	2	
8	Tap 2"		13	23	5	

Figure 11.11 SOR sheet for Problem 2.

2. Operations are performed on eight machines as follows:
   - 1. Face: lathe #609
   - 2. Drill-Tap: machine #501
   - 3. Drill: machine #101
   - 4. Tap: machine #301
   - 5. Drill: machine #102
   - 6. Tap: machine #302
   - 7. Drill: machine #103
   - 8. Tap: machine #303

   The times for these operations are shown on the SOR sheet in Figure 11.11.
   a. Reproduce a copy of Figure 11.11. On the copy draw in all times as illustrated in Figure 11.4; determine the actual cycle time.
   b. The lathe #609 is a large but somewhat movable machine. The drill-tap machine #501 is very large and cannot be moved. All of the other drill machines and tap machines are small and movable. Sketch the layout of the machines in a workcell such that one worker could easily walk between them in the 2 to 5 seconds allotted. Show the worker's path and locations of stock areas for incoming and outgoing material.

3. Refer to Table 11.1. Assume all eight operations have been replaced with machines that have completion times per unit of 45 seconds (including 10 seconds for setup and handling). Assume also that demand for the product has increased such that the resulting takt time is reduced from 95 seconds to only 49 seconds. The machine B at operation B1 is more than twice the size of the other machines, which are somewhat small, mobile, and can be located about 2 seconds apart (walk time). Determine the number of workers required, and discuss the necessary changes to the SOS sheet in Figure 11.10. Make your own assumptions about walk times.

Chapter 12

Quality at the Source and Mistake-Proofing

The man who makes no mistakes does not usually make anything.

—**Bishop W. C. Magee**

It is the nature of man to err, but only the fool perseveres in error.

—**Cicero**

The key to discovering and eliminating defects are inspection procedures that give accurate and timely information about causes of defects. Statistical process control (SPC) is one such kind of inspection procedure: whenever a potential problem is detected, the place in the process where the problem likely originated is identified so the cause can be diagnosed and corrective action taken. There are, however, other kinds of inspection procedures and, argues Shigeo Shingo, the production expert, if the goal is to **eliminate defects**, these other kinds too, not SPC alone, must be employed.

The four concepts for eliminating defects discussed in this chapter are 100% inspection, Jidoka, source inspection, and pokayoke.[1] These concepts are part of the total quality toolkit, and they are applied in concert along with the SPC and design for manufacture practices, mentioned in Chapter 4. It takes all of them, not one or two alone, to eliminate defects and to elevate quality to the level mandated by world-class competition.

SPC Limitations

There is no worker, machine, or process that does not, on rare occasions, do something incorrectly. Given the random nature of errors and mistakes in production systems, there is the probability that a defect will occur at some time other than during sampling inspection. Since SPC relies on sampling, it can miss occasional problems from sources that are ephemeral. Even when shifts occur in process parameters and are detected by sampling, the time lag between when the process shift occurs and is first detected allows time for a substantial quantity of items to be affected and for the cause of the shift to disappear. Further, to achieve very small defect levels using SPC, it is necessary to employ variables inspection. Yet there are cases where variables inspection is inappropriate and where defects can be detected only through human sensory inspection or automated inspection of attributes.

Besides limitations from sampling, there are cases where SPC is simply not practical or feasible, for example, when items are produced only in small batch quantities or in quantities too small to establish guaranteed process stability, and in processes where no defects or process shifting is allowable. This chapter discusses forms of inspection systems and related measures applicable when SPC alone is inadequate or cannot be employed, or where the goal is not to just reduce defects but to eliminate them.

100% Inspection (Screening)

As mentioned, the two fundamental drawbacks of SPC are that it relies on sampling and that it allows a sometimes substantial delay between when a problem originates and is corrected. So, to minimize the chance of overlooking defects or of missing random problems that have fleeting causes, it is necessary to do 100% inspection.[2] Additionally, to minimize the time lag between when a problem occurs and is remedied, it is necessary to combine inspection, analysis, and corrective action with the original work task. To this end, wherever feasible, these duties should be given to the people doing the work tasks, the frontline workers. This is accomplished via self-checks and successive checks.

Self-Checks and Successive Checks

Self-Checks

After a worker performs a task, he checks the result. If he detects a problem and if the solution is within his capability, he immediately fixes it. Self-check inspection and correction is the most rapid kind of informative inspection system. Potentially, however, self-check objectivity and accuracy can be poor if workers are biased in judging their own workmanship or if they forget to inspect things. One way to increase inspection objectivity and reduce oversights is with successive checks.

Successive Checks

With this method, the next worker in the process inspects the previous worker's output. In Figure 12.1, worker B checks worker A's output, worker C checks worker B's output, and so on. Whenever a worker detects a problem that originated upstream, he passes the item back to the

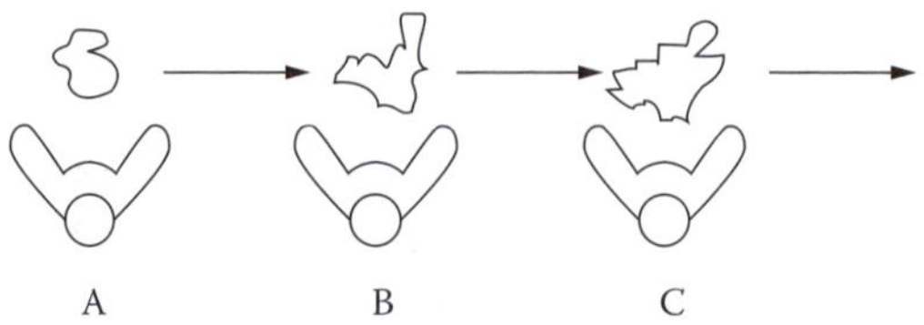

A B C

Figure 12.1 Self-checks and successive checks.

responsible worker, who then corrects the problem and does whatever is necessary to prevent it from reoccurring. For exceptionally important items there can be a double check: both worker B and worker C check the output of worker A.

If it is impossible for the next worker in the process to perform the inspection, then the worker at the next nearest subsequent operation does it instead. Whenever inspection calls for particular knowledge, skill, or judgment, an additional **specialized check** is introduced wherein a worker with the requisite skill and knowledge also does checking. After each day's work, the special inspector meets with the worker whose output she inspected and other workers doing successive checks to discuss the results. The purpose of the discussion is to point out defects or problems the workers missed, and, eventually, to transfer inspection skills from the special inspector to the workers.

Despite the potential for lower objectivity, self-checks, conducted properly, can be more effective than successive checks. Quick feedback is essential to achieving zero defects because unless identified immediately, problem sources vanish, only to reappear and cause problems later on. Also, given that most people prefer to find their own mistakes versus others finding them first, self-checks do not garner worker animosity or resistance.

In the same vein, successive checks are more effective than are checks by supervisors or staff inspectors. Successive checks result in somewhat-quick feedback about quality problems, though beyond that, workers tend to be more accepting of mistakes pointed out by their colleagues than by managers and staff members. An immediate colleague can point out a particular mistake made only moments earlier, whereas a supervisor or inspector will more likely speak in generalities about something that happened possibly hours or days before.

Requirements for Self-Checks and Successive Checks

Some minimal requirements for effective screening with self-checks and successive checks include setting check targets, enabling quick feedback and action, and management showing consideration and providing support to workers.

Check Targets

Requiring workers to check too many things on every item is counterproductive. They either forget to check some of them or they get sloppy with whatever they do check. Each worker should inspect only two or three check targets. Features that are critical in terms of safety, performance, appearance, and so on, or are subject to frequent problems should always be checked. Other features should be checked based upon statistics maintained on defect occurrences as identified by workers or customers. Every few weeks defect statistics should be reviewed to identify the most frequent problems, and these should be considered as check targets until a permanent solution is found to eliminate them.

Feedback and Action

Whenever a problem or defect is spotted, the worker responsible for the problem (if he is not the worker who spotted it) is notified immediately. If the problem is chronic, the worker stops the process to remedy it before further items are affected. When the worker cannot remedy a problem quickly, other workers and staff specialists are called in for assistance. Colored lights called **andons** located at each workstation, and **status boards** located centrally above a line or workcell for everyone to see, are used to signal each workstation's status or need for assistance. For example, a green light (or light off) means "situation normal"; yellow means "working to resolve a problem"; red means "difficult problem, send help." Yellow and red usually suggest a temporary stoppage of work at the workstation; if the process is synchronized, then every station stops. Problems in manufacturing processes often stem from inappropriate procedures and inadequate worker training, and with andons it is easy to identify the stages where workers need training, procedures need changing, and so on.

Consideration and Support for Workers

Frontline workers are obviously at the core of self-checks and successive checks, so there is no way this process will work without worker dedication. There are several important interrelated points here:

1. Workers must know (and management must ensure) that the inspection process is not used as a tool to evaluate them. If they think the purpose of inspection is to gather data to evaluate them, the workers will collectively undermine the process by overlooking problems. Some human error is inevitable and is expected; the purpose of self-checks and successive checks is to identify places where inadvertent or procedural errors occur, and to improve procedures to reduce errors.
2. Workers must be given a time allowance to improve quality. They must be given extra time during the production day to permit line stoppages and correct interim problems, as well as at the end of the day or week to discuss problem causes and permanent solutions. Providing this time is one way management shows its commitment to finding long-term solutions to quality problems.
3. In most operations, workers have relatively little control over the major factors that influence quality, so simply giving them additional responsibility is not enough. Joseph Juran, the quality guru, distinguished between worker-controllable and management-controllable situations. In terms of preventing or discovering and eliminating defects
   - A worker-controllable situation is one where workers are provided with (1) knowledge of what they are supposed to do, (2) knowledge of what they are actually doing, and (3) a process that is capable of meeting specifications.
   - If any of these three conditions is missing, the situation is management controllable.

   Juran estimates that about 80% of defects are management controllable regardless of industry, that is, situations where, simply, workers do not have adequate knowledge or an adequate process to do quality work.[3] Competent inspection and problem solving by workers is predicated on their being well trained in inspection, data collection, and problem solving. Besides training, management must provide whatever technical resources are necessary to make the production process capable (e.g., redefining specifications or altering the process). Worker effort alone is often insufficient to improve process capability.
4. Preexisting quantity-oriented quotas and piece rates must be eliminated before workers can be expected to take seriously any responsibility for self-checks and successive checks.[4] Workers paid according to production quotas or piece rates will overlook defects because identifying and fixing defects will reduce their volume of output.

Automation

Although it is often believed that automated inspection is better than human inspection, in general, it is not. When 100% manual inspection is employed using self-checks or successive checks, the rate of error detection can be almost as good as with automated devices. Only when it is very difficult or physically impossible for humans to do an inspection, when the unit cycle times are very short and when the inspection accuracy must be very high is automated inspection necessary and more effective. Generally, however, the overriding advantage of automation is not accuracy or reliability of measurement but cost. Replacing manual inspection with automated inspection tends to reduce inspection costs in the long run, especially when items are produced in large quantities and have a large number of inspection targets.

Cycle Time

The typical contention about using self-checks and successive checks, especially with screening, is increased cycle time. Shingo argues that, although the cycle time increases initially (workers need additional time to do checks and remedy problems on the spot), the increase is often small (sometimes only 10%) and, eventually, drops to only a few percent or less. As the inspection procedures become more familiar, the workers integrate them into their preexisting tasks. As more defects are spotted and problem sources eliminated, the frequency of errors and work stoppage decreases. Even when the ultimate cycle time does not return to the preexisting value, the overall average production lead time is reduced since final inspection and rework are virtually eliminated. In processes where the cycle times are very short and where inspection time would have a significant impact, less than 100% of items, say, every second, fifth, or tenth item, can be done instead. As mentioned, automated inspection is also a consideration.

Pursuit of Perfection: Limits of Inspection

Screening with self-checks and successive checks can improve quality levels beyond sampling inspection by increasing both the comprehensiveness of screening and the speed of feedback; however, even it cannot achieve perfection. Just as inadvertent mistakes exist in any production task, so too they exist in inspection. Inadvertent inspection mistakes are inevitable, and no amount of measurement calibration, inspection experience, or training can eliminate them. Even when the reliability and consistency (degree of error variation) of inspection can be improved by automation, some error will still be present.

The accumulated effect of small inaccuracies in inspection can be significant, even when the process defect rate (rate prior to inspection) is already small. For example, suppose an operation with an average defect rate of 0.25% (.0025) employs 100% inspection with both self-checks and successive checks. Further, suppose the checks are each 95% effective at finding defects. At the operation itself, where a 0.25% defect rate translates into 2500 defective ppm (parts per million), the self-check will identify 95% of them, or 2,375 ppm. Thus, 125 defective ppm will remain undetected and go to the next operation. At the next operation, the successive check will identify 95% of the undetected defects, or 119 ppm more. Thus, 6 defect ppm will remain undetected—an error rate of 0.00007%.

Although a defective rate of 6 ppm might seem inconsequential, consider that a typical product is comprised of numerous components assembled at many stages and that the cumulative effect of defects in every component on the final product is exponential. In the example, suppose the final product consists of 100 components, every one of which has a final defect rate of only 6 ppm. Since each part is 999,994 ppm good, the overall effect on the final assembled product of 100 parts is 0.999994^{100} or 0.9994002, the equivalent of 999,400 ppm of final product with no defective parts. This leaves 600 ppm that contain defective components. This example illustrates the difficulty and virtual impossibility of achieving zero defects, even when using 100% inspection and double-checking items that were pretty good to start with.

Jidoka

Jidoka is a Japanese term that in one sense refers to automation in the usual way, but in another sense refers to automatic control of defects, that is, a process that has built-in mechanisms that prevent it from proceeding whenever a defect or abnormality is detected. Shigeo Shingo, the originator of the single-minute exchange of dies (SMED) setup reduction method, formalized the concept, but Sakichi Totoda had applied it in 1902 when he invented a loom that would automatically stop whenever a thread broke.[5] This automatic shutdown of a process is also called **autonomation** and the concept applies to manual processes as well as automated ones.

Autonomation

One form of autonomation is **line stop**, which, as mentioned, refers to a worker's responsibility for stopping a process when a problem is discovered. Line stop should also occur when a worker is unable to complete a task in the required cycle time. For example, if the required cycle time is 15 minutes and a worker needs 15.5 minutes, then the process will be held up 30 seconds while the worker finishes. A situation like that requires analysis of the task, as well as a change in procedures and times on the standard operation routine. Autonomation can also be incorporated into automated, mechanical processes. An example of mechanical autonomation is a sensor on a machine that stops the machine whenever (a) it has completed one machining cycle, (b) the number of parts between it and the next machine (intermachine buffer) reaches a maximum, or (c) the machine's performance begins to change.

Managers often resist giving line-stop responsibility to workers because it interrupts production schedules, increases cycle times, and idles facilities. Of course, line stop responsibility should cause interruptions, especially initially, otherwise there will be no indication that quality problems are being identified and resolved. As problems are eliminated and fewer remain that necessitate stopping the line, the frequency of interruptions should drop to very few. As a Toyota manager stated, however, if the line is stopped too infrequently (fewer than fives times a day at Toyota) that means that defects are getting by and the process is losing effectiveness.

The importance of stopping the process in a manually controlled system to correct a defect or take time to do something right even if it exceeds the takt time must be impressed on workers, who, often, are hesitant to slow production or draw attention to themselves. However, as Monden points out, even when workers are motivated to conform to the line-stop concept, mechanical means might be required to assist them. For example, at Toyota a worker walks next to a moving assembly line for a certain distance within which she is expected to be able to complete her tasks. If the worker walks farther than the expected distance, she steps on a mat that stops the line. The same line has a hand drill connected to a rail running next to the line. A worker using the drill to put lug nuts on wheels walks next to the line, but if the drill connection passes a certain point on the rail, the line stops.[6]

Andons

Andon lights, described earlier, are an important part of autonomation. It is not enough to give workers responsibility for line stop; there must also be mechanisms to inform everyone in the process of the location of the line stop and the nature of the problem. For example, besides a defect problem, the line can be stopped because of a machine problem, a shortage of material, a required setup, or because the required amount has been produced. When everyone knows the workstation origin and source of the line stop, they can take appropriate action—continue working, stop working and go to the station needing assistance, or stop working and wait for instructions.

Source Inspection and Pokayoke

Shingo asserts that since traditional inspections are aimed at finding problems in outputs (parts, material, products), any conclusions drawn from such inspections will serve to remedy defects or restore process parameters, but not to prevent the defect or process changes in the first place. As a result, according to Shingo, given that every defect or process change has a cause or source, the only way to eliminate defects is to discover the *conditions* that give rise to defects or process changes, and eliminate them. This is what Shingo refers to as **source inspection**.

Most often, defects occur because of the following situations:

1. Inappropriate work processes or operating procedures (e.g., unsuitable heat-treatment temperature, incorrect assembly of machining procedure).
2. Excessive variation in operations (e.g., excessive play or wobble in a machine) resulting in machine wear, vibration, out-of-adjustment operation, or lack of specificity in procedures.
3. Damaged or defective raw materials.
4. Inadvertent errors by workers or machines (e.g., an occasional random oversight by assembly workers or a machine occasionally jamming because of dirt buildup).

In general, the first three of these problematic situations can be resolved, respectively, by

1. Improving work processes and procedures, and using standard operation routines
2. Exercising good housekeeping, preventive maintenance (PM), and constant monitoring of equipment
3. Working with suppliers to ensure no defective materials ever enter the process

Nevertheless, even with perfect procedures, perfect equipment, and perfect raw materials, defects occur because of situation 4, inadvertent errors. Screening will catch most defects from inadvertent error, but some defects will still get by.

Before discussing the solution to the problem of defects from inadvertent errors, consider the following five examples of situations where no amount of worker training or vigilance would eliminate defects.

Example 1: A worker in an assembly cell spot welds frames at eight locations. The cell produces three kinds of frames; the three frames all require eight welds, but the welds are in slightly different locations for each frame. The problem is that sometimes the worker misses a weld.

Example 2: Five similar but different products move along the same assembly line (in a mixed-model assembly process). The different products require different fittings that workers select from bins and attach as the products move by. Occasionally workers select and attach the wrong fittings to a product.

Example 3: Small batches of items arrive in small containers at a paint department, and each is painted the color specified by a code number on a kanban ticket attached to the container. Sometimes the wrong paint is selected and every item in the container is painted the wrong color.

Example 4: Metal panels are formed by inserting rectangular sheets into a stamping press. Occasionally the worker inserts the sheet upside down.

Example 5: A worker installs three switches on a control panel, and for each switch she must first insert a spring. Sometimes she forgets to insert the spring. Without the spring, a switch will not work.

Source Inspection

In all of the previous examples, subsequent inspection of the part or product will likely reveal the defects, although no amount of inspection will ever prevent the defects from occurring again. Since complete inspection will ultimately reveal most of the defects, the concern here is more over the waste incurred by the defects than with the fact that they occurred in the first place. In example 4, the defective metal sheets will have to be scrapped, and in all the other examples the defective parts will have to be disassembled or reworked.

The one commonality in all the examples is that in each case the source of the defects is known. That being the case, these sources can be monitored and eventually eliminated. Eliminating the source of a defect is equivalent to eliminating the opportunity for the defect to ever occur again.

Shingo refers to schemes or devices used in source inspection as **pokayoke**, a Japanese term that is roughly equivalent to error proofing or mistake proofing.[7] The following gives examples of pokayoke solutions for eliminating mistakes and defects in the five examples.

Example 1: Each time a frame is welded, it is first placed on a frame bed where two clamps automatically drop over it (Figure 12.2). The spot welder is attached to a counter, which in turn is attached to the clamps. Only after the number of spot welds reaches eight will the clamps release. No frame with fewer than eight welds can be released from the frame bed.

Example 2: Attached on the side of each product is a large card with a code. Just before the product arrives at the station where the fittings are to be attached, a scanner reads the card and automatically turns on lights on all bins containing parts for that product (Figure 12.3). The normal time required for a worker to withdraw and attach parts to each product is 40 seconds. Every bin has a motion sensor, and unless parts have been withdrawn from all the lighted bins within 30 sec, lights flash at the bins where parts have yet to be withdrawn. A buzzer sounds and the line stops until the parts are withdrawn.

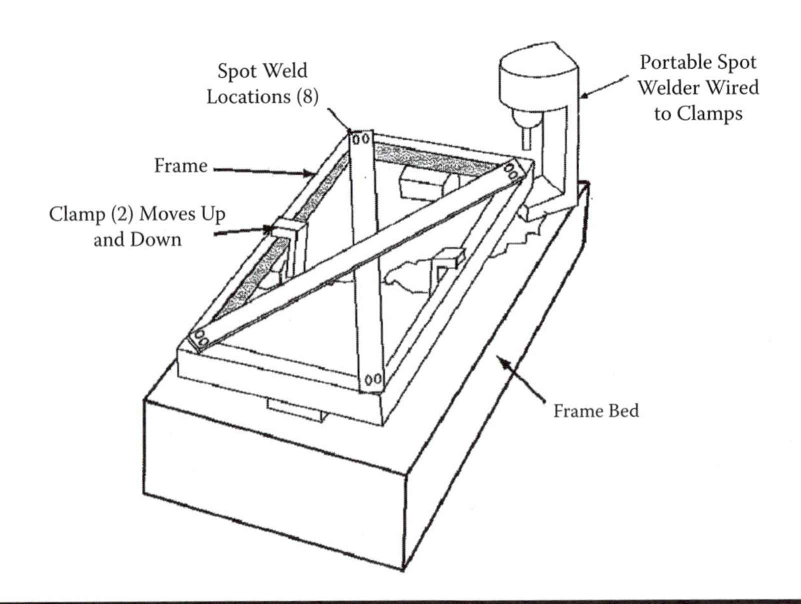

Figure 12.2 Example 1. Pokayoke to ensure correct number of welds.

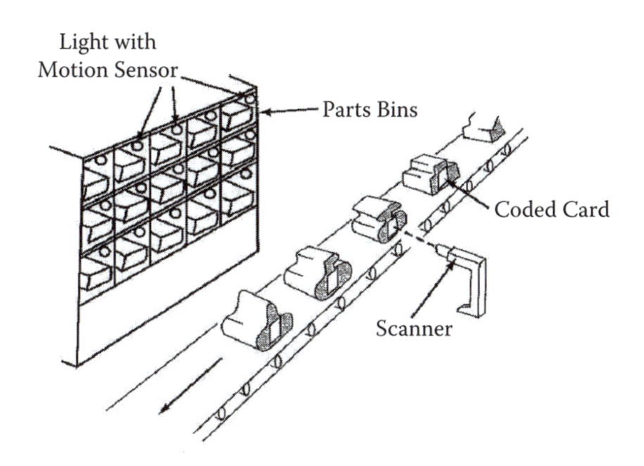

Figure 12.3 Example 2: Pokayoke to ensure correct parts selection.

Example 3: Similar to Example 2, attached to each container of items to be painted is a kanban card that specifies the paint color. The worker scans the card with a card reader, which activates a light on the paint container located on a large rack. In addition, a photocell by each paint container monitors for two motions: removal of the container from the rack within a prespecified period of time and then return of the container within a prespecified time period. If the second motion is not detected, a buzzer sounds. It should be noted that the purpose of the photocell is not to pace the worker but to make sure that the right paint container is removed for use and then is returned to the right location in the rack.

Example 4: In every metal sheet to be formed, small holes are drilled as part of an earlier operation. In the lower die of the press, pegs have been inserted at locations to correspond exactly with the locations of holes in the sheets (Figure 12.4). If a sheet is inserted upside down, the pegs and holes do not align. This prevents the sheet from sitting properly on the die, a clue to the operator that the sheet is inserted incorrectly.

Example 5: One step was added to the worker's task (Figure 12.5). Upon starting on a new control panel, the worker (a) first withdraws three springs from the supply bin and puts them in a small bowl. She then (b) withdraws the springs from the bowl and inserts them in the switch locations. Finally, she (c) installs the switches. This little step eliminated the problem of the worker overlooking a spring on a switch.

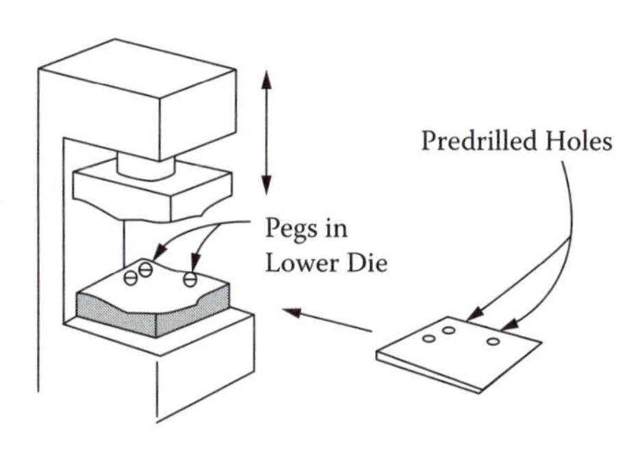

Figure 12.4 Example 5: Pokayoke to ensure proper placement of sheet in press.

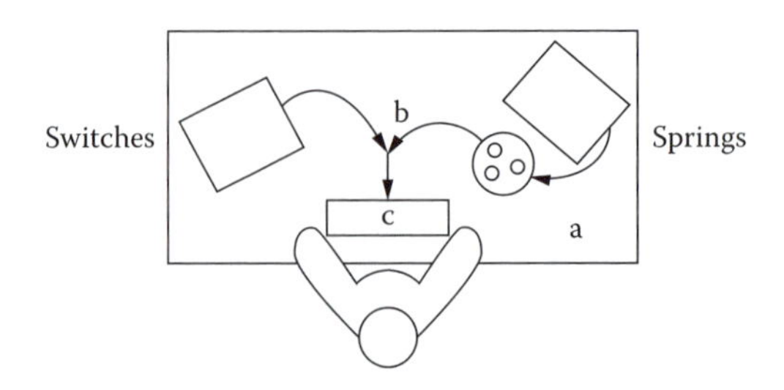

Figure 12.5 Example 5: Pokayoke to eliminate missing springs.

Pokayoke Functions

Any kind of system or mechanism that prevents defects from happening can be called a pokayoke. There are two broad functions that pokayokes serve: regulatory and setting.

Regulatory Pokayokes

These are devices that either control a process or give warning about it. A *control pokayoke* is a device that shuts down an operation whenever it detects an abnormality, thereby preventing defects in a succession of items. An example is a sensor on a drill press that detects erratic motion and automatically stops the machine. Example 1 is a control pokayoke: as long as fewer than eight welds have been made (an abnormality), the clamp prevents a frame from being released and moved to the next stage of the process. The line stop in example 2 is also a control pokayoke.

A *warning pokayoke* is a device-activated light or buzzer that only signals an abnormality. The photocell-activated buzzer in example 3 is a warning pokayoke. In a continuous production process, warning pokayokes are not as effective as control pokayokes since they allow the abnormality to continue until someone heeds the warning and stops the process.

Setting Pokayokes

These are devices that check for or ensure proper settings or counts in a process. They apply any place where positioning or orientation is important. Placement of pegs in the lower die in example 4 is a form of setting pokayoke; it guarantees that a sheet is properly set into the press. The device in example 1 is a count pokayoke; it guarantees that no less than eight spot welds are made on each frame. Example 5 is another count pokayoke; it helps ensure that the worker will always insert a spring in a switch before installing the switch on the panel.

Many pokayokes use automatic switches attached to electronic counters, proximity and photoelectric sensors, motion and vibration sensors, and pressure and temperature sensors; other pokayokes, however, are but simple fixtures or procedures that require nothing automated or electronic (examples 4 and 5).

The pokayokes in all the above examples are intended to prevent inadvertent human mistakes. Pokayokes can also be attached to machines to prevent even fully automated processes from deviating. For example, a piece of equipment that maintains certain operating parameters (temperature, pressure, revolutions per minute, etc.) can be monitored and the process stopped if the required limits on parameters are exceeded.

The concept of pokayoking can be applied anywhere that mistakes can happen (which means, literally, everywhere). Applications include pokayokes installed on equipment to prevent mistakes in equipment changeovers and setups, and in periodic equipment PM procedures.

Most pokayoke solutions are somewhat simple in concept, easy to implement, and inexpensive. Shingo, in what is perhaps the best reference on the subject, details numerous examples of pokayoke solutions, most that use devices or materials that cost a few hundred dollars or less.[8] Technology offers other, though more expensive pokayoke solutions. An example is an electronically coded radio frequency identification (RFID) tag affixed to a part. As each step in a procedure is performed on the part, information denoting that the step has been completed is recorded on the tag. A reader tracks information on the tag, and only after all steps have been completed is the part released to the next operation.

The following example further demonstrates the range of pokayoke applications.

Case in Point: Pokayoke at a Military Retail Supply Operation[9]

The U.S. Army retail supply system connects all military units with storage depots that stock repair parts. Part of the system, the supply support activity (SSA) at Fort Carson, Colorado, maintains a stock of repair parts for equipment used at the base. SSA orders, tracks, and controls the parts.

SSA faced many problems caused by inadvertent errors caused by both customers and SSA operators:

- Customers sometimes placed an order that contained outdated or erroneous information, so the SSA system could not process the order.
- Customers often did not pick up parts when the parts arrived, which caused a shortage of storage space at SSA and problems with inventory control.
- SSA performed batch processing daily, weekly, and monthly that required operators to perform a series of intricate procedures to compile and send information to suppliers. The processing required operators to perform a series of intricate procedures. The procedures resulted in many inadvertent errors that prevented the system from processing order requisitions. The system never flagged operators about errors, and subsequently it would be impossible to determine who had made an error. Ill will grew between operators and customers over who was to blame for order errors.
- Part-number information for arriving parts was occasionally entered incorrectly into the SSA system. The system would ignore the mistake and take no further action. In effect, the order was lost: the customer did not receive the item, the supplier did not get paid, and the item had to be reordered.

SSA has since adopted several pokayoke procedures to cover errors at all stages of customer ordering, batch processing, and order receipt. To solve batch-processing errors, operators are now forced to perform procedures properly and sequentially. Operators are given strict guidance and a warning list. A computer program checks inputs and does not allow an operator to continue processing data until the right procedural sequence is followed.

A bar-code reader ensures accurate processing of incoming parts. All incoming parts have bar-code labels that are read upon receipt at SSA. A bar-code printing system has also been installed to prepare labels for inventory parts tracking and locating all items held in stock.

Customers are now required to load the current parts catalog onto their computers before placing orders. This prevents outdated or erroneous parts information from ever appearing on orders. Customers are also required to pick up any ordered parts that have arrived prior to placing new orders.

The following improvements have resulted:

- Monthly adjustments to dollar amount of inventory holdings dropped from $3,000 to $250
- Inventory accuracy (percentage of items in the correct inventory location) rose from 65% to 98%
- Percent of receipts with some form of error dropped from as much as 90% to 0%
- Batch processing errors dropped from 15 to 20 per month to 0
- Errors in information from customers dropped from 22% to 0%
- The average request processing time (difference between date when a supplier receives a request and the original date on the request) fell from 12.5 days to 1.6 days

The main cost for the pokayokes was the time and effort to isolate the problems and determine appropriate mistake-proofing devices. The dollar investment was less than $1,000.

Pokayoke Ideas

Many pokayoke ideas originate with frontline workers, particularly ideas for how they can eliminate their own mistakes. To implement these ideas requires assistance from supervisors, engineers, maintenance staff, machinists, toolmakers, and so on. They team up with shop workers to share concepts, refine ideas, consider ways to implement the ideas, and, where technology is involved, to analyze and work out the details. The McDonnell & Miller case in point is an example of workers, team leaders, and supervisors teaming up for a pokayoke idea.

Case in Point: Error-Proofing at McDonnell & Miller[10]

The following notice was posted on the shop bulletin board:

> Recently a letter from a customer alerted us to a problem that existed in our diverter valve assembly. The customer had received an order of our diverter valves that contained a unit that was incomplete in its assembly.

The Focused Factory No. 4 team found that an operator had inadvertently forgotten to include a spring and thermostat into the unit's body. Upon investigation, it was found that it was quite easy for an operator to forget these parts.

A team was formed to address the problem and find a solution. Mr. Amaro, the group leader in the diverter area, drew up a plan to install a limit switch to the holding fixture that would not allow the operator to remove the valve body from the fixture if any of the internal components were missing.

The focused factory supervisor and team leader listened to Mr. Amaro's idea and agreed that the switch would work. The approval to go ahead was given, and within one week the fixture was retrofitted with the limit switch. The cost for this work was minimal.

Tests were performed by the operators. The results were excellent. The limit switch can sense the weight (or lack of weight) of the spring and thermostat. If any parts are missing in the body, the switch will not let the operator remove the assembly from the fixture. This feature assures us that no incomplete assembly will leave the work area and be sent to our customers.

Due to the action taken by Mr. Amaro, research will begin to see if this idea can be adapted to other areas of the plant.

Of course, other people who help the frontline workers implement pokayoke ideas have pokayoke ideas of their own that might or might not be related to the workers' ideas. Pokayoke ideas can come from anywhere, and everyone must be encouraged to "think pokayoke." To this end, for example, product designers should think of ways to build features into products or components that will preclude any possibility of workers making mistakes in assembling them. Product engineers can design parts to have subtle differences that would be transparent to the customer but would prevent assembly workers from ever confusing the parts. Toolmakers, machinists, and process engineers can design equipment, fixtures, procedures, or entire systems such that human mistakes are virtually impossible or such that abnormalities in worker or equipment functioning can be quickly detected and the process stopped before the output is affected. Many of these pokayoke ideas should be addressed early, while design issues are still being resolved in product development. Such issues are best dealt with by a cross-functional team engaged in concurrent engineering of the product and process designs.

Improvement should not be put off because of the desire to do further analysis (paralysis from analysis). Simple solutions to reduce quality problems should be implemented immediately as stopgaps until more effective and robust measures are found and implemented. Ideally, these shop-floor centered quality efforts support quality-of-design and design-for-manufacture efforts by feeding data back about process control and defect problems to concurrent engineering design teams.[11]

Continuous Improvement

The concept of pokayoke includes never-ending improvement: not only can the pokayoke for a particular application always be improved (made simpler, more reliable, more cost effective), but new pokayokes must continually be developed for new processes, new applications, and new circumstances. In example 4, the pegs in the lower die preclude mistakes, but only for one kind of stamped part; for other parts made with other dies, other solutions are needed to prevent the metal sheets from being inserted the wrong way.

In some of the previous examples the pokayokes themselves introduce the possibility of error: in examples 2 and 3, coded cards on each product or container indicate the right parts or paint color, but there is nothing to prevent the wrong card from being attached to the product or batch. The counter on the spot welder in example 1 ensures every frame gets eight welds; however, if a seven-weld frame were added to the product mix, then a way would have to be found to ensure that the frame got seven welds and that it did not get confused with the eight-weld frames. In all these cases, what is needed is a *meta-pokayoke*, a device or procedure that will prevent mistakes in the application of each pokayoke itself.

In a particular system, whether existing or conceptual, it is seldom necessary to muse about where problems or defects might originate. Experience and accumulated data from workers performing self-checks and successive checks, statistics on defects, customer complaints, and warranty claims provide information about defects that can be traced to the defect-producing sources. What is necessary, however, is that good data be collected about products and processes, and that the data be used to refine existing products and processes and to design new ones. The data will indicate where existing systems need changes to eliminate defects and will guide designers of new products and processes to minimize defects. Nonetheless, even a new, so-called mistake-proof system will never be completely mistake proof. Any operational system must be monitored, and the resulting data used to determine what needs changing, either in terms of fundamental modification or by simple pokayoking.

Most every process can be continually improved by adding pokayokes, but eventually, as the process itself is modified, the pokayokes themselves become outmoded or inappropriate, and must be replaced by new or different ones. As part of continuous process improvement, the process of improving pokayokes must be continuous, too.

Summary

There is no worker, machine, system, or process that does not on rare occasions perform incorrectly, and, given the random, fleeting nature of these performance lapses, no amount of statistical sampling will detect them. In addition, many human-caused errors are inadvertent, that is, mistakes that no amount of training or diligence can prevent. To preclude these and other sources of rare quality problems, measures beyond SPC must be employed. One measure is 100% inspection (screening) with worker self-checks, which will usually detect the random, rare defect missed by sampling inspection. Screening also provides immediate feedback so the source of the defect can be found and remedied. In cases where inspection error allows some defects to go undetected, then successive checks and special checks can also be instituted.

An important part of ensuring defect elimination is to prohibit any defect from progressing further in the process. This is the concept of Jidoka. Either the defect is fixed immediately, or the defective item is set aside for later analysis and rework. Process-related problems are resolved as soon as they are discovered, if possible; otherwise, workers have authority to stop the process or production line, which is the concept of autonomation or line stop. Autonomation also refers to electronic or automated sensing systems that stop the process upon detecting a defect or operational abnormality. Workers throughout the plant are kept apprised on the status of each machine and the overall process by signal lights on every machine (andons) and overhead displays (status boards).

At some point, even all that checking is not enough to prevent some (albeit a very tiny percentage) of the defectives from getting through. Shingo asserts that to achieve zero defects it is necessary to monitor and then eliminate the sources of defects. This is the concept of source inspection. Sources of defects are abnormalities in materials, machines, and procedures, so if those sources are screened, then inspection of outputs is unnecessary. An example of source inspection is monitoring equipment so equipment will shut down when it starts to perform abnormally.

Many defects are the result of human mistakes in procedures, such as a worker occasionally overlooking a step or doing it incorrectly. The solution is to identify how the step affords opportunity for error and to alter the step or incorporate other steps that will preclude the opportunity. Pokayoke, or mistake-proofing, involves adding steps, or simple manual or automated checking devices to eliminate inadvertent errors. Pokayoke devices and procedures are also used to shut down an operation to prevent an impending error or to warn about abnormalities that could lead to errors. Most pokayoke ideas are relatively simple, inexpensive to implement, and come from worker suggestions.

Pokayokes for mistake-proofing are especially important in lean factories where no defects are tolerated and in pull production systems where inventories are small and stoppages anywhere soon halt the entire process. It should be obvious, however, that the quality-enhancing benefits from source inspection and pokayoke are realizable in all production and service processes, not just in pull systems or even manufacturing.

Notes

1. Much of the material for this chapter is adapted from Shigeo Shingo, *Zero Quality Control: Source Inspection and the Pokayoke System* [trans. by A. Dillon] (Cambridge, MA: Productivity Press, 1986).
2. Inspection accuracy is seldom perfect, so even 100% inspection will not find every defect. Still, assuming high accuracy in the inspection, the percentage of defects missed will be much less than with random inspection.
3. See J. M. Juran and F. Gryna, *Quality Planning and Analysis*, 3rd ed. (New York: McGraw-Hill, 1993), 348–360.
4. This is precisely the message of the eleventh of Deming's 14 points: eliminate numerical quotas. In a quota system such as piecework, people are paid for the number of units they produce, whether or not they are defective. See M. Walton. *The Deming Management Method* (New York: Perigee Books, 1986).
5. P. Dennis, *Lean Production Simplified* (New York: Productivity Press, 2001), 89.
6. Y. Monden, *Toyota Production System*, 2nd ed. (Norcross, GA: Industrial Engineering Press, 1993), 227–229.
7. Pokayoke is not a word but is a derivative of *bakayoke*, the Japanese term for "idiot proofing" or "fool proofing" and refers to methods aimed at eliminating errors due to mistakes (and not necessarily originating from idiots or fools).
8. Shingo, *Zero Quality Control*.
9. T. Snell and J. Atwater. Using pokayoke concepts to improve a military retail supply system. *Production and Inventory Management* 37, no. 4 (1996): 44–49.

10. Courtesy of Avi Soni, Manager of Manufacturing Engineering, ITT McDonnell & Miller.
11. C. Dyer, p. 73, in G. Taguchi and D. Clausing, Robust quality, *Harvard Business Review,* (January/February 1990): 65–76.

Suggested Reading

M. Baudin. *Working with Machines: The Nuts and Bolts of Lean Operations with Jidoka.* New York: Productivity Press, 2007.

Questions

1. Why is it not possible to eliminate defects by solely using SPC?
2. In self-checks and successive checks, what happens when a worker spots a defect?
3. Why do managers resist giving line-stop responsibility to workers?
4. Before workers can be expected to take on responsibility and the authority associated with self-checks and successive checks, what assurances and resources must they have?
5. In generic terms, what are the typical sources of defects? Give examples of each.
6. In 10 words or less, what is the principle of source inspection?
7. Of the different possible sources of defects, what sources do pokayoke address?
8. Describe the difference between control pokayoke, warning pokayoke, and setting pokayoke. Give an example where you might apply each in your work or daily experience (allow yourself freedom to be creative on this).
9. Word processing software contains many kinds of pokayokes. Where?
10. What is Jidoka? What is autonomation? Describe the relationship between Jidoka and autonomation.

PROBLEMS

1. A product is made of an assembly of 10 parts, each produced at a different workcell. The average defect rate of the parts before inspection in each cell is 5%. Before leaving the cell, every part passes through a two-stage inspection process, each stage of which identifies 95% of the remaining defects. No identified defects are allowed to proceed without being corrected.
    a. What is the defect rate of parts following the inspections? What is this rate expressed as parts per million?
    b. Suppose a defect in any of the 10 parts results in a defective assembled product. What is the defect rate of assembled products in parts per million?
    c. If a mistake occurs while assembling 1 in every 1000 products (assuming the assembly uses the 10 parts described earlier) what then is defect rate in parts per million?
    d. Assume the maximum acceptable defect rate is 250 ppm. Assuming no change in the quality of the 10 parts used in assembly, what must the mistake rate during assembly be reduced to?

2. Boxed-Goods Inc. packages replacement-parts kits. The company packages 10 kinds of kits. Each kit has between 4 and 8 kinds of parts, and the total parts count for a kit is between 20 and 55. As an example, a kit with 1 part J, 9 part Ks, 3 part Ls, 14 part Ps, and 7 part Rs would have a total

parts count of 34, and be comprised of five different kinds of parts. Parts range in size from an average 1 inch on a side to 4 inches on a side. Kits are hand-packaged in batches according to customer order. A typical order is for 10,000 kits. The packaging rate is a function of kit size: roughly 1.5 seconds per part, plus 20 seconds; thus, for example, a 20-piece kit takes 50 seconds.

The problem facing Boxed-Goods is that occasionally a worker makes a mistake resulting in an incorrect parts count or wrong kind of part in a kit. Suggest at least two pokayoke procedures that might eliminate the parts-count problem, the wrong-part problem, or both at once. State any assumptions about the parts necessary for the procedures to work. Discuss the pros and cons of each procedure in terms of time, cost, quality, and flexibility.

LEAN PRODUCTION PLANNING, CONTROL, AND SUPPLY CHAINS

Previous parts of this book addressed lean concepts, methodologies, and tools for continuous improvement, waste reduction, and total quality as applied to manufacturing. This part focuses on how the lean production concepts and techniques of Part II fit within the broader context of manufacturing planning and control.

A recurring topic is pull production, introduced in Chapter 8. A quick look at the key features of pull production—kanban cards and containers—tells a great deal about the purpose of pull production, which is to authorize and control work on the shop floor. Although pull production is also a tool for continuous improvement, that improvement happens through pull production achieving ever-better control over aspects of factory work.

Control implies the existence of standards, goals, or plans to which system performance is compared and upon which corrective action is taken. No control technique can be fully understood without reference to the broader system that it is supposed to control, and that is also true of pull production. In this part of the book we take a big step back and look at the broader production system, a system that besides physically producing the product also forecasts demand, accumulates customer orders, and translates forecasts and orders into broad plans, detailed plans, daily schedules, and authorizations to procure materials and execute work that, ultimately, lead to the final product. After looking at techniques for planning and scheduling work and for balancing and synchronizing stages of the production process, we look at a broad framework that integrates these techniques with the pull production and lean techniques discussed earlier. We then again take a step forward to look more closely at what is necessary to adapt material requirements planning (MRP)-based production planning and control systems to pull production.

In lean production, product design, development, and manufacture are viewed as a single process. Matters relating to speed, quality, and quantity are addressed from the top down. Emphasis is on identifying and improving the process elements that add value and eliminating others that do not. That process perspective extends beyond the factory walls to include suppliers. An organization trying to succeed at lean production and total quality management will have a difficult time unless it sees suppliers as forward elements of its process and as partners in its bid to gain competitive market advantage. Because suppliers are viewed as elements of the greater process, concepts like customer–supplier relationships and supply chain management are especially important in lean production.

The chapters in this part of the book are:

Chapter 13: Uniform Flow and Mixed-Model Scheduling
Chapter 14: Synchronizing and Balancing the Process
Chapter 15: Planning and Control in Pull Production
Chapter 16: Lean Production in the Supply Chain

Chapter 13

Uniform Flow and Mixed-Model Scheduling

The future influences the present just as much as the past.

—Friedrich Nietzsche

De do do do, de da da da.

—The Police, "De Do Do Do, De Da Da Da"

A characteristic of an efficient production system is that jobs and materials flow smoothly through the system, most production lead time is value-added processing, and jobs hardly ever have to wait. Since production schedules dictate the frequency and level of changes in products and output volumes, smooth production flow is largely a matter of scheduling.

This chapter discusses ways to prepare production plans, master production schedules, and final assembly schedules that meet customer demand, yet minimize production disruptions and changes in production capacity. One way to do this is to use **production-leveling** techniques to hold the frequency and size of changes in production schedules to a minimum.

The chapter also describes how final assembly schedules are prepared in pull production systems. Pull production requires somewhat uniform demand, and, as such, the concepts of production leveling and pull production scheduling are related. In addition, the chapter discusses the broader topic of production scheduling in three manufacturing environments: make to stock, make to order, and assemble to order. Ways to minimize costs and problems from scheduling disruptions in nonpull and make-to-order systems are also discussed.

Production Leveling

Plants that make a variety of different products typically use batch production. The frequent changeovers between different products and different size batches result in variation in the daily workload of every department and workcenter. Figure 13.1 represents the production schedule for three products:

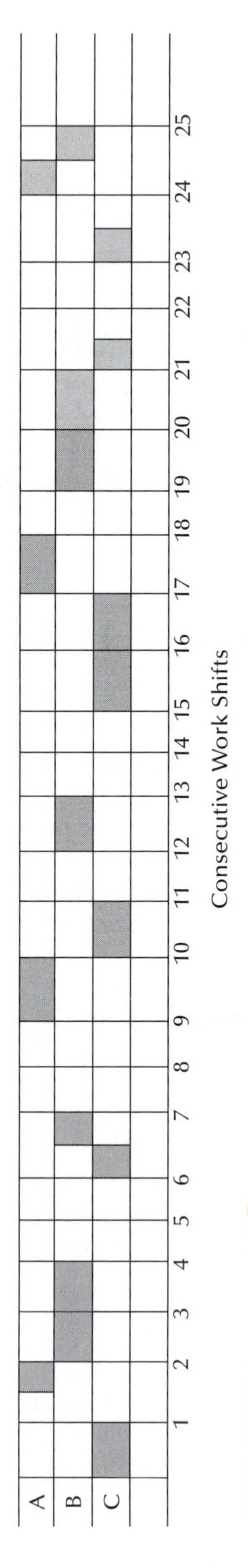

Figure 13.1 Production schedule for products A, B, and C.

A, B, and C. Each shaded block represents one or more work shifts scheduled to produce a batch of one product. The schedule shows large variation in two ways—in the size of the production runs (batch size) and in the frequency of the runs (batch interval). If other products besides A, B, and C must also be produced, they will have to be fit in the schedule somewhere between the batches shown in Figure 13.1, so the schedules for them will also share high variation in terms of batch size and interval.

Variation in scheduled batch sizes and batch intervals results from management's attempts to match production levels with demand. Demand levels vary because of sales fluctuations, promotions, end-of-month quotas, and so on, and management adjusts the sizes and intervals of production batches to compensate. Variation in production schedules also stems from variability inherent in the production process. Even when production is scheduled using fixed time periods (the schedule in Figure 13.1 uses half-shift minimal time blocks), the output quantity in any specified time period will vary due to material shortages, worker absenteeism, and equipment problems. These factors reduce output expectations, and the shortfall must be filled in subsequent time blocks (other shifts or overtime). Conversely, output sometimes exceeds the planned amount, and production in subsequent time blocks must be reduced.

Leveling Production with Buffer Stocks

Any variation in scheduled production for a finished product has a ripple effect on the production and delivery schedules of every upstream operation and supplier. A common way for the upstream stages of the production–distribution chain to absorb this variation is to carry buffer stocks of raw and in-process material. Buffer stocks provide some degree of certainty, and during periods of slack demand they keep materials flowing and workcenters productive. Since managers and supervisors are sometimes evaluated on workcenter utilization, to prevent idle time or underutilized resources they try to retain a backlog of work. Thus, in-process buffer stocks and work backlogs are one tactic for leveling production.

There are of course drawbacks to the tactic. Recall discussions from earlier chapters of the consequences of high inventories, and then consider the vast quantity of inventory that accumulates plantwide when an organization relies on buffer stocks to level production. Though removing variation from the production output is desirable, the question is, Does it have to come at such great expense?

The answer is no. Another, sometimes better way to minimize variation is simply to **level the production schedule**, that is, establish a master production schedule (MPS) where every product is produced on a regular basis and in a fixed-size batch. A level production schedule provides the same benefits as WIP buffer stocks without the drawbacks.

Leveling Production with Uniform Schedules

With a uniform, **level production schedule**, the same quantity is made in each production run for a product, and the production runs occur at regularly scheduled intervals. In other words, the batch size and batch interval for a given product are constant. Figure 13.2 shows two uniform, leveled schedules for products A, B, and C. For each product, the same quantity is produced in each batch, and the production runs are evenly placed. With a leveled schedule, every operation in the process follows such a pattern. This uniform pattern introduces sameness into the daily work of every workcenter, and sameness is easier to handle than change.

The smaller the batch size and the interval, the more level the schedule and smoother the flow of materials plantwide. In Figure 13.2 schedule b is more level than schedule a and would result in smoother flow of material, which means less backlogs, less WIP inventory, and shorter lead times than schedule a.

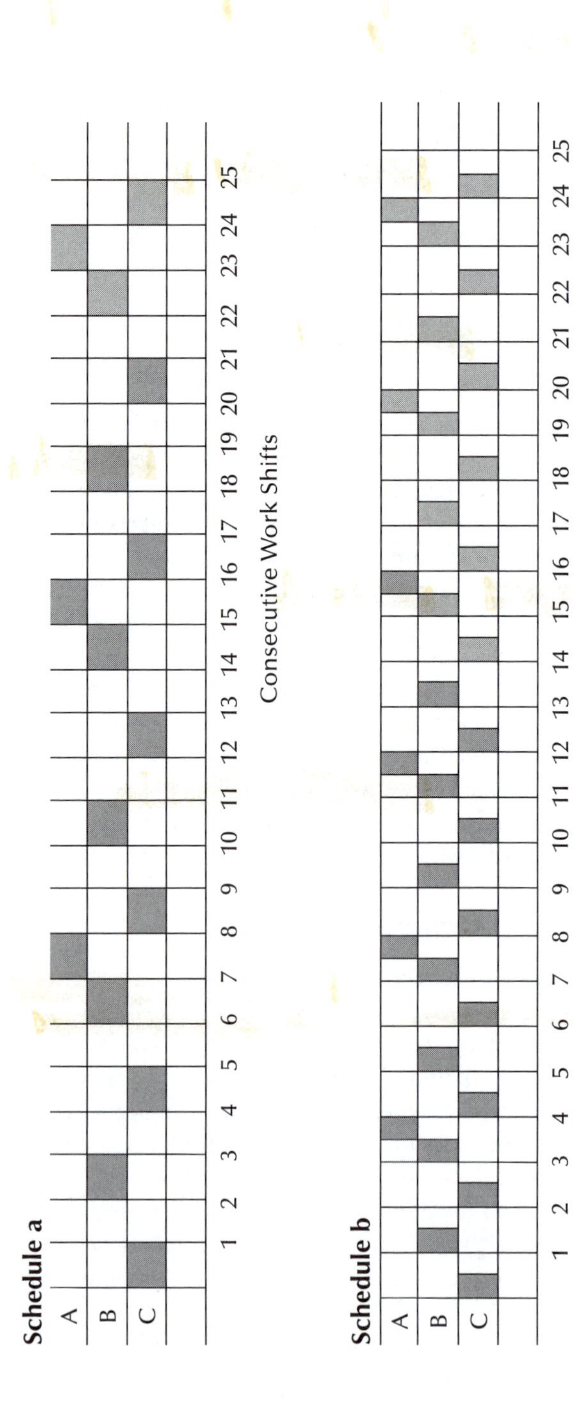

Figure 13.2 Two uniform, level schedules. Note that for each schedule the batch size and batch interval for each product stay constant.

Besides the time, cost, and quality benefits of smaller WIP, level schedules give supervisors and workers more time to focus on the work at hand. With fewer distractions from changing workloads, and fewer problems from work expediting and slowdowns, workers have more time to identify areas of the process that need improvement and to develop solutions and implement them.

Requirements for Leveling Production Schedules

For production leveling to be practical, three requirements, or conditions, must be met. The first requirement pertains to stable product demand, something that a company might or might not be able to control. The other two are things a company can and must do: shorten setups and match production with demand.

Continuous, Stable Demand

To maintain a level production schedule in the presence of fluctuating demand, some quantity of finished product must be held as buffer stock. This buffer stock protects the entire production–distribution process from product demand variability, and allows upstream stages to operate on a somewhat level basis. This is illustrated in Figure 13.3.

Now, keeping a buffer stock of finished product is practical only if demand is somewhat continuous; it is not practical with sporadic demand, since the buffer stock can sit there for a long time, possibly indefinitely. As a practical matter, demand does not have to be large, a few units a week, every week, might be enough. When demand is continuous, then the higher cost of carrying some finished goods inventory is offset by lower costs of carrying smaller WIP inventories everywhere else.

Of course, recommending inventory anywhere is a contradiction to lean principles. To keep that inventory small, the company must manage the demand so as to minimize variation. One way to do that is to segregate customers into tiers:[1]

- First tier: High volume, common processes (no special setups or operations)
- Second tier: Substantial volume and much process commonality
- Third tier: Low volume, sporadic orders with little process commonality

The first two tiers represent the best or bread-and-butter customer accounts; the last tier represents the lower profit and harder-to-satisfy accounts. Often, the first two tiers comprise 20% to 40% of total customers but 70% to 90% of sales. It is often possible to work with first-tier customers and even second-tier customers to circumvent abrupt changes in demand. These customers

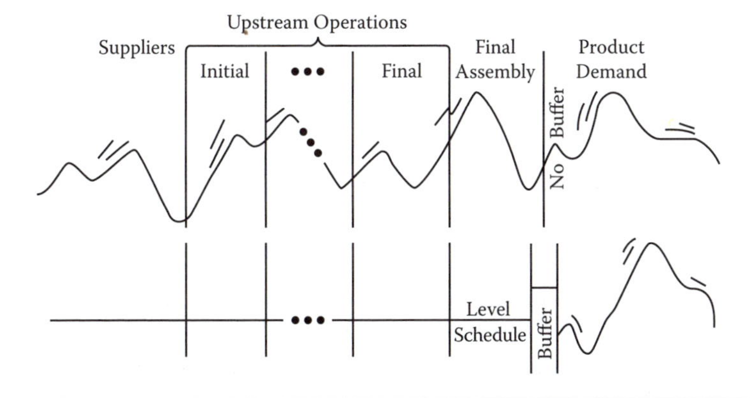

Figure 13.3 Effect of level schedules and finished goods buffer on upstream operations and suppliers.

are encouraged to inform the manufacturer of anticipated needs far enough in advance so that it will be unnecessary to carry finished goods stock to service them. Production schedules for these customers can be then gradually raised or lowered to meet demand.

During peak demand periods, the focus should be on satisfying first-tier and second-tier customers, and possibly even turning down orders from third-tier customers. During slow periods, orders from third-tier customers are accepted because production schedules have ample excess capacity and orders are not disruptive.

Short Setup Times

The time to change over a process from one product to another must be short. As discussed in Chapter 6, lengthy changeovers consume too much of the workday and leave relatively less time for actual production. *Ceteris paribus*, they also mandate larger batch sizes (e.g., see schedule a in Figure 13.2), which restrict the allowable degree of production leveling. The ideal changeover is a one-touch procedure that takes only a few minutes, although changeovers as long as 30 to 60 minutes might also be practical, depending on the number of products, length of workday, and demand.

Production = Demand

The frequency and batch size of production runs in leveled production should roughly correspond to the average actual demand for the product. This means that the batch sizes and production run intervals should be based upon product demand and not something arbitrary, like a daily work shift or half shift. The goal of each production run should be to achieve the scheduled output quantity, not to maximize output.

In addition, the scheduled production quantity (batch size) should be periodically readjusted to match the average demand level over a specified time period. The more often the demand level shifts, the more often the leveled production schedule must be readjusted. This will be discussed later.

The demand figure used to plan production levels for future periods is typically based on forecasts. When the goal is to level the production schedule of a number of different products, the forecasts should be for the **aggregate** demand for all the products in a product group or product family; it should not be for the demand of each individual product. This is an important point. A forecast for aggregate demand is more accurate because the numbers in the aggregate forecast are larger than for individual forecasts, and patterns for large numbers are inherently more stable than patterns for small numbers.[2] In this context, more stable means more predictable. Also, estimates for individual products are always high or low, though in the aggregate the high estimates and low estimates cancel out each other, giving better accuracy.

Leveling Focus

In organizations that produce a variety of products, the focus should be on leveling the schedules for the highest-volume products. This goes back to what was said before about tier 1 and tier 2 customers. A company that produces 60 different product groups wherein, say, five customers account for 50% of the production volume can dramatically reduce WIP levels and lead times plantwide just by leveling the production schedules of those five. Even when most of the other products continue to be scheduled and produced in the usual erratic fashion, leveling the schedules of the few highest-volume ones will substantially reduce workload variation and uncertainty for most workcenters. Product-quantity analysis, described in Chapter 9, is useful for assessing which product groups have the highest volumes and for which schedules should be leveled.

Leveling the Master Schedule

The way to force leveled production on a product is to level the MPS. The concept of leveling implies selecting a production quantity and producing that same quantity on a periodic basis. Leveling production, a common topic in classical aggregate planning,[3] is reviewed here as a precursor to the techniques described later. We start with the simplest case, leveling the production schedule for a product group, then look at leveling production schedules for multiple products simultaneously.

Leveling One Product Group

One consideration in choosing the level of production is the time period over which that level is to be maintained. Although it is desirable to keep the production level fixed for as long as possible, it is evident that demand changes will require periodic changes to the production level. Occasionally the level must be changed. Whether the change is every 3 months, every month, or sooner depends upon changes in the demand pattern. The solid line in Figure 13.4 represents the anticipated demand pattern over a 2-year period. Demand variation appears to be seasonal and the overall trend downward. The dotted line represents the planned uniform level of production.

Leveling production over a 2-year period like this is inappropriate for at least three reasons. First, since the production level during the first year is less than average demand, an initial stock of inventory must be available to meet demand, and the inventory investment for the first year would likely be high. Second, the production level throughout most of year 2 exceeds demand; this plus the fact that demand appears to be declining will likely result in a large stock of inventory remaining at the end of the second year. Reason three is that demand forecasts become more unreliable with time, so planning the first year's production largely based on the second year's demand is chancy.

The plan in Figure 13.5 considers just the first year (the second year shown only for reference), and sets production at two levels that, together, more closely align with the demand pattern than the one level in Figure 13.4. Closer alignment means fewer inventories necessary to meet the initial demand peak and resulting from buildups during slack periods.

It should be noted, and is clear from the example, that the chosen level of production cannot simply be the average demand over some time horizon. The level chosen must account for preexisting stock and be able to satisfy periods of peak demand.

Although the intent of production leveling is to keep production uniform for as long as possible, if forecasted demand exhibits considerable variation between seasons or months, then the level of production should be adjusted seasonally or monthly as needed (production = demand); see Example 1.

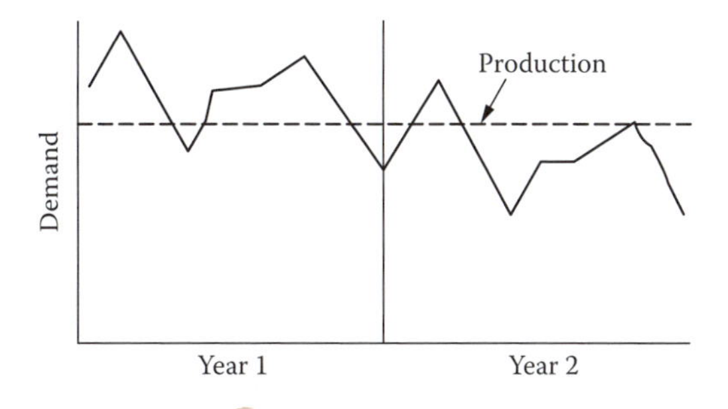

Figure 13.4 Leveled production over 2-year period.

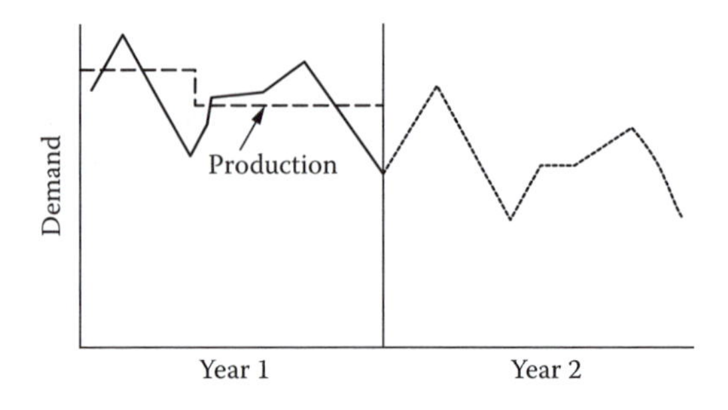

Figure 13.5 Leveled production over first year only.

Example 1: Level Production Schedule with Seasonal Variation

Demand for the seasons is forecast as follows:

Winter	12,000
Spring	48,000
Summer	60,000
Fall	42,000

One way to level the production schedule is to divide the seasonal forecasts by 3 to get monthly demands, then divide the monthly demands by 4 (assuming 4 weeks per month) to get weekly demands. The resulting production schedule would be as follows:

Dec.–Feb.	4,000/month	1,000 units/wk
Mar.–May	16,000/month	4,000 units/wk
June–Aug.	20,000/month	5,000 units/wk
Sept.–Nov.	14,000/month	3,500 units/wk

To avoid the extremes (going from 1,000 to 4,000, and from 5,000 to 3,500) the schedule might be modified to 3,500 per week in the first and third quarters, though that would require carrying inventory from the first to the third quarter.

Leveling Multiple Products

Next consider the procedure for leveling the production schedule for several products at once. Though we are now looking at multiple products, the procedure is similar for leveling one product group. For each product, we seek to produce about the same amount every period. The size of the period can be a month, week, or day, though the smaller the period, the smoother (more level) the production schedule. For example, if the plan indicates that 4,000 units of a product must be produced every month, then one approach would be to produce all 4,000 in the first week of the month. Though the monthly schedule would look level, the weekly schedule would look choppy because in each month it specifies 1 week on (making the product) and 3 weeks off (not making it), a pattern that is repeated over and over. Had the plan specified instead 200 units a day (assuming 20 days/month), then production would be uniform throughout the month.

Example 2: Leveling Production of Three Products

Assume the production requirements for one month are

Product	Quantity
Product A	4,000
Product B	2,000
Product C	1,000

Other products are made too, but products A, B, and C are the high-volume ones.

Schedule a in Figure 13.6 shows one possible MPS for the products. Although the production schedule looks somewhat level (it remains at 2,000 units for 3 weeks and changes only once during the month) it is not. Because of the large size of the batches (2,000 or 1,000), production involves correspondingly large batches of components and subassemblies, which imposes large week-to-week variation in the workloads of upstream operations supplying these parts. Products A, B, and C are different, and equal quantities of them do not equate to equal production times, types and quantities of components, or any other measure of resources and capacity required to produce them.

Now, if changeover times throughout the production system are long, then schedule a might be the only feasible schedule. If, however, the changeover times are small, then possibly some amount of every product could be produced every week. This is shown in schedule b in Figure 13.6. Since in schedule b the same amount of each product is produced every week, then the requirements imposed on upstream operations will remain the same, week by week. At least for production of parts and components going into products A, B, and C, the schedules for upstream operations will be uniform and entirely predictable. Uniformity translates into a weekly routine, which means simplified weekly work planning and scheduling for upstream operations.

Even better than a weekly level schedule is a daily level schedule. If changeovers can be reduced to one-touch procedures, then the same amount of every product could be produced every day, shown in schedule c. With this schedule, every upstream operation every day produces the same volume of every kind of part needed for the three products.

Three MPS Alternatives

	Week 1	*Week 2*	*Week 3*	*Week 4*
a	2,000A	2,000A	2,000B	1,000C
b	1,000A	1,000A	1,000A	1,000A
	500B	500B	500B	500B
	250C	250C	250C	250C
	Day 1	*Day 2*	*Day 3*	*Etc.*
c	200A	200A	200A	200A
	100B	100B	100B	100B
	50C	50C	50C	50C

Figure 13.6 Three master production schedule alternatives.

In a **daily leveled schedule** the production volume for each product is set at 1/20th the monthly requirement (assuming 20 working days/month). If monthly product demand is 4,000 units, then daily production would be 200 units. If the demand variation within the month is large, then half-month demands can be used instead, and the daily production for each half-month set at 1/10th of the half-month amount.

To take this a step further, suppose demand fluctuates significantly from day to day. Does that mean that the scheduled production level should also fluctuate day to day? The answer is both no and yes. No, because a production schedule that changes every day is obviously not level. It lacks daily routine and requires every upstream operation to carry buffer stock or readjust capacity (setups, inputs, outputs) to meet fluctuating workloads. This is a common predicament, but one which production leveling seeks to avoid. On the other hand, yes, but only if the daily change in production represents a relatively small adjustment to a base schedule that is somewhat level. The allowable degree of adjustment will depend on how much variation upstream operations are capable of absorbing on short notice. Toyota, for example, uses a level 10-day production schedule as a base, but adjusts the schedule daily to incorporate the most recent customer order information. That daily adjustment, however, is limited to ±10% of the base-scheduled amount. Still, even with such a small adjustment limit, this provides enough flexibility to vary day-to-day production by as much as 23%.

Case in Point: Level Production at an Electronics Plant

Park[4] performed a simulation study for a large Korean electronics manufacturer to answer management's question as to whether level production would be good for the company. He investigated the effect of level production on various sublines and suppliers. Figure 13.7 is an example of one month's demand imposed on a supplier when the electronics firm used traditional nonlevel production schedules. Though daily demand averaged about 9,000 parts, because it fluctuated between 12,500 and 5,000 parts, the supplier had to operate at a

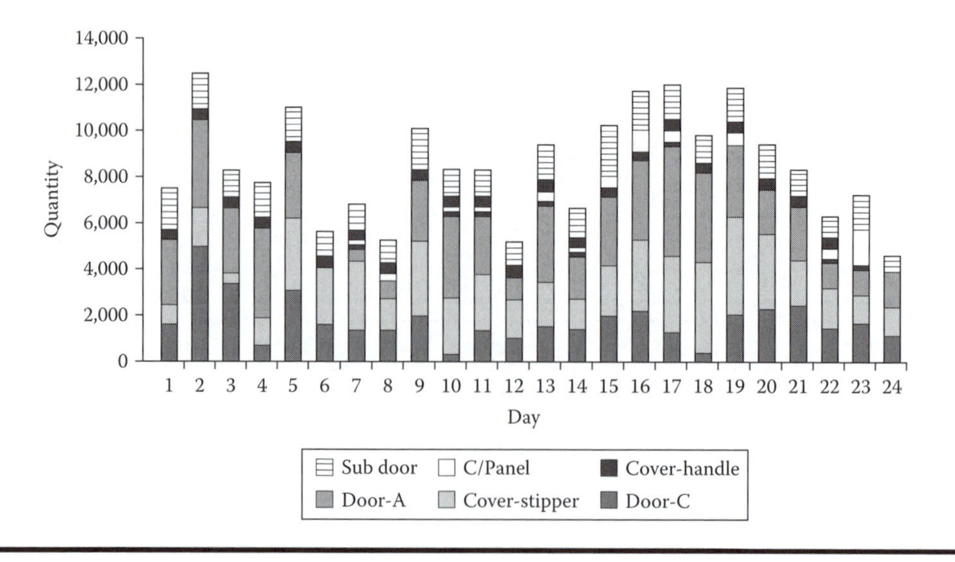

Figure 13.7 Supplier demand before level production.

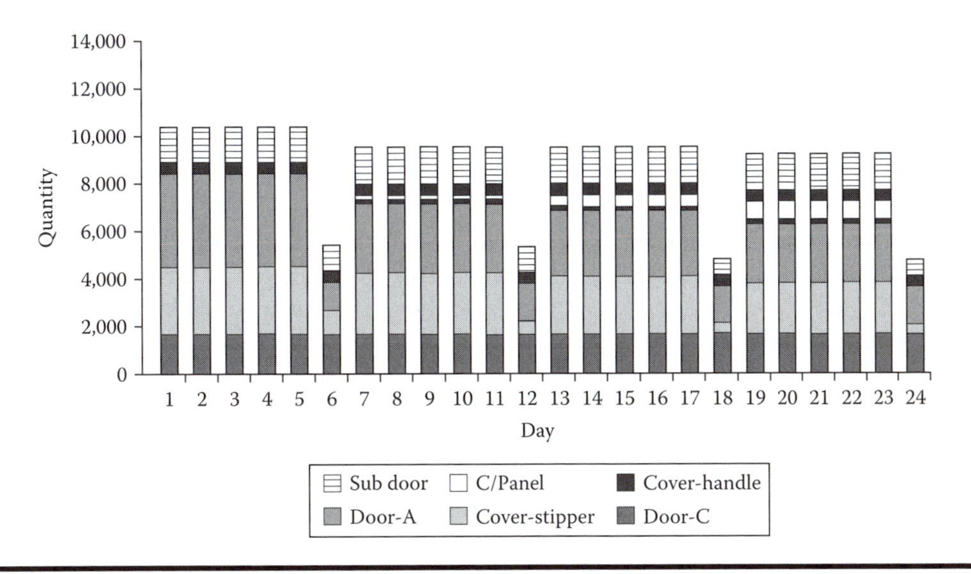

Figure 13.8 Supplier demand with level production.

capacity of 11,000. Even with that, it was necessary for the supplier to preproduce several days in advance and carry 3 days inventory to meet peak demand. Similar fluctuations were experienced by many of the firm's 200 vendors. Within the manufacturer's own plant, sublines carried 3 days average inventory plus an additional day of inventory as a safety stock to handle schedule fluctuations.

Figure 13.8 shows the demand for the same supplier assuming the electronics firm switched to level production (the half-load days are Saturdays). With level demand, the supplier no longer has to preproduce. Prior to leveling, the supplier needed 1 week to fill orders; after leveling, 3 days. Inventory level on sublines in the plant dropped 50%.

Based on the results of the simulation analysis, the electronics firm implemented level production on several of its lines. In many cases the simulation predictions proved correct. Workers reacted positively to the changes, and productivity and product quality increased. Vendors like leveling too, but for the certainty it provides. Order changes have been practically eliminated, and delivery dates and quantities are fixed a month in advance. The electronics firm implemented leveling in steps, first by cutting original batches in half, then by producing to fill shipping containers, then by leveling daily production (as in Figure 13.8). Leveling for some items has been extended to mixed-model production, a topic we cover later.

Leveled Schedules: A Cooperative Effort

Preparing leveled production schedules is not something production planners do occasionally and behind closed doors. Schedules are constantly refined as new, more accurate demand information becomes available.

Practical, feasible scheduling requires involvement from people in sales, marketing, engineering, production, and finance. The production levels selected must be able to satisfy actual, firm customer orders and projected demand requirements, yet do so with the existing capacity and minimal buffer stock. Thus, planning the production level must take into account manufacturing's ability to adjust capacity, as well as sales's ability to provide quick order information and accurate sales forecasts.

The marketing department must coordinate promotional schemes with manufacturing to ensure that production output can keep up with possible sales increases. Marketing must be sensitive to the goal of production leveling, and avoid doing anything that amplifies demand variability. Sales must know the impact that each customer order will have on production schedules and profits, and be willing to forgo orders when the return is questionable, especially in periods of peak demand. This is an important point, because sales and promotional efforts can have a major influence on demand fluctuation. When production is busy, additional demand generated from occasional, marginal customers can result in degraded service to regular customers. By having finance involved, the dollar ramifications of stimulating or quelling demand and of adjusting capacity to meet demand can be determined.

Achieving level production also requires that the parts and components of different products be somewhat interchangeable. For effective production of multiple products in uniform small batches, design engineers must be mindful of the effects their designs have on production processes and changeovers. Among other matters, design for manufacture and assembly (DFMA) efforts must consider the effect of product design on production leveling.

Level Scheduling in Pull Production

Production leveling can be beneficial in any production environment that meets the requirements listed earlier. For pull production systems, however, production leveling is itself a requirement. Pull production systems function well only when demand for materials at every stage is smooth and steady. For that demand to be level, production schedules must also be level.

Mixed-Model Production

Final Assembly Schedule

Unlike a push production system, which requires a schedule for every operation, a pull production system utilizes only one schedule, a schedule for the last stage of the process. As shown in Figure 13.9, each operation produces in response to orders from downstream. Since orders move

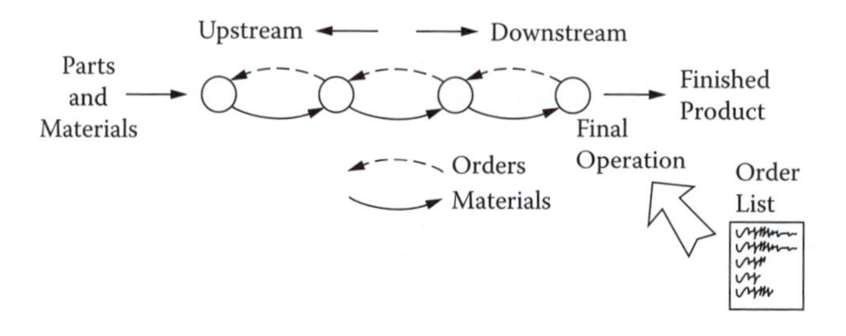

Figure 13.9 Process initiated by orders from the final operation.

upstream, the process is initiated by orders that come from the final operation. Whatever orders the final operation must meet, so must the operations upstream. The final operation is often an assembly operation—the place where all the components and subassemblies produced upstream are finally put together. Thus, the schedule or order list at the final operation is referred to as the **final assembly schedule,** or **FAS**.

Heijunka: Mixed-Model Production

A pull system that produces multiple kinds of products requires a uniform production schedule for each of them. Ideally each product has its own line and final assembly operation that produces according to its own final assembly schedule. But usually it is not cost effective or feasible for every product to have its own line or final assembly area, especially when the number of products is large and demand for each is relatively small. In the usual case, several different products will be produced on the same production line or same final assembly station.[5] Production of multiple kinds of products on a repetitive basis, in a mixed fashion (some As, some Bs, and so on) on a single line or assembly station is referred to as **mixed-model production** (MMP) or mixed-model assembly. The Japanese term for MMP is **heijunka**, which at Toyota refers to "distributing the production of different [product types] evenly over the course of a day, a week, or a month."[6]

In heijunka or MMP, batch production of different products is avoided; instead, products are interspersed together and produced in a mixed sequence. Mixing different products in sequence like this results in smooth, steady demand for the upstream operations that supply components and materials.

Batch Size

As stated, the mixing of different products in MMP is done in systematic fashion that avoids batching of consecutive products. For example, referring to Figure 13.2, if products A, B, and C are produced in the same final assembly area, then schedules a and b represent a form of MMP because production of the three products is mixed. The schedules, however, are a rather crude form of MMP because they result in production of somewhat large batches (in Figure 13.2, each production run in schedule a extends for an entire shift and in schedule b half a shift) with somewhat long intervals between batches. For each batch of product, the required materials and component parts must be produced in advance by upstream operations, and when the batch is large, the materials that go into it are usually produced in large batches also.[7] In general, large batch production at final assembly imposes large batch production on immediate upstream stages, which imposes large batch production on earlier stages, and so on. The irregularities resulting from large batches such as in Figure 13.2 might be more than a pull production system can handle.[8]

In pull production, working with small WIP is the modus operandi.[9] Given that the size of the batches at final assembly somewhat dictates the size of batches everywhere else in the production system, then to keep WIP small everywhere, the batch size at final assembly must also be small.

MMP and Production Smoothing

Producing in mixed small batches smoothes the demand requirements everywhere in the process. This is shown in Figure 13.10 using the three schedules from Figure 13.6. Product A is highlighted to illustrate how reducing the size of the production batch results in a smoother production

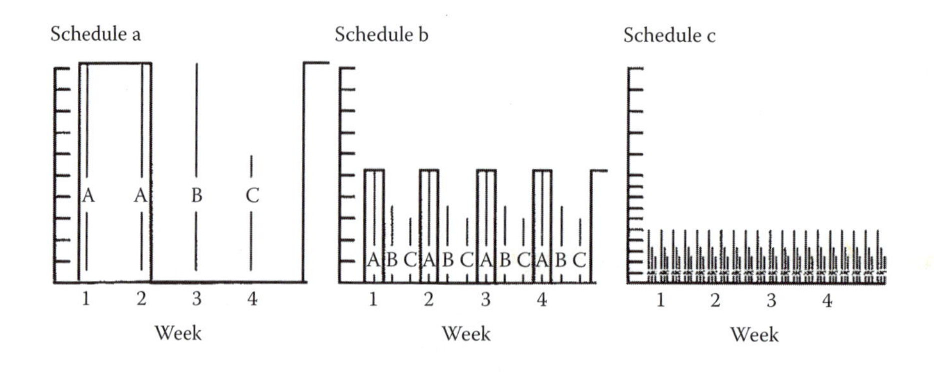

Figure 13.10 Effect of smaller batches on production smoothing.

schedule. Smoothing the production schedule results in smoother flow of orders going upstream, and, hence, smoother flow of materials going downstream and everywhere else in the process.

The MMP Schedule

Theoretically, maximal smoothing is achieved by scheduling production at final assembly in batches of size one. A batch of size one means that, ideally, one unit of A is produced, then a unit of B, a unit of C, and so on, and the resulting production sequence is ABCABC.... This repeating sequence is called an **MMP schedule**. In pull production, this MMP schedule is the FAS. The sequence tells workers at final assembly what they must produce and in what order. This is the only schedule required in pull production.

Now, looking more closely at the MMP schedule, if it specified ABCABC..., then the result would be units of A, B, and C produced in equal quantities. If the demand for the three products is different, as would be expected, then the schedule should indicate the different products being produced in the desired proportions. The frequency that each product appears on the MMP schedule should be proportionate to the demand for that product. The following example gives a procedure for determining the MMP schedule that will yield the desired quantity of each product.

Example 3: Establishing an MMP Schedule

Monthly demand for products A, B, and C is 4,000, 2,000, and 1,000 units, respectively. Assume the products involve similar components, technology, and processing steps, and thus can all be assembled at the same final assembly stage. There are three steps to setting the MMP schedule.

Step 1: Determine the daily production requirements. Assuming 20 working days per month, the daily production requirements are:

Product A	4,000/20 = 200
Product B	2,000/20 = 100
Product C	1,000/20 = 50
All products	7,000/20 = 350

Step 2: Determine the repeating sequence. The dual goal of MMP is to produce units (a) in conformance with production requirements, and (b) in the "most mixed" fashion. To achieve both we need to determine the appropriate MMP **repeating sequence**, which is the particular ordering of the different products to be repeated throughout the day. To find the repeating sequence:

1. Determine the largest integer that divides evenly into the daily requirements of all the products. The requirements are 200, 100, and 50, so the largest integer divisor of the three is 50.
2. Calculate the minimum ratio among the three products by dividing the daily requirement of each product by the integer divisor. For the integer 50, the ratio is 4:2:1. The sum of the ratio numbers is 7, so each repeating sequence will have 7 units: 4 product As, 2 product Bs, 1 product C. Note that this sequence must be repeated 50 times a day to produce the required amounts for every product.

Step 3. Determine the product ordering within the repeating sequence. One possible ordering is AAAABBC. This ordering satisfies the ratio requirement of 4 As, 2 Bs, and 1 C but is still somewhat choppy since it batches all the As and all the Bs together. A less choppy (more mixed) sequence would be AABABAC. This is probably the most mixed ordering for this example.

Assuming this ordering is adopted, the production process would repeat the sequence AABABAC 50 times a day. The only other information necessary to round out the FAS is to specify the takt time (required cycle time [CT]). If the work day is 8 hours, then, on average, the takt time must be 8/350 = 0.02285 hr, or about 82.3 sec/unit. The matter of applying takt time to synchronize the process is described in the next chapter.

Maximally mixed ordering gives the smoothest flow, though for some purposes maximal mixing might not be best. If, to continue the example, the takt time is very short and the changeover times between products are somewhat long, then the ordering AAAABBC might be preferred over AABABAC because it requires fewer changeovers from product A to product B, then to product C, and so on. Or, if the products are packaged two to a container, then AAAAAAAABBBBCC (repeated 25 times) might be better because items could then be packaged as soon as they are made (in AABABAC units have to wait; production of successive B units is interrupted by an A unit, and C units can be packaged only in alternate sequences).

Seldom is the demand in nice round numbers as in the example, so the procedure for determining the minimum ratio requires making ad hoc adjustments. For example, if the daily demand for product A were 213, that figure might first be rounded to 200 so a repeating sequence could be easily determined. The remaining units (13 here) would be added to (or deleted from, if rounding were upward) the schedule at the start or finish of the day, or uniformly throughout. When the number of different product models is very large, finding the most mixed MMP sequence is more difficult, and in that case manufacturers use computational computer algorithms to develop the daily sequence.[10]

Requirements for MMP

The minimal requirements for MMP are the same as for level production schedules: somewhat continuous demand, small setup times, and demand-driven production. As with level production schedules, trying to implement MMP when demand is discontinuous or one time is impossible or impractical. Maximal mixing in MMP can require almost continual changeover, so setups must be simple, one touch, and time minuscule. MMP schedules are demand driven, but the demand is a composite of incoming customer orders and leveled production schedules. Details of this are explained in Chapter 15.

Mixed-model production has three additional requirements, all familiar aspects of lean production: flexible workers, total quality assurance, and small-lot deliveries of material.

Flexible Workers

Workers must be cross-trained and equipment adaptable to work on every product in the MMP sequence. Ideally, besides being able to perform a variety of tasks at a given workstation, workers should be able to perform tasks at multiple workstations. That way a bottleneck anywhere caused by a quality problem, too high CTs, or insufficient capacity can be tackled by shifting workers between operations.

Effective Quality Assurance

Moving workers between operations increases flexibility, but it can also increase quality problems since workers must remember many different tasks and operations. Standard operations, poka-yoking, source inspection, and line stop are thus essential to preventing workers from skipping steps, taking incorrect steps, or using wrong parts. At the Toyota assembly plant in Georgetown, Kentucky, use of autonomation and line stop causes an average of 1,700 shutdowns per shift. The attitude of management is that the 6% production downtime is well worth it because of all the problems being discovered.[11]

Small-Lot Material Supply

Since MMP involves producing a variety of products all at once, variety in the materials used can be large. To minimize the amount of in-process inventory and confusion about what parts go into what models, materials must be synchronized to arrive at the final assembly stage in the exact quantities and at the exact times needed. The process of synchronizing upstream parts production and delivery with MMP final assembly is described in the next chapter.

Advantages of MMP

MMP has all the advantages of level production schedules, including low variation in production schedules, low WIP inventory, reduced lead times, and ability to meet demand with lower average production capacity. Other advantages of MMP are discussed in the following sections.

Elimination of Losses Due to Line Changeover

On a traditional assembly line, the entire line is shut down for changeover between production models, and the shutdown can take days, weeks, or longer depending on the extent of differences between models. With MMP an entire line is rarely shut down for changeover. Unless the changeover is for a new product that requires substantial process retooling, the line continues to function during changeovers, which occur almost continually. In automobile production, for example, a single line that produces cars, SUVs, and small trucks can readily be adjusted to meet changing demand for each vehicle type without ever stopping the line. The adjustment is made simply by altering the proportion of each vehicle type in the MMP sequence.

Process Improvement

Workers rotating through a variety of tasks and operations in MMP are more aware of problems and motivated to eliminate them than are workers assigned to only one model. In large batch production where, say, one model is produced monthly between changeovers, process problems tend to be patched and then forgotten; months later when the model is run again, the problems recur, and workers again learn the process and patch the problems. In MMP, a problem that is not fixed appears over and over, which motivates workers to find a permanent solution.

Balanced Work Loads[12]

MMP on an assembly line results in an even allocation of work throughout. In the usual single-model line, tasks at different stations require different times, say, 40 seconds, 45 seconds, and 55 seconds on a three-station line operating at a 60-second CT. Not only do workers with long task times find this inequitable and aggravating, but workers at stations with short task times have no incentive to rotate to other stations or to contribute to finding ways to reduce the process CT. With MMP and rotating assignments, the workloads among stations are more balanced, and workers have no vested interest in the task times at any one station.

Fewer Losses from Material Shortages

In the event of a material shortage in MMP, only particular models requiring that material are affected. Work continues on other models, and more of those models are produced until the material arrives. When the shortage ends, production of the interrupted models resumes and is temporarily increased until the deficit is filled.

Production Planning and Scheduling in Different Circumstances

The fewer the number of final products and the larger, more stable the demand for each, the simpler it is to do production planning and scheduling, including for MMP. Conversely, the greater the number of products and the smaller, less stable the demand, the more difficult it is.

One way companies prepare plans and schedule production to accommodate different product variety and product volume combinations is to modify their definition of an **end item** to whatever works best. In general, an end item is whatever item a company prepares, plans, and schedules to produce. It is the item that appears in the MPS and at the top level (0 level) of the bill of materials. Beyond this general definition, however, what a company chooses to call an end item can vary depending on the complexity and volume of the product and on the production philosophy to which the company subscribes.

Production Philosophy

Consider the three most common production philosophies, make to stock, assemble to order, and make to order.

Make to Stock

Make to stock (MTS) companies make products in anticipation of demand. Usually these products go into finished goods stock before being withdrawn to fill customer orders. Relatively few products (say, less than 100) are produced in large volume, and each product typically contains a large number of components. In MTS companies, the end item is a finished product or group of products that are identical except for minor features. Televisions are an example. An MRP system, if one is used, maintains a separate bill of materials (BOM) for each end item and creates a separate MPS for every product. Since products are made in advance of actual orders, the quantities in production schedules are based on forecasts.

Assemble to Order

Assemble to order (ATO) companies produce subassemblies according to forecasts and then combine the subassemblies into unique combinations as requested by customers. A large variety of different products can be produced by combining different combinations of relatively few kinds of subassemblies. Companies that produce computers and automobiles are examples. In ATO companies, the end item is the option for a kind of subassembly. For example, say a company makes golf carts by assembling engines, drive trains, and chassis, and the drive trains and chassis each come in different models or options—the drive trains come in standard and heavy-duty versions, and the chassis come in medium and stretched versions. The focus of the MPS is on the production of subassembly options, not the final product, golf carts, and these subassembly options are produced in advance of customer orders. Similarly, the BOMs maintained by the MRP system are for the subassembly options, not for final products. The use of options, also called modules, in production scheduling is described later.

Make to Order

Make to order (MTO) companies produce products in response to actual customer orders, so they carry little finished goods inventory. They can produce many kinds of products (many hundreds) each in small quantity by using different combinations of relatively few kinds of components. Pharmaceuticals are an example. Because of the large number of potential products and possible small demand for each, it is impossible to forecast demand for products. Also, it is impractical to maintain the BOM for every product. (In a related philosophy, engineer to order, products are not only produced but are designed to meet requirements of particular customers. Before the customer order, the product does not exist, hence, neither does the BOM.)

If the number of products is not too large, say, less than 100, then all products can be included in the MPS. To prepare a master schedule solely from customer orders, there must be a backlog of orders. From this backlog, orders are chosen and inserted into the schedule according to order priority and production capacity constraints.[13]

When the number of products is large, it is difficult to handle all of them on the master schedule. In that case, the end item in the master schedule can represent a group of products such as a product family. Requirements for parts and components for each product within the group are derived from percentages of the anticipated mix of products within the group. This procedure is discussed later.

The three philosophies represent three fundamental types of product structures, as shown in Figure 13.11: (1) few products, each potentially comprised of many kinds of parts; (2) many products, each made from a different combination of subassembly options and where each subassembly itself might consist of many parts; and (3) many products, each consisting of different combinations of a relatively few kinds of parts. In general, items at the narrowest part of the product structure should be the items upon which requirements planning (in MRP) and master production scheduling are based. This is because this part of the structure represents the point of greatest commonality among all products, and items at that level are the simplest to plan and schedule. Production

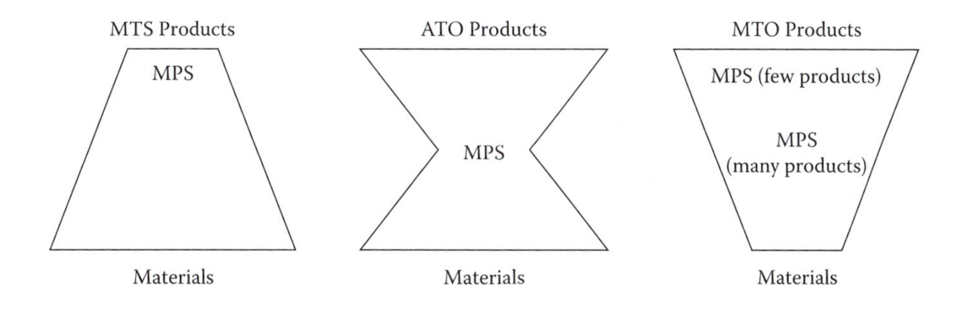

Figure 13.11 Different kinds of product structures and level at which master production scheduling (MPS) is performed.

planning and scheduling for MTS and ATO usually focuses on items at the top and middle of the product structures, respectively. For MTS, master scheduling focuses at any level of the structure depending on the number of final products, though for simplicity it should be on items somewhat low in the structure—on materials, parts, and subassemblies of relatively common items.

Final Assembly Scheduling versus Master Production Scheduling

It is important to distinguish between scheduling for MPS and scheduling for final assembly. The difference is one of immediacy: whereas the MPS often portrays anticipated orders or forecasted demand, the FAS usually portrays actual orders. Whereas the former drives material procurement and preparation to do work, the latter drives the actual execution of work. Paraphrasing Mather:[14]

> Think of the MPS as the means for bringing in raw materials and purchased items to the plant and processing them. The FAS then takes over and converts the items and materials into finished goods. The analogy is a relay race where the baton is passed from the MPS to the FAS. The objective is to make the FAS responsive to market conditions while simultaneously keeping the MPS stable.

The MPS must look far enough into the future so that material and capacity requirements can be anticipated, ordered, and on hand when needed. As a result, production and procurement lead times are the determining factor in setting the time horizon for the MPS. In contrast, the time horizon for the FAS need cover only the time between when all components or subassemblies become available for production of an item and when the production of the item is completed.

When the number of kinds of final products is very small (for example, aircraft), the FAS is the MPS. In such cases, no product is made without an existing product order, and actual customer orders drive the MPS.

In MTS companies, the MPS drives the FAS, that is, final assembly happens in anticipation of demand. Even then, however, the FAS must take into account existing (actual) orders and the most recent changes in anticipated demand. If current demand falls off substantially, or if inventories reach the maximum desired levels, then production should be stopped.

When the number of final products is very large, the MPS focuses on subassemblies or options, not on the individual products. The FAS specifies the particular products to be made through various combinations of subassemblies and components, whereas the MPS assures that those components and subassemblies will be available at final assembly in sufficient quantities to meet the FAS.

The remainder of the chapter discusses particulars for smoothing production schedules (FAS or MPS, whichever is appropriate) under each of the three manufacturing philosophies.

MTS: Uniform Load Production Schedule

Products that have large, stable demand can be readily scheduled using MMP as described earlier. However, even products with somewhat small demand can be scheduled using MMP, providing that the demand is somewhat continuous. For example, suppose product-quantity analysis is used to single out from hundreds of products the 11 with the highest volume. One way to level the production schedule is to produce some of the first four products every day, some of the next three every week, and all of the last four every month, once a month.

Example 4: Creating a Uniform Production Schedule

Figure 13.12 shows the monthly demand and production breakdown for the 11 products. According to the figure, it should be possible to satisfy demand for all 11 products with a daily average production rate of 475 units. Therefore, in creating the daily production schedule, the goal should be to maintain an average of 475 units/day.

One possible production schedule for the 11 products is shown in Figure 13.13. Daily production in that schedule varies from a high of 495 to a low of 445. The largest day-to-day variation is

Demand Volume and Scheduled Production Volume

Product	*Monthly Volume*	*Daily Production*	*Weekly Production*	*Monthly Production*
A	4,000	200		
B	2,000	100		
C	1,000	50		
D	760	38		
E	540		135	
F	500		125	
G	300		75	
H	150			150
I	100			100
J	80			80
K	70			70
Total	9,500	388	355	400

Average daily production = 9,500/20 = 475.

Figure 13.12 Demand volume and scheduled production volume.

Level Daily Production Schedule

	Production Schedule													
	Week 1 (days)					*Week 2*			*Week 3*			*Week 4*		
Product	*1*	*2*	*3*	*4*	*5*	*	*4*	*5*	*	*4*	*5*	*	*4*	*5*
A	210	210	210	190	180									
B	150	150	50	50	100									
C	60	60	60	60	10									
D	60	60	60	20	20		Same			Same			Same	
E			135											
F				125										
G					75									
H				50	100									
I							25	75						
J											80			
K														70
Total	480	480	485	495	485	*	470	460	*	445	455	*	445	445

Note: Production schedule for days 1–3 remains the same every week.

Figure 13.13 Level daily production schedule.

in week 3 where production goes from 485 in day 3 to 445 in day 4. Since the amount of variation (40/445 = 0.089, about 9%) is small, the process should be able to adjust to it. In actuality, the lowest production rate of 445 per day would never happen because many other products besides the 11 listed must also be produced, and production of them would be scheduled during periods of low production activity—days 4 and 5 during weeks 3 and 4.

The identical MMP schedule is used on days 1 and 2 to produce products A to D, and a different schedule is used on day 3 to include product E. Once the MMP schedules for these three days have been set, they remain the same week after week. For days 4 and 5, different MMP schedules are used on different weeks to accommodate products F to K. Unless the overall level or mix of demand changes, once set, all of these MMP schedules will remain about the same, month after month.

The MMP schedule represents anticipated production, and the actual daily FAS will differ somewhat to reflect recent demand changes and actual customer orders received.

Assemble To Order

When a company makes numerous kinds of products (1,000s, 10,000s, or 100,000s of combinations of styles, options, features), it is impossible to anticipate demand and plan production for each of them. It is also infeasible to maintain a detailed BOM for each of them. A way to deal with the problem of too much variety is to sidestep it and instead deal with something for which there is much less variety.

Modular Bills

In ATO practice, a BOM is maintained for each subassembly or component option but not for individual final products. The BOM for such an option is called a **modular bill** of materials.

Figure 13.4 shows the modular bill for product family X, where product family X represents all possible variations of a product. Each column represents a subassembly (body, engine, etc.) and shows the available options. Together, different combinations of options yield a possible 54 variations (3 bodies × 2 engines × 3 covers × 3 instrumentations) of product X. Since ordinarily the BOM for a product shows the particular combination of subassemblies used in a product, 54 BOMs would be required to show the same information as expressed in Figure 13.14.

As Figure 13.14 indicates, a modular bill shows all the possible options, but not particular combinations of them. Each box in the figure is a **module** and represents an option. Since there

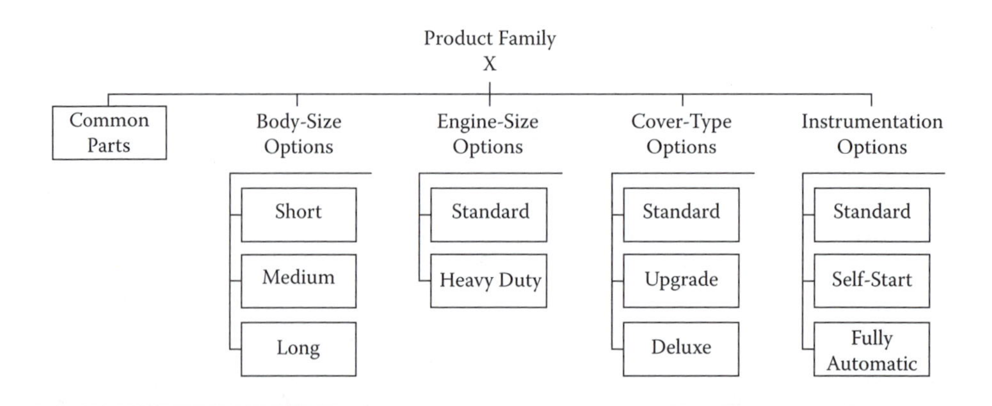

Figure 13.14 Modular bill showing available options for product family X.

are 11 options (3 bodies, 2 engines, 3 covers, 3 instrumentations), then product X can be fully represented by 11 BOMs, one for each option. (In Figure 13.14 there are 12 BOMs; the additional one represents common parts, or parts used in all versions of product family X.) In a modular bill, the options are elevated to level 0 status in the tree structure, and the BOM for final products (products in family X) does not exist.

Since many final-product variations result from different combinations of options of relatively few subassemblies, the demand for each option is always larger than the demand for each variety of final product. As a result, it is easier to estimate demand for modules than it is for particular products. Although demand for the products in which the subassemblies are used might be erratic and unstable, the requirements for the subassemblies themselves, aggregated over all products in which they are used, might be stable and predictable.

Figure 13.14, for example, represents 54 possible kinds of products, but only two sizes of engines. The demand for each of the 54 products might be small and hard to predict; however, the aggregate demand for all 54 kinds of products will, obviously, be larger and thus easier to predict. Once this aggregate demand is forecast, the requirement for a kind of engine is determined by multiplying the forecast by the percentage expected to require that engine (the percentage is usually derived from the percentage of past sales having that engine). This is explained later.

The use of modular bills presumes that the BOMs have been structured in terms of the modules from which final products are made. In many cases, however, preexisting BOMs are for final products, not for modules. The next section discusses how to create modular BOMs from a group of BOMs for final products.

Modularization Procedure[15]

To restructure the BOMs for a group of product models into modular bills, first look at the level 1 components in the BOMs for all the product models to determine which components are common to all models and which are unique to only certain ones. Then, cluster the components into the categories they have in common. Some components will not be assignable to categories, in which case the level 2 components should be investigated for commonalities.

Example 5: Modularization of Products

Consider a product that comes in the following variations:

Body size: big, small
Clutch torque speed: high, medium, low

There are six possible variations or models for the product. Suppose the level 1 BOMs for each of them is as shown in Figure 13.15. The level 1 components share the following categories:

Common to all models: J, S
Big models: L, Z
Small models: K, W
High-speed models: C, O, R2
Medium-speed models: B, N, R3
Low-speed model: A, M, R4

The remaining components (D1, D2, E1, E2, and so on) are unique to each model. Now, to determine whether these components can be modularized, look at the level 2 BOM and observe

Possible Model Combinations for Final Products

Body:	Big	Big	Big	Small	Small	Small
Clutch:	High	Medium	Low	High	Medium	Low
Level 1 Components	J	J	J	J	J	J
	C	B	A	C	B	A
	L	L	L	K	K	K
	D1	E1	F1	I1	H1	G1
	S	S	S	S	S	S
	D2	E2	F2	I2	H2	G2
	O	N	M	O	N	M
	Z	Z	Z	W	W	W
	R2	R3	R4	R2	R3	R4

Figure 13.15 Components in level 1 BOMs for six product models.

the parts breakdown of each component. Suppose the level 2 breakdowns are as shown in Figure 13.16. Next, refer back to the original categories (common, big, small, high, medium, and low) to see whether it is possible to classify the level 2 components into the same categories. As it turns out, it is possible, with the following result (level 2 components in bold):

Common: J, S, **H3**
Big: L, Z, **C4**
Small: K, W, **C5**
High: C, O, R2, **C1**, **G5**
Medium: B, N, R3, **C2**, **G3**
Low: A, M, R4, **C3**, **G4**

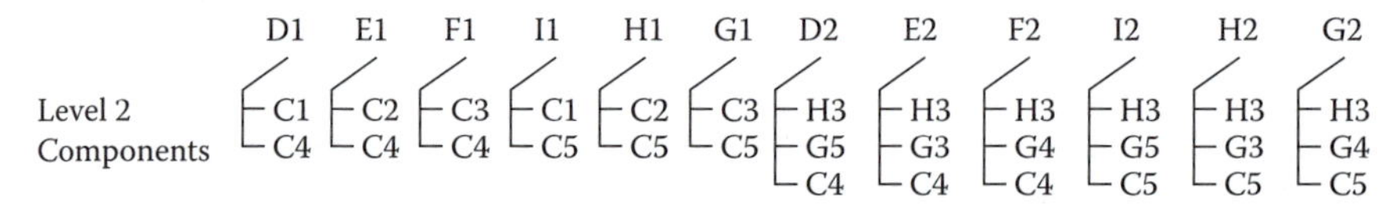

Figure 13.16 Level 2 BOM.

Thus, the six final product BOMs can be replaced by six modular bills: common, big, small, high, medium, low. In this example, the number of modular bills is the same as the number of product models, six, though as explained, when the number of possible product models is very large, the number of modular BOMs will be much smaller.

Usually, a module is a subassembly that is produced and temporarily held in stock until a customer order arrives that calls for its use in a product. The production lead time required to assemble modules into the final product is one determinant of the level of the item chosen for modularization. If the time to combine modules into a final product exceeds the customer-expected lead time, then the level of the subassembly in the module might have to be raised from a simple subassembly to an assembled group of modules, or a major subassembly. As a rule, it takes longer to assemble many simple subassemblies than a few major subassemblies (groups of already combined simple subassemblies). Thus, the acceptable lead time at final assembly is an important criterion for choosing the appropriate level of subassembly to call an end item a module.

Planning Bills

Modular bills elevate lower-level items in the BOM to level 0 end item status and eliminate the need for the former (final product) level 0 item. In the example for product X, each subassembly module is considered an end item in itself.

When the number of possible final products is very large, a type of modular bill called a **planning bill** is used for production scheduling. Planning bills greatly simplify production scheduling and improve the ability of production schedules based on forecasts to satisfy actual customer demand. As a case in point, in the example for product X in Figure 13.14 there are 54 variations. Whereas it would be difficult to accurately anticipate demand for each of the possible variations, it is relatively easy to forecast the percentages of products that will involve different options from sales records and conversations with customers. These percentages are precisely what a planning bill provides. The planning bill does not specify the exact amount of each product to produce, but it does specify the amount of materials and subassemblies needed to fill actual orders in MTO and ATO factories. Following is an example.

Example 6: Production Planning Using Planning Bills

Figure 13.14 illustrated the modular bill concept. Suppose past sales of products in product family X are evaluated for customer preference, and these preferences are reflected by the relative percentages for options as shown in Figure 13.17.

Suppose for an upcoming month, forecasted demand for all product models is 2,000 units. A level production schedule is used, and the MPS specifies a daily rate of about 100 units. The schedule, however, does not specify the exact mix of the 54 models since that will not be known until orders are compiled and the FAS is prepared.

Without knowing the actual individual product models, the planning bill makes it possible to go ahead and estimate the number of each type of component that will likely be needed. Assume an MRP system orders materials using weekly time buckets, 4 weeks per month. Based on the planning bills in Figure 13.17 and assuming a 5-day workweek, the weekly requirements for the modules will be Percentage × 5 days × Daily rate, or

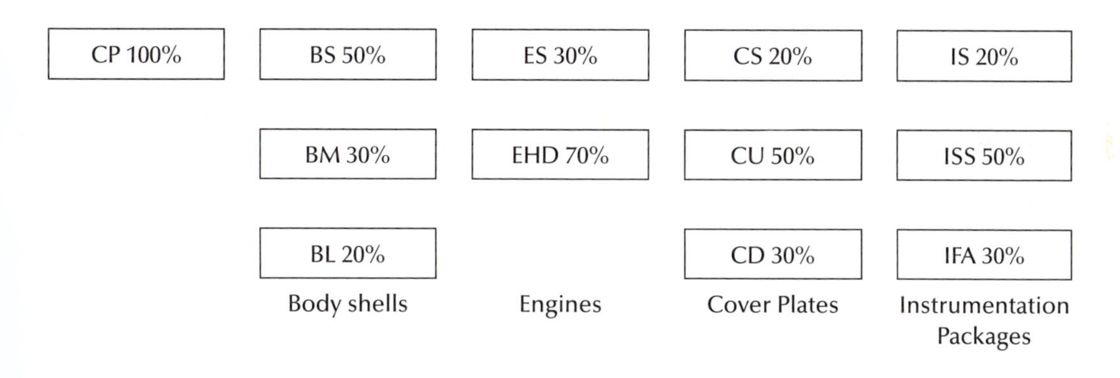

Figure 13.17 Planning bills showing percentage breakdowns.

CP	500	**BL**	100	**ES**	150	**CS**	100	**IS**	100
		BM	150	**EHD**	350	**CU**	250	**ISS**	250
		BS	250			**CD**	150	**IFA**	150

These numbers represent the quantities of each option that should be available to final assembly each week to meet anticipated demand for all varieties of the product.

To illustrate how planning for modules is tied to MMP, suppose the same company (that produces 54 models) also produces the body shells (BL, BM, BS) and the instrumentation packages (IS, ISS, IFA) on two separate pull production lines. For each line, the FAS can be expressed in terms of a daily MMP sequence:

Body shells: Sequence of 2 BLs, 3 BMs, and 5 BSs (e.g., BS-BM-BS-BL-BS-BM-BS-BM-BS-BL) repeated 10 times a day, 5 days a week.

Instrumentation packages: Sequence of 2 ISs, 5 ISSs, and 3 IFAs (e.g., ISS-IFA-ISS-IS-ISS-IFA-ISS-IFA-ISS-IS) repeated 10 times a day, 5 days a week.

The MMP sequences for body shells and instrumentation packages also determine the requirements for parts and components that are packaged by upstream workcells.

As with assembly of the final product, the FASs for body shells and instrumentation are modified daily to reflect actual customer orders as received. Benetton, the Italian clothing manufacturer does a twist on this, producing 25% of its output undyed, then coloring it at the last minute based on current sales data. The other 75% of output is produced with color according to sales forecast.

The percentages in the planning bills must be periodically checked and adjusted to reflect the most recent product order mix. Since actual requirements for final assembly will likely vary from the amounts estimated by the planning bill, safety stock for some items might be necessary. It is OK to carry an excess of inexpensive parts but not so for expensive parts, so close tabs must be kept on the latter. To avoid stockout of items controlled by kanban, kanban links should be established with upstream operations and suppliers (i.e., material supply links with guaranteed short production lead times for items produced in-house and short purchasing lead times for vendor-supplied items).

Alternative to Planning Bills

An alternative to planning bills is to forecast demand for each option directly. Using time series of the historical requirements for each option and taking into account the anticipated trends, the requirement for each option is forecast separately. Whether this gives more accurate requirements forecasts than planning bills is an open question; the answer is to try both methods, then pick the one that gives the best results.

Role of Concurrent Engineering

Simplified production planning and scheduling follows the principle "plan for the few, but produce for the many." Helping production conform to this principle is yet another area where concurrent engineering plays an important role.

Example 7: Simplified Planning through Simplified Design

A manufacturer produces 10 kinds of electrical motors. For each kind of motor, it also produces a mount and cover. Each cover has an access hole, and to accommodate different customers' requirements the hole comes in eight sizes. Therefore, the total number of combinations of covers and hole sizes is 80. Besides these covers, the material requirements include 10 sizes of motors and 10 sizes of motor mounts.

Now, suppose the design team redesigns the motor mount such that one size mount can be used with any size motor without loss of functionality. Also, suppose with only one size mount there need be only one size cover. The material requirements are now 10 sizes of motors, 1 size of mount, and 8 covers (one for each hole size). Note that there are still 80 possible final product combinations (10 motors, 8 possible hole sizes in each cover).

As a further simplification, a team of product and manufacturing engineers determines that the hole in the cover can be drilled at the time of final assembly according to the customer order. The hole size is thus specified on the FAS, and now the number of possible covers is reduced to one.

Simplifying products to simplify planning requires the combined effort of people in marketing, engineering, production, and purchasing. Marketing determines from customers the kinds of options they want and then engineering determines ways to design the product to meet the range of wants. Marketing, production, and engineering discuss how to achieve the wants through possible combinations of options, then zero in on combinations that will simplify planning and scheduling, minimize production costs and lead times, and be feasible given constraints of production costs, capabilities, and lead times. The end result is something more pleasing to both customers and producers. At one time, General Motor's midsized cars were produced in 1,900,000 combinations, though that number has since been reduced to fewer than 1,000.[16]

Make to Order

When demand is sporadic or every order is somewhat unique (the true job shop), then level production is impossible. If, however, product demand is uniform and somewhat continuous, then level production is possible.

Suppose demand is continuous, but is also highly variable. The only way to level production in that case is to maintain a backlog of orders. Given a sufficient backlog, then the production process has a pool of orders from which to continuously draw at a uniform rate. Of course, any backlog increases the time that customers must wait, though the actual wait also depends on how long orders are held before being released to production, and how long setup and production take. Thus, even with a backlog, lead times can be shortened if (1) marketing sends to production only confirmed orders (orders from paying customers), and (2) production minimizes the time to execute the orders (minimizes times for setup, worker reassignments, routing, and so on).

Scheduling with Backlogs

Briefly, the procedure is as follows. Upon receipt of an order, marketing enters the order into the backlog. In about a week, marketing confirms the order and, at that time, gives production notice of all confirmed orders in the backlog. About a week later, the list of confirmed orders is transferred to production; at that time, production starts fitting the orders into the production schedule for, say, 2 weeks later. Given a 2-week lead time and backlog, it might be possible for production to prepare a somewhat-level production schedule.

The total lead time in this example (time between orders being first received and being filled) is 3 to 4 weeks. The size of time buckets (time blocks for each stage of order receipt, order confirmation, order transmittal to production, and so on) is assumed to be 1 week and that contributes to longer lead times. With experience, the size of the time buckets at different stages should be reduced, which will reduce the time customers have to wait.

Minimizing Scheduling Problems

The worst case for production scheduling is when the number of products is large and the demand for each is low or one of a kind. In this case, level production schedules and pull production are impossible. Nonetheless, the goal is still to reduce waste, and that is done in the following ways:[17]

1. Simplify the BOMs. Reduce the number of levels in the BOMs to a minimum; this will reduce the number of parts to be managed, the numbers of transactions to be processed, and the time and cost of processing.
2. Use group technology and standard parts. Design products so that almost anything a customer would want could be achieved by assembly of similar or identical components. For procured components and parts, stick with standard items that are less expensive to procure and stock, and that can be procured on short notice. At the same time, try to keep WIP inventory to a minimum, and use simple visual controls to track inventory.
3. Make only what is needed. Do not use estimated lead times to anticipate what will be needed, and do not keep backlogs of work just to keep everyone busy. A simple form of visual kanban can be used to signal when a workcenter should produce more to meet demand from downstream workstations. When a workcenter becomes a bottleneck, workcenters upstream should stop and provide help to relieve the bottleneck. Items that must be produced in advance should be produced only in small batches and on a periodic basis, say, four times a year or once every month. The latter is a crude start toward repetitive production.
4. Produce in lot sizes that are small and easy to count. Produce large jobs in small batches and use small transfer batches (for an order of 100 units, process and move it along in batches of 10 at a time). This is done to avoid waiting at every operation until the entirety of a large batch is produced, which increases WIP inventories and lead times everywhere in the plant.
5. Use simple visual control systems. Replace sophisticated, remote tracking systems with visual systems. Described elsewhere in this book, they include production schedules posted where everyone can see them; signals such as andons and status boards to display production status (normal and malfunction); production procedures and standard operations charts; and charts showing goals versus performance (production, quality, worker skill proficiency). Other visual means to facilitate production control include kanban cards, standard-sized containers to limit inventory, and shop-floor layouts that allow people to see things important to their jobs (the status of upstream and downstream workcenters, equipment functioning, stock levels, etc.).
6. Do not overload the shop or particular operations. Overloading the shop floor increases WIP and lead times everywhere. Keep the workload at an amount that the shop can handle. If downstream operations experience problems, then upstream operations should slow down or stop. Keep the amount of rework at each operation below a prespecified limit; if the limit is reached, stop the operation until the rework is completed.
7. Use days or hours for planning lead times. Products that are custom made, one of a kind, or that have erratic demand cannot be effectively produced using pull production, and traditional means for planning and scheduling as provided by MRP systems are more appropriate. Although long lead times and high WIP inventory are hallmarks of MRP systems, both can be reduced if the procurement and process run times used in scheduling accurately reflect reality. Wherever possible, the MRP system should be modified to use daily, or even hourly lead times, and the times should be continuously checked and updated. Following is an example of detailed work scheduling in a push system using short time buckets.

Example 8: Scheduling in a Push System

Alpha Company makes customized furniture for corporations, churches, temples, and museums according to customer requirements. Rarely is a design used by more than one customer, so every job is unique. Both the frequency and the size of job orders are highly erratic.

Suppose an order is received to make 10 units of a kind of a wooden table. The table is an assembly of a top, four legs, and support pieces for the legs. The top and support pieces must be cut and sanded; the legs must be cut, milled, and sanded. After the top, legs, and supports are assembled, the table is given a coat of primer, then four coats of paint. The steps of the process (represented by the product **bill of activities**) and the lead time at each step are shown in Figure 13.18.

To minimize the production lead time, each component (legs, supports, and top) is scheduled separately at each of the operations (cutting, sanding, etc.). For example, cutting the legs, the top, and the supports are considered as three jobs, and each job is scheduled separately. (This contrasts with lumping all the components together and scheduling them as one cutting job.)[18]

The scheduling process begins by assigning a due date for shipment to the customer. This is the time by which the last operation, package, must be completed; if the shipment occurs midday, then it would be possible to schedule packaging for early the same day. Moving down the bill in Figure 13.18, the next operation is paint, which is scheduled to start 4 days earlier. The procedure of scheduling the start of an activity backward from the time when the activity must be completed is called **back scheduling**. This procedure is repeated for the next activity, primer, and for all remaining activities.

When the schedule shows more than one job at an operation, a decision must be made as to which job gets priority. For example, the top, legs, and supports must all be sanded, so a decision must be made about which goes first, which second, and which last.

Figure 13.19 shows the resulting back schedule for manufacturing the 10 tables. The dotted lines in the figure represent waiting times, for example, in the sand operation the legs must wait until the tops are sanded, and the supports must wait until the legs are sanded.

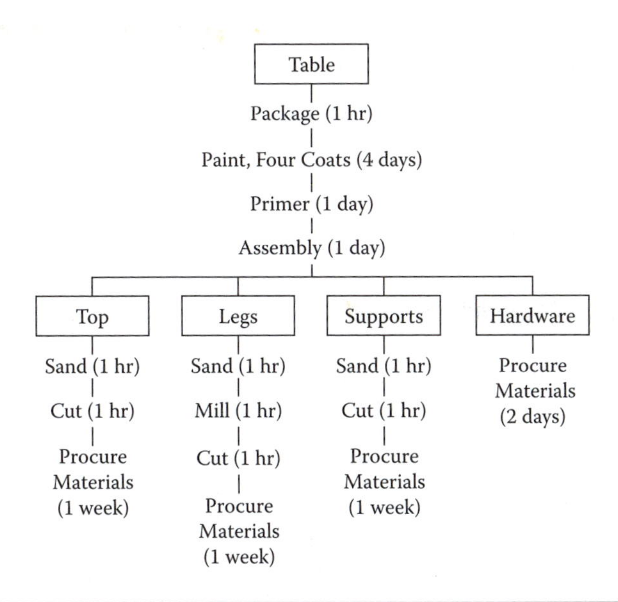

Figure 13.18 Bill of activities for table manufacture. Lead times (shown in parentheses) are for producing 10 tables.

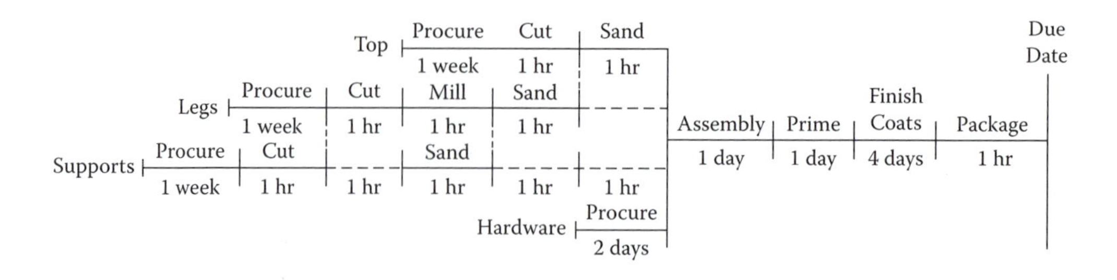

Figure 13.19 Back schedule for table. Lead times are not shown to scale.

The total lead time for the entire 10-table order is the length of the longest lead time branch on the back schedule. In Figure 13.19 this is the supports branch, the length of which, starting with procurement and ending with packaging, is

$$5 \text{ days (1 wk)} + 5 \text{ hr} + 6 \text{ days} + 1 \text{ hr} = 11 \text{ days} + 6 \text{ hr}$$

Assuming that batches of components for the tables must wait no longer than the times shown in Figure 13.19, the order should be completed within 12 days.

With an MRP system capable of scheduling in daily or hourly time buckets, production lead times can be held to a minimum. Of course, two concomitant requirements are that shop-floor operations are tightly controlled so the lead times are, in fact, the same as those used by the MRP system' and that the system is immediately updated to reflect changes in lead times. Small time-bucket planning combined with small batch-size production minimizes the time wasted in jobs waiting.

Hybrid Systems

Some companies handle high product variety by using a combination of ATO and MTO production activities. Each kind of product is assembled on an order-by-order basis and involves common modules and subassemblies, as well as unique, one-of-a-kind components. The modules are produced in repetitive fashion using pull production methods and MMP schedules; the unique components are produced order by order and are back scheduled. In these hybrid cases, daily scheduling of workcells and autonomous workstations is done by a combination of kanban cards for the repetitive pull jobs, and MRP-generated orders for one-of-a-kind push jobs.[19] Following is an example.

Example 9: Hybrid Scheduling Process

Beta Company makes furniture for national department store chains. Although the models and styles produced for the chains are different, the main components for all of them are similar. Thus, the aggregate demand for components for all of the products is somewhat stable and predictable.

Since Beta Company, unlike Alpha Company in the previous example, makes products from different combinations of identical components and since each component has stable demand, the components can be produced repetitively using leveled schedules and pull production, even though the final product assemblies cannot. The final product assemblies, as well as any operations that are unique to particular orders, are scheduled with an MRP-type back scheduling system.

Among the company's products are tables. The company makes tables in different dimensions, woods, and finishes, though the feature that most distinguishes a table is the table top: each is made to customer requirements. Otherwise, every table the company produces uses the same hardware, leg supports, and one of three styles of legs—A, B, or C.

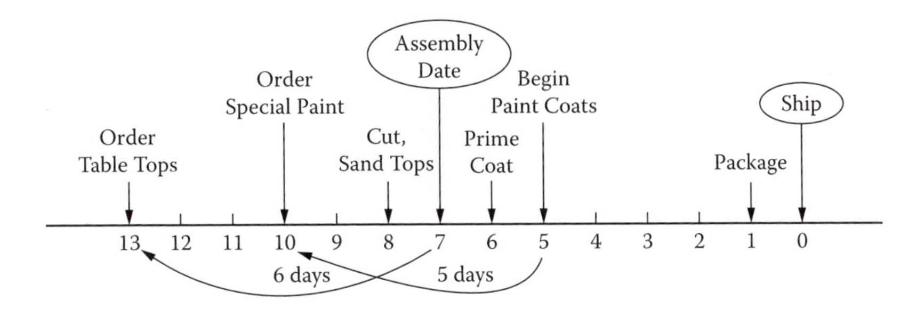

Figure 13.20 Hybrid system: pull and back scheduling.

Demand for legs for all tables produced is large and somewhat stable, so legs can be produced using a pull system—either in small batches of A, B, or C, or in an MMP sequence, whichever works best. Thus, whenever the company receives an order to produce a particular kind of table, only the tops must be back scheduled. The back schedule includes lead times for procuring the tops and for every operation performed on them. If additional steps must be performed on the tables after assembly, then those steps must be back scheduled also.

Suppose an order for 10 tables is received and suppose the lead time for procuring, cutting, and sanding the 10 tops is 6 days. Thus, the tops must be ordered 6 days prior to table assembly. Further, suppose after the tables are assembled they are to receive one primer coat and four paint coats, and then be packaged. If the coating operations are specific to the customer order, then they also must be back scheduled. If the paint finish is special and not carried in stock, then the paint procurement must be back scheduled, too. Say we allow the following lead times: 5 days to procure paint, 5 days to apply primer and four coats of paint, and 1 day to package the tables. Based on those lead times, the production schedule will be as shown in Figure 13.20.

The legs, supports, and hardware will arrive at the assembly station 7 days before the ship date; these materials will have arrived there using kanban cards or other pull signals. The assembly date, 7 days before the shipping date, has been set to allow 1 day for assembly, 6 days for applying the primer and paint coatings, and 1 day for packaging. The overall lead time is thus 13 days.

Summary

Variation in production schedules contributes to longer production lead times and higher costs, overages, and defect rates. Conversely, smooth, level production—producing items at a uniform rate and in small batches—minimizes uncertainty, reduces costs, improves capacity utilization, and yields better product quality and customer service. In many companies the Pareto principle dominates demand: relatively few products account for the highest volume. These products are the primary focus for production leveling. In level production the ideal case is to produce some of every item every hour, though for low-volume items the alternative is to produce some of every item every day or some every few days or weeks.

Achieving level production is a cross-functional effort since imbalances between demand and production often result when marketing, distribution, and production work independently and at cross-purposes. A multifunctional planning team with members from marketing, sales, finance, production, and other areas coordinate decisions and information about product launch, promotion strategies, customers' demand, and planned levels of production.

Level production is a prerequisite for pull production. Pull production requires somewhat uniform demand, which is achieved through a level production schedule for the final stage of the process. Since items are pulled through the process, if the schedule at the final stage is level, demand everywhere upstream will be level, too. When a number of different items is produced in the same process, uniform demand is achieved by intermixing products. This is called mixed-model production (MMP) and refers to the mixed sequencing of different items as they move through the process. The more uniformly that items are intermixed in the sequence, the more uniform the demand for materials everywhere in the process. MMP requires flexible operations, cross-trained workers, strong emphasis on quality assurance, and small-lot delivery of incoming materials. MMP reduces losses from line changeovers and material shortages, and enhances process improvements.

Level and MMP strategies are easily implemented in make-to-stock operations where demand is stable and predictable. They can also be implemented in make-to-assemble situations for production of subassembly options (modules), though usually not for the final products in which the options are used. Since demand for options or modules tends to be more stable than demand for the differentiated products comprising them, the modules can be produced at a uniform rate and then combined into final products to satisfy last-minute customer orders. The ability to modularize products requires forethought about simplicity and commonality of parts in product design.

The most difficult production planning and scheduling situation is make to order where product variety is high and demand for each product is sporadic or one time. In that situation the Pareto principle is applied to determine which, if any, products could be made to stock or made to assemble, for which cases production leveling and mixed-model strategies can be used. For products that must be custom produced on an order-by-order basis, the emphasis must be on achieving high quality, good service, and low cost by ways discussed elsewhere in this book, including producing in small lots, avoiding overproduction, simplifying product design, using a visual shop-floor tracking system, and employing accurate lead-time information.

Notes

1. R. Schonberger, *World Class Manufacturing: The Next Decade* (New York: The Free Press, 1996), 147–150.
2. This principle is called the law of large numbers.
3. See, for example, N. Gaither, *Production and Operations Management,* 6th ed., Fort Worth: Dryden Press, 1994), Chapter 9. R. Chase and N. Aquilano, *Production and Operations Management*, 7th ed. (Chicago: Irwin, 1995), 529–531.
4. P. Park, Uniform plant loading through level production, *Production and Inventory Management* (Second Quarter 1993):12–17.
5. The concept of mixed model production is frequently used in continuous and repetitive processes such as manufacture of automobiles and electronics. The concept also applies, however, to production of small batches of products in a family in focused job shops and production workcells.
6. *The Toyota Production System* (Toyota City, Japan: TMC International Public Affairs Division and Operations Management Consulting Division, 1992), 2.
7. Actually, stages feeding into final assembly could produce materials in smaller batches if they could guarantee that a sufficient number of the smaller batches arrived at final assembly in time for the large batch production. However, the scheduling and coordination of production and delivery of different size batches at different stages of the process—all to ensure that an adequate amount of materials is on hand for batch production at later stages—is rather cumbersome, particularly when production involves batches of different kinds of products (requiring different kinds of input materials). As a result, it is common to schedule production of batches of roughly equivalent size at all stages throughout the process.
8. Pull production with large batch final assembly is possible, but it requires building large stocks of material at each stage of the process, possibly too large to make sense. The large stocks of materials negate one reason for pull production, which is to keep stock low and reveal problems in the system.

9 Again, a pull system will work with large stocks, though, given the emphasis in pull production on reducing the size of stocks as much as possible, it makes little sense to try to use pull production while also maintaining big inventories.
10. Monden, for example, describes a rather sophisticated method used by Toyota for mixing assembly of automobiles that have a large variety of product features and options. The method results in maximal smoothing of the final assembly sequence. See Y. Monden, *Toyota Production System*, 2nd ed. (Norcross, GA: Industrial Engineering and Management Press, 1993), Chapter 17 and Appendix 2; for another method, see T. Vollmann, W. Berry, and D. Whybark, *Manufacturing Planning and Control Systems* (New York: Irwin/McGraw-Hill, 1997), 484–487.
11. B. Coleman and M. Reza Vaghefi, Heijunka (?): A key to the Toyota Production System. *Production and Inventory Management* (Fourth Quarter 1994): 31–35.
12. Ibid.
13. R. Hall, *Zero Inventories* (Homewood, IL: Dow-Jones Irwin, 1983), 69–72.
14. H. Mather, *Competitive Manufacturing* (Englewood Cliffs, NJ: Prentice Hall, 1988), 86–87.
15. S. Narasimhan, D. McLeavey, and P. Billington, *Production Planning and Inventory Control*, 2nd ed., (Englewood Cliffs, NJ: Prentice-Hall, 1995), 313–323.
16. Schonberger, *World Class Manufacturing*, 119.
17. R. Hall, *Attaining Manufacturing Excellence* (Burr Ridge, IL: Irwin, 1987), 208–209; R. Schonberger, World-Class Manufacturing (New York: The Free Press, 1987), 208–209.
18. Why scheduling three jobs results in shorter lead time than scheduling one job should be obvious: when scheduled separately the three jobs can each be processed separately (legs, top, supports), then moved to the next step (cutting, milling, sanding, whatever); when scheduled as one job, enough time must be allowed for all the components (legs, top, supports) to be processed.
19. For discussion of hybrid shop-floor control systems, see B. Williams, *Manufacturing for Survival* (Reading, MA: Addison-Wesley, 1995), 281–286.

Suggested Reading

O. Maimon, E. Khmelnitsky, and K. Kogan. *Optimal Flow Control in Manufacturing Systems: Production Planning and Scheduling* (Applied Optimization). Norwell, MA: Springer, 1998.

A. Smalley. *Creating Level Pull: A Lean Production-System Improvement Guide for Production-Control, Operations, and Engineering Professionals*. Cambridge, MA: Lean Enterprise Institute, 2004.

J. Vatalaro, and R. Taylor. *Implementing a Mixed Model Kanban System: The Lean Replenishment Technique for Pull Production*. New York: Productivity Press, 2005.

Questions

1. What are the major sources of variability in production schedules?
2. In 10 words or less define *level production*.
3. Why level production? What advantages are offered over nonlevel production?
4. Discuss the requirements for level production schedules.
5. Suppose a monthly production schedule shows that 1,000 product Ws, 2,000 product Ys, and 1,500 product Zs are to be produced every week. Since the same quantity of every product is produced every week, is this an example of level production? If so, why? If not, what must be done to make it level?
6. What is the contribution of sales and marketing, and product engineering to level production?
7. Explain how MMP results in production leveling.
8. Referring to the products in question 5, what would the sequence of products be if an MMP procedure was used?

9. In a pull production system, what is the relationship between the amount of WIP required between upstream stages of the process and the amount of leveling in the production schedule?
10. For what reasons would a sequence other than the maximally mixed MMP sequence be employed?
11. Describe the following three production philosophies: make to stock, make to order, assemble to order. In each case, where is the end item for production planning purposes?
12. Contrast the relative roles of the final assembly schedule (FAS) and the master production schedule (MPS).
13. Describe how a bill of materials and a modular bill of materials differ. How is each used? Why use one instead of the other?
14. Explain the pros and cons of holding in stock higher-level subassemblies (assemblies of subassemblies) versus lower-level subassemblies (subassemblies or modules).
15. Explain what a planning bill is. How is it used?
16. Discuss how concurrent engineering efforts ultimately affect production scheduling procedures.
17. What are some rules for reducing waste and improving quality and service in make-to-order production environments?
18. Give some examples (other than from this chapter) of companies or situations that would use a combination of pull production with mixed model FASs and MRP-generated push schedules.

PROBLEMS

1. Forecasted annual demand for a product line is given below. Assume a minimum batch size of 50 and a 250-day work year.

Product	*Forecast*
A	150,000
B	75,000
C	50,000
D	37,500
E	25,000
F	7,500
G	2,500
H	1,250
I	1,250

a. Develop a level schedule for the production line assuming minimum size product batches and no time required for changeover.

b. The line produces at a rate of 160 units/hr. Is the current 7-hour workday adequate to produce the quantity in the leveled schedule? If not, to what time must the workday be modified?

c. The company wants to produce the daily batches in a mixed-model sequence. Suggest such a sequence.

2. The LAN's End company produces peripheral devices for computer networks. Among its products are three styles of modems: LEB, LEM, and LEX. The FAS, which is fixed 4 weeks in advance of production, specifies producing one LEB, four LEMs, and six LEXs every 20 minutes throughout the day in a 7-hour workday.
   a. What is the daily total production for each of the three products?
   b. By trying to maintain a 7-hour workday, the company finds that on average it falls short by 11 units/day. What is the problem? What are alternatives to resolve it?

3. The LAN's End line assembles three kinds of switching boxes: SBA, SBN, and SBX. Demand for SBA is half that of SBN, but is the same as SBX. Assembly of SBA takes 6 minutes; assembly of SBN and SBX, 2 minutes each.
   a. Develop a mixed-model sequence for the three products. How often does it repeat every 420-minute workday, and what is the daily production for each kind of box?
   b. Repeat (*a*) except suppose that demand for SBA is three-fourths that for SBN, and that demand for SBX is one-fourth that for SBN.

4. What is the mixed-model repeating sequence for meeting a daily production requirement of 48 model As, 24 model Bs, 12 model Cs, 36 model Ds, and 24 models Es?

5. A product comes in three models, J, K, L, and the demand for each relative to the total is 10%, 30%, and 60%. What is the mixed-model repeating sequence?

6. A schedule calls for producing 40 As, 20 Bs, and 30 Cs. Process time is 5 minutes per A, 8 minutes per B, and 10 minutes per C.
   a. Develop the mixed-model repeating sequence. Assuming 60 seconds changeover between products, how long will the scheduled production take?
   b. Assume no more than 700 minutes a day is available to produce these products. Suggest a way to reduce the required production time to meet this constraint.

7. Following is a level 1 breakdown of the parts for four models of a product:

	Commercial		*Military*	
	Standard (CS)	*Deluxe (CD)*	*Standard (MS)*	*Deluxe (MD)*
Small parts	CP	CP	CP	CP
Body shell	BSS	BSM	BSL	BSL
Engine size	ESS	ESHD	ESHD	ESHD
Cover type	CTS	CTD	CTU	CTU
Instrumentation	IS	IFA	IS	IFA

   a. Modularize the components. How many planning bills are required? Can the components be completely modularized?
   b. Suppose the level 2 breakdown for the three body shell components shows:

Subassembly	*Parts*
BSS	F, SS, BF
BSM	F, SM, BF
BSL	F, SL, BF, J

Revise the planning bills from (a).

c. Suppose on average total demand for all models is expected to be about 2,000 units/month. Further, based on historical sales, the following sales breakdown is expected: CS, 40%; CD, 20%; MS, 30%; MD, 10%. What is the percentage breakdown of commercial and military products, and standard and deluxe products? What is the monthly demand for each?

d. Assume the percentage breakdown given in (c) and that production is leveled to 100 products/day. From the planning bills, determine the daily material requirements for all components.

8. Juarez Inc. manufactures telephone answering machines in the following configurations:

Options	*No. Choices*
Colors	6
Cord/cordless	2
Memory capability	3
Features switch	2
Tone select	2
Volume select	2

a. If there is a BOM for each possible configuration, how many BOMs are required for all configurations?
b. If modularization is possible, what number of BOMs is required?
c. The Juarez company sells 10,000 machines/week. Considering only choices of tone select and volume select, the breakdown is as follows:

Tone select (T)	30%
No tone select (NT)	70%
Volume select (V)	60%
No volume select (NV)	40%

d. What number of each possible combination should the company plan to produce? Can you determine the number of each possible combination (e.g., both T and V)?

9. To simplify the items at level 1 of a product BOMs, Julia Megan & Co. has started putting small, loose parts for its products in kits. For four of the products, the number and kind of small parts is very similar; in particular,

Part	*Product A*	*Product B*	*Product C*	*Product D*
Nut 2403	6	6	7	5
Washer 7403	6	6	7	5
Bolt 6403	6	6	7	5
Nut 2614	3	3	4	2
Bolt 6614	3	3	4	2

O-Ring 4320	1	2	2	1
Belt 2118	2	2	1	1
Seal 18J	1	1	1	0

One kit is to be used for all four products (on the product's BOMs, the part numbers above will be replaced by a single number, that of the kit). List the contents of the kit.

10. The following figure shows the level 1 BOMs for different options offered in a product. Can the components be completely modularized? List the planning bills of the components that can be modularized.

Options	*Products*							
Chassis	St	St	St	St	Del	Del	Del	Del
Motor	Big	Big	Sma	Sma	Big	Big	Sma	Sma
Carriage	Hi	Lo	Hi	Lo	Hi	Lo	Hi	Lo
Components, Level 1, BOM	L J C I A N G E	L J D I A O G F	L K C I A P H E	L K D I A Q H F	M J C I B N G E	M J D I B O G F	M K C I B P H E	M K D I B Q H F

11. For problem 10, suppose the level 2 breakdown shows the following:

Component	*N*	*O*	*P*	*Q*
Parts, Level 2 BOM	R T W V	R U V	S T V W	S U V

Is it now possible to completely modularize the parts and components for the products in problem 10? List the planning bills.

12. Zemco Dynamics Inc. makes tables with the following options:
Top size: small, medium, large
Top material: cherry, maple, oak, Corian
Legs: spindle, ornate, Corinthian
Leaves (offered on medium and large only): 0, 1, or 2

a. Assume legs and leaves are matched to the tables by material and size (e.g., a large table with a cherry top will be matched with large cherry legs and large cherry leaves). What is the number of possible configurations of tables?
b. If complete modularization were possible, how many planning bills would be needed? List them.
c. Suppose, historically, annual demand has been for 12,000 tables with the following percentage breakdown for options:

Size	*Material*	*Legs*
Small 30%	Cherry 20%	Spindle 50%
Medium 50%	Maple 30%	Ornate 30%
Large 20%	Oak 40 %	Corinthian 20%
	Corian 10%	

Oak Leaves		
	Large	*Medium*
None	30%	20%
1	60%	50%
2	10%	30%

Assume uniform production throughout the year. For any month, can you determine the number of large oak tables produced?

d. Suppose 80 large oak tables must be produced. Can you determine the number of spindle legs and the total number of leaves needed for these tables? If you can, give the answer. If you cannot, discuss why.

Chapter 14

Synchronizing and Balancing the Process

Order is Heav'ns first law.

—Alexander Pope

Timing is everything.

—Anonymous

Uniform flow is a key aspect of pull production. To achieve uniform flow, the master production schedule must be leveled and the size of discrete batches produced for final assembly must be small. But uniform flow at the rate necessary to meet requirements of the final stage takes more than leveled master production schedules and mixed-model assembly schedules. Although these schedules establish the necessary production level, material requirements, and production sequence, there remains the matter of production **timing**.

In pull production, the rate of production of every stage must correspond to the rate of final assembly. In mixed-model production (MMP), all materials should move through the process in a uniform fashion and arrive at the final stage at the time and in the quantity needed and sequence specified by the final assembly schedule (FAS). Aligning the production rates and sequences of all upstream workcenters so that everything arrives as needed at the final stage, is called **synchronization** and is the subject of this chapter.

And achieve smooth flow and uniform production, the capacity of every operation must be adjusted such that, on average, all of them are capable of processing materials at the required rate. Achieving smooth, uniform form by equalizing the capacity or workload of all stages is called **balancing** and is the other subject of this chapter.

Synchronization

In make to stock (MTS) processes and portions of assemble to order (ATO) processes where products are manufactured on a repetitive basis, the production process should be synchronized such that every upstream operation produces at the rate required to satisfy demand. In a pull system, that rate is set by the FAS. To visualize a synchronized process, imagine a conveyor system that moves material continuously from one operation to the next. The speed of the conveyor system is set according to the demand rate of the final product. To enable the entire system to meet demand without delays or work in process (WIP) buildup, every operation along the conveyor must produce at about the same rate—the rate at which the conveyor system is moving. Such is a synchronized process.

But here the conveyor system is just a useful analogy, since such a system is not necessary to achieve synchronous flow of materials in a process. To enable somewhat synchronous flow in a process, it is only necessary that the rate of production and transfer of items at upstream stations roughly match the rate of demand for these items at downstream stations. Extending this concept, the flow of materials throughout the entire process is synchronized by setting the production of every operation (and the transfer of materials between operations) equal to the rate of demand at the final stage of the process. That demand rate is, in turn, determined by the production schedule at final assembly.

Synchronized Cycle Times

In a pull system, synchronization is achieved by setting the *cycle times* (CTs) at every upstream operation according to the cycle time at final assembly. As mentioned, in a pull system only the last stage of the process has a daily schedule. If production is to involve repetitive manufacture of multiple, similar products, the daily schedule will be an MMP schedule. This MMP schedule is used to determine the CTs of all products, and these product CTs are used to derive the maximum CTs of every upstream operation that supplies parts for the products. Briefly, each operation is synchronized to the overall process by setting its CT as a multiple of the required CT. The following example illustrates this.

Example 1: Setting CTs to Synchronize the Process

Suppose the daily requirements for three products are

Product A: 200 units
Product B: 100 units
Product C: 50 units
All products: 350 units

Assuming 420 minutes a day for production, the required product CTs are

Product A: 420/200 = 2.1 min (60 sec/min) = 126 sec
Product B: 420/100 = 252 sec
Product C: 420/50 = 504 sec
All products: 420/350 = 72 sec

In other words, the final assembly station must complete a product A every 126 seconds, a product B every 252 seconds, and a product C every 504 seconds. Overall, the final assembly station will complete a product every 72 seconds on average. This 72-second CT is the **drumbeat** of the process. Note that everywhere in the process the required CT is the **takt time,** as mentioned earlier.

Consider just product A. To synchronize its production every upstream operation supplying its parts must produce and transport a part once every 126 seconds, on average. If two of a kind of part are needed for each Product A, then the operations making that part must produce and transport one part every 63 seconds, on average. The same rationale applies to parts and materials produced for products B and C.

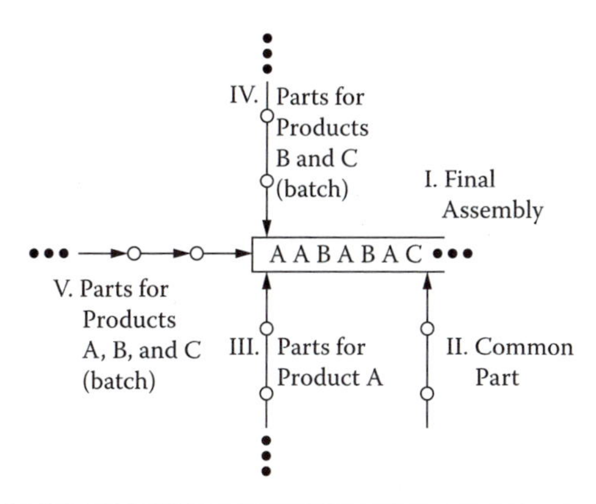

Figure 14.1 Final assembly and subprocesses for producing products A, B, and C.

Suppose the process for manufacturing the three products consists of five major subprocesses, described next and shown in Figure 14.1.

Subprocess I: This is the final assembly line for products A, B, and C. Assume the FAS specifies an MMP sequence of AABABAC, repeated 50 times, and one product every 72 seconds, on average. As Figure 14.2 shows, with this sequence and a 72-second average CT, the required CTs for all three products are indeed achieved.

Subprocess II: This subprocess continuously produces a part that is used in all three products. Since 350 units of this part must be made, the CT for every operation in this subprocess must be the same as the CT at final assembly, 72 seconds.

Subprocess III: This subprocess continuously produces parts that are used only for product A; thus, it must produce 200 units of this part per day. The CT of every operation in this subprocess must be, at most, the same as the CT of product A at final assembly, 126 seconds.

Subprocess IV: This subprocess produces in alternate batches a part for product B and a part for product C. Suppose the size of the batches is 10 units, the same as a standard container. Also assume that production of each batch is preceded by a changeover (from a batch of parts for B, to a batch of parts for C, and vice versa) that has an internal setup time of 10 minutes.

Each day subprocess IV must produce 100 parts for product B, and 50 parts for product C. Since the batch size is 10, the corresponding mixed-model batch sequence will be BCBBCBBCBBCBCBB (where each B and C represents a batch of 10). Thus, the daily number of average setups is 10.

The time consumed by 10 setups is 100 minutes, which leaves 420 – 100 = 320 min a day for production. Thus, the CT for every operation in this subprocess must be, at most, 320 min/150 units = 2.133 min, or 128 sec.

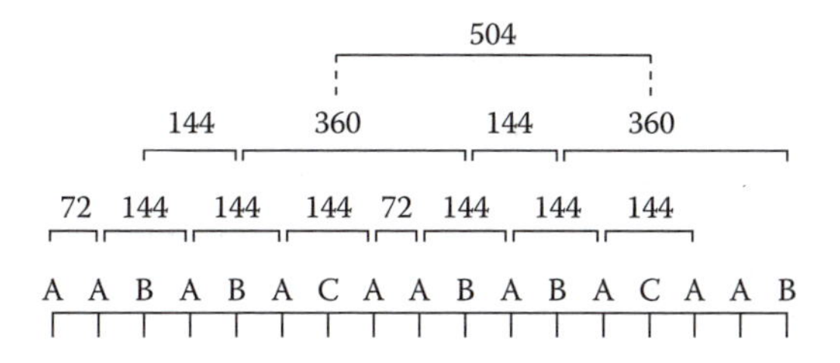

Figure 14.2 Correspondence between 72-second interval and product CT.

Subprocess V: This subprocess alternately produces parts for product A, product B, and product C. Determining the CT is done in the same way as for subprocess IV. Daily, this subprocess must produce 200 parts for product A, 100 parts for product B, and 50 parts for product C. Assuming again a batch size of 10 units, the corresponding batch sequence will be identical to the sequence shown in Figure 14.2, except in this subprocess each A, B, and C represents a batch of 10 units. This sequence, AABABAC, must be repeated five times a day. Since there are six changeovers between batches of different parts in a single mix-model sequence (AA, B, A, B, A, C), there are 30 setups a day. With 10 min/setup and a 420 minute workday, 420 – (30 × 10) = 120 min remain for production. Thus, the maximum CT must be 120 min/350 units = .342 min = 20.5 sec.

Note that in both subprocess IV and subprocess V where setups are involved, if the required CTs are too small to be attainable, then the batch size must be increased. For example, doubling the batch size to 20 units at subprocess V cuts the number of setups to 15, which allows a maximum CT of [420 – (15 × 10)]/350 = .771 min, or about 46.2 seconds per unit.

The FAS provides the basis for setting CTs for all components and subassemblies; in turn, these CTs serve as the basis for setting CTs of parts that feed into the components and subassemblies. A similar rationale can be continued back through the process and to parts procured from external suppliers. It is of no use for an upstream operation or supplier to provide parts at a rate slower than, on average, that required by demand. Neither is it good for operations to produce at rates much faster than required, since, then, WIP inventory grows and operations must periodically stop to enable final assembly to catch up.

After product CTs have been set, the operations at all upstream processes are structured to produce roughly at those times. If the operations include linked workcells, as described in Chapter 10, the actual CT of each cell is adjusted by assigning workers and refining tasks such that it does not exceed the required product CTs (takt times). Further, these CTs are posted on the standard operating routines (described in Chapter 11) and become the goal toward which process improvement is directed.

Given all the bother to establish product CTs and to synchronize processes around them, the MMP and final assembly schedule must be held somewhat frozen for at least a few weeks. Substantial changes in CTs are permitted only with advance notice. This is the topic of the next chapter.

The Essence of Cycle Time

The concept of CT in synchronous production does not mean exactly the same thing as production rate, even though, mathematically, one is the inverse of the other. For example, a process with a CT of 60 seconds has a theoretical production rate of 1/60 unit/sec, which is the same as 1 unit per minute or 60 units per hour.

Yet even if every operation were to produce at the rate of 60 units/hr, the process would not necessarily be synchronized. In a synchronized process, a required CT of 60 seconds means one thing: one unit of a finished product (or part or component) will be produced every 60 seconds, uniformly, throughout the day. In contrast, a production rate of 60 units/hr can mean many things, including 120 units/hr produced once every 2 hours, 480 units/hr produced once every 8 hours, and so on. Any of these ways results in an average production rate of 60 units/hr, yet unless all operations in the process are coordinated, there will sometimes be WIP buildups or shortages between stages of the process. However, when every operation produces according to the same required CT, material flows smoothly, and there is minimal WIP buildups and no material shortages along the way.

Bottleneck Scheduling

An assumption in synchronizing a process around the final assembly schedule is that all operations have adequate capacity to meet that schedule, otherwise the process would not be able to keep up with the required pace. In essence, this is like saying that if there is a bottleneck in the process, then it is at the final assembly stage. In a process that is somewhat long running and stable, it is possible to adjust capacity everywhere such that, in fact, all operations are able to meet the requirements of final assembly with only minimal adjustment.

In many cases, however, the product mixes and volumes keep changing, and as a consequence, bottlenecks appear at places other than at final assembly. In such cases, the FAS no longer is the determinant of what flows through the system nor can it set the pace. Scheduling a process based upon the bottleneck constraint is called **bottleneck scheduling**.

Principles

Much of the current awareness about and practice of managing a process from the bottleneck is based on the pioneering work of Eliyahu Goldratt.[1] Given that the bottleneck constrains process throughput, efforts to increase throughput must start at the bottleneck. To increase bottleneck throughput, the setup time must be reduced or the size of the process batch must be increased. To minimize the lead time when the process batch is large, the transfer batches from the bottleneck should be small. Synchronizing a production process by managing the bottleneck involves the following principles.

Throughput Pace

Given that the throughput of a process is restricted by whatever the process bottleneck can handle, it makes sense to set the pace of the process according to the capacity of the bottleneck. This is similar to the manner in which the FAS sets the drumbeat when the process everywhere has sufficient capacity.

Buffer Stock

Goldratt says that an hour lost at the bottleneck is an hour lost everywhere in the system. To ensure that the bottleneck is never without work, a buffer of jobs should be maintained ahead of it. If upstream operations are interrupted, the buffer will permit the bottleneck operation to continue working. The buffer is established by releasing jobs into the process so that they arrive at the bottleneck operation ahead of when that operation is expected to be ready to process them.

Process Scheduling

It does not make sense to release more jobs into the system than can be processed through the bottleneck since the excess jobs will queue up as WIP inventory. Thus, the timing of jobs released into the process should be predicated on the bottleneck's capacity to process those jobs. The jobs should be released upon receipt of order signals from the bottleneck. In essence, this is the same as starting at the bottleneck and pulling jobs into and through the system by a logistical rope in the form of schedules or kanbans.

Drum–Buffer–Rope

The concepts of (1) setting the drumbeat for the process based on the bottleneck, (2) establishing buffer stock ahead of the bottleneck, and (3) pulling material into the process from the bottleneck

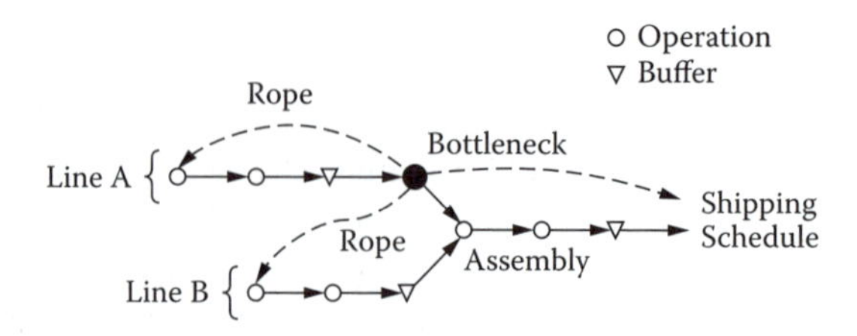

Figure 14.3 Drum–buffer–rope system.

are illustrated in Figure 14.3. Application of these principles together is referred to as the **drum–buffer–rope** system, or **DBR**. In Figure 14.3 jobs are pulled through lines A and B upon release of orders (or a scheduled derived) from the bottleneck. Operations downstream from the bottleneck are faster than the bottleneck, so they do not need a schedule.

In addition to the buffer before the bottleneck, Figure 14.3 shows two other buffers, one before assembly to ensure that interruptions in line B will not interrupt the assembly process and one before shipping to ensure the integrity of the shipping schedule. The shipping schedule for a job is based upon the time the job is expected to leave the bottleneck plus the time it will take to move through the remainder of the process. For details about locating and sizing buffers, as well as setting schedules based on the bottleneck, see the references in the Notes.

Any operation can become a bottleneck. If a large batch is released to an operation on short notice, the operation will not be able to handle it. The easiest ways to locate a bottleneck is to look for an operation that consistently has jobs waiting and to ask shop workers.

Increasing process throughput starts with improving the bottleneck, though as soon as the bottleneck is improved enough, a new bottleneck will form at another operation (what was formerly the next slowest operation). If a process can be reengineered to change the location of the bottleneck, the bottleneck should be placed as close as possible to the start of the process. That is because in a sequence of interdependent operations, operations earlier in the sequence are less influenced by variability in other operations and hence are easier to control.

Pull from Bottleneck

The rope in DBR can be a kanban card issued from the bottleneck operation to the first stage of the process. When materials enter the process at multiple places, shown as A and B in Figure 14.3, cards must be sent to all of them. This procedure works well in processes where the bottleneck is stable and where the times through the process routing are constant.[2]

Balancing

Once the CTs in a synchronized process are set, the production capacity of all workcenters must be adjusted to conform to those CTs. The adjustment can involve, for example, increasing capacity at bottleneck operations, decreasing capacity at nonbottlenecks, or reconfiguring tasks at workstations so they require less time or more time.

Balancing refers to the procedure of adjusting the times at workcenters to conform as much as possible to the required CT (takt time). A **balanced process** is one where the actual CTs at all stages are equal. Strictly speaking, the goal of achieving a completely balanced process is appropriate only in processes that are **paced**, that is, where material moves on a conveyor or chain at a constant speed past workstations. Such is the case of product layouts and MMP lines. In other processes more similar to job shops and with many products and routings, the goal of a balanced process is inappropriate. Seeking to balance a process around one product or routing makes it more difficult to adopt operations to accommodate many products and routings.

In this section, the focus is primarily on balancing tasks in paced production lines, though the procedures discussed pertain to any coordinated sequence of operations, including cellular layouts. The procedures balance workstation times by manipulating assignments of tasks and workers/machines to workstations.

Line Balancing

As an introduction to the concept of balancing, consider first a simple example from traditional line balancing. Line balancing refers to assigning tasks (elemental units of work) to a workstation or operations sequence such that

1. The CT of the combined sequence of workstations satisfies the required CT (product CTs as described before).
2. The tasks are assigned in the right order.
3. The assignment is as efficient as possible.

The first point states that the output of the sequence of workstations meets demand. For that to happen, the CT of the slowest workstation in the line (the bottleneck) must not exceed the required CT (takt time). The second point states that the assignment of tasks to workstations meets precedence requirements. Whenever a number of tasks are to be performed, there is a logical sequence or ordering that must be followed. Certain tasks take precedence over others, so they must be done before the others can be started.

The third point states that the resultant number of workstations or operations in the line is the minimum possible given the required CT and precedence relationships. The following example illustrates a simple line-balancing procedure.

Example 2: Simple Line Balancing Using the Longest Operation Time Rule

Suppose the assembly procedure for a product involves the following seven tasks:

Task	*Time (min)*
1	0.7
2	0.5
3	0.4
4	0.6
5	0.5
6	0.8
7	0.3
Total	3.8

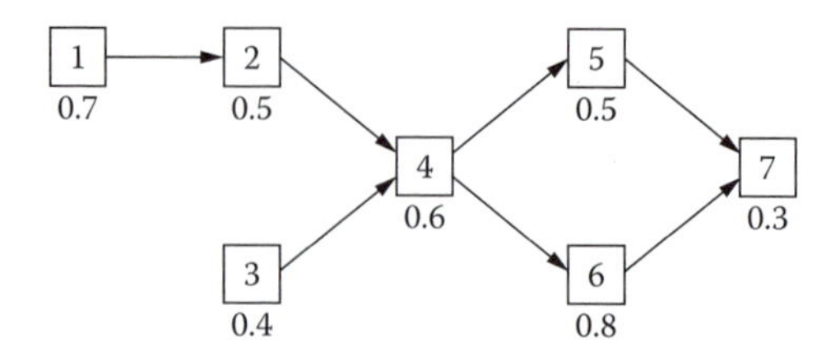

Figure 14.4 Precedence diagram for seven assembly tasks.

The total time of 3.8 minutes is the **work content** or actual work time necessary to make one unit of product.

Suppose demand for the product is 420 units/day. The product is produced on an assembly line that operates 7 hr/day (420 minutes), so the required CT is

$$\text{Required CT} = \text{Operating time} \div \text{Demand} = 420 \text{ min/day} \div 420 \text{ units/day} = 1 \text{ min/unit}$$

Suppose the seven tasks must be performed according to the relationships shown in Figure 14.4, called the **precedence diagram**. The figure, which can be interpreted by looking at it backward, says that before task 7 can be performed, tasks 5 and 6 must be completed; before tasks 5 and 6 can each be performed, task 4 must be completed; before task 4 is performed, tasks 2 and 3 must be completed; before task 2 is performed, task 1 must be completed.

Now, to assign the tasks to a sequence of workstations on an assembly line, look at the precedence diagram to see which tasks can be assigned to the first workstation. For this purpose, look at the precedence diagram starting on the left. Figure 14.4 indicates that either task 1 or task 3 can be assigned; the question is which. There are different ways to decide between tasks when there is a tie. One way is to select the task that requires the **longest operation time** yet does not exceed the available remaining time at the workstation. Therefore, assign task 1 to station 1. Task 1 takes 0.7 minutes, which is 1.0 – 0.7 = 0.3 minutes less than the required CT. Thus, 0.3 minutes remain available at station 1.

Next we can assign tasks 2 or 3. Since both exceed the available time at station 1, the next assignment must be to station 2. Pick task 2 because it takes longer than task 3. This leaves 1.0 – 0.5 = 0.5 minutes available at station 2.

The next task to be assigned must be task 3, because the precedence diagram indicates that tasks 3 and 2 must both precede task 4. Task 3 requires 0.4 minutes, which fits into the available time at station 2. With both tasks 2 and 3 assigned, station 2 has 0.1 minute available.

Task 4 comes next. The time of 0.6 minutes exceeds the available time at station 2, so the task must be assigned to station 3. This leaves 1.0 – 0.6 = 0.4 minutes available.

The procedure continues in this fashion until all remaining tasks have been assigned. The assignment procedure results are summarized in Table 14.1. The resulting line will have five workstations as shown in Figure 14.5.

In general, the efficiency of a sequence of workstations is computed as

$$\text{Efficiency} = \text{Work content}/(\text{Number of workstations} \times \text{Required CT})$$

The efficiency for the line in Figure 14.5 is

$$3.8/(5 \times 1.00) = 0.733, \text{ or } 73.3\%$$

For high efficiency, as many tasks as possible must be packed into each workstation within the constraint of the required CT and restrictions of the precedence relationships. For this example, the efficiency is the feasible maximum.

Table 14.1 Procedure of Assigning Tasks to Stations

Station	*Time Available (min)*	*Tasks Eligible*	*Will Fit*	*Assign (time in min)*
1	1.0	1, 3	1, 3	1 (0.7)
	0.3	2, 3	None	—
2	1.0	2, 3	2, 3	2 (0.5)
	0.5	3	3	3 (0.4)
	0.1	4	None	—
3	1.0	4	4	4 (0.6)
	0.6	5, 6	None	—
4	1.0	5, 6	5, 6	6 (0.8)
	0.2	5	None	—
5	1.0	5	5	5 (0.5)
	0.5	7	7	7 (0.3)

A line that achieves the required output and is as efficient as possible is considered a balanced line. There are many heuristic and optimal-seeking methods for assigning tasks to stations on a line, though they are beyond the scope of our discussion.[3] Our interest is on assigning tasks to an assembly line, workcell, or other process for MMP.

Balancing for MMP

The individual task times used for traditional line balancing are determined for whatever particular product is being produced. In Example 2 the task times are for such a product. In MMP, however, more than one product is being produced, so for each product the time for a given task might be different. Unlike traditional line balancing, which considers the production of only one product, MMP balancing must consider simultaneous production of multiple different products.

In general, MMP balancing requires assigning tasks to a sequence of workstations such that

1. The CT time for each product at each workstation satisfied the required product CT.
2. The assignment for all products is as efficient as possible.

Workstation	1	2	3	4	5
Tasks Assigned	1	2, 3	4	6	5, 7
Task Time	0.7	0.5, 0.4	0.6	0.8	0.5, 0.3
Required CT	1.0	1.0	1.0	1.0	1.0
Idle Time	0.3	0.1	0.4	0.2	0.2

Figure 14.5 Resulting task assignment to five workstations.

One way to assign tasks to workstations in MMP such that the line produces according to the required CTs for all the products is to use the **weighted average time** rule. The weighted average time is the amount of time, on average, required at a workstation to perform tasks. So that every workstation in MMP can produce the required amount of every product, this rule states that the weighted average time at each workstation cannot exceed the required CT for all product models.

The weighted average time can be written as

$$\Sigma\Sigma q_j t_{ij}$$

where

q_j = the proportion of product j in the MMP repeating sequence, (j = 1, 2, …, P), = (production of j)/(production of all products)

t_{ij} = the time to perform task i on product j, where i is a task assigned to the workstation

The weighted average rule can be expressed as

$$\Sigma\Sigma q_j t_{ij} \leq \text{Required CT}$$

where

Required CT = Available production time/Production quantity for all products

The following example illustrates this concept.

Example 3: Assigning Tasks Using the Weighted Average Rule

Suppose products A, B, and C are to be assembled on a line. The assembly of each product requires doing four tasks, but the time for each task varies by product. Table 14.2 shows task times t_{ij}, where i = 1, 2, 3, 4; j = A, B, C. Suppose daily demands for products A, B, and C are 240, 120, and 60, respectively. The total demand is 420.

To find an assignment, first determine the product proportions $q_j = D_j/\Sigma D$. Since ΣD = 420, the proportions are

$$q_A = 240/420 = 0.571,\ q_B = 120/420 = 0.286,\ q_C = 60/420 = 0.143$$

Next, determine the required CT for all products, where

$$\text{Required CT} = \text{Available production time}/\Sigma D$$

Table 14.2 Task Times in Minutes, t_{ij}

	Product j		
Task i	*A*	*B*	*C*
1	0.7	0.6	0.5
2	0.6	0.5	0.6
3	0.5	0.4	0.4
4	0.6	0.6	0.7
Work content	2.2	2.1	2.1

Table 14.3 Weighted Average Times

Workstation	*Tasks (i)*	*Weighted Average Times (min)*
1	1	$0.571(0.7) + 0.286(0.6) + 0.143(0.5) = 0.643$
2	2, 3	$0.571(0.6 + 0.5) + 0.286(0.5 + 0.4) + 0.143(0.6 + 0.4) = 1.029$
3	4	$0.571(0.6) + 0.286(0.6) + 0.143(0.7) = 0.614$

Assume available time is 420 minutes, then the required CTs for the three products are

Product	*D (daily)*	*Required CT (min)*
A	240	1.75
B	120	3.5
C	60	7.0
All products	420	1.0

The "all products" CT of 1 minute represents the required CT of the overall process; it specifies that one product will be produced every minute, on average.

Suppose, as a trial, we consider a line similar to that in Figure 14.5, but to simplify the problem, look only at the first three workstations. To conform to the weighted average rule, the weighted average task time $\Sigma\Sigma q_j t_{ij}$ at each workstation must not exceed the required CT. The weighted average task times at the three workstations are shown in Table 14.3. As defined earlier, these times indicate the time required on average for each workstation to process the multiple kinds of products. To illustrate, assume that the MMP repeating sequence is AABABAC. Figure 14.6 shows the actual task times at each workstation to process the products in one repeating sequence. Note that the average of the times at a workstation in Figure 14.6 is the same as the weighted average task time in Table 14.3.

This particular assignment does not conform to the weighted task time rule since the weighted time at the bottleneck (1.029 minutes at workstation 2) exceeds the required CT of 1 minute. It will thus be necessary to try other assignments, for example, combining tasks 1 and 2 at workstation 1, or tasks 3 and 4 at workstation 3, but they too would result in weighted average times being too

Workstation	1	2	3
Task	1	2 + 3	4
A	0.7	1.1	0.6
A	0.7	1.1	0.6
B	0.6	0.9	0.6
A	0.7	1.1	0.6
B	0.6	0.9	0.6
A	0.7	1.1	0.6
C	0.5	1.0	0.7
Sum	4.5	7.2	4.3
Average	0.643	1.029	0.614

Figure 14.6 Workstation average times, three products.

large. If we insist on the assignment in Figure 14.6, then we must reduce the weighted average time at the bottleneck.

Focusing on workstation 2, the times for tasks 2 and 3 among products A, B, and C must be reduced such that the weighted average time becomes 1 minute or less. For example, (and you can verify this) shaving 3 seconds (0.05 minute) off task 2 or task 3 for just product A will bring the weighted average time at workstation 2 down to 1 minute.

With the weighted average principle, balancing in MMP is achieved by pairing tasks of more time-consuming and less time-consuming products together. Thus, it is OK if the task times at a workstation for one product exceed the required CT, as long as that product can be paired with other products that have smaller task times, such that the resultant weighted average time of all tasks does not exceed the required CT. Of course, the pairing is not arbitrary but depends on q_i, which depends on the product demand.

Other Ways to Achieve Balance

The assignment of tasks to workstations to achieve a balance can be done in different ways. Besides the aforementioned general approach are two alternatives: dynamic balance and parallel lines.

Dynamic Balance

In the example in the previous section, the balance was **constant** since workstations always perform the same tasks, regardless of the product. That is, for example, workstation 1 performs task 1 on every product, workstation 2 does tasks 2 and 3 on every product, and so on. Another way to achieve balance is to change the balance, depending on the product. The balance is **dynamic,** referring to the fact that the mix of tasks can be changed with each product. For example, a workstation might perform, say, task 1 on product A, but then do tasks 1 and 2 on products B and C. Assuming no technical or skill constraints (and that workers are able to keep track of the changing tasks), dynamic balance is another way to conform to the weighed average rule.

Parallel Line

When task times for different products vary considerably, or when the task times cannot be reduced so that the weighted average time satisfies the required CT, then another way to meet the required CT is to split the line into two or more parallel lines. That is, at workstations on the line where the weighted average task time is too large, simply add a parallel workstation.

Example 4: Balancing with a Parallel Line

Refer back to the line in the previous example. Suppose the task times at workstations 1 and 3 are the same, but at workstation 2 the times are now

Product A: 1.3 min
Product B: 0.9 min
Product C: 1.0 min

Assuming the same product proportions q_i; as in the previous example, the weighted average time at workstation 2 is then

$$0.571(1.3) + 0.286(0.9) + 0.143(1.0) = 1.143 \text{ min}$$

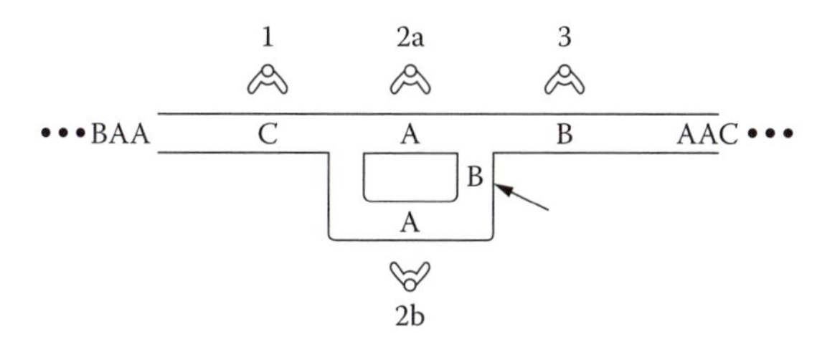

Figure 14.7 Parallel line for lengthy tasks.

This time exceeds the required CT of 1 minute. Now, if some tasks could not be assigned to other workstations, then a second workstation might be added in parallel to workstation 2, as shown in Figure 14.7. With these parallel workstations, whenever a part arrives, it goes to either workstation 2a or 2b, depending on whichever is available first. After the part is completed at workstation 2a or 2b, it rejoins the main line.

In general, the effect of having two parallel workstations at a stage of the line is to cut the total task time at that stage in half (given a task time at each workstation of 1.143 minutes, the effective task time at stage 2 of the process would be 1.143/2 = 0.572 minutes). In actuality, the average task time from parallel lines in MMP will be somewhat larger than the mathematical average because a part leaving either parallel workstation sometimes has to wait until after a part leaves the other workstation in order to retain the original MMP sequence. In Figure 14.7, for example, the arrow points to a part B that has been completed at workstation 2b but which cannot immediately reenter the main line because doing so would change the original MMP sequence of AABABAC. Part B must wait there until after part A at workstation 2a has been completed and moved to workstation 3.

Balancing for Synchronous Flow

The previous discussion on balancing focused on a particular line, such as a final assembly line. If other areas of the plant feed parts into the line, and if those areas are focused factories or workcells dedicated to these parts, then, ideally, the entire process (parts production and final assembly) should be balanced. This involves balancing tasks at all upstream operations feeding into final assembly as well as tasks on the final assembly line. To achieve balance, the task groupings at every operation, workcell, and machining department must be adjusted so that the resulting weighted average task time at each of them is as close as possible to the required CT. This is done by adding or subtracting tasks at operations, moving workers between tasks and operations, and adjusting the time to perform work tasks.

In synchronous production every operation produces just enough to satisfy downstream demand. If a Kanban system is employed, the demand is signaled by cards. The goal of workcells and operations everywhere in the process is to match the required CT time and do so with minimum waste. As Monden explains, balancing is a matter of continuous improvement and worker reallocation, a process that involves three steps:[4]

1. Eliminate wasteful tasks.
2. Reallocate tasks.
3. Reduce the workforce; return to step 1.

The next section illustrates the process.

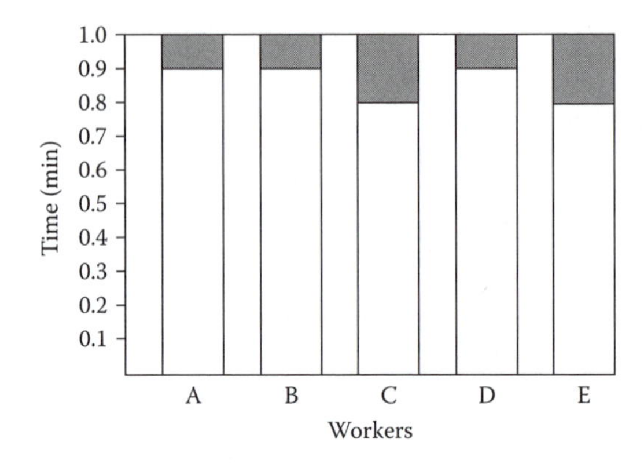

Figure 14.8 Initial worker assignment.

Balancing through Worker Reassignment

Suppose the required CT is 1 minute, and tasks and workers in a cell are being reassigned to meet that CT. Suppose the initial assignment results in five workers at five workstations, with the times shown in Figure 14.8. Workers A, B, and D are the bottlenecks of the cell since they require the longest time. Since that time, 0.9 minutes, is less than the required CT, the cell output is adequate. However, because every workstation has times that are lower than the required CT, every station cell must be idled occasionally so the cell does not produce in excess of the required output. This idle time is represented by the shaded areas in Figure 14.8.

The time for each worker is comprised of the times for multiple separate tasks that can (theoretically) be modified or shifted between work stations. Suppose by a combination of eliminating wasteful motion and reallocating tasks, the tasks can be rebalanced so that every worker takes 0.86 minutes, as shown in Figure 14.9. This at first seems to be an improvement because everyone now takes the same amount of time (the cell is perfectly balanced); however it is not, because the improvement cannot be realized. In fact, given that the cell CT must be held to 1 minute, the percent idle time in the cell remains unchanged by this so-called improvement.

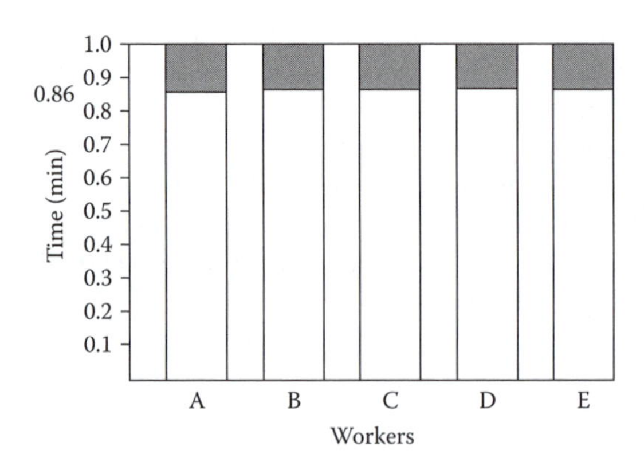

Figure 14.9 Possible alternate worker assignment.

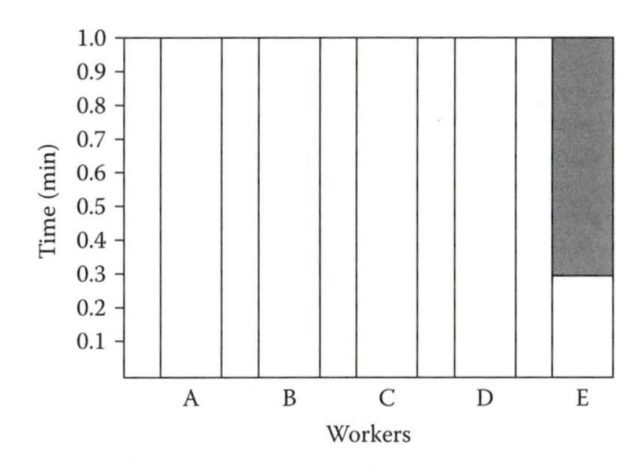

Figure 14.10 Preferred worker assignment.

Since one goal of balancing is to achieve the required CT with the greatest efficiency (i.e., minimum waste), a better solution would be something like the assignment shown in Figure 14.10. There, the tasks have been shifted so that four workers each have 1 minute of work, while the fifth worker has 0.3 minutes. Although the cell idle time is still the same, the assignment is better because now all effort can be focused on improving tasks such that, ultimately, worker E's tasks can be reallocated among workers A to D, and worker E can then be transferred out of the cell. The cell can then operate with only four workers.

Balancing operations plantwide is a continuous process. Changes in required CTs and from delays or shutdowns at individual operations are constant threats to synchronous production. As countermeasures, plants rely on[5]

- Visual signals (andons, charts) to keep workers informed about problems at all stages of the process
- Multiskilled operators, flexible operations, and short changeover times to enable quick adaptation to changes in product demand and product mix
- Less than full-capacity scheduling to allow for delays or unanticipated demand changes

Workcells usually have some capability to adapt to CT changes without stopping work (i.e., without reshuffling workers or tasks). In pull production systems, the adaptability is usually within ±10% of the nominal CT. As long as production schedules (rate of production and mix of products) do not change greatly, no capacity planning is necessary for operations to adjust to changing CTs.

Maintaining Synchronization

In a synchronized process, whenever the final production schedule is changed, so must the CTs of upstream operations. Changes are inevitable, although maintaining level final production schedules reduces the frequency and magnitude of the changes. If the change in schedule represents but a small daily or weekly adjustment to a level baseline schedule, then a simple way for upstream operations to keep pace is to lengthen or shorten the workday. This is far simpler than having every operation readjust to new product CTs, which requires shifting worker assignments, changing machine configurations, and so on.

It should be evident that a delay or interruption anywhere in a synchronized process has potential to desynchronize the entire process. For that reason **andon** lights are located at each operation (or a **signal board** located centrally over the entire process) and are used to display the status of operations. When a light for an operation flashes red, supervisors and adjacent workers rush over to resolve the problem. In a truly synchronized process, a red light at one operation is a signal for every operation to stop. Only after the problem has been resolved (green light) do all operations resume, and they all resume simultaneously.

Adapting to Schedule Changes

A major consideration in setting production schedules is the ability of the shop floor to achieve production levels as specified by the required CT.

Returning to our three-product example, suppose daily production has been proceeding according to the schedule shown in Table 14.4, column (1). On average, a product is produced at a rate of 1 every 72 seconds, so 72 seconds is the CT to which upstream operations have become accustomed. This CT corresponds to an average production rate of 50 units/hr (3,600 sec/[72 sec/unit]).

Now, suppose the demand for product B increases from 100 to 150 units/day. This change results in an increase in the total daily production requirement from 350 units to 400 units (column (2)). There are two ways the production process can meet this increase: increase the production workday or decrease the CT (increase the production rate). In general, altering either the production workday or production rate, or some combination, is the way production processes adjust to changes in demand.

Alter the Production Workday

Altering production requires lengthening the production workday. Since going from 350 units to 400 units represents an increase of 14.3% in the required output, the workday must be increased to (1.143 × 7 hours) 8 hours. Instead of 7 hours, the workday will be 8 hours.

In addition, the MMP repeating sequence must be changed to reflect the increase in demand for product B. Using the procedure described in the preceding chapter, the sequence before the change is AABABAC; to reflect the required increase in B, the new sequence must now include three B's with four A's and one C. Suppose the sequence of ABABABAC is adopted. Repeating this sequence 50 times a day will meet the requirements for every product.

Table 14.4 Schedule Change by Adjusting Production Time

	(1)		*(2)*		*(3)*
Product	*Original Schedule*	*CT (sec)*	*Revised Schedule*	*CT (sec)*	*% Change in CT*
A	200 (57%)	126	200 (50%)	144	14.3
B	100 (29%)	252	150 (38%)	192	–23.8
C	50 (14%)	504	50 (12%)	576	14.3
All	350 (100%)	72	400 (100%)	72	0

Notice from column (3) in Table 14.4 that although the CTs for each product change somewhat, the overall CT (all products) remains the same. Although every operation must adjust its production rates for the different products by the percentages shown in column (3), the changes might not be too difficult since the CT percentages for product B are offset by CT increases for products A and C. Overall, the hourly capacity of the process will remain unchanged.

Alter the Production Rate (Adjust Cycle Time)

Suppose the current 7-hour workday is to be maintained, in which case to meet the product B demand it will be necessary to increase the overall production rate. In the example, going from 350 units a day to 400 units for product B will require that the process CT be reduced from 72 seconds to 63 seconds (420 min/400 units). The average production output rate must be increased from 50 units/hr to over 57.14 units/hr (400 units/7 hr).

As shown in Table 14.5, columns (2) and (3), the increased production will require a significant decrease in CT for product B (33%), which in the short run will probably be difficult to achieve because there is no compensating increase in the CTs for products A or C (output of B must be increased with no decrease in output for A and C). Unless every operation that provides parts for product B can find ways to decrease the CT by 33% (without lengthening the CTs for parts for A and C), the process will not be able to adjust. In that case, the first alternative (increase the length of the work day) must be adopted, at least until ways are found to reduce the CT for product B.

In Practice: Adjusting to Schedule Changes

There is of course a third way to meet changing product demand and product mix: alter both the length of the workday and the product CTs. Initially the workday length would be increased; then, as ways are found to reduce CTs, the workday can be decreased, eventually back to normal.

In pull production operations, supervisors, and worker teams on the shop floor assume much of the responsibility for deciding how they will satisfy CT requirements. Management does the long-term capacity planning (rough-cut capacity planning), but the daily details are left to the shop floor. Management determines the overall CT (takt) requirements for a process by setting the MMP schedule and workday length, but then the shop floor determines how to balance tasks and assign workers to achieve those CTs.

Since it can be difficult to adjust to large changes in CTs, even with flexible equipment and multiskilled or temporary workers, large changes in production requirements are ordinarily not

Table 14.5 Schedule Change by Adjusting Only Cycle Times

	(1)		*(2)*		*(3)*
Product	*Original Schedule*	*CT (sec)*	*Revised Schedule*	*CT (sec)*	*% Change in CT*
A	200 (57%)	126	200 (50%)	126	0.0%
B	100 (29%)	252	150 (38%)	168	–33.3%
C	50 (14%)	504	50 (12%)	504	0.0%
All	350 (100%)	72	400 (100%)	63	–12.5

attempted in successive months, much less in successive weeks. Rather, the overall average production level is set, and small adjustments are made to the level on a daily or weekly basis. Though the overall average CT might have to be changed monthly or more frequently, operations on the floor are given adequate time to plan for the changes. Also, through rough-cut capacity planning, management knows what schedule changes the shop floor is capable of absorbing. It then proposes schedule changes far enough in advance so the shop floor has adequate time to adjust. The exact procedure is the topic of the next chapter.

The simplest way to adjust output in the short run is to alter the length of the workday (say, between 7 and 10 hours).[6] During slow periods, production time is shortened and remaining time is used for problem solving and PM. This is simplest only in terms of adjusting output to meet demand, for it is anything but simple from the point of view of workers, many of whom rely on regular work hours for scheduling transportation to and from work, child daycare, and so on. Any considered changes in the workday should take into account potential worker hardships and allow workers adequate time to make personal arrangements.

Adding shifts is another way to adjust the workday. For example, Prince Castle Inc., mentioned in Chapter 3, ordinarily runs a full day shift plus two reduced night shifts with skeleton crews. The skeleton crews on the second and third shifts can be brought up to full force on short notice with temporary help during periods of unusually high or spiked demand.

Summary

There is a saying that timing is everything, and although in pull production it is not quite everything, it is still very important. Ideally, materials move through the production process in a smooth, uniform fashion. The schedule of the final stage of the process determines the rate at which materials from upstream are consumed and, hence, the rate at which they should move downstream. Everywhere in the process materials should be produced and moved at the same rate. The CT of products going out the end of the process (determined by the rate of demand) should theoretically determine the CT of materials coming from upstream. The procedure of setting the CTs of all upstream operations to meet the CT of the final operation is called synchronization.

The concept of CT rather than production rate is used in pull production because it implies regularity, uniformity, and periodicity in output—a leveled, uniform flow—whereas production rate implies a quantity produced within a period of time without regard to uniformity.

The pace of a process is determined by the slowest operation, the bottleneck. When the bottleneck capacity is less than demand, the release of jobs into a system should be predicated on the capacity of the bottleneck to process those jobs. In a production system that utilizes the drum–buffer–rope concept, the bottleneck sets the pace of production everywhere in the process, a buffer of work protects the bottleneck from upstream interruptions, and jobs are released into the system based on a schedule or work backlog at the bottleneck.

Related to synchronization and CT is the topic of balancing. In a balanced process, every stage has about the same work content, and the work content time is equivalent to the required CT. When the times to perform tasks at different workstations are different, the process is unbalanced, which implies that some workers do more work than others. Since process output must be constrained to the required CT, workers whose tasks take less time will be idle during part of every cycle. In MMP, balancing is based upon the weighted average task times for the product models in the mixed-model sequence.

The process of balancing is ongoing and as such another aspect of continuous improvement. Once tasks have been assigned to workstations, supervisors and workers scrutinize them for improvement opportunities, and then rebalance the tasks such that most of the idle time in the process is on one worker. Eventually, as tasks continue to be reassigned, all the work content is shifted to some workers away from others, who will then be reassigned elsewhere and whose workstations will be eliminated.

Two ways to meet changing demand requirements are to alter the length of the production workday and to alter the rate of the production process. The latter is inherently more difficult since it involves changing the required CT, and, potentially, rebalancing tasks everywhere in the process. Thus, altering the length of the workday is the favored method, at least in the short run, and modern union contracts often specify a range of times, say, 7 to 10 hours, as a normal workday. Such changes, however, can cause hardship for workers, and thus should be somewhat infrequent and only follow advance notice. Long-term shifts in demand can initially be handled by changes in the workday or by overtime; through continuous improvement, the CTs should be adjusted so that, eventually, the workday can be brought back to normal.

Notes

1. E. Goldratt, *The Goal,* (Millford, CT: North River Press, 1986); see also E. Goldratt and R. Fox, *The Race* (Croton-on-Hudson, NY: North River Press, 1986; M. Umble and M. Srikanth, *Synchronous Manufacturing* (Cincinnati, OH: South-Western Publishing, 1990).
2. What happens in plants where the process consists of multiple possible routings that share a common bottleneck? This is handled using a variant of bottleneck scheduling and a procedure called CONWIP, described in W. Spearman and M. Hopp, *Factory Physics: Foundations of Manufacturing Management* (Chicago: Irwin, 1996), 444–477, 478–483.
3. See, for example, R. Askin and C. Standridge, *Modeling and Analysis of Manufacturing Systems* (New York: John Wiley, 1993), Chapter 2. An excellent reference on the topic of sequencing and balancing, including evaluation of some common line-balance methods, is E. Buffa and J. Miller, *Production-Inventory Systems: Planning and Control,* 3rd ed. (Homewood, IL: Richard D. Irwin, 1979), Chapter 9.
4. Y. Monden, *Toyota Production System*, 2nd ed. (Norcross, VA: Industrial Engineering and Management Press, 1993), 179–183.
5. J. Black, *The Design of the Factory with a Future* (New York: McGraw-Hill, 1991), 152.
6. R. Hall, *Zero Inventories* (Homewood, IL: Dow Jones Irwin, 1983), 62–63.

Suggested Readings

J. Y-T. Leung, and J. Anderson. *Handbook of Scheduling: Algorithms, Models, and Performance Analysis.* Boca Raton, FL: Chapman & Hall/CRC, 2004.

B. Rekiek, and A. Delchambre. *Assembly Line Design: The Balancing of Mixed-Model Hybrid Assembly Lines with Genetic Algorithms.* Brussels, Belgium: Springer, 2005.

Questions

1. Explain the concept of synchronization in a process. What is necessary to synchronize a process consisting of a sequence of operations?
2. How is a process synchronized when the completion times per unit of the operations in the process are different?

3. Explain (a) the relationship between production rate and cycle time, and (b) their interpretation in synchronous production.
4. Regarding the production capacity of all operations in a process, what is necessary to base the CT of those operations on the required CT at final assembly?
5. Explain the meaning of the terms *drum*, *buffer*, and *rope*.
6. In bottleneck scheduling, why is it desirable to maintain a buffer of work ahead of the bottleneck?
7. What does it mean for the workstations in a process to be balanced?
8. Discuss the meaning of the term *line balancing* as it applies to an MMP process.
9. Comment on the following: A process has eight workstations in sequence; the workstation with the greatest work content takes 94 seconds; therefore the CT for the entire process should be set to 94 seconds per unit.
10. Discuss ways to reduce the effects of disruptions in a synchronized process.
11. Discuss the ramifications (relative pros and cons, ease or difficulty) of adapting production to schedule changes by (a) altering the length of the workday and (b) altering the production rate.

PROBLEMS

1. At Feebo Company, components U, V, W, and X are assembled in a cell in daily quantities of 70, 140, 280, and 35, respectively.
   a. Assuming a 480-minute workday, what is the average overall component CT of the cell?
   b. What is the MMP sequence?

2. Assume the level 1 BOM breakdown of parts for the components in problem 1 are as follows (parenthesis indicates number of this part needed):

	U	*V*	*W*	*X*
Parts	B	A	A	C
Level 1	D	D	B	E
BOM	F	G	E	G(2)
	H(2)	H(2)	F	H
			H	

These parts come from the following places: parts H and E are supplied by a vendor, and the other parts are produced in the Feebo plant—parts A and D on another line, parts B and F in one cell, and parts C and G in another cell.

a. What is the MMP sequence for the line that produces parts A and D? What is the takt time of the line?
b. The cell that produces B and F does each in batches of 50. Changeover time between batches is 3 minutes. What must the CT be at the cell for it to satisfy demand for parts going into components U and W?
c. The cell that produces parts C and G does it in mixed-model fashion. What is the MMP sequence? What is the average required CT?
d. The supplier delivers parts H and E directly to their points of usage in the plant at 8 a.m. and noon. How many of each part, on average, should be in each delivery?

Table 14.6 Task Time in Minutes t_{ij}

Task i	*Product j*		
	X	*Y*	*Z*
a	2	3	2
b	1	2	1
c	3	3	3
d	2	1	4
e	3	2	2
f	4	2	1
g	1	2	1
h	2	2	2

3. Average daily demand for a product is 95 units. A production workcell has four workstations and runs 400 min/day. The time to complete tasks at the workstations is 210, 180, 205, and 160 seconds, respectively. Discuss the goal and general strategy for balancing tasks in this process.

4. Figure 14.11 is the precedence diagram for an assembly process. Table 14.6 shows the task time in minutes t_{ij} for i = a, b, c, …, h; j = X, Y, Z. Suppose the daily demand is 30 units for X, 15 units for Y, 60 units for Z. Assume a 480-minute workday.
   a. What is the required CT for X, Y, Z, and overall?
   b. Compute the weighted average task time for each task.
   c. Using the weighted average task times and the precedence diagram, assign tasks to a line using the longest operating time rule.
   d. What is the efficiency of the line?
   e. What is the potential maximum CT of this line? What would be the daily output of the line at this CT for each of the three products?
   f. If you wanted to increase the output of the line beyond the maximum determined in part e, at what workstation should you direct your improvement effort? If you were to direct all your improvement effort at this workstation, what would be the most improvement that you could make (i.e., further improvement at this workstation would not increase the output of the line)?

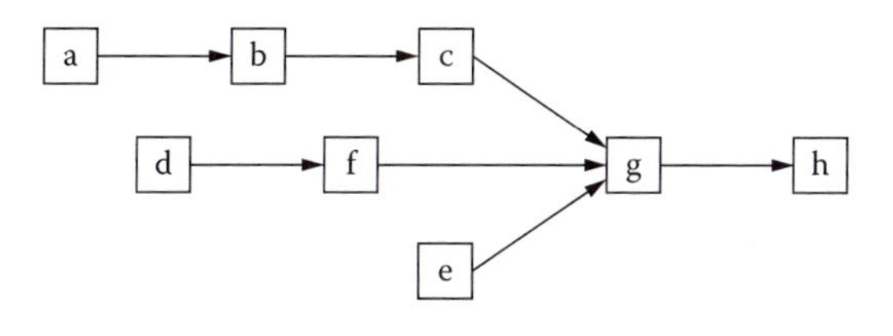

Figure 14.11 Precedence diagram.

Chapter 15

Planning and Control in Pull Production

Let us not go over the same old ground. Let us rather plan for what is to come.

—Marcus Tullius Cicero

Want to see God laugh? Make a plan.

—Anonymous

This chapter describes a framework for preparation and execution of the pull production plans and schedules described in previous chapters. The framework is a simplified version of the Toyota Production System. Toyota originated pull production and has the longest experience with all aspects of it; as a result, Toyota has become the standard for integrated planning, scheduling, and control in pull production.[1] But Toyota's planning process is complicated, largely because Toyota is a huge corporation with millions of customers, and its planning process must deal with information from a vast network of dealerships, suppliers, and production facilities worldwide. The discussion here avoids the complicating details of the Toyota system and focuses on those aspects of most relevance to other industries.

Today most large and medium companies use enterprise resource planning (ERP) systems with embedded manufacturing resources planning (MRP II) systems for production planning and control. Thus, in adopting pull production they are not starting with a clean slate, but have preexisting procedures for planning, scheduling, and controlling production. Some functions of ERP/MRP systems remain the same even with pull production, in particular, functions for demand management, rough-cut capacity planning, master scheduling, and aspects of requirements planning. Other functions of these systems, however, especially for detailed scheduling, shop-floor control, and inventory updating are unnecessary and must be modified for pull production. This chapter addresses the differences and the necessary changes to make MRP-based systems compatible to pull production.

The Whole Enchilada

Production planning, scheduling, and control in pull production are conducted within a **production planning and control** (PPC) framework that has centralized and decentralized components. We start with an overview of the system and of the two main components, shown in Figure 15.1.

Centralized System

The **centralized** part of the system comprises staff personnel and systems that perform order entry, demand forecasting, rough-cut capacity planning, master production scheduling, and material requirements planning. The role of this part of the PPC system is to accumulate demand information and formulate production plans and schedules. In a multiplant company, it subdivides demand information into product categories and prepares master schedules for every plant. It develops medium-range demand forecasts and accumulates short-range and current demand figures for firm customer orders. Demand information is continuously compared to existing and

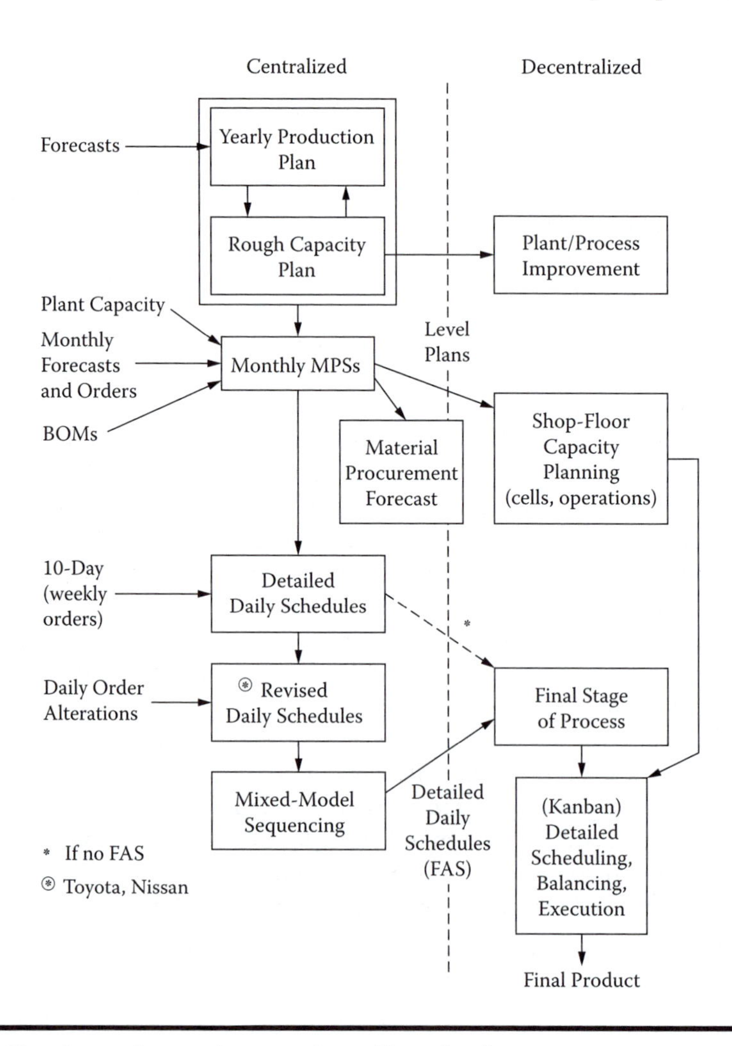

Figure 15.1 Planning and control system for pull production.

short-term production capacity to determine the best level of production, that is, the level that meets demand and is within or near the capability of existing capacity and inventories. This forecasting and capacity planning is performed by ERP/MRP systems in traditional ways.

Preparation of level master production schedules (MPSs) begins months in advance of actual production. As execution of the MPS draws near, production departments, workcenters, and suppliers are apprised of anticipated changes in production levels and material requirements.

Besides creating forecasts, rough capacity plans, and MPSs, the centralized part of the system also prepares and issues the daily production schedules for the final stage, or the final assembly schedule (FAS). As described earlier, this is the only schedule needed to determine CTs and material requirements in a pull process.

The centralized system creates daily schedules by incorporating the most recent customer order information into the level schedule. Although recent order information is essential to providing good customer service, it creates variation in daily production requirements. Since a pull system requires a somewhat level daily schedule, the centralized system has a dual goal: to incorporate recent customer orders yet keep day-to-day production variation to a minimum.

Decentralized System

The role of the **decentralized** part of the system is to oblige the production requirements as specified in the MPSs and daily schedules. Because the decentralized part is comprised of departmental managers, shop-floor supervisors, and teams of workers, it is primarily a shop-floor-based system. Plant and department managers receive MPS preplans from the central system, which they use to try to determine how to produce products in the mix and volume specified by the plans. If a plan is difficult or impossible to meet, the centralized system is notified and the plan is revised.

With the exception of the final stage of the process, every department and workcell in the process does its own daily work scheduling and control. In a synchronized pull system, daily scheduling is largely a matter of each operation or workcell responding to orders from downstream. Supervisors and worker teams make most decisions about execution and control. When a job produced in a batch size exceeds one container or when multiple kanbans arrive at once, the decision about the batch size or the job priority of the kanban or the batch size are often made by the supervisor at the operation. The supervisors and worker teams also handle necessary schedule changes due to equipment breakdowns, material shortages, and other disruptions.

Though responsibility for planning, scheduling, and execution of work is split between the centralized and decentralized systems, the two systems work together. In setting production levels the centralized system takes into account the production capacity of the shop floor. The shop floor sends the centralized system frequent, updated information about production rates, setup times, equipment reliability, and so on. The centralized system specifies on the plan what the shop floor will do but far enough in advance so the plan can be modified if the shop says it cannot do it. Execution and control of the production quantities specified in the MPSs and daily schedules are left almost entirely to the decentralized part of the system.

In the traditional ERP/MRP-based PPC system, everything from long-range planning to detailed capacity planning, scheduling, and control is done by the centralized system. But in any business, it is a given that things change and that no plan or schedule will remain completely valid for very long. One drawback with centralized planning, scheduling, and control is that it is almost impossible to quickly revise plans and schedules to account for all the work and priority changes and interruptions happening everywhere in the plant. By the time information gets to the system and schedules are revised, the information and, hence, the schedules are obsolete. Because of this, most traditional PPC systems end up having a decentralized part, which, de facto, does much of the real detailed work scheduling and control. The decentralized part consists of frontline workers and supervisors on the shop floor who constantly struggle to overcome shortfalls in the schedules they are supposed to follow. Their actions, however, are informal, sometimes spontaneous and ad hoc, and often not totally effective.

In a pull production system, the role that frontline worker teams and supervisors play in PPC is formalized. Supervisors and worker teams are trained to be competent to schedule and control work to meet requirements as set by level MPSs. Since it is impossible for a centralized system to get up-to-the-minute reports on everything, it makes sense to decentralize as much of the planning, scheduling, and control as possible (this applies to any production process, pull system or not).

Having said that, I now say this: Most pull production systems still rely on MRP II systems for aspects of production planning and scheduling. After all, ERP/MRP systems do many things, and some of them (forecasting, rough-cut capacity planning, master scheduling, material requirements planning, and ordering) must continue to be done even in pull production.

We now look more closely at the major components within the centralized and decentralized systems.

Centralized Planning and Control System

The centralized system anticipates demand, then sets in motion the actions for procurement, organization, and utilization of resources necessary to meet demand. The main functions of the centralized system are monthly planning of the MPS, daily scheduling of final production, and materials forecasting and procurement.

Monthly Planning

The centralized system prepares monthly demand forecasts for each product, product group, or other end items, and then creates the MPS for each. Figure 15.2 illustrates this process.

Planning MPSs for Future Periods

Each month the demand over some planning horizon is estimated. If, say, the length of the planning horizon is 3 months, then demand for the current month and each of the next 2 months

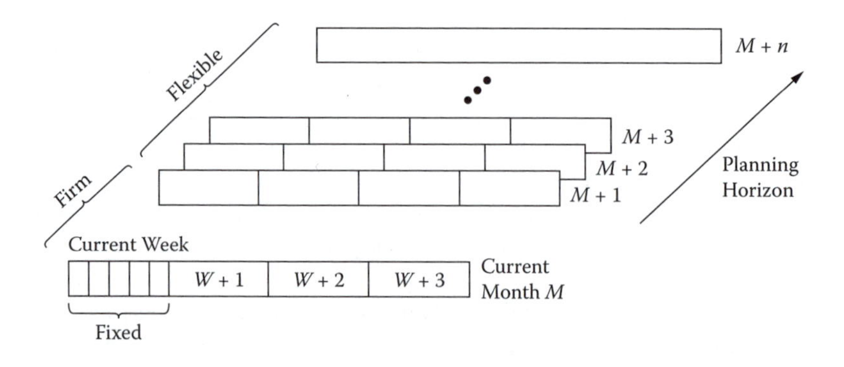

Figure 15.2 Monthly planning process.

is forecast.[2] Master production schedules are then prepared for each of the months, taking into account forecasts, preexisting production capacity, inventory levels, and known customer orders. The MPSs are leveled as described in Chapter 13.

The tentative MPSs for 2, 3, or more months into the future (M + 2, M + 3, etc., in Figure 15.2) are the basis for estimating production capacity requirements. Because no actions or commitments for those MPSs will have as yet been made, the schedules are flexible and easily revised. However, as shown in Figure 15.2, the MPSs prepared for the current and next month are considered firm or fixed because actions for them have already been taken. Once purchase orders for raw materials are sent out, additional workers have been hired or assigned, and contracts for outsourced items have been set, the commitment to follow through on an MPS increases, and the schedule becomes somewhat fixed.

MPSs for Shop Floor Planning

Tentative MPSs for future months (M + 2 and beyond) are circulated at a predetermined time (usually around the middle of the month) to managers and supervisors in production lines and production departments. From the tentative MPSs and the number of predicted working days in the month, the required average daily production level for a product can be determined. For planning like this an MRP II system is a useful tool.

Given the average daily production level for every product, the production department can estimate the number of workers needed, equipment requirements, and the length of workdays necessary to meet the level MPS. By midmonth every operation in the process is able to estimate what it must do to meet the production requirements for the following month.

Daily Scheduling

The MPS for the current month is broken down into weekly schedules, then into daily schedules, incorporating in each week the most recent information on forecast demand and customer orders. Such information is used to develop, first, an average daily production level and then eventually a final, fixed daily production schedule for the final stage of the process (final assembly or otherwise). If the final stage does mixed-model assembly, the daily schedule is the FAS as described in Chapter 13. In effect, the size of the time bucket in this scheduling system is 1 day.

Incorporating the most recent order information into the final schedule causes the production level to vary somewhat day to day. As a consequence all operations everywhere experience some variation from the planned baseline amount. But since each produces only what is authorized from downstream, the entire system readily adapts: slightly higher demand at the final stage results in a slight increase in the rate of demand signals sent upstream; slightly lower demand results in a slight decrease in the signals. As mentioned, a pull system can readily absorb a variation of about ±10% from the baseline amount without needing to change to the number of cards or containers in use.

Integrating Recent Demand Information

The process of refining daily plans to reflect the most recent order information is illustrated in Figure 15.3. Every week the centralized system receives customer order information for a 10-day period.[3] This information includes firm customer orders and the most recent sales forecasts from

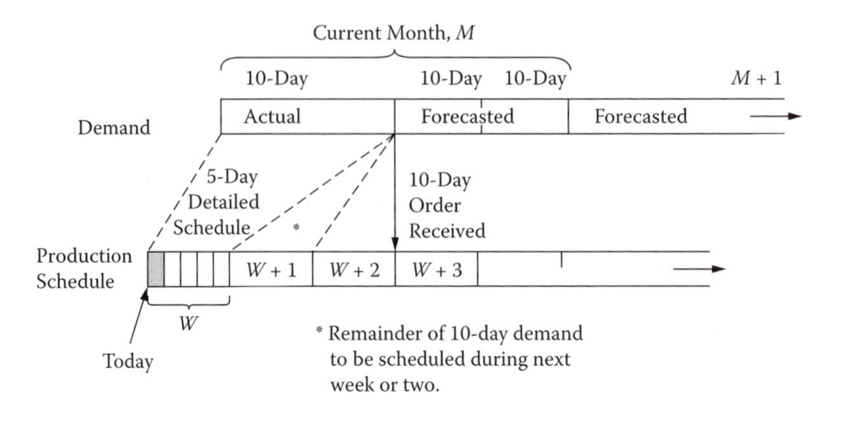

Figure 15.3 Daily planning process.

sources such as dealers, sales people, and distributors. The total demand for three successive 10-day orders (i.e., 30 days demand) must fall within the planned production volume of the level production schedule for the month. If it exceeds the planned volume and cannot be accommodated by overtime, then the excess is included in the production schedule for the following month. If it is apparent that demand is on the rise and will continue to exceed current capacity levels, then the overall planned capacity level is ratcheted up.

The central system receives the 10-day demand information at least 1 week prior to the 2-week time block that the information represents. For example, customer orders to be filled in the first 10 days of May should be received 1 week prior to that, or by April 24. Note that although the system requires 10-day order information a week before production, much of the information might still be forecast, leaving room to insert actual orders that arrive later. As a consequence, the time between the arrival of a customer order and production to fill that order is often as little as 4 to 10 days. This is explained in the next section.

Having the 10-day demand for all products allows time to sort out and sequence products so the final production schedule for the first week closely fits the baseline level of the current MPS. The system then uses this information to estimate the daily production rate for each product or end item. Out of 10 days of demand, only 5 days can be satisfied in the first week (5 working days) of production. The remaining orders from the 10-day plan are then scheduled for production in the following week or, at the latest, within the week after that.[4]

Daily Order Alterations

At companies like Toyota and Nissan, the centralized system performs one additional step in setting the daily production schedules. Besides information on 10-day demand, the system incorporates information received every day from sales sources about order alterations. Alterations include new orders, canceled orders, and changes to orders affecting the most recent 10-day period. The daily schedule is originally prepared using 10-day demand figures is then updated to account for these alterations. The updated version becomes the final daily production schedule. Of course, finalization of that daily schedule happens a few days or more in advance of the day the schedule will be used. How far in advance of the production day it is finalized depends on lead times for materials procurement and manufacturing setup.

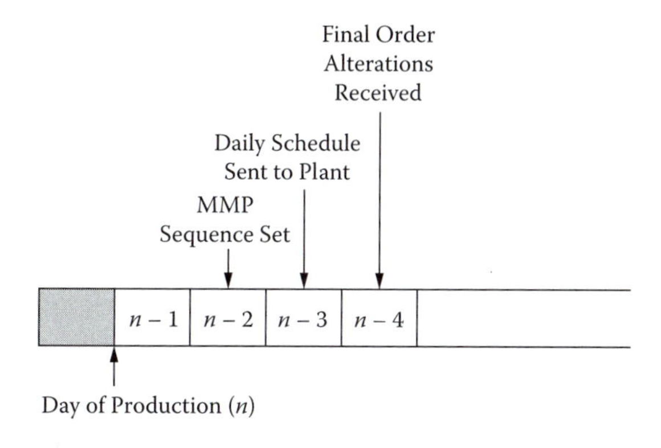

Figure 15.4 Incorporating daily demand into daily schedules.

At Toyota the daily final assembly schedule is prepared 4 days in advance of the production day (shown in Figure 15.4). While setting the final daily schedule, the system also sets the requirements for all materials, parts, and assemblies for the day of production, 4 days later. Three days before production the centralized system transmits the finalized daily schedule to the plant. Two days before production the plant determines the MMP sequence for the schedule. On the day of production, the MMP sequence is displayed at the final stage of the process, where it is executed. Assuming that the production lead time for final assembly is 1 day, then products coming off the assembly line on any given day will reflect customer order information received as little as 4 days earlier.

For large products like automobiles, major components such as engines and transmissions are each assembled on their own production lines. Each of these lines also has a level daily production schedule and MMP sequence, which is derived from the daily FAS for the automobile assembly line.

Some manufacturers allow salespeople access to the daily production schedule. Since some of the models produced in the MMP sequence are for forecast demand (not firm orders), these are available for immediate sale. The salesperson can inform a customer about these models (the ones available for quick delivery) and also about how long they will have to wait for other models not on the current daily schedule.

Material Procurement Forecast

The quantity of materials and parts for all products is estimated from the monthly MPS and bill of materials (BOM). For each product or other end item, requirements for procured parts and components are determined using the same logical procedures as employed in MRP systems.[5] At the start of each month, information about the anticipated daily material requirements for that month is sent to suppliers. Forecast material requirements for the following month or two are also sent to allow suppliers time to adjust to changes in requirements levels.

Kanban Supplier Link

Many pull production companies have direct **Kanban links** to their suppliers. Supplier Kanban systems, described in Chapter 16, work very much like internal pull systems except the upstream operations to which orders and empty containers are sent are external suppliers. Since a level MPS will result in material requirements that are also somewhat level, suppliers can expect the requirements (quantity and order frequency) to remain somewhat constant over a given period of time. Although the actual daily requirements might be somewhat variable (just as the MMP sequence and daily production schedule are somewhat variable), the variation, if kept small, can readily be absorbed by the supplier Kanban system. If the variation is anticipated to exceed what the supplier Kanban system can absorb, then the parameters of the supplier Kanban system (delivery cycle time [CT], number of supplier kanbans, delivery frequency, and number of units per container) are revised. Daily production schedules, which are set a week in advance (at Toyota, 4 days), are used to determine daily material requirements. Each week the parameters of the supplier Kanban system are reviewed and adjusted as necessary to reflect demand for the upcoming week.

MRP Supplier Link

Not all pull production companies use Kanban for daily materials procurement. At Nissan, for example, daily material ordering is done by a centralized MRP system that uses daily time buckets. The order process combines the finalized daily production schedule for 4 days later with the BOMs for the products to be assembled. To meet daily requirements, suppliers deliver eight times a day. Eighty percent of supplied parts are ordered in this way. For common, standardized parts, orders are placed less frequently, typically about every 2 weeks using 10-day order information. For specialized parts for low-volume products, orders are placed only monthly.

Decentralized Planning and Control System

The biggest difference between pull systems and traditional push systems is that in the former the functions of detailed planning, scheduling, and control are decentralized, whereas in the latter they are centralized. In pull production, shop-floor supervisors and work teams assume much of the detailed capacity planning, scheduling, and work control. In many instances they also handle procurement of purchased materials.

Detailed Capacity Planning

Initial Capacity Planning

Each month the production output of a cell or independent station might have to be altered to meet changes in the monthly MPS. This requires production capacity flexibility, which is achieved through adjustable work hours, flexible workcell configurations, short changeover times, cross-trained workers, and so on.

If the capacity of workcells or stages of a process cannot be increased to meet the MPS and if the shortage cannot be made up through overtime, inventory, or outsourcing, then the capacity of that workcell or stage, not the original MPS, is the process bottleneck; that capacity dictates the output rate for the entire process. If the rate is less than the level specified in the MPS, and assuming there are no other means by which to reach the level of the MPS, then the MPS must be revised downward.

The worker team at each cell or department has its own personal computer for performing capacity planning, quality analysis, inventory control, and reporting. Information about the MPS and BOMs is retrieved from the centralized PPC system. Information about current capacity and inventory is sent to the centralized system.

Capacity Fine-Tuning

Even after the MPS and the production capacity of every operation have been mutually adjusted to a feasible level, adjustments continue as more recent information becomes available. The initial MPSs, the preplans prepared 1 or 2 months in advance, provide only forecasts and rough estimates of demand, so any plans for the workcells and departments to meet those schedules are considered provisional. As late as 1 or 2 weeks before production is to begin, the plans for workcells and departments might have to be revised to incorporate the most recent demand information.

Even as the FAS is being executed, workers continue to fine-tune each operation. Through experience and continuous review of standard operations routines (SORs; described in Chapter 11) frontline teams seek ways to alter work procedures, number of kanbans, worker assignments, and so forth, to achieve better balance between available work capacity and required CTs.

Shop-Floor Control

In traditional companies the sole purpose of data collection in the plant is to provide the centralized PPC system with information for accounting and planning. But in lean companies the purpose of data collection is also to provide frontline workers with information for scheduling and controlling their work.[6]

Visual Management, Again

In lean companies, much of the information that workers rely on for daily scheduling and control is generated and displayed by the **visual management system**, so-called visual because people know the status of the system and what to do simply by looking at kanban cards, containers, and workplace signals. With a visual system it is not necessary to look hard or look far to see what needs to be done. Other elements of visual management described in this book include information post-its (posted measures about quality, throughput, setup times, and machine performance); andons signaling the status of a process, workcell, or machine; posted schedules, quality procedures, and control charts; posted SORs; and pokayokes for signaling and preventing defects. Visual management simplifies tracking of jobs and system status and pinpoints areas on the shop floor having problems or needing improvement.

Role of Worker Teams

Part of the role of worker teams in the decentralized system is to track performance measures such as production rates, CTs, lead times, product quality, and equipment availability and effectiveness. Daily, hourly, and by event, workers update charts about defects, breakdowns, setup times, and so forth. Supervisors prepare summaries of this information and send it (say, via an intranet) to the centralized system, which uses it for updating estimates of production capacity levels and setting MPSs and daily production schedules. Up-to-date information from the shop floor keeps the centralized planning staff in touch with reality and helps ensure that MPSs and daily schedules are realistic.

Adapting MRP-Based PPC Systems to Pull Production

Adapting an MRP-based system to pull production requires fundamental changes to shop floor control procedures and to aspects of the centralized PPC system. Assuming that the product demand and the production process meet the requirements of pull production as described in Chapter 8, a company can begin transitioning to pull production by undertaking changes as described in this section.

Simplified Bills of Materials

A primary goal of lean companies is to manufacture products with the least amount of waste. Associated with this goal is the principle of simplification: accomplish the same ends, but in a less complex, more basic way and with fewer steps. In manufacturing that translates into producing products with fewer parts and using processes that require fewer steps.

In reducing the number of steps in a process, nonvalue-added steps such as storage are eliminated first. In pull production, work in process (WIP) levels are progressively reduced, sometimes to the point where they become nonexistent. Eliminating a place where inventory is stored is tantamount to eliminating a place where material is tracked and scheduled. Pull production can thus simplify a manufacturing process simply by eliminating steps that involve administrative procedures, even if no processing steps are eliminated.

Since upstream stages of pull systems do not require work schedules, the centralized PPC system (MRP or otherwise) does not need to generate requirements lists or schedules for all stages of the process. The centralized system need only prepare a schedule for the final stage of the process and order releases for procured materials from outside, non-Kanban- linked suppliers. In a pull production system, an MRP II system has much less work to do than in a push system.

Software developers have created add-on modules for adapting MRP II computer systems to pull production. These systems assume (or mandate) that BOM structures are expressed in the simplest terms possible and that places in the process requiring scheduling and tracking are minimized. One of the first things a company must do when converting from push to pull is to simplify the BOM for each end item.

The typical MRP system uses the BOM in many ways. Besides the materials and parts that must be procured, the BOM identifies manufacturing and assembly operations that must be scheduled and monitored. For example, the BOM in Figure 15.5 shows that item R is comprised of items S and Y, both of which must be individually manufactured and then combined. In traditional push manufacturing, the MRP system maintains inventory records for every item at every stage of the production process and tracks inventory levels and production status. In reference to the BOM in Figure 15.5, inventory records are thus maintained for the end product (the top-level, item R),

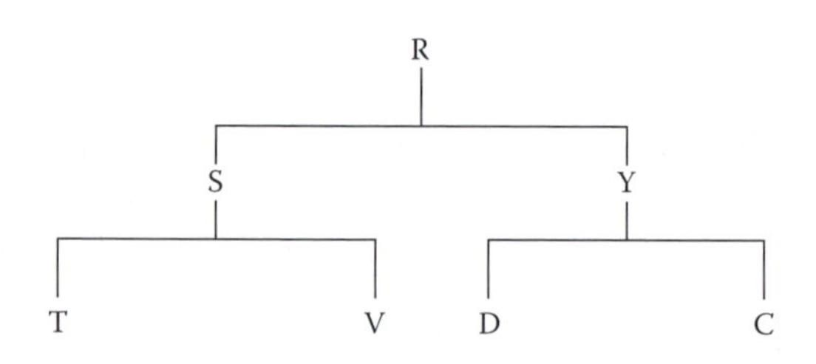

Figure 15.5 BOM structure: three levels, seven records.

all raw materials and purchased parts (bottom level; items T, V, D, C), and all WIP (intermediate level, items S and Y), a total of seven records.

Each record corresponds to a place somewhere in the plant (a shop-floor location, a stock room, etc.) where raw material, WIP, or finished items are stocked. For plants that deal with many jobs simultaneously, this recordkeeping is considered necessary so jobs do not get lost or mixed up. Although recordkeeping is often necessary for legal purposes and financial reporting, most recordkeeping is wasteful and unnecessary.

It should be obvious that the greater the number of levels and branches in the BOM, the greater the number of records the MRP system must maintain. The greater the number of records, the higher the recordkeeping cost, and the more difficult it is to keep the records current for purposes of scheduling and control. Conversely, the fewer the records, the lower the recordkeeping cost and the easier it is to maintain record accuracy.

Since a pull-production system controls work and materials throughout the process by delimiting the number of containers, any other form of control mechanism at stages other than at the final one is unnecessary. This is equivalent to saying that, for purposes of scheduling and control, all records for intermediate levels of the BOM are unnecessary. Eliminating these records is referred to as **flattening the BOM**. Once flattened, a BOM has only two levels, one for the end item and one for the low-level parts that go into it.

Flat BOMs

An MRP system ordinarily maintains records for every location where material is stocked. If instead of being held in stock, a material is moved directly from one stage of a process to the next, or if it is held relatively briefly at a stage and in a small quantity (as in pull production), then there is no need to track it at that stage. Lack of need to keep track of an item at a stage corresponds to removing the item from the BOM and removing the inventory record for it from the MRP system. Hence, just smoothing production flow and eliminating storage places in the process flatten the BOM, even though the number of parts in the product and stages in the process remain the same. (This assumes continuous or repetitive material flow, accompanied by shop-floor visual management.)

Cellular manufacturing (the topic of Chapter 10) also leads to flattening of BOMs. Material in a cell moves directly from one workstation to the next and is held so briefly between stations that it need not be tracked. Thus, whether a product is made in a workcell instead of a job shop (where it is held in WIP inventory between every stage of the process) or in a process with stages linked by a repetitive pull system, the BOM structure for it can be flattened.

Example 1: Flattening the BOM with Cellular Manufacturing

Suppose product X is ordinarily produced in a job shop and requires nine production steps. After each step, it is held in inventory until it can be processed by the next step. The process is represented by the 10-level BOM in Figure 15.6. The lowest level in the BOM represents raw material inputs for the first step, and successively higher levels represent WIP and raw material inputs for subsequent steps. Each letter in Figure 15.6 represents a stock record; hence, the MRP system maintains 19 records for this product. Every movement of materials into or out of a stock represents a transaction and a stock record that must be updated. Suppose now that the first two steps in Figure 15.6 (combining parts J and K, then combining A and L) are done in one workcell, the next three steps (combining B and R, then C and S, then D and M) are done in another workcell, and the last four steps (combining E and N, F and O, G and P, H and Q) are done in a third workcell. An inventory control rule of thumb says that as long as a material keeps moving it is not necessary

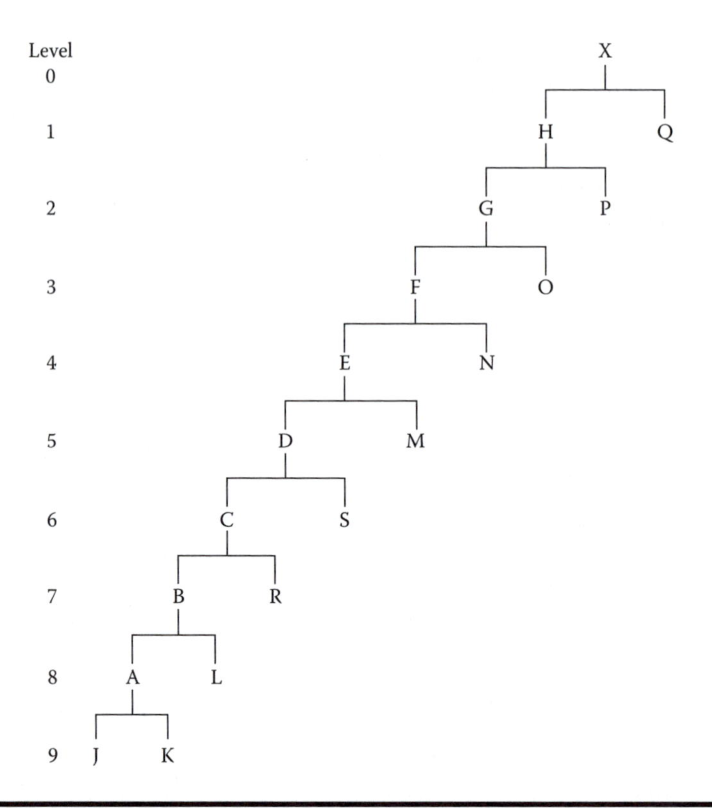

Figure 15.6 BOM structure for nine-step process.

to track it; the material need be accounted for only at places in the process where it is detained. In a workcell, material flows continuously from one workstation to the next, so tracking within the cell is unnecessary. The only places in a workcell where material is held are at the input and output stock areas. Thus, the BOM for a product made in a workcell can have as few as two levels.

In the present case, where the process has been converted into three workcells, the BOM can be collapsed from 10 levels to just 4 levels, shown in Figure 15.7:

- Level 3 represents stocks of material going into cell 1.
- Level 2 represents stocks of material going into cell 2, including item B, the output of cell 1.

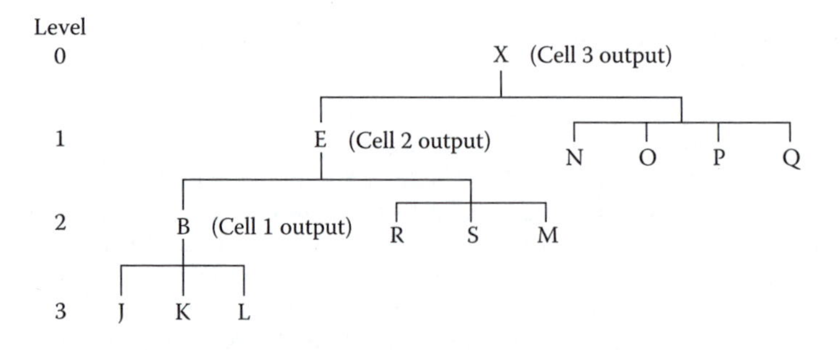

Figure 15.7 Flattened BOMs for nine-step process and three workcells (original BOM in Figure 15.6).

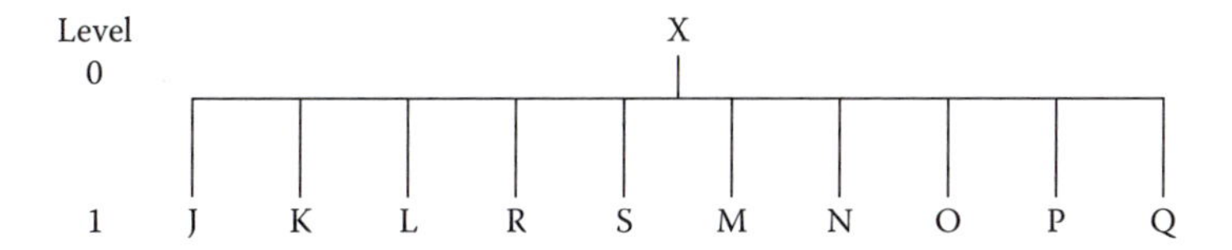

Figure 15.8 Flattened BOM for nine-step process in one workcell.

- Level 1 represents stocks of material going into cell 1, including item E, the output of cell 2.
- Level 0 is the stock of output from cell 3, end item X.

As another example, suppose all nine steps are performed by just one cell. The BOM would then be reduced to the two levels shown in Figure 15.8.

When a production process and the corresponding facility layout are modified for pull or cellular production, many locations where WIP material was previously held are eliminated. As a consequence, the preexisting BOM for the product must be altered to reflect the new process. Phantom records are a way to eliminate those once-held items from the BOM.

Phantom Records

Suppose a plant is converting its production process from a job shop to pull production or cellular manufacturing. The preexisting BOMs of the end items manufactured in the process can be flattened by removing records for items at intermediate stages of the process or workcell. An alternative, however, to physically removing the records is to transform them into **phantom records**. A phantom record represents a material that never actually goes into storage, but is in a momentary transitional state (a component or subassembly) and is en route to the next stage of the process. By using phantom records, the BOM for an end item can be simplified without altering the basic structure of the BOM or of the inventory data file. This is preferable when the cost of restructuring a data file is large, when the current BOM structure is needed for accounting purposes, or when intermediate items in the structure are used for purposes besides end-item production and must be tracked. In those cases, a phantom record is simply substituted into the existing BOM structure for a preexisting record. The phantom record retains the location of the original record in the BOM structure and can be ignored, if desired; if tracking and accounting information are needed, the record can be activated and accessed.

Example 2: Simplified BOM Using Phantom Records

Suppose the nine-step process described in the previous example is performed by a sequence of three workcells. Figure 15.9 shows how the BOM from Figure 15.6 would look if this change were made and if phantom records were inserted in the BOM wherever material was once, but will no longer be, held in stock. Actually, the phantom records represent materials between stations within a workcell. Since the items flow somewhat continuously, they do not need to be tracked, and, given that they are listed as phantoms in the BOM, no transactions (additions or withdrawals) are posted against them. Essentially, the phantom records are ignored by the MRP scheduling system. (Notice, if you erase the phantom records in the structure in Figure 15.9, you get a structure identical to that in Figure 15.7.) Since phantom records are not intended to represent steps in the process, lead times for them are considered to be zero. If, however, it becomes necessary to track these intermediate items, the phantom records for them are simply activated.

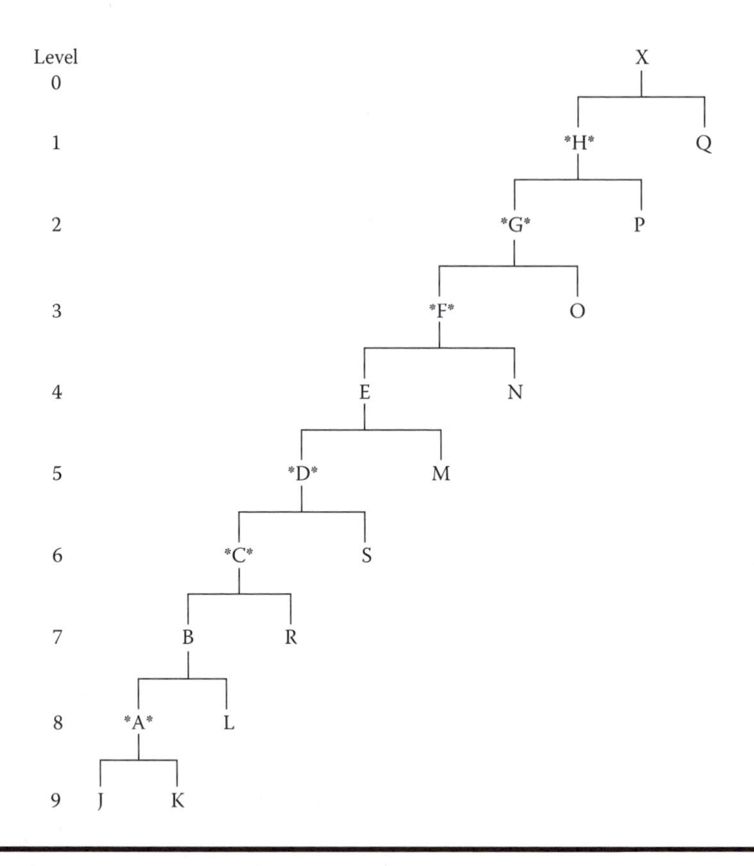

Figure 15.9 BOM using phantom records (letters with asterisks) for 9-step process performed in three workcells.

Phantom records are also used for another purpose: to represent items that are held in stock, but for purposes other than production. That is, they can represent items that are produced at a stage of the process but are then separated out and not used in the end item. For example, some quantity of a component produced in a process might be siphoned off (diverted from the rest of the process) for the purpose of a reserve for service parts or for testing and marketing. The phantom record for this component enables inventory of this component to be monitored, controlled, and maintained apart from inventory of the identical component used in production of the end item. Although phantom records for these spare and test parts are not treated like other MRP records, the records for parts at the lower levels used to make the service and test parts are treated like other MRP records. In other words, the net requirements for these lower-level parts must be planned and transactions against the MRP records for them must be posted.

Stock Areas and Point of Use

Corresponding to each MRP record (including phantom records for items momentarily held in stock) is a **stock area**, a physical location where material is held for use by a workcenter, workcell, or workstation. The stock area is the place where items represented by MRP records and phantom records are located. To minimize handling time and cost, the stock area is ideally located at the place where the material is used—the **point of use**. Material at the point of use is ordinarily

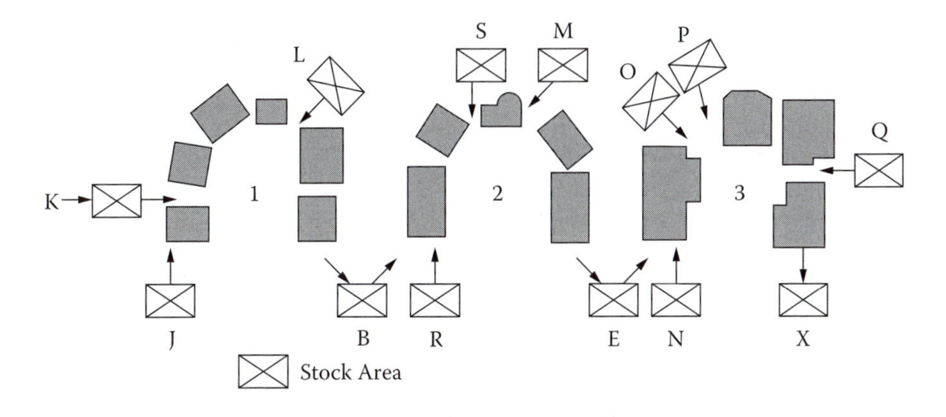

Figure 15.10 Three-cell process showing point-of-use stock areas.

within the close reach of workers. Ideally, a stock record exists for each point of use such that if the same part is used at multiple locations throughout a plant, the amount at each can be known.

For the three-cell process in the previous examples (represented in Figure 15.7) there is a stock area for each kind of raw material and for the output of each of the three cells. If the stock areas are located at their points of use, then they would be located at places in the three cells as shown in Figure 15.10.

Every production process can be conceptualized as a series of stock areas through which material moves on its way to becoming a finished product. Referring to Figure 15.10:

- In cell 1, workers remove materials from the respective stock areas for items J, K, and L to produce a unit of B, which they place in the stock area for item B.
- In cell 2, workers remove materials from the respective stock areas for items B, R, S, and M to create a unit of E, which they placed in the stock area for item E.
- In cell 3, workers follow a similar process to produce a unit of item X.

In a pull production system each stock area is the location for a production card (P-kanban) or the destination for a withdrawal card (C-kanban). In fact, any place in the system where there is a **kanban mailbox** or equivalent for posting cards is usually also a stock area for keeping containers and material.

Although, ideally, the stock areas are located at points of use, that might be impractical when the number of kinds of material needed at a workstation is large. The next best alternative then is to keep some of the material in a reserve stock location as near as possible to the point of use and for workers or material handlers to retrieve the material as needed. Another alternative is to hold the incoming material in the receiving area or staging area. The receiving area serves as a staging area and can be treated as a stock area with firm limits on all quantities held.

Companies first implementing pull production often have much inventory and cannot locate it all at points of use. As they gain experience and reduce the inventory, however, they will be able to locate most if not all material at the points of use; in addition, deliveries from external suppliers can be made directly to those areas to eliminate the need for storerooms and warehousing. This last point is described Chapter 16.

Postdeduct and Deduct Lists[7]

Even in pull production it is necessary to keep track of the quantities of completed end items and of procured parts and materials that need replenishment. In some PPC systems this tracking and updating is done through a procedure called **postdeduct** or **backflushing**. For every completed end item, the quantity on the MPS (or final schedule) is decreased by a unit, and the on-hand inventory balance of all parts going into the end item is decreased by the amounts used. When a new container of material is first accessed, or when the quantity of material reaches a minimum level, a replenishment transaction is initiated either electronically by a purchasing module, or by a shop worker or supervisor phoning or faxing the supplier. Either way, the transaction might also involve determining the replenishment quantity, adding that quantity to the stock area record (assuming short lead time on delivery), and notifying accounts payable. Such a transaction is called a **purchasing kanban** or **blanket transaction**.

The postdeduct procedure occurs everywhere in the process designated as a **deduct point**. At each deduct point, the inventory records to be updated are specified on a **deduct list**. As items move past a deduct point, the inventory records of items listed are automatically updated. Often, automatic or handheld radio frequency identification (RFID) scanners are used to simplify data entry to the recordkeeping system. As each item passes the deduct point, a worker scans a label on the item, and the record for the item is automatically updated.

In traditional MRP systems, inventory records are updated whenever items are moved from one operation to the next.[8] Every move, which usually involves a large batch of materials, requires an order or requisition. With deduct points, work requisitions are unnecessary, and updating occurs even if only a single unit is produced. The following example illustrates the procedure.

Example 3: Postdeduct for the Three-Workcell Process

Product X is produced in the three-workcell process described earlier. The product BOM is shown in Figure 15.7 and locations of stock areas in Figure 15.10. Suppose three deduct points are used, one at the end of each workcell process. The three deduct points function as follows:

1. Deduct point 1 is the completion of product X at cell 3. Thus, each time a unit of product X is completed, the inventory balance for X is increased by a unit (or the MPS for X is decreased by a unit), and the inventory balances for all parts going into X are decreased by a unit. The parts whose inventories will be decreased are specified on the deduct list, which, for product X are items E, N, O, P, and Q. (The deduct-list items are indicated as level 1 in Figure 15.7.)
2. Deduct point 2 is the completion of subassembly E. Whenever a unit of E is completed, the inventory balance in the record for E is increased and the balances in records for B, R, S, and M are decreased (level 2 in Figure 15.7).
3. Deduct point 3 is the completion of item B. For each increase in a unit of B completed, the balances for J, K, and L are decreased (level 3, Figure 15.7).

Sometimes only one deduct point is needed—at the end of the entire process. Upon completion of an end item, one unit is subtracted from the MPS and the on-hand inventory balances of all parts in the end item's deduct list are reduced by the number of parts used in the end item. This procedure is called **superflush**. Referring to Figure 15.7 and the three-cell example, each completion of end item X would be accompanied by a reduction in the balances of items J, K, L, R, S, M, N, O, P, and Q all at once. There would be no transactions for items B and E.

With only one deduct point, however, inventory balances tend to be less accurate since there is a time lag between when items are withdrawn from a stock area and when the same items within the completed end item pass from the final stage of production. In the interim, before they reach the deduct point and before the postdeduct procedure, inventory records will not reflect that the items have been removed. When the CTs are small and the schedules are level, errors in the records will be minimal. When, however, the CTs are long, using only one deduct point at the end of the process can result in significant inaccuracy in inventory records since material consumed at various stages of the process will not be accounted for until much later. In such cases it is necessary to use several deduct points located throughout the process.

In the postdeduct procedure, system software scans the end item BOM, starting at the top level. The software logic moves through the BOM, bypassing phantom records until it encounters nonphantom, "real" records. It then updates those records. Thus, decisions about whether intermediate stages of a process should be represented by phantom records are partly dictated by the throughput rate of the process and the level of inventory accuracy desired. If the throughput rate is fast, phantom records can be used at all intermediate levels of the BOM. If it is slow, some real records should be left at intermediate levels to correspond to places in the process where the postdeduct procedure should be performed.

Rate-Based Master Schedules

Traditional MRP systems generate weekly schedules for every operation, though, as mentioned, pull production requires only daily schedules and only for the end-item. The conversion starts with the assumption that the MPS is somewhat level. As discussed in Chapter 13, this means that with, say, a monthly planning time horizon, the forecasted demand for the month can be met by producing at a uniform level throughout the month. If a traditional MRP-generated MPS with weekly time buckets is used, the production quantity specified would be about the same every week.

To convert the MPS into the equivalent daily time buckets, monthly volume is translated into a **daily production rate**, which is the rate that will be maintained throughout the month (or whatever period the MPS covers). Then, instead of issuing authorization orders for each day, the MRP will issue only a single authorization order on the day the daily production rate is to commence. No further authorization is issued until the rate is changed. Example 4 shows how the daily production rate is determined.

Example 4: Daily Production Rate

Refer to Figure 15.11. Row 3 (anticipated monthly demand) is the sum of row 1 (projected sales and promised deliveries) and row 2 (production shortages from prior months).

To determine the daily production rate (row 5), divide the anticipated demand by the number of working days (row 4), and round the result up or down. Available units (row 8) are the production units (production rate × number of working days), plus any stock on hand (row 7). Units left over in any month (ending units, row 9) become the on-hand stock (row 7) for the following month; negative ending units (shortages) become the backlog for the next month (row 2).

The example presumes a monthly time frame for the MPS. The general procedure for setting the production rate would be identical for shorter time frames of, say, 10 days, as described in Chapter 13, and the shop floor is apprised in advance of the estimated CTs so workcenters can plan for changes to achieve the necessary capacity and balance. Once the production rate has been set, it can be altered on a daily basis, as described earlier, to accommodate the most recent orders.

Row	*April*	*May*	*June*	*July*
1. Forecast	3,900	4,100	3,800	3,700
2. Backlog	90*	0	0	16
3. Anticipated demand	3,990	4,100	3,800	3,716
4. Working days	20	23	22	21
5. Production rate	200	178	172	177
6. Production units	4,000	4,094	3,784	3,717
7. On hand	0	10	4	0
8. Available units	4,000	4,104	3,788	3,717
9. Ending units	10	4	(16)	1

* Shortage for months prior to April.

Figure 15.11 Determination of daily production rate.

Implementing Pull Production with MRP PPC

Implementing pull production requires changes to both the centralized and decentralized systems. Many of the changes, including modifications necessary to make an MRP II system compatible with pull production, have been addressed in this chapter; other changes involving planning and scheduling philosophy and practice were covered in preceding chapters. At the shop-floor level, the necessary changes are very broad in scope and encompass almost every aspect of lean production addressed in Part II of this book. All told, the changes necessary to implement pull production might appear overwhelming. For that reason it must be done in a carefully planned, systematic fashion.

Implementing pull production should begin with a pilot project for a product and production process where the known difficulties are few and where success is likely. As discussed in Chapters 3 and 8, implementing lean philosophy and methodologies, including pull production, requires the support of top management, high-level coordination of a steering committee, and the dedicated effort of a committed, enthusiastic shop-floor team. Ideally, the pilot product and process should be relatively autonomous from other processes in the plant. Operations that require working on products beyond the pilot project should be avoided. Implementation must consider housekeeping, layout, process routing, setup reduction, equipment reliability and maintenance, process capability, quality control, and training, in addition to the planning and control topics of this chapter. Personnel in engineering, maintenance, quality assurance, planning, and other support areas must work with the shop-floor team and learn (from then on) to coordinate their efforts. Many of the problems, issues, and steps in implementing pull production are similar to those for implementing cellular manufacturing as discussed in Chapter 10.

Some or significant progress in setup reduction, equipment PM, quality checking, inventory reduction, and workplace organization must occur before pull production procedures are introduced. Suppliers too must be involved in implementation because pull production requires frequent, small-batch deliveries of high-quality materials. Such commitment from suppliers often requires their participation in setting delivery schedules, defect levels, and even product specifications. These topics are covered in the next chapter.

Besides all of this is the matter of transforming the centralized MRP PPC system and associated shop-floor procedures to enable pull production.[9] The following steps summarize this transformation.

Step 1: Create a Logical Flow; Improve Material Handling

Use of stockrooms should be discontinued and all stock should be located on the shop floor near points of use. When that is not possible, as much stock as possible should be located at points of use and the rest kept in an overflow stockroom.

All areas of the shop floor should be linked with a material handling system, either an automated system or a team of special material handlers. This handling system replaces the MRP procedure of extracting inventory from the stockroom.

Information should be included in the MRP system to show the new locations of all materials. MRP systems often do not show the physical locations of stock, so when the same material is located at multiple points of use, a way is needed to keep track of it (either modify the MRP or adopt a scheme that keeps one record but accounts for multiple stock points).[10]

At this stage the MRP system continues to schedule and release production orders as before. Meantime, efforts progress to improve setup times, machine operations, preventive maintenance, and quality control. With enough improvements, it is possible to begin producing in smaller batch sizes and reduce stock levels.

The overall production process is adjusted so the final stage is able to produce at a uniform rate and follow a level schedule. If several models are to be produced, then the length of production runs for each must be reduced to eventually allow for MMP. Start with daily runs (one model per shift), then twice-daily runs, then hourly runs, and so forth, working for the goal of single-unit model production.

The layout of the process, including routing, material traffic flow, and space requirements should be scrutinized: Does the routing flow make sense? Does the location of machines, facilities, tools, equipment, and stock areas make sense? Revise the routing and layout to minimize waste (handling, inventory, waiting, etc.). Once the process is able to produce in small batches and follow a somewhat level schedule, the next step can begin.

Step 2: Introduce the Pull System

Given that setup times, product quality, and equipment reliability are sufficiently improved, a pull system using containers, cards, deduct points, and so on, as described in this chapter and Chapter 8 can be phased in. Usually it is easiest to phase in a pull system stage by stage, starting at the final stage of the process and working backward. After the container size is determined, the number of cards can be estimated, and stock areas and kanban signal methods (e.g., kanban mailboxes) can be set up. Initial trials will usually reveal problems with machine changeovers and setups, quality, and work layout (the "rocks," or problems hidden, covered by inventory described in Chapter 3). These are handled by holding extra stock everywhere, though as the problems are resolved, extra stock is no longer needed.

Pull production cannot handle defective components and materials. If the defect rate is high, the defective items must be first sorted out. Says Hall, the rule of thumb is that the defect rate should be less than 1% and almost never higher than 3%.[11] The goal is zero defects, but companies first adopting pull production rarely have all the PM, setup, pokayoke, product design, and other procedures in place to achieve near-zero defect rates. Eventually, however, through quality of design and mistake-proofing programs, they should be able to reduce defect rates to well below 1%.

At this stage the MRP system is still authorizing work orders and placing orders from suppliers. The BOMs are flattened (usually by coding the intermediate items as phantoms) and the system no longer does detailed capacity requirements planning or issuing of detailed schedules to individual operations.

Step 3: Create a New Layout; Reduce Reliance on MRP

Once the pull system for a product family or mixed-model group has been balanced and demand has been deemed large and somewhat stable, a new layout—a flow line or linked workcells—can be created. As stock areas are reduced in size or eliminated, operations can be moved close together.

The role of the centralized MRP II system continues to be transformed; by the end of this stage the sole functions of the system will be to (1) accumulate forecasts and customer order information and generate the MPSs, (2) release daily production rates (FAS or other daily production schedule) to the final stage of the process, (3) generate and release orders for materials to external suppliers that are not Kanban-linked, and (4) update the remaining inventory records in the BOMs through the postdeduct procedure.

Step 4: Continuously Improve Processes

Everything is expanded on, including improvements to setups, operations, maintenance, workcell layouts, and so forth. With time and experience, the pull system can be expanded to larger portions of the pilot process, and then to other products and processes.

Summary

Production planning and control (PPC) in pull production is handled jointly by centralized and decentralized systems. The functions of forecasting, order entry, rough-cut capacity planning, master scheduling, and requirements planning are performed by a centralized staff in much the same way as in traditional MRP-based PPC systems. Based on demand forecasts and current orders, the centralized system develops monthly MPSs. These are shared with suppliers and with departments and workcells on the shop floor so each can plan ahead to meet the schedules. Preplans are sent months in advance, though final daily schedules might incorporate order information received as late as a week before production.

In pull production, much of the short- and near-term planning and virtually all of the shop-floor control is handled at the factory level. Though the centralized system prepares the level MPSs, supervisors and frontline work teams determine staffing levels and worker assignments. The only daily schedule prepared by the centralized system is a schedule for the final operation. As a result, shop-floor operations perform largely on their own, and, consequently, they are largely self-regulating. This eliminates much of the need for centralized tracking and control systems, which are often ineffective anyway. Materials procurement from suppliers can still be done using an MRP batch-ordering system, though another way is to link suppliers directly to the shop floor with kanbans; this, in effect, extends the pull process beyond the factory.

Companies with ERP/MRP PPC systems can adapt the systems to pull production by modifying product BOMs and inventory records in the computer database, altering the inventory update procedure, and switching from batch-oriented to rate-based MPSs. Since material moves somewhat rapidly through a pull system, the only places where materials are really stored are places where they enter the process or leave the process. WIP, because it is small and transitory, often does not need to be tracked. As a result, the BOM can be flattened, sometimes to include only two levels: raw material and final product.

A simple way of updating inventory records while reducing the number of transactions is to use postdeduct or backflushing: upon completion of a product, the inventory balances for all parts in the product are reduced by the number of parts used in the product. To eliminate double handling and inventory transactions, material is held only in stock areas next to the place where it will be used, the point of use.

Successful implementation of pull production involves matters beyond the shop floor, beyond the PPC system, and even beyond the company. Suppliers too must subscribe to lean philosophy and become partners in the lean production process, the topic of the next chapter.

Notes

1. The most widely referenced materials are Taiichi Ohno, *Toyota Production System: Beyond Large-Scale Manufacturing* (Cambridge, MA: Productivity Press, 1988); and Y. Monden, *Toyota Production System: An Integrated Approach to Just-in-Time*, 2nd ed. (Norcross, GA: Industrial Engineering and Management Press, 1993). Other authors have described in general terms a process for smoothing production schedules and for integrating schedules into a pull production system; it is clear, however, that the model they use is the Toyota process as described by Ohno and Monden.
2. The length of the planning horizon depends on the anticipated month-to-month demand variation. When variation is relatively little, a 2-month horizon might suffice; for products with wide demand swings throughout the year, the horizon should be at least 6 months. This reflects the fact that the greater the anticipated change, the longer it takes the production system to adjust. But there is a paradox here: the farther into the future you look, the less reliable your forecast and, thus, the harder it is for you to plan. Hence, as you plan farther into the future, be cognizant of the fact that your plans will have to be revised, perhaps repeatedly, by the time they are executed. This is simply acknowledging the reality of forecasting and planning.
3. The number of days used by Toyota and Nissan is 10. The longer the time between when customers place orders and when the orders are registered in the order entry system and the more erratic the anticipated customer demand is, the greater the number of days required in the planning period (say, 20 or 30 days instead of 10). However, the problem with increasing the number of days is that relatively fewer actual orders from later days will be included. Thus, in a 30-day order period, there might be few actual orders for the last 10 or 20 days of that period; so as not to underestimate production requirements for those days, it is necessary to pad the demand with forecasts of anticipated orders.
4. Notice that this system presumes the presence of a large backorder. If a large backorder is not accumulated over a 10-day period, then (1) the number of days must be extended (as suggested in note 3), or (2) a make-to-stock philosophy must be adopted. Of course, the more you extend the number of days (option 1), the more of the planning horizon you must base on forecasts (again, note 3), hence, the more you will make to stock. Option 1 becomes option 2, de facto.
5. In fact, as described later, the material requirements estimate might be made in MRP fashion by the centralized system.
6. H. Steudel and P. Desruelle, *Manufacturing in the Nineties: How to Become a Lean, Mean, and World-Class Competitor* (New York: Van Nostrand Reinhold, 1992), 301. This book is a good reference on aspects of pull production, especially capacity planning, detailed cell planning, scheduling, and control; see pp. 261–311.
7. For further discussion of postdeduct and other procedures for inventory accounting and replenishment, see W. Sandras, *Just-in-Time: Making It Happen* (Essex Junction, VT: Oliver Wight Publications, 1989), 198–208.

8. Actually, updating may occur only at discrete times, such as when the MRP system software is run.
9. From S. Flapper, G. Miltenburg, and J. Winjgaard, Embedding JIT into MRP, *International Journal of Production Research* 29, no. 2 (1991): 329–341; R. Hall, *Zero Inventories*. (Homewood, IL: Dow Jones-Irwin, 1983), 271–284; W. Sandras, *Just-in-Time: Making It Happen*, 195–212; B. Williams, *Competitive Manufacturing* (Reading, MA: Addison-Wesley, 1996), 282–284.
10. Sandras, *Just-in-Time: Making It Happen*, 98–200.
11. Hall, *Zero Inventories*, 275.

Suggested Readings

J. Clement, A. Coldrick, and J. Sari. *Manufacturing Data Structures: Building Foundations for Excellence with Bills of Materials and Process Information*. New York: Wiley, 1995.

H. Hirano. *JIT Implementation Manual—The Complete Guide to Just-In-Time Manufacturing: Volume 3. Flow Manufacturing—Multi-Process Operations and Kanban*, 2nd ed. New York: Productivity Press, 2009.

D. Sheldon. *World Class Master Scheduling: Best Practices and Lean Six Sigma Continuous Improvement*. Fort Lauderdale, FL: J. Ross Publishing, 2006.

J. D. Viale. *JIT Forecasting and Master Scheduling: Not an Oxymoron* (Crisp Fifty-Minute Series). Menlo Park, CA: Crisp Learning, 1997.

T. Vollmann, W. Berry, D. C. Whybark, and F. R. Jacobs. *Manufacturing Planning and Control Systems for Supply Chain Management: The Definitive Guide for Professionals*, 5th ed. New York: McGraw-Hill, 2004.

Questions

1. Explain the procedure of preparing tentative production plans for pull production. Describe what happens as these plans become more firm and, eventually, fixed. What are the purposes of longer-range flexible plans versus shorter-range firm plans? What role does ERP/MRP II serve in all of this?
2. Explain the procedure for setting daily schedules based upon initial production plans and incoming customer orders.
3. What is the purpose of sharing tentative production plans and schedules with suppliers?
4. For each of the following kinds of procurement systems, discuss applications and relative advantages and disadvantages:
   a. Supplier Kanban links
   b. MRP systems for daily (or more frequent) order releases
   c. MRP systems for less-than-daily order releases

5. Discuss the relationship between preparing the MPS and shop-floor capacity planning. What role does the agility of the shop floor have in this relationship? How is such agility achieved?
6. Describe the purpose of the visual management system in shop-floor control. Review the ways that such visual management is achieved.
7. What effect does pull production have on product BOMs? Why do products made in pull systems need fewer-level BOMs than the same products produced in MRP-type push systems?

8. Explain the purpose of phantom records.
9. Explain the concept of postdeduct (backflushing). What is a deduct point? What is a deduct list?
10. Explain the difference between a typical MPS and a rate-based production schedule. Why in a pull production system is one preferable over the other?

PROBLEMS

1. Shown below is the BOM for a large motor assembly.

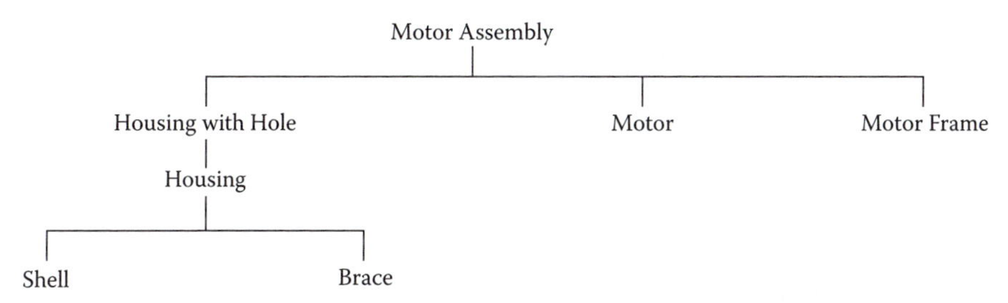

An MRP system is used to schedule each step. Currently for motor assemblies the lot size is 100 units, the lead time is 1 week, and 80 motors/wk are assembled. There are currently 110 units in stock. The MRP record for the motor assembly is as follows:

	Week					
	0	*1*	*2*	*3*	*4*	*5*
Gross requirement (GR)		80	80	80	80	80
Scheduled receipts (SR)						
On hand (OH)	110	110	30	50	70	90
Net requirement (NR)			50	30	10	
Planned order release (POR)		100	100	100		

Assembly of the housing and drilling of the hole in the housing are done in two separate areas of the plant. The final motor assembly is done in a third area. It is proposed that the brace and shell assembly, drilling the hole, and final assembly of finished motors all be done in a single workcell.

a. What would the BOM reflecting this change look like?
b. Assuming now lot-for-lot production and lead time of 0, what would the aforementioned MRP record look like? Use the same tabular format. If the motor housing with hole were phantomed, where in the BOM would the phantom be located?
c. What would the MRP record for the brace look like? Assume a lot size of 30, on hand inventory of 80, and lead time of 1 week.

2. At Gornigs Corp., pull production with visual shop-floor controls is being implemented for two products, ARM and CARM. Following are the BOMs for the products under the existing push production system:

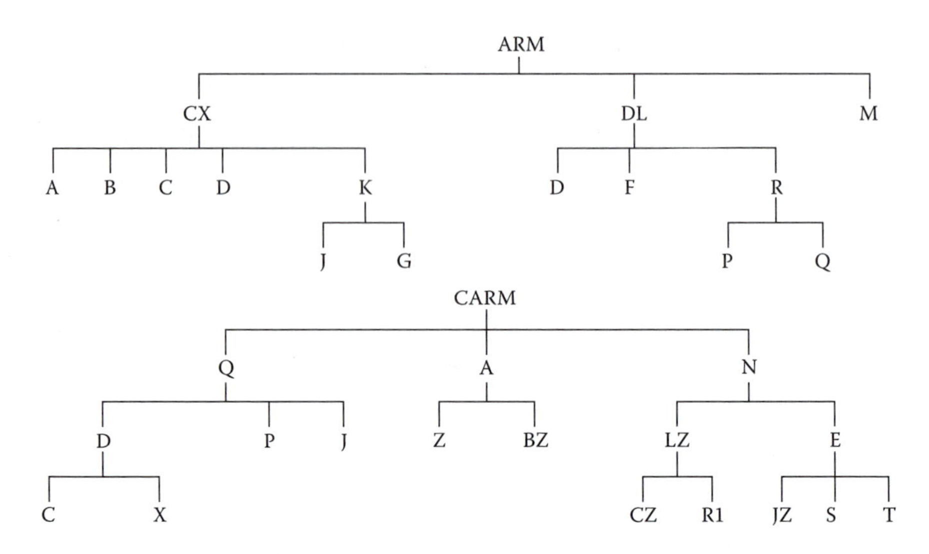

a. Assuming both products will be produced repetitively with short CTs, explain how the material should be tracked through the system.
b. Reduce the BOMs to reflect the tracking system you proposed in (a).
c. For spare parts, the DL component in the ARM product must be produced in excess of the amount required for ARM. Describe how this should be handled by the material tracking system and what effect it would have on the BOM you created for ARM in part (b).

3. Following is the BOM for product X.

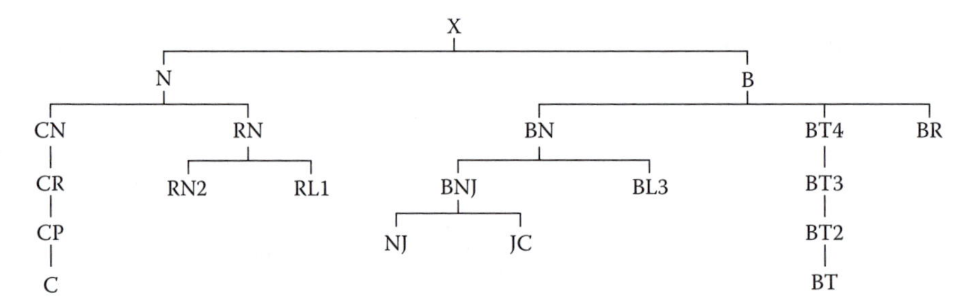

The plant is converting to cellular manufacturing, with stock areas of the cells linked by kanbans. The switch to cells will affect the following operations:
- All operations to convert item C into item CN will be performed in machining cell 1.
- All operations to convert item BT into BT4 will be done in machining cell 2.
- The assembly operations that create part B will all be done in assembly cell 3.
- The assembly operations that create part N and final product X will all be done in assembly cell 4.

a. Draw a flow diagram illustrating the movement of materials and evolution of the items in the BOM as they go through the cells.
b. Draw a diagram illustrating the floor plan of the new cellular system. Show the location of the four cells, the operations in each cell (use intermediate levels on the BOM as surrogate operations), all places where inbound and outbound stocks of material are held, and places where materials are transferred from one cell to another. The cells should be arranged so as to minimize intercell distances for transferring material.

c. Draw the new BOM to reflect the cellular/pull process.
d. Suppose the original BOM structure is to be retained. Show where phantom bills should be placed to give the effect of a flattened BOM.

4. The July rate-based MPS, and the demand for August are as follows:

Row	*July*	*Aug*
1. Forecast	3,700	4,000
2. Backlog	37	?
3. Anticipated demand	3,737	4,020
4. Working days	21	20
5. Production rate	177	?
6. Production units	3,717	?
7. On hand	0	0
8. Available units	3,717	?
9. Ending units	(20)	?

Determine the production rate for August and complete the MPS.

5. The rate-based MPS for December and demand for January are as follows:

Row	*Dec*	*Jan*
1. Forecast	15,600	12,400
2. Backlog	990	?
3. Anticipated demand	16,590	12,400
4. Working days	15	21
5. Production rate	1,026	?
6. Production units	15,390	?
7. On hand	1,200	?
8. Available units	16,590	?
9. Ending units	0	?

Determine the production rate for January and complete the MPS.

6. Suppose the demand forecast for February is 17,000 units, and 860 units are on backorder.
a. Assuming nineteen 8-hour working days in February, what daily production rate is required?
b. What is the required CT of the process? Assume 480 min/day.

c. The bottleneck operation in the process is a manual assembly procedure that takes 2 minutes. How many workers should be assigned to that operation?
d. By what amount would the workday length have to be increased, starting on the sixth working day, then held the same for the remainder of the month, such that at month's end there would be no backorders? Company policy is to give workers 10 working days notice before increasing the length of the workday. Given the 10-day policy, can they implement this change?
e. Suppose that actual new customer commitments (excluding backorders) as of February 1 indicate for the month a daily demand of 850 units for the first 7 days, then 940 units for the next 7 days, then 1,100 units for the remaining 5 days. To retain the existing process CT, by how much must the current 8-hour working day be increased to meet demand for the remainder of the month? Assume the same 10-day policy as stated earlier, and that the change will not go into effect until the eleventh working day, but then will stay the same for the duration of the month. Assume no backorders remaining at the end of the month.

Chapter 16

Lean Production in the Supply Chain

I am he as you are he as you are me, and we are all together.

—The Beatles, "I Am the Walrus"

All organizations rely on suppliers for inputs, and an estimated 60% to 70% of the final cost of manufactured items is from purchased materials, components, and services. Manufacturers depend on suppliers not only for materials, but also for machines and tools, spare parts, and services that keep machines and processes functioning.

The performance of the customer–supplier network, called the **supply chain**, affects the competitive advantage of every company. Each company is constrained by the other companies in the chain, even when it comes to improvement efforts. Potential cost and lead-time savings and quality improvements at a company can be easily outweighed by the cost, lead time, and defect increases of its suppliers and distributors.

Lean production acknowledges the importance of the supply chain to competitive advantage and stipulates that improvement efforts must not be confined to a company, but must be extended to its suppliers and distributors. Extending improvement efforts beyond a company requires that managers take a new look at the costs and tradeoffs of working with suppliers, and that customer and supplier companies adopt new ways of working together.

This chapter covers a range of topics related to supply chain management: new customer–supplier relationships, the changing role of the purchasing function, and application of lean concepts throughout the supply chain. It ventures into the territory of **logistics management**, which is managing the movement of materials from suppliers to customers and at distribution points between them. Logistics management is undergoing a revolution of sorts, partly because of the growing numbers of customers that are becoming lean/just-in-time (JIT) companies, and, consequently, the growing quality and service requirements being imposed on suppliers and third-party carriers.

Produce Versus Buy

An organization relies on suppliers for parts, products, and services. While some manufacturers produce many or most of the parts and components that comprise their products, other manufacturers produce very few or none, and simply assemble parts and components made by suppliers. While potentially a manufacturer has much to gain by outsourcing production of parts and components, it also has something to lose.

Relying on Suppliers

Purchased materials are a major source of variability in the cost, quality, and delivery of finished products. They account for 50% of the quality problems in manufactured items and as much as 75% of all warranty claims.[1] Clearly, a manufacturer cannot provide its customers with high quality, low cost, and quick delivery of its products or services without receiving the same from its suppliers.

In the past, manufacturers tried to control part variability by producing (rather than buying) the needed parts themselves. An extreme case is Ford and the monstrous River Rouge plant that Eijia Toyoda visited in the 1950s. It produced virtually 100% of a car's parts, even the steel and glass for those parts. But the fact is, a small company that is highly specialized in only one or a few products can usually produce parts better and cheaper than a large company that tries to produce everything. Why?

Breadth sacrifices depth. Trying to produce everything requires huge resources, and few companies, including even the most resourceful, can muster the expertise and resources necessary to produce everything well. On the other hand, a supplier that focuses on just one kind of product has relatively more resources to devote to producing it and to perform the research and development necessary to keep improving it. Also, the supplier likely produces parts for many customers and can achieve economies of scale that one company producing for itself cannot. A reputable supplier may have learned to manufacture an item for low cost and high quality, the result of expertise that has taken years of hard work to accumulate. A customer who outsources to such a supplier reaps the benefits of that expertise and years of hard work.

Core Competency

Although much can be gained by outsourcing production of parts or components to valued suppliers, a producer must be careful not to outsource its own core competency, its raison d'être, that is, whatever its own customers most recognize it for. Automakers, for example, are recognized as being just that—makers of automobiles. But making an automobile is not the same as making the thousands of parts that go into it, and core competency in auto making lies in the design and assembly of the overall auto and of major systems such as engine/fuel, suspension, electronics, frame, and body. The automaker must have competency both in design innovation and applications of new materials and technologies for these systems, and in the integration of those systems into the overall vehicle design. By entrusting to suppliers the design and production of subcomponents and parts not close to its core competency, the automaker reserves more of its own resources to devote to developing critical skills and capabilities necessary to keep enhancing its core competency.

Again, Toyota is an example. Although it outsources production for 70% of its vehicle components, it works hard to retain competency in the technology of vehicle production and of some of the key parts that it outsources. If a technology is considered essential to the vehicle, then Toyota believes it should master it. For example, a critical component in the engine for the Prius hybrid car is a semiconductor switching device. Toyota believes that hybrid technology is the wave of the future, and even though it had no prior expertise in semiconductors, it built a new plant to produce the component rather than outsource it. Toyota believes that to effectively manage suppliers it is necessary for it to stay at the forefront of the technology of those suppliers, and only after it has mastered a technology will Toyota seek to outsource it.

In 1988 the company opened an electronics plant because over 30% of its vehicle content is electronics-related and, the company believes, to maintain its destiny as a leader it must stay abreast of that technology. Management felt that too heavy a reliance on electronics suppliers is risky and, hence, the need to accumulate the competency in developing innovative electronics technologies and applications for Toyota vehicles.[2]

The underlying theme of all this is focus. The more focused you are, the better job you do in your area of focus. Just as manufacturers do a better job by subdividing their production facilities into focused factories, so they reap benefits by procuring inputs from suppliers who are focused on what they do. The crucial issue is knowing the difference between what to focus on and what to let others focus on.

Supply Chain Management

The supply chain is the multitiered system of suppliers through which materials and information flow, ultimately reaching the final manufacturing customer (illustrated in Figure 16.1). Suppliers that provide materials and information directly to the final manufacturing customer are called

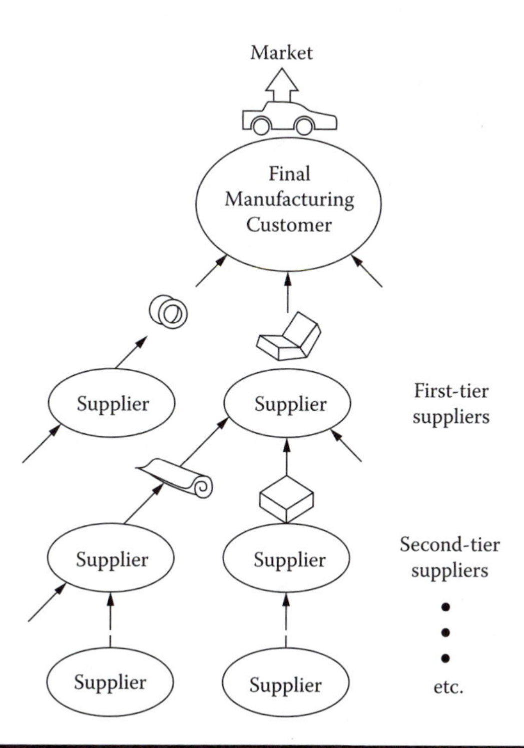

Figure 16.1 The supply chain.

first-tier suppliers. Each first-tier supplier is itself a customer supplied by companies from the next lower tier. Timely, right-quantity and right-quality deliveries to customers at every tier depend on the performance of suppliers at the tiers below them. Consequently, the only way a company anywhere in the chain can meet its schedule, quality, and quantity requirements is for suppliers at every lower tier to meet their requirements also. Such is the thrust of **supply chain management**: coordinating the activities of suppliers to meet the requirements of the customers above them in the supply chain.

Process and Customer Focus

There are two key features to a lean supply chain. First, every company in the chain recognizes that it is processes, not isolated functions, that create value in products and services. The manufacturing process extends beyond the producer's factory doors and includes the processes of its suppliers, its suppliers' suppliers, on down to suppliers at the lowest tier of the chain. Each supplier is an **external factory** whose processes must be coordinated with its customer's processes.

Second, every company in the chain is customer focused, which means it understands the wants and needs of its customers—an understanding that goes beyond knowing its delivery, quality, quantity, and cost requirements and includes the markets, processes, culture, and constraints of the customer. This understanding includes recognition of who the customer is, because the real customer is not only the company with whom the supplier does direct business, it also is its customer's customers, customers above that, on up to the customer at the top of the supply chain. Every supplier has a hand in contributing value to the final product at the top of the chain, and when customers of that final product are not satisfied, the implications ripple down to impact companies everywhere in the chain. Obviously, loss of business at the top of the chain means loss of business to everyone.

Process orientation and customer orientation is the result of teamwork. At successive tiers up and down the supply chain, companies integrate their processes and develop shared goals, plans, and schedules that mutually benefit them and the companies they supply. Companies in the chain come to think of themselves as part of an industry group that competes with other industry groups for market share. Industry groups able to coordinate the design, manufacture, and logistics of products for the final market have great advantage over groups that do not work together. The evidence is the early competitive successes in U.S. markets by Japanese automobile and electronics companies, each of which is part of a Japanese system for that control financing and material flow through tight customer–supplier relationships. The system, called *keiretsu,* gives big customers large control over suppliers, but also ensures that suppliers' needs are provided for. Today, many of the principles of supplier–customer relationships practiced worldwide derive from customer–supplier relationships as practiced by Japanese companies.

Customer–Supplier Relationships

Lean production philosophy recognizes that for customers to acquire the best purchased items, it is often necessary for them to work with suppliers to help make them the best. Working with suppliers means many things, but it especially means joint problem solving, practicing quality at the source, and exchanging information.[3]

Joint Problem Solving

Joint problem solving takes many forms. Suppliers and customers participate in product design and concurrent engineering, setup reduction and preventive maintenance (PM), and implementation of cellular manufacturing and pull systems. Companies solicit from their suppliers comments and suggestions about the final product and sometimes give them considerable leeway in parts design by providing only functional requirements, then entrusting them with the entire detailed design and manufacture of the part. To reciprocate, the customer works with the supplier to help the supplier resolve cost or quality problems, and find better ways to meet the requirements.

Quality at the Source

Quality at the source begins with quality at the supplier. When a supplier guarantees 100% quality of incoming materials, the customer can eliminate inspection of arriving materials. Incoming materials can be moved directly to points of use, and buffer inventories to cover defects can be eliminated. High quality requires high process capability, and the customer helps its supplier achieve high capability by sharing its own experience and expertise (assuming it has already achieved its own high capability). As described later, companies continue to work with only the few suppliers who are able to continuously meet tough requirements.

Information Sharing

Companies and their suppliers mutually exchange information about long-range plans, production schedules, design changes, and problems. A company gives advance notice about changes in product mix and product demand so its suppliers have time to adjust. Suppliers give advance notice about quality or delivery problems so the customers can help them find solutions and develop contingencies.

For customers and suppliers to embrace these changes requires nothing short of fundamental change in the way they customarily relate to each other. It requires moving from an adversarial-type relationship to a more trusting, partnership-type relationship.

Partnership Relationships

Adopting a process orientation and customer orientation requires moving away from traditional arms-length, short-sighted, customer–supplier relationships. In the old system of production, customers were distrusted and detached from suppliers. For every part it needed a customer had several current or potential suppliers. The singular purchase criterion was lowest price, and the customer played one supplier against the other. In such an environment, suppliers and customers are adversaries, and suppliers feel no obligation to provide anything beyond the minimal required quality and service. The system keeps prices down but tends to drive up inventory, ordering, purchasing, and quality costs; beyond that, it mitigates against improvements in the quality and delivery of parts, services, and the final product. Such a system cannot compete with a system wherein customers and suppliers develop relationships that provide for mutual gain. That kind of relationship, called a **partnership**, completely changes the nature of business. Differences between this partnership form of relationship and the traditional, adversarial relationship, listed in Figure 16.2, are discussed in the following sections.

	TRADITIONAL	*PARTNERSHIP*
Purchase Criteria	Lowest bid	Competency, ability, capacity, and willingness to work with customer to improve price, quality, and delivery.
Design Source	Customer	Customer and supplier
Number of Suppliers	Several for each item	One or few for each item or commodity group
Customer Business Volume per Supplier	Limited: multiple suppliers share business	High: one or few suppliers get all of the business
Type of Agreement	Purchase order; contracts to meet immediate requirements	Contract plus agreement about working relationship
Terms of Agreement		
Duration	Short-term, or as needed by customer	Long-term, multiple years
Price/Cost	Lowest bid, inefficiencies, and waste keep prices/costs high	Negotiated price/cost savings from supplier improvements shared with customer
Quality	Variable; customer relies on incoming inspection	High; quality at the source; supplier uses SPC, TQM, etc.
Shipping: Frequency/ Size/Location	Infrequent/large/dock or stockroom	Frequent/small/point-of-use
Order Mechanism	Mail or phone	FAX, phone, EDI, or kanban
Customer–Supplier Interaction	Formal information exchange, limited to customer requirements; no teamwork; supplier service limited to minimal requirements	Frequent formal and informal exchange of plans, schedules problems, ideas; teamwork and mutual commitment based on trust; cooperation to resolve problems and improve supplier's products and processes

Figure 16.2 Customer–supplier relationships.

Purchase Criteria

Price will always be a major criterion for purchase decisions, but beyond price, additional criteria must be scrutinized for the long-term cost implications. Even though a supplier might not have the lowest price, other factors can make it the lowest *cost* supplier. These include the supplier's core competency and skills, current and future production capacity and capability, ability to meet immediate and projected quality and delivery requirements, and willingness (win–win attitude) to work at continuous improvement and partner in design and production.

Design Source

A major source of continuous product improvement is improvement of product parts and components. The traditional relationship stunts such improvement; for example, because parts are designed by the customer, the supplier merely has to produce them to specification. It does not matter whether the supplier knows a way to improve the design; reduce its cost; or increase its durability, reliability, appearance, or manufacturability. In the traditional relationship, the customer is not interested in the supplier's ideas.

In a partnership, the supplier is recognized for competency in design as well as production. The customer develops the requirements and specifications for the part, then expects the supplier to do all or some of the design for the part. Suppliers are included in concurrent engineering teams to capture their ideas during early product development. A case in point is Ford, where major body and interior components, lights, carpeting, and plastic parts are designed and produced by suppliers. One carpet supplier pointed out that Ford was using five different colors of carpet in the trunk, although color did not matter to most customers and 7% could be saved on carpet cost by using one color.[4]

Number of Suppliers

Traditional purchasing practice is to use multiple suppliers for every kind of part needed. The practice stems largely from distrust or inability of any one supplier to provide parts in the specified time, quality, and quantity; as mentioned, it also keeps prices low through supplier competition.

Multisourcing, however, prevents any one supplier from getting the lion's share of customer business, which prevents it from realizing savings in economies of scale, some of which could be passed to the customer. Further, it inspires low supplier confidence about long-run, sustained business with the customer, which discourages the supplier from doing anything extra to improve equipment, processes, or technology that might benefit the customer. In addition, suppliers to companies using multiple sources tend to provide parts that serve the needs of all of them, rather than specialized parts that would better serve particular needs better. Although the parts are suitable for a broad range of customers, they are not as good as if the supplier had focused on the needs of just one customer.

Another problem with multisourcing is that it increases the overall variability in quality. As Figure 16.3 shows, even when the variability in quality of parts from each supplier is small (dashed lines), variability of the parts from all suppliers (heavy line) can be large. This concept also applies to variability of other things such as delivery schedules, delivery quantities, and supplier responsiveness.

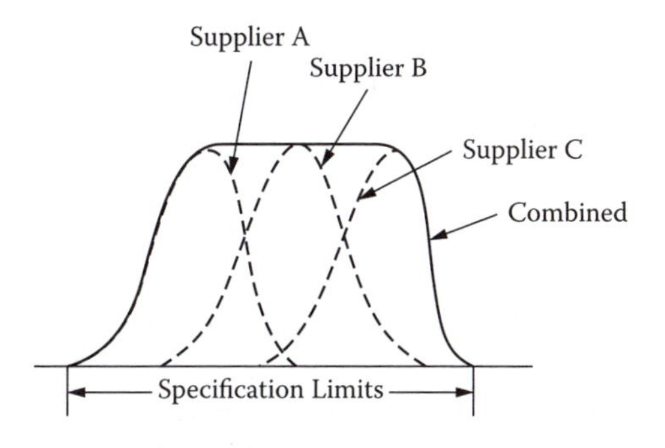

Figure 16.3 Variability from multiple suppliers.

Further, as mentioned earlier, when a customer discovers a nonconforming or defective part it is difficult to pinpoint from which supplier the part came. A company with many suppliers for a part becomes preoccupied with the part's conformance to specification. When it is already difficult to enforce conformance to specifications, improvement within specifications is unlikely.

In a partnership the number of suppliers is reduced until only the best few remain; the goal, ultimately, is to reduce the number to a single or **sole source**, the one supplier recognized as being most capable of supplying a particular part. To guard against interruptions in deliveries, two or more suppliers with similar capabilities are retained. Each of these is considered as a **single source** in that, under normal conditions, one supplier provides, say, parts A, B, and C, and the other provides parts D, E, and F; both, however, have similar capability, in which case business can be shifted entirely to one supplier if the other has an emergency. Honda in France does this by ordinarily using one supplier for right taillights and another for left taillights.[5] Companies that make the same products at multiple plants also use single sourcing wherein each plant has but one supplier for a part, but the supplier is different for every plant.

Since it is likely that workers in the plant of one supplier are in the same union as workers in the plant of the two or three alternate suppliers, a strike will shut down all of them. When a strike is imminent, there is still the old standby: build inventory.

Both customers and suppliers benefit from single-source relationships. The supplier gets *all* of the customer's business, and that is the incentive for the supplier to work harder to retain that business. If the agreement is a long-term contract, the supplier has the incentive to invest in the technology and processes to make parts so good as to virtually guarantee contract renewal.

Type of Agreement

Traditional purchase agreements are short term, apply to a single order, and are limited to definition of price, specifications, quantity, and delivery date. For each purchase, a list of suppliers is reviewed, one is chosen, and the order is sent. To keep suppliers on the list on their toes, orders are rotated among them. The practice is continued even with **blanket purchase-orders**, which are agreements that cover purchases for an extended period of time, by holding similar agreements simultaneously with multiple suppliers.

One way that partnership agreements differ from traditional contracts is that the customer and the supplier mutually agree on aspects of the relationship. The agreement is separate from a procurement contract for a particular commodity. It lays down ground rules the parties must follow, mutual expectations, considerations, and duties as part of the relationship. The agreement can be far reaching in scope; for example, it might state that the supplier agrees to continuously improve its quality, service, and productivity and reduce its inventories and delays. Expectations regarding price, quality, and delivery are also covered, though broadly since the emphasis of the agreement is on working relationships, not conditions for particular orders.

Terms of Agreement

The major differences between partnership agreements and traditional contracts are in the time period covered and expectations about product pricing, quality, deliveries, and order placing.

Duration. Partnership agreements are long-term. Unlike procurement for a one-time order, a partnership extends for at least 1 year. One year is actually too short because it is not enough time for parties to develop the relationship and bend to accommodate each other's needs. Agreements of 3 to 5 years are becoming more common. Suppliers are monitored for performance and the agreement terminated if expectations are not met.

Price/cost. Prices are negotiated. Savings achieved in the supplier's operations are shared with the customer. A proposed price increase by the supplier is not accepted without explanation. Schonberger[6] mentions the case about a Polaroid supplier that asked for a price increase to cover higher-cost materials. Polaroid buyers went to the supplier's plant and were able to find ways to reduce the supplier's process cost to offset the higher material cost.

Quality. Partnerships practice quality at the source. The supplier guarantees 100% quality, which obviates the need for incoming inspection by the customer. The customer assists the supplier in the pursuit of quality and improvement: it sends its own quality experts to develop, certify, and coach the supplier in quality methods, it furnishes training (plan–do–check–act [PDCA] and statistical process control [SPC], for example), and it provides technology (pokayoke devices, automatic inspection, and control equipment). Motorola's suppliers participate in a series of Motorola-run courses on topics such as basic problem solving, design for manufacture and assembly (DFMA), JIT, and Motorola's Six Sigma and Total Customer Satisfaction programs for total quality management (TQM).

Shipping. Although the traditional practice of shipping infrequently and in batches to fill a truck keeps unit shipping costs to a minimum, it results in higher inventory holding costs for the customer. In a partnership, the supplier makes frequent, small shipments to conform to the customer's immediate needs. In theory, a pull production plant should be supplied on demand with small shipments from suppliers as an extension of the pull process.

To reduce material handling time and cost, customers redesign their receiving areas with multiple docks so incoming material can be quickly moved to points of use. In many cases, supplier drivers are authorized to unload their shipments onto carts and move them to the points of use.

Even for the nonpull customer, frequent deliveries in small quantities are good because they reduce variability in delivery times, incoming inventory, and the need to carry safety stock. For example, the variability in shipping time for a supplier that promises two deliveries, one between 10 a.m. and 12 noon and another between 3 p.m. and 5 p.m., will be much less than the variability for a supplier that promises one shipment between 10 a.m. and 5 p.m. Hewlett Packard stipulates that suppliers must meet a daily delivery window; if they miss the window more than three times a year, they risk losing the contract.

So that small deliveries will be economical, the supplier is expected to adjust aspects of its operations, such as to locate a plant closer to the customer, use smaller vehicles, or schedule deliveries along multisupplier/multicustomer routes (milk runs). These concepts are discussed later.

Ordering and vendor-managed inventory. Often, the supplier is expected to keep track of the customer's on-hand quantity and to determine the amount to ship each delivery. The concept of the supplier managing its customer's inventory is called, appropriately, **vendor-managed inventory**. Empty Kanban containers are one mechanism for order placing: When a driver arrives with a shipment, he checks for empty containers, which he is authorized to replenish on the next delivery. At a higher level, called **electronic data interchange (EDI)**, supplier and customer computers are linked to enable direct ordering, billing, and even funds transfer for payment between them. The supplier can determine at any time the customer's most recent sales and demand projections, and know exactly how much to deliver.

Customer–Supplier Interaction

Partnerships foster interaction between people at all levels of the customer and supplier organizations. Frontline supervisors, shop workers, and managers in the customer plant are encouraged to meet their counterparts in the supplier plant. The concept of people dealing with people is exploited to diminish the mind-set of one company dealing with another company. When customer and supplier organizations see each other's company in terms of the people who work there, they become more sensitive and responsive to each other's ideas, needs, and constraints. A delivery driver who makes daily trips to the point of use in a plant becomes a potential source for suggestions about ways the supplier might improve its service or ways the customer might take better advantage of that service.

Likewise, when shop-floor workers from the customer company meets their counterparts at the supplier company each group develops an understanding about the other group's needs, which helps them determine ways to better produce and use the parts.

Sometimes a supplier devotes a focused factory entirely to one customer, and workers and managers there deal exclusively with workers and managers in the customer's plant they supply. The supplier's process is a direct extension of the customer's process, and the people in the two processes work together. When the customer business is large enough in terms of shipments, volume, and commitment, the supplier might locate a small plant near the customer. This is a common practice for high-volume suppliers in the automobile and aerospace industries. The ultimate is when the supplier locates a plant within the customer's plant to produce components or subassemblies for immediate use. Such a concept is now underway on a large scale in a Volkswagen truck making plant in Brazil. The following case illustrates some of the aforementioned points.

Case in Point: Supplier partnerships at Xerox[7]

Xerox's North American operation began an initiative with suppliers of sheet metal parts focused on improving parts quality and delivery performance and reducing parts costs through better utilization of supplier assets. Sheet metal parts represent about 6,000 part numbers and an annual cost of $100 million. The cost of a part depends heavily on the overhead burden allocated to it, and in many cases overhead is 40% to 60% of the total cost. One of Xerox's goals was to increase its supplier's business so overhead costs could be spread over a higher volume and the unit costs could be reduced. Higher volume would come from increased business from both Xerox and other customers. Xerox prefers that its share of a supplier's business not exceed 50% and, ideally, be 30% to 40%.

Xerox had 40 sheet metal suppliers. As a first step, it formed a team with the task of cutting that number to 10. To decide which 10, the team obtained data from all the suppliers on labor and overhead costs, parts quality, management attitude, delivery performance, and technical capability. It rank-ordered the suppliers, then met with the top 10 to discuss their willingness to participate in a partnership program. As part of the program, Xerox would increase its business to the suppliers, but the suppliers would be responsible for increasing their non-Xerox business, both to reduce dependency on Xerox and to increase the volume over which to spread overhead costs. Two of the suppliers declined, so Xerox invited suppliers 11 and 12, both which accepted. The remaining 30 firms were phased out as the older Xerox products that used their parts were dropped.

Another feature of the program is that Xerox provides suppliers with improvement tools ranging from simple checksheets to activity-based costing software. It runs seminars for the suppliers on topics ranging from leadership to benchmarking, and routinely visits the supplier plants. Xerox also sets performance goals for the suppliers. The following goals are typical:

- Guarantee fewer than 300 ppm (defective parts per million)
- Reduce average annual price 5%
- Deliver newly order products within 7 weeks
- Meet JIT delivery schedules
- Participate in continuous improvement programs
- Use standardized containers for materials shipped

Each year Xerox evaluates supplier performance against best practices and adjusts the goals to meet the best-practice levels of its competitors in the marketplace.

It Doesn't Come Easy

Some experienced manufacturing executives think that establishing partnerships is the hardest part of lean production. Although many suppliers readily embrace the idea or start out ahead of the customer in terms of understanding its merits, others reject it out of distrust or resistance to outsider meddling with their business. As Robert Hall says,

> The relationship between the companies must be built with people-to-people bridges at several points—line management, engineering, quality organization, [and] it is a long developmental journey to this state of thinking if companies are just emerging from order-at-time haggling and expediting.[8]

Richard Dauch, former executive VP of manufacturing for Chrysler, said that his own experience in forming new supplier relationships was not all milk and honey:[9]

> There was strong resistance ... We instituted a system of chargebacks for vendor failures to reinforce our recommendations, proposals, requests, and requirements. We did meticulous cost studies to calculate the total expense to Chrysler when a vendor failed to perform [and caused such problems as] line stoppages, ... inventory holding costs, record keeping, sorting, reworking, unpacking, repacking ...We grabbed the supplier's attention by subtracting these costs from submitted invoices. Once we had their attention, we got action!

Given the typical adversarial relationship between customers and suppliers, it is not surprising that many companies are skeptical about forming partnerships. One party cannot force another into a partnership—although, as Dauch implies, big companies can exert pressure on their suppliers. But they have always been able to do that. The difference to a supplier in a partnership is that, although they still get pressured, suppliers each end up with a bigger share of the customer's business.

Suppliers must be willing to develop measures to meet customer requirements, and the customer must be willing to assist suppliers and adjust requirements occasionally until the suppliers can meet them. As Hall states, "The bottom line to [a partnership is] an appreciation that suppliers and their manufacturing customer are not really in competition with each other. The real competition is for the customer who uses the end item produced."[10]

Small-Customer, Big-Supplier Partnership[11]

Not every manufacturer is a big company like Toyota or Xerox that has flocks of suppliers lining up to form partnerships. Small companies with relatively little market power are likely to hear their suppliers say, "This is what we do. Take it or leave it." Placed in that situation, how does a small company motivate a supplier to team up with it and make frequent, small deliveries of reasonably priced, defect-free items?

Part of the answer, it appears, is to use multiple sources, which is contrary to recommended partnership practice. In a survey of 70 small lean manufacturers, 77% had no intention of reducing the number of suppliers in the next 3 years.[12] Being small customers, they felt, competitive bidding is still the best way for them to obtain good quality, delivery, and price. These companies feared that single sourcing and high dependency on one supplier would reduce their negotiating power with each supplier.

There are, however, two strategies that small companies can use to increase the chances that their big-company suppliers will be willing to meet their lean requirements. First, they can use suppliers that advocate and practice lean/TQM, or are industry or ISO certified. A supplier that is itself practicing lean will be more sensitive and responsive to a customer that is trying to do the same. Even if a company does not use lean practices, being ISO or industry certified makes it more likely to be capable of meeting a customer's quality requirements than one that is not certified. Of course, ISO certification is no guarantee, and every company must be evaluated individually. A big supplier that is certified might be unwilling to enter into a full-scale partnership with a small customer, though it might agree to a limited partnership. The second strategy is for the small company to maintain long-standing relationships with all its suppliers. Many big companies sustain themselves by serving a multitude of small customers, and they value customers that give them regular business, even though the business is small. A longstanding relationship allows the customer more opportunity to learn about the supplier's strengths and weaknesses, to get to know the supplier's representatives, and, through the representatives, to coax the supplier a little to adopt ways that improve its responsiveness to the customer's needs.

Supplier Selection

Suppliers are selected based upon two broad measures: product design and manufacturing process. The supplier provides the customer with a design, model, or prototype of the item it proposes to make, and the customer then tests this design to determine if it meets requirements. For the supplier to be selected, however, passing design tests is usually not enough since the design or prototype alone says nothing about the supplier's capability to produce the product. As a result, the supplier selection process usually puts heavy weight on the supplier's process capability. Since it is easier to improve a product design than to revamp a company's manufacturing process or service philosophy, a supplier with an inferior design but a superior process is often selected over a supplier with a better design but a poorer process.

Certification

The procedure for assessing a supplier's capability to meet delivery, quality, cost, and flexibility criteria is called **certification**. The emphasis in certification is on processes and existing or proposed programs for continuous improvement, variance and waste reduction, and implementation of lean practices. Ford designates suppliers as Q1 to indicate they are certified. A supplier that has been certified is put on the company's short list of suppliers. Buyers and managers using the list have the assurance that suppliers will provide incoming parts that do not need inspection and service that meets minimal standards.

Certification by Customer[13]

A customer team comprised of members from purchasing, manufacturing, engineering, and accounting decides certification status. The team reviews a supplier's past performance on the same or similar products using information from buyer organizations and industry, trade, and consumer groups. It also surveys the supplier's technological and process competency, and managerial and financial strengths and weaknesses using, at one extreme, a simple questionnaire, or, at the other, a team visit to the supplier's site.

As an example, a producer of medical devices used the following questions to survey potential suppliers:[14]

- Have you received the product requirements and do you agree that they can be fully met?
- Are your final inspection results documented?
- Do you agree to provide the purchaser advance notice about production problems, schedule changes, or product design changes?
- Describe the air-filtration system you use.
- What protective garments do your employees wear to reduce contamination?

The most comprehensive form of survey is when the team visits the supplier's plant, observes its processes and procedures, reviews documentation, and interviews workers and managers. Depending on the product and industry, the team might look at the following specifics:[15]

- Management: Philosophy, organization structure, commitment to continuous improvement
- Product design: Design organization, design systems used, merit of specifications, use of change control methods, developmental and testing laboratories
- Manufacturing: Physical facilities, PM and machine availability, special processes, process capability and flexibility, production capacity, lot identification and traceability, PPC systems, setup times, transfer distances, lead time and inventory reduction techniques, employee skills and attitudes
- Quality assurance: Organization structure; quality assurance personnel; quality planning; methods of quality analysis and corrective action; disposition of nonconforming items; training, motivation, and involvement of workers; control over subcontractors and suppliers
- Data management: Facilities, procedures, effective usage of reports
- Performance results: Performance levels attained, prestigious customers, prestigious suppliers and subcontractors, reputation among existing customers

The on-site survey should be comprehensive enough to enable a conclusion about the supplier's competency to meet product requirements. Surveys that focus solely on organization structure, documentation, and procedures are inadequate.

Upon completion of the survey, the team reports to a management group its findings about the supplier's strengths and weaknesses, effectiveness and need for assistance, and whether the supplier should be certified.

Certification by Industry Standard or Award

The assessment for certification process can be time consuming and burdensome to a supplier, especially to one that is being surveyed by multiple customers using multiple assessment criteria. A supplier trying to comply with a dozen customer surveys can get bogged down, as can a customer who is surveying a dozen suppliers. An alternative is for the customer to rely on industry certification standards, for example, ISO 9000 and ISO 14000, which are standards for quality management and environmental management, respectively. Professional societies also certify companies; for example, the Society of Manufacturing Engineers offers Lean Gold, Silver, and Bronze certifications.

With industry certification, the assessment is performed by a third party (ISO, SME, etc.), not a customer, and all suppliers meeting the assessment criteria appear on a published list. Rather than assessing and certifying each supplier, a company accepts the assessment criteria of the ISO registrars, SME award examiners, and so forth, and chooses suppliers only from those that have been certified or received an award.

Evaluation

Companies periodically evaluate their suppliers and provide them with feedback about performance, problems, opportunities, and areas needing improvement. The evaluations are done quarterly, monthly, or sooner if the supplier has problems. The evaluations rate a supplier with respect to minimal customer requirements or against other suppliers. Evaluations that use quantitative ratings are the best because they specify customer expectations in ways that are easily communicated and measurable. Following is an example.

Example 1: Supplier Evaluation Scheme

A supplier is assessed for three criteria: quality, delivery, and service. The maximum rating for quality is 100 and points are subtracted when a supplier fails to meet minimal requirements. Fifty points each are assigned for on-time delivery and service, with points subtracted for deviations from requirements. Thus, there are 200 total possible points for rating the supplier. If the total drops below, say, 160, the supplier must allow the customer to perform an on-site evaluation audit. If the supplier plant is overseas, the audit is contracted to a consultant.

Periodic evaluations provide a performance record that clearly shows upward or downward trends. Performance that is mediocre and not improving (say, 130 on the aforementioned scale) is justification to terminate the relationship. Such decertification implies that the supplier is discontinued as a source until it makes certain improvements.

Suppliers that score well on supplier evaluations are sometimes given special recognition and awards. Top-rated suppliers are awarded trophies, and their managers and workers are given parties or luncheons. Some corporations take out full-page ads in publications listing their best suppliers.

Suppliers rated as superior (through evaluation by prominent customers or in third-party reviews) do not have to wait for new customers to come around. Being certified has marketable value, and certified suppliers seeking more business often solicit potential customers by advertising the areas in which they are strong, admitting to areas where they are weak, and promising to keep improving in both places.

Purchasing

In many companies the only function that deals with suppliers is purchasing. Traditionally, the role of this function has been to process and purchase orders and requisitions, reconcile information about incoming items, solicit and evaluate bids, and prepare contracts. The role has included dealing with problems and complaints about unmet needs from internal and external customers of procured items. Buyers on the purchasing staff are sometimes specialized to handle procurement of a particular commodity, component, finished good, or service, and deal with suppliers in particular technologies or industries. In general, the more advanced or complex a manufacturer's product and its components, the more essential it is for buyers to have specialized expertise.

Whereas buyers used to operate independently of other functional areas, in lean they participate as members of concurrent engineering teams. Buyers know about materials and components, their sources and costs, and can provide engineers with comments and suggestions about ways to improve product quality and manufacturability. Participation in concurrent engineering is one of many ways the purchasing role is changing.

Evolution of Purchasing[16]

In the days before MRP systems a planner would break down a product's build schedule into the product's constituent parts, then a buyer would locate and negotiate with suppliers to make those parts. When MRP arrived, the parts breakdown was computerized, but buyers still negotiated with suppliers and determined the size and timing of parts orders. With MRP II, detailed orders and their release dates were determined automatically but were sent to suppliers that buyers had identified and contracted with.

In a lean facility, managers and buyers in the purchasing function still handle supplier contracts, but individual replenishments are signaled by Kanban and actual orders are placed by frontline workers on the shop floor. A pull production system connects the customer's production area to the supplier's production area, and only exception orders are handled by the purchasing function. Figure 16.4 illustrates this by showing a Kanban label with replenishment information and a frontline associate pointing a handheld scanner at a UPC on the label. The device transmits an order signal that is relayed directly to the supplier. In similar fashion, replenishment orders to suppliers can be signaled directly using RFID kanban tags on containers or on the individual parts themselves.

Clearly, with the advent of MRP II and pull production much of the responsibility for actual material ordering and receipt has shifted from the purchasing area to manufacturing. Nonetheless, the role of the purchasing function is expanding as companies put more energy into concurrent engineering and supply chain management and expect purchasing to be active in both.

Role of Purchasing in Lean Production

In lean companies, the primary role of the purchasing function lies in three areas: specifying requirements, selecting suppliers, and managing supplier relationships. Purchasing is responsible for clarifying specifications imposed on the supplier's production processes and quality control procedures. These requirements are essential when the customer's products and components must meet tough requirements as, for example, in pharmaceuticals and sophisticated electromechanical systems.

	F.F. # 3	
	265507	
CIRCUIT BOARD UNI-MA TCH 24 VOLT		
	Vendor: Beltronles	
Reorder Point	720 Pes.	
	12 Boxes	
Kanban	960 Pes.	
Lot Size Qty.	16 Boxes	265507
Quantity Per Box	60	
Kanban card. After scanning return this card to part location.		

Figure 16.4 Associate scanning a Kanban card (shown, upper) to initiate an order.

Purchasing communicates to suppliers the product specifications, and selects candidate suppliers. For specific contracts, purchasing negotiates the terms for pricing, quantity discount prices, return policies, warrantees, delivery conditions, and penalties for failure to meet quantity, quality, and delivery terms. For companies seeking partnerships, purchasing seeks suppliers willing to work with the company, then assists in final selection. Purchasing sometimes also assists a supplier in developing capabilities it currently does not possess.

Finally, purchasing is the major information conduit between the customer and suppliers. It provides suppliers with production schedules so they are able to plan deliveries, it assesses supplier performance, and works with suppliers about areas in need of improvement. In many companies purchasing also evaluates suppliers' processes and capabilities and certifies suppliers.

Although purchasing's first responsibility is to the customer (the company to which it belongs), it also has responsibility to the supplier. When a supplier has a valid complaint about, for example, the customer making frequent changes to product designs or order schedules—changes that affect the supplier's ability to meet customer needs—purchasing will often represent the supplier and argue its case. One U.S. automaker used to send a supplier a 6-weeks advance schedule on production requirements, but the advance notice was useless because the actual schedule was changed daily.[17] The supplier never knew for sure how much it had to produce until the schedule arrived at 6 a.m. each day. Purchasing presented the problem to the customer's production managers, who worked out a procedure to reduce the fluctuations in the daily schedules.

Lean in the Supply Chain

Whether applied to the distribution of final end products or the transport and handling of raw materials and parts between tiers, application of lean production methods reduces waste and improves service everywhere in the supply chain. Customer–supplier partnerships spread the use of these methods to companies at every tier and improve performance of the entire industry group.

The usual handling–transport–distribution process includes numerous steps, most of which, as shown in Figure 16.5a, are nonvalue added. If the transport part of the process involves multiple carriers or rerouting and redistribution at intermediate places between supplier and customer, it too will have numerous nonvalue-added steps. One function of lean in the supply chain is to eliminate these steps so the process looks more like Figure 16.5b. This section illustrates lean methods for making such improvements.

Facilities Layout

Materials shipping, handling, and receipt are big sources of supply chain waste. Materials sit idle and require double handling because of poor layout in receiving and shipping areas. Such waste can be reduced by replacing the usual single set of docks with multiple docks at the start and end of every process. Figure 16.6 shows the layout of a large plant for an appliance manufacturer. The plant is divided into focused factories that each produces a family of appliances. The plant has few walls to obstruct the movement of materials; equipment and point-of-use storage areas can be easily relocated, as needed. Each focused factory has it own receiving and shipping docks located at the front and back ends of the process. Materials unique to each product family arrive at one dock, and shipments of finished appliances depart from the other. Commonly, materials arrive at the

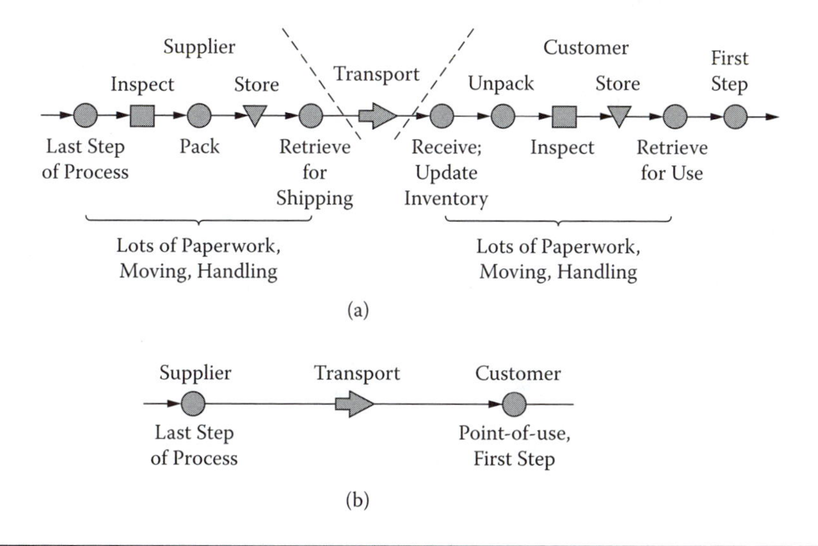

Figure 16.5 Waste at the handling–transport–distribution process.

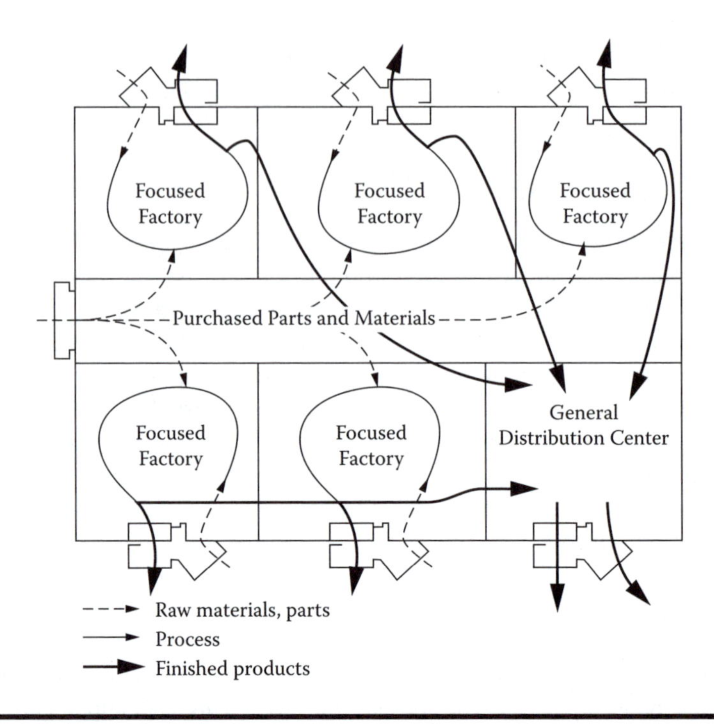

Figure 16.6 Plant layout organized for minimal waste in shipping, receiving, and handling.

central receiving dock in small containers and are moved by handcart to points of use. Finished products to be shipped in mixed-product groups are routed to the general distribution center where they are grouped by customer order and loaded onto trucks. Incoming deliveries of parts and materials and outgoing shipments of finished products are made in small, frequent amounts. Except in the distribution center, there is little inventory in the plant.

The time and cost of storage and handling can also be further reduced by having delivery trucks drive into the plant and to points of use. The example in Figure 16.7 shows items being off-loaded from a truck onto a mobile gravity conveyor chute, on which they slide directly to small storage areas next to the assembly line where they are needed.

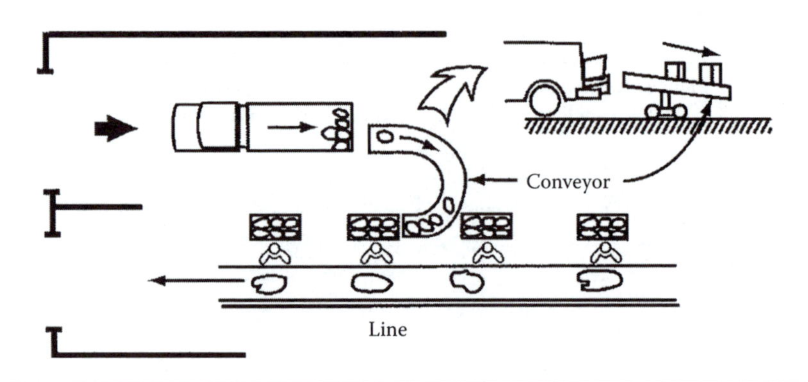

Figure 16.7 Truck delivery at points of use.

Teamwork

It used to be that a truck driver would drive a truck, then stand around waiting while dock workers loaded and unloaded the truck and dispatchers prepared freight bills. That is now changing at some companies where drivers are responsible for preparing freight bills and helping load and unload items (depending on the item). At one distribution company, drivers are responsible for calling customers in advance about weights and destinations of daily pickups. The company holds a Saturday breakfast four times a year for drivers to meet and voice ideas and opinions.[18]

Docks and distribution centers are frequently scenes of confusion and sources of waste and costly errors. Trailers newly loaded are mistakenly unloaded, are loaded with the wrong items, are sent to the wrong destinations, or sit ready to roll but waiting on paperwork. The confusion often stems from dispatchers, schedulers, dock workers, and drivers working autonomously and not as a team. Like the factory floor, docks and distribution centers can benefit from teamwork and workers sharing opinions about ideas and problems.

Setup Reduction and Small-Batch Shipping

The cost of fuel, tolls, and a driver's wages is roughly the same for a given route whether a vehicle is fully loaded or partially loaded, so full-load shipments are preferred over partial loads because the per-unit hauling cost is less. Thus, analogous to lengthy equipment setups and large batch sizes, it would seem that shipping large quantities on an infrequent basis makes more economic sense, even if the resulting greater on-hand inventory at the destination does not.

To make small-quantity, frequent deliveries economical, there are two general approaches: reduce the fixed cost of hauling or schedule enough customer–supplier stops to fill a truck. The fixed hauling cost is a function of distance and vehicle operating cost. To minimize distance-related costs, some companies only contract suppliers located nearby—within a maximum shipping radius of, say, 100 miles. To reduce operating costs, haulers use vehicles of size tailored to daily shipping requirements. Small vans are used for multiple, small-quantity, short trips, while large trailers for large, infrequent or long-distance loads. With a small vehicle it can be more practical to make five daily shipments instead of one large weekly shipment.

The major cost in hauling is the driver, and that does not change much if the truck is large or small; thus, there are obvious advantages to using large trucks and filling them up, especially for long routes. Companies employ several strategies to fill trucks and, at the same time, satisfy their customers' demands for small, frequent deliveries and pickups. Figure 16.8 shows examples of these **milk runs** or routes for daily deliveries.

- In (a), the supplier fills a truck, which then makes deliveries to multiple customers.
- In (b), the truck is filled from a cluster or string of multiple suppliers for delivery to a cluster or string of one or more customers.
- In (c), the truck is filled from a cluster of several suppliers for delivery to a remote, common customer.
- In (d), a truck going to a customer meets a truck coming back from the customer at the halfway point, where drivers switch trucks and return back home. This approach, which can be incorporated into (b) and (c), is good for long distance routes because drivers do not have to spend nights away from home.

Customers that receive multiple daily shipments from many suppliers can also direct the shipments into a hub for freight consolidation. At the former NUMMI automotive plant in Fremont,

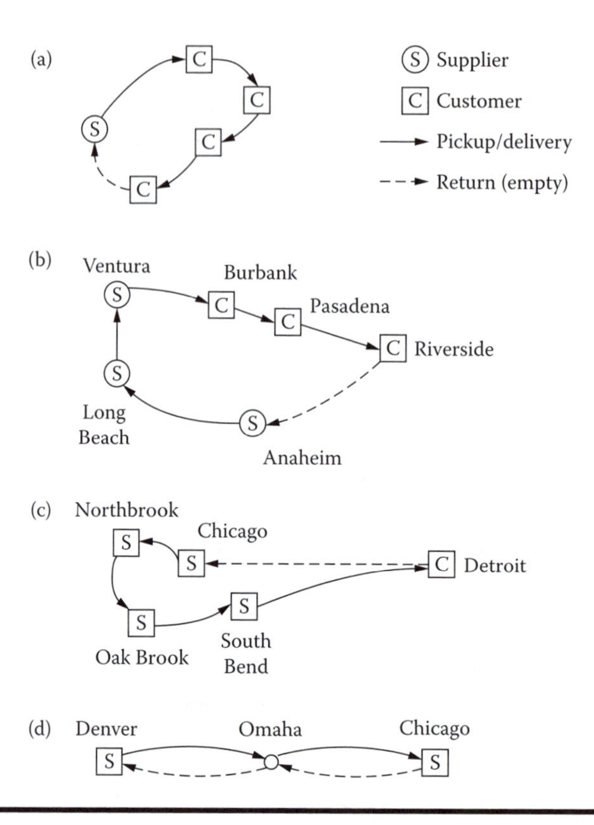

Figure 16.8 Configurations of daily milk runs.

California, shipments from Midwest suppliers were directed to a rail hub in Chicago for daily drop-off. The drop-offs are consolidated for shipment by train and arrived in Fremont 70 hours later.[19]

When trucks or rail cars from multiple suppliers make deliveries to multiple customers, items must be loaded in a sequence such that they can be unloaded at each destination without wasting time shuffling items to access the right ones. One approach is to load the items such that an open path remains to items that must be unloaded first. Another is to use vehicles with roll-up sides so items can be loaded and unloaded in any sequence. The same sequencing consideration applies especially to mixed items for delivery to a single customer. For example, seats delivered to an automaker should be arranged so they can be unloaded (using a system like in Figure 16.7) in the sequence (color, fabric, etc.) that matches the mixed-model sequence on the assembly line of cars in which they will be installed.

Case in Point: JIT Distribution at Wal-Mart

Wal-Mart application of cross-docking and hub-and-spoke distribution are examples of supply chain waste cutting. With **cross-docking**, a concept developed by Wal-Mart, goods are continuously delivered from suppliers to Wal-Mart warehouses where they are picked, repackaged, and dispatched to individual stores. The warehouses use laser-guided conveyor systems that read bar codes on incoming goods and direct the goods to trucks going to the right destinations. On average, goods spend only 48 hours or less at a warehouse—the time it takes to move them through the process from one loading dock to another.

Wal-Mart warehouses are strategically located to serve stores within a 150 to 300 mile radius. This **hub-and-spoke** approach allows a single truck to resupply two or three stores each trip. Each warehouse serves as the distribution hub for about 175 stores, and 85% of all Wal-Mart goods are routed through this hub-and-spoke system. The process is very effective: whereas industry-wide distribution costs average 4.5% to 5% of sales, Wal-Mart cost is less than 3%.

When production CTs are short, loading and unloading of trucks is the bottleneck that holds up shipments. Increasing the number of trucks but keeping the number of drivers the same can remove the bottleneck. For example, Figure 16.9 shows an empty truck (6) waiting at the customer and a loaded truck (3) waiting at supplier B. When trucks 2 and 5 arrive at supplier B and the customer, respectively, their drivers switch to trucks 6 and 3. By the time these trucks get to suppliers A and C, trucks 1 and 4, respectively, will have been loaded and are ready for the arriving drivers, who switch trucks and drive them to supplier B and the customer. The approach not only reduces overall transportation time, but it also increases the flexibility of the system to satisfy varying levels of customer demand. The situation is analogous to workcells: during low customer demand, one driver cycles through the entire route by switching trucks; for greater demand, two or more drivers switch trucks.

On most shipping routes there is the problem of **backhaul**, which is that the vehicle must return to its point of origin, which doubles the one-way time and cost of every delivery. The problem is solved when the vehicle carries a billable load on the return trip. Circular routes are desirable (Figure 16.8a) because they reduce the length of the return route, though another alternative is to contract customers that need goods shipped to destinations along the return route. Since trucks returning from lean/JIT customers must carry empty kanban containers, the containers should be "nestable" or collapsible to allow room for goods to be delivered on the way back.

While it is common for large customers to maintain their own fleets of vehicles for making daily deliveries and pickups, the capital equipment costs and logistical complexity of lean/JIT networks (examples in Figure 16.8) often mandate that shipments be contracted out to experience

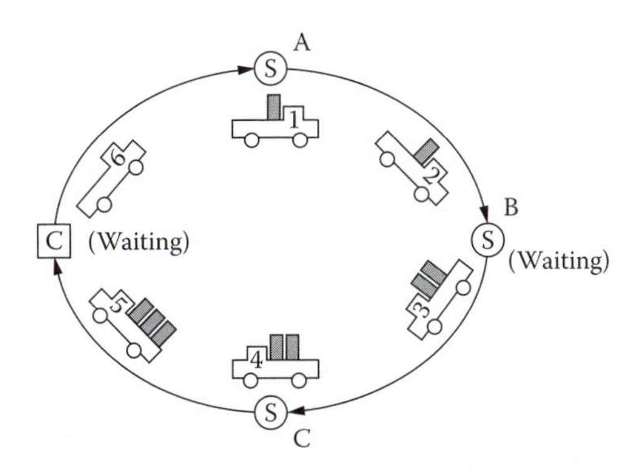

Figure 16.9 Decreasing delivery times: six trucks, two drivers.

carriers that have preexisting shipping and distribution networks. As in choosing suppliers, carriers should be selected on the basis of ability to meet requirements and become certified. Also like suppliers, the number of carriers should be kept to a minimum so each gets a larger share of the business and gives better service in return.

Preventive Maintenance

The role of preventive maintenance and total productive maintenance (TPM) in the supply chain is to eliminate breakdowns of vehicles (trucks, forklifts, rail equipment) during operation, increase vehicle safety, and tailor vehicles for better service (quicker loading and unloading, better loading organization, quicker freight processing, and quicker repairs). Many TPM concepts from the factory floor can be applied to transportation, such as prescheduling preventive and predictive maintenance during nonwork hours; training drivers to enable them to perform basic vehicular inspection, simple maintenance tasks, and basic roadside repairs; and improving the design of truck beds, doors, and lift gates based on suggestions from drivers, customers, and suppliers.

Kanban

The Kanban procedure can readily be extended to external suppliers. Whenever the customer empties a container of parts, that container and attached kanban card signal an order to the supplier. Empty containers picked up by a driver on each daily delivery are returned to the supplier, filled, and delivered to the customer on the return trip. The supplement at the end of this chapter gives one procedure for determining the number of kanbans in a supplier-linked pull system.

When the travel time between customer and supplier is long (more than a day or so), an electronic Kanban system can be used. When a container is emptied, the customer scans the UPC code on the kanban card that transmits information to the supplier about what is needed, how much, and when (Figure 16.4). Electronic kanbans are especially appropriate for high-volume, repetitive-use items.

Harmon and Peterson give an example of a supply chain connecting an auto assembly plant and four tiers of suppliers: seat assembler, seat cover manufacturer, fabric mill, and thread suppliers. Between the seat-cover manufacturer and the fabric mill is a supplier-linked Kanban process, illustrated in Figure 16.10:

> [The] seat cover plant starts the [kanban] process by taking a roll of fabric from focused factory storage for cutting and sewing [1] … the bar coded [kanban card] on the roll is read [2] … this information is accumulated daily and transmitted to the fabric supplier [3], where the information is used to withdraw the required number of rolls from storage for shipment [4]. The [kanban cards] on the outbound rolls are read … and are transmitted to [the seat cover plant] to update its inventory status [5].[21]

Information from the kanbans on outbound rolls also signals the supplier about rolls withdrawn from its outbound stock that must be replenished. The fabric suppliers are located at a 30-hour drive from the seat cover plant. Before Kanban, shipments were made weekly and, thus, inventory at the seat cover plant averaged half the weekly demand. With electronic Kanban, requirement updates and deliveries are made daily, so the inventory average is one-half the daily demand.

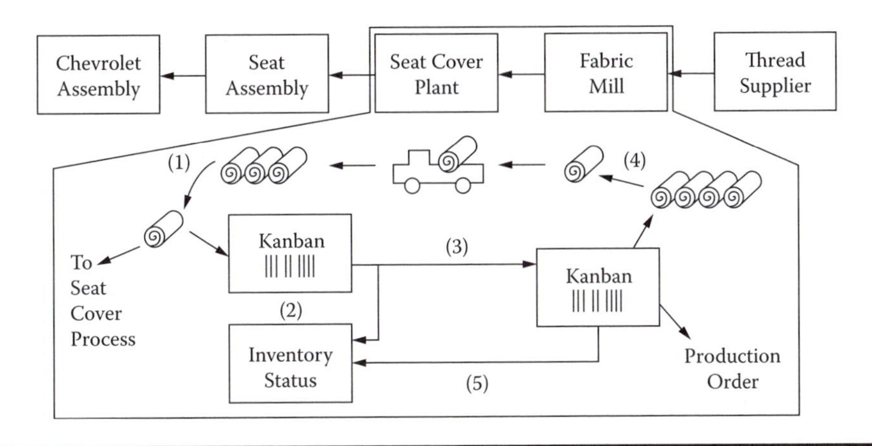

Figure 16.10 Electronic Kanban for replenishment orders.

The size and shape of Kanban shipping containers are jointly determined by the customer and supplier to facilitate ease of use, handling, storage, and shipment. The containers are usually small (the guideline being no more than one-tenth the daily demand) for easy handling and standardized in size and shape so they can be used with many different kinds of parts. If the containers are partitioned, then counting the parts in a container is trivial, and the total quantity in a shipment can be determined just by looking at the number of containers. Figure 16.11 shows Kanban containers for incoming material arranged for visual inspection. Upside-down containers signal authorization for replenishment.

Figure 16.11 Vendor-managed stock area arranged for simple visual assessment of on-hand inventory.

Before suppliers are able to convert their own operations into small-batch, quick-response production, they must, in effect, carry in finished goods inventory stock that formerly was the customer's raw materials inventory. Though this might appear as a bad thing from the supplier's point of view, from a lean supply-chain point of view it is not so bad for a number of reasons, described by Schonberger:[22]

- Transportation costs and return costs are avoided when orders are canceled.
- Materials need not be warehoused by the customer but can be delivered directly to the point of use as needed.
- Goods are less likely to be damaged at the supplier's warehouse than when sitting on the shop floor prior to use.
- The cost of the supplier holding the stock is an incentive for the supplier to develop its own lean ways to reduce inventories.
- The amount the supplier must carry will be small if the customer maintains stable order schedules and gives the supplier adequate advance notice about changes. This point is elaborated upon next.

Communication and Scheduling

Supplier production schedules are driven by a customer's monthly, weekly, or less-frequent orders. A supplier that values a customer's business and does not want to risk running short must maintain a stock of finished goods items to absorb fluctuations in customer orders. The greater the fluctuations, the more stock the supplier must carry. In a lean environment, unless the supplier knows customer demand in advance or can count on somewhat uniform demand in the short run, it will have to carry even more inventory. This, however, is not what you call good supply chain management: the supplier is not being treated as a partner and the system still has inventory waste, only it is at a different place (at the supplier). To enable a supplier to meet customer demand without carrying inventory requires three things: (1) the customer sharing its plans and schedules, (2) production leveling, and (3) point-to-point communication.

Share Plans, Maintain Uniform Schedules

Pull production in a factory requires that operations and cells in a factory be notified in advance about the anticipated production level and product mix so they have time to adjust the required level of capacity. After that, the schedule is held somewhat uniform. Daily fluctuations occur, but every attempt is made to minimize them.

Extending the pull process beyond the factory, the same protocol should apply to scheduling orders for suppliers. The customer notifies the supplier of the estimated mix and volume of orders 1 to 2 months in advance, depending on the lead time for the supplier to acquire materials and adjust its own capacity. This advance notice gives the supplier time to start doing whatever necessary to meet the requirements—add workers or hours, transfer workers, cut back or build inventory. Once the customer commits to a monthly requirement, it sends orders that are somewhat uniform and coincide with the supplier's uniform production schedule. There is, in fact, an interesting dynamic here: If the customer does not first level its order pattern (if its orders remain erratic), then the production and delivery pattern of the supplier will be erratic, too, and that will make it more difficult for the customer to level its production—a circular process. As Hall says,

If the supplier—one of many—does not deliver or if its material is unusable, either the customer process stops completely or a schedule change goes out to many other suppliers to try to keep everything running. In this chicken-or-egg situation, a uniform load schedule comes first. Without it, there is no pattern to work against, however poor the actual performance to schedule might be.[23]

Point-to-Point Communication

The customer sends to its suppliers monthly forecasts, plans, and anticipated daily level schedules from its centralized planning system. Variations from level daily orders, however, as well as quality complaints, emergent production problems, or suggestions for improvements are handled locally. Workers or supervisors at an operation or workcell in the customer's plant deal directly with workers in the cells or operations that supply them.

One way to facilitate communications between the customer and the supplier is to link their respective planning and control computer systems. Direct-link electronic data interchange (EDI) enables each party to access portions of each other's systems, which eliminates delays in sharing schedule updates.

Milliken and Company, the textile and chemical manufacturer, synchronizes everyone in the supply chain, starting with bar-code point-of-sale information from retailers to schedule its production. By using point-of-sale data from sample retailers to schedule production, the lead time between a retailer store ordering fabric and receiving it has been reduced from 18 weeks to 5 weeks. Walmart has its own version of this kind of system, called Retail Link, an EDI system that connects stores, warehouses, and suppliers, and processes all purchase orders, changes, and confirmations. Retail Link is a proprietary system that enables any Walmart supplier to determine at any time, 24 hours a day, the number of its products being sold. Retail Link eliminates much of the guesswork for suppliers in planning for production quantities and cuts days off delivery times.

Point-to-point communication not only reduces information time lags but also information distortion. Updates in forecasts and plans for the final customer (top tier) are communicated *directly* to suppliers at every tier of the chain, which can then begin to make immediate adjustment for them. The alternate—the usual stepwise, tier-to-tier downward communication—results in a progressive worsening of distortion in supplier requirements at lower tier levels, shown in Figure 16.12. Even when the schedule at the top level is relatively uniform, just one tier later it becomes choppy, and two tiers down, very choppy. The increasing distortion of order information and resulting production schedules at successive tiers of the supply chain is called the **bullwhip effect**.

According to Harmon and Peterson, the factors in the supply chain that amplify this distortion are:[24]

- Safety stocks
- Lead time allowances in excess of actual process time
- Inaccuracies in inventory and requirements records and transactions
- Lot sizing
- Effects of scrap and rework
- Planning/scheduling systems with large time buckets (weekly)

They describe a GM division that was able to largely eliminate supply chain distortion by adopting daily scheduling time buckets and electronic Kanban between some tiers in the supply chain.[25] Because direct communication between the top of the chain and all suppliers eliminates

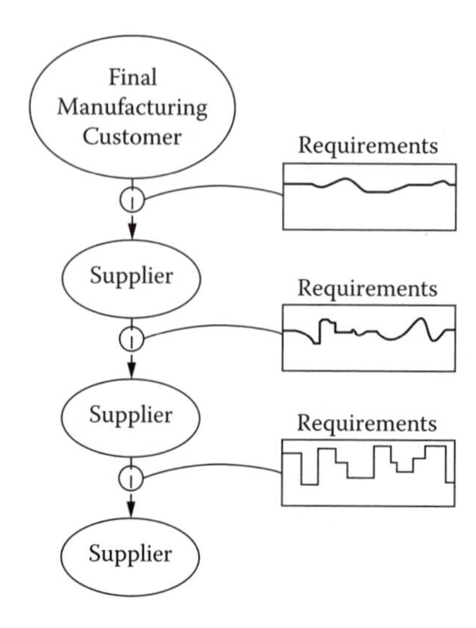

Figure 16.12 Production requirements as passed tier to tier.

most of the planning uncertainty for suppliers, it obviates the need for suppliers to rely on safety stocks and safety lead times—two big sources of requirements distortion.

Electronic communication, however, does little to facilitate local-level, person-to-person contact. For that purpose, less sophisticated and less expensive technologies (fax, telephone, e-mail, in person) are adequate. In fact, relying solely on sophisticated EDI and not establishing direct links at lower levels can be an expensive mistake. Technology alone is not the answer. Partnerships require building relationships, without which the most advanced communication technology will be for naught.

The following case exemplifies lean applications in a supply chain partnership.

Case in Point: Supply-chain JIT at Dow Chemical Company

Cook and Rogowski describe a customer–supplier partnership initiated by Dow Chemical Company with one of its customers.[26] As part of a corporate initiative to adopt lean methods in its operations, Dow invited the customer to participate in a pilot effort to reduce supply chain waste.

The product Dow supplied to the customer is anhydrous hydrochloric acid (AHCl). Prior to the partnership, the customer determined its requirements for AHCl by the usual process of forecasting demand and developing a master production schedule (MPS). By referencing BOMs and inventory status, the requirements were translated into purchase orders that were released to Dow via EDI. Upon receipt of the order Dow would pump the requested volume of AHCl into Dow-owned tank cars.

Shipments between Dow's plant in Texas and the customer plant in Michigan involved four railroads and three switching points to transfer tank cars from one railroad to another. Both Dow and the customer tracked the shipments. Although demand at the customer's plant averaged two tank cars a day, an average of 16 cars was held as buffer. The value of the excess 14 tank cars of AHCl was $280,000; the value of the 14 cars themselves, $1,540,000.

Dow and the customer formed a team to find ways to reduce the uncertainty and variability in demand forecasts and shipping lead time. The team changed the customer's forecasting procedure and increased the accuracy of its BOMs and inventory status files. It improved the flow of sales order information through the customer's departments and Dow's own marketing department and increased the reporting frequency of information to adjust AHCl requirements. As a result, 3-week forecast accuracy was increased from 50% to 70%; 2-week accuracy, from 60% to 82% percent; and 1-week accuracy, from 75% to 95%.

Before the partnership, average delivery lead time was 8 days with a standard deviation of 3.1 days. Much of this time was spent at rail switching points. With help from the railroads the team identified a shorter route (less miles) that involved three railroads instead of four. This eliminated one switching point and shortened the average lead time to 6 days and the standard deviation to 1.4 days. Also, the team assigned tracking exclusively to Dow, which eliminated the customer expense of tracking. Dow agreed to send the customer weekly reports about tank car locations and expected arrival times.

Because of improved forecast accuracy and decreased lead time, the customer was able to reduce its inventory buffer from 16 tank cars to 6 and save over $880,000 in working capital. The team is now trying to reduce the buffer at the customer plant from six tank cars to two. Dow has improved its responsiveness and delivery reliability to the customer and has a much better understanding of the customer's markets, operations, and concerns.

Getting Started: Begin at Home

Suppose a company intends to adopt supply chain management concepts and expects its suppliers to adopt lean techniques like those described. A word of warning: First the company had better make sure that it, itself, is already far along in having adopted lean philosophy and implementing lean techniques, especially regarding things it expects from its suppliers. Not doing so makes no sense. Deliveries of defect-free materials to a company that itself has poor quality procedures will make little difference in the final product. Frequent small-batch deliveries of parts to a company that produces in large batches will only increase the buildup of incoming parts, paperwork, and material handling costs. Besides, frequent or daily deliveries are ineffective if order lead times are long. Far better then daily deliveries for orders placed a month earlier are weekly deliveries for orders placed 1 week prior. Customers that are struggling with their own lean efforts will understand this; nonlean customers won't.

Summary

The supply chain is the network of customers and suppliers wherein each provides the other (directly or indirectly) with materials, components, services, and orders. Since every organization relies on suppliers, efforts to improve competitiveness must extend beyond the organization and to its suppliers.

By outsourcing a company can often get goods and services that are higher quality and lower cost than if it tried to produce them itself. But the benefits of outsourcing depend in part on the nature of the relationship between the customer and the supplier. By adopting a partnership-type relationship, both the customer and the supplier have something to gain. In a partnership, customers can expect more from suppliers in terms of a better, faster, cheaper, and more agile response to their demands. The partnership may include sharing information with suppliers about plans and products, or giving suppliers more say in design decisions. When a company develops a partnership, it also reduces its suppliers to a relatively small number. Many companies have only one or two suppliers for each item needed: thus, the supplier in a partnership is rewarded with longer-term, higher-volume business. Good customer–supplier relationships begin with both the customer and the supplier realizing they are not competing with each other but are together in competing against other customer–supplier groups.

The purchasing function, the traditional linkage between customer and supplier companies, serves an important role in supply chain management. Formerly restricted to finding new suppliers, processing purchasing orders, and soliciting and evaluating contracts, the role of the purchasing function has been expanded to include participation in concurrent engineering teams and assessment and certification of suppliers.

Most of the lean principles and techniques for reducing waste in factories can be applied to most places in the supply chain, including the distribution channels linking customers and suppliers. Part of the purpose of partnerships is to spread the technology of continuous improvement and waste reduction to every tier of the chain. Concepts such as focused factories, employee involvement, setup reduction, small batches, teamwork, preventive maintenance, and Kanban offer opportunities to improve logistics and distribution performance everywhere.

For customer companies to receive JIT service they must provide suppliers with information about demand forecasts, projected orders, and current and projected production requirements. At the least, a customer must share information about anticipated future demand so its suppliers are able to adjust their capacity. In the best case, companies at all tiers of the supply chain receive demand information directly from the customer at the top of the chain. Direct transmission reduces the information time lags and distortions that occur when orders move through the supply chain, company by company, tier by tier. Direct information from the final customer allows all tiers of the supply chain to work together in a synchronized fashion and to everywhere minimize lead times and reduce the costs of moving, handling, and storage.

Appendix: Supplier Kanban[27]

An **S-kanban** or **supplier kanban** serves a purpose similar to the conveyance or withdrawal kanban (C-kanban) discussed in Chapter 8. It authorizes withdrawal of containers from an outbound buffer to replenish a customer's inbound buffer, though here the outbound buffer represents finished goods at a supplier's plant. A difference between a C-kanban and an S-kanban is that, timewise, the former authorizes replenishment on an as-needed basis, while the latter authorizes replenishment at fixed, periodic time intervals as negotiated by the customer and the supplier.

Although this system relies on fixed-interval deliveries, it can accommodate some variation in demand through variation in the number of containers authorized for replenishment each delivery. Ordinarily, the number of containers needing to be filled in a given period depends on the number of containers consumed in the previous period.

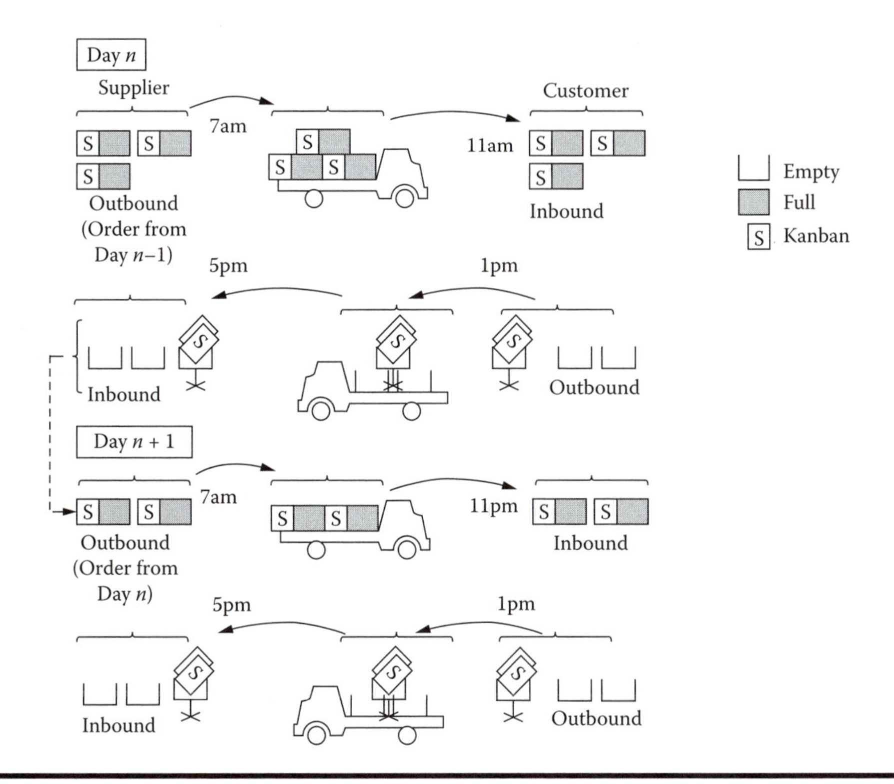

Figure 16.13 Example of S-kanban.

To illustrate, suppose the customer and supplier negotiate a delivery interval of 24 hours. As shown in Figure 16.13, each morning (say, at 7 a.m.) a truck departs the supplier plant and delivers full containers and their S-kanbans (based on the *previous day's* order) to the customer's plant (arriving, say, at 11 a.m.). Later that day (say, 1 p.m.) the truck departs the customer plant with empty containers and attached S-kanbans and drops them off at the supplier (5 p.m.). The next day the process repeats. The number of full containers delivered is based on the number of empty containers dropped off the day before.

If a supplier replenishes just one kind of item, there is no need for a kanban card since the containers themselves can serve as order cards. But when it replenishes more than one kind of item with the same containers, separate cards are necessary to specify the number of containers of each item to replenish.

The number of S-kanbans is determined in a similar way to the other kinds of kanbans, namely,

$$K_S = (D/Q) \times C \times [(1 + I)/N]$$

where

C = round-trip **travel CT** between the customer and the supplier, including loading and unloading time, rounded up to the nearest full day

N = number of daily conveyances (withdrawals and deliveries from/to the customer),

I = supplier production lead time expressed in terms of the number of delivery intervals between conveyances

Example 2: Computation of Number of S-Kanbans

Suppose the travel time each way between customer and supplier is less than one working day. The supplier needs one day to fill any order. The number of negotiated deliveries is three per day. Daily demand is 60 units and standard container size is 5 units.

During the one day required to fill each order, three deliveries occur, so I = 3. Rounding travel time up to the nearest full day, C = 1. Therefore,

$$K_S = (60/5) \times 1 \times [(1 + 3)/3] = 16 \text{ S-kanbans}$$

At any point in time, 4 of the kanbans will be at the customer's plant (1/3 of the daily requirement), while the other 12 will be en route or at the supplier plant (the supplier is on average working on three, 2-kanban batches). The 16 S-kanbans allows no safety margin, so another 2 kanbans might be added for a total of 18.

From a supplier's perspective, S-kanbans are like P-kanbans because each authorizes production to fill a container. The time to fill any order, 8 hours in the example, is the time allowed to fill any number of containers ordered (it is the rough equivalent of production lead time). Though the number varies from delivery to delivery, in the long run the average order quantity is equivalent to the average demand, and, knowing the latter, the supplier is able to compensate for small variations. The assumption, however, is that the customer is attempting to comply with a somewhat uniform schedule, and orders to suppliers will also be somewhat uniform.

Although this kind of delivery system requires a supplier to commit to regular, short-interval deliveries, it affords the supplier a degree of certainty in predicting order requirements. If the customer expects demand to change dramatically, then it must inform the supplier far enough in advance so that production lead times can be renegotiated and the number of S-kanbans readjusted.

Notes

1. J. Juran, and F. Gryna, *Quality Planning and Analysis*, 3rd ed. (New York: McGraw-Hill, 1993), 313.
2. J. Liker, *The Toyota Way* (New York: McGraw-Hill, 2004), 209–10.
3. S. Melnyk and D. Denzler, *Operations Management: A Value-Driven Approach* (Chicago: Irwin, 1996), 419.
4. Lessons from Detroit: Get Supplier Involved Early, *Purchasing*, (Oct. 19, 1995): 38–42.
5. R. Schonberger, *World Class Manufacturing* (New York: The Free Press, 1986), 161.
6. Ibid., 157.
7. J. Blocher, C. Lackey, and V. Mabert, From JIT purchasing to supplier partnerships at Xerox, *Target* 9, no. 3 (May/June 1993): 12–18.
8. R. Hall, *Attaining Manufacturing Excellence* (Burr Ridge, IL: Irwin, 1987), 233.
9. R. Dauch, *Passion for Manufacturing* (Dearborn, MI: Society for Manufacturing Engineers, 1993), 100.
10. Hall, *Attaining Manufacturing Excellence,* 231.
11. Portions of this section based on M. Abeysinghe, Vendor Relationships of Companies with Low Bargaining Power (Loyola University Chicago, February 1996).
12. L. Ettkin, F. Raiszadeh, and H. Hunt, Just-in-Time: A Timely Opportunity for Small Manufacturers, *Industrial Management* (Jan/Feb 1990): 16–18.
13. For more on supplier certification see U. Akinc, Selecting a set of vendors in a manufacturing environment. *Journal of Operations Management* 11, no. 2 (June 1993): 107–122; K. Bhote, *Strategic Supply Management: A Blueprint for Revitalizing the Manufacturer-Supplier Relationship* (New York: AMACOM, 1989); J. Lewis, *The Connected Corporation: How Leading Companies Win through Customer-Supplier Alliances* (New York: The Free Press, 1995); L. de Boer, E. Labro, and P. Morlacchi, A review of methods supporting supplier selection. *European Journal of Purchasing & Supply Management* 7, no. 2 (June 2001): 75–89.

14. Juran and Gryna, *Quality Planning and Analysis,* 318–319.
15. Juran and Gryna, *Quality Planning and Analysis,* 319; Melnyk and Denzler, *Operations Management,* 643.
16. Juran and Gryna, *Quality Planning and Analysis,* 314–321; Melnyk and Denzler, *Operations Management,* 610–611.
17. Melnyk and Denzler, *Operations Management,* 611.
18. S. Gibson, S. The Con-Way Express cross-dock operation (Loyola University Chicago, February 1996).
19. W. Sandras, 1989. *Just-in-Time: Making It Happen* (Essex Junction, VT: Oliver Wight Publications, 1989), 175.
20. P. Ghemawat, WalMart stores discount operations, *Harvard Business Review* (May 1989): 1–9; R. Halverson, Retooling retailing via information technology, *Discount Store News* (May 15, 1995): 73–75.
21. R. Harmon and L. Peterson. *Reinventing the Factory* (New York: The Free Press, 1990), 263.
22. Schonberger, *World Class Manufacturing,* 160–161.
23. Hall, *Attaining Manufacturing Excellence,* 238.
24. Ibid., 261–262.
25. Ibid., 262.
26. R. Cook and R. Rogowski, Applying JIT principles to continuous process manufacturing supply chains, *Production and Inventory Management Journal* (First Quarter 1996): 12–17.
27. Adapted from Y. Monden, *Toyota Production System,* 2nd ed. (Norcross, GA: Industrial Engineering and Management Press, 1993), 287–290.

Suggested Readings

J. Bossert, J. Brown, and R. Maass. *Supplier Certification: A Continuous Improvement Strategy.* Milwaukee, WI: ASQ Press, 1990.

M. Christopher. *Logistics and Supply Chain Management: Creating Value-Adding Networks,* 3rd ed. London: FT Press, 2005.

C. Cordon, and T. Vollmann. *The Power of Two: How Smart Companies Create Win-Win Customer-Supplier Partnerships that Outperform the Competition.* New York: Palgrave Macmillan, 2008.

J. Cunningham, O. Fiume, and E. *Adams. Real Numbers: Management Accounting in a Lean Organization.* Durham, NC: Managing Times Press, 2003.

J. Martin. Lean Six Sigma for Supply Chain Management. New York: McGraw-Hill Professional, 2006.

R. D. Nelson, P. Moody, and J. Stegner. *The Purchasing Machine: How the Top Ten Companies Use Best Practices to Manage Their Supply Chains.* New York: Free Press, 2001.

Questions

1. In 10 words or less, define what constitutes the supply chain.
2. What role does the supply chain play in the competitive position of any one company?
3. What is the role of suppliers and supplier relationships in the Toyota Production System and concurrent engineering? To answer this question, review Chapters 1 and 4.
4. Discuss the tradeoffs facing a manufacturer in deciding between designing and producing a product itself versus outsourcing design and production.
5. Describe briefly the philosophy of supply chain management (SCM).
6. What is the role of supplier–customer teamwork in SCM? How is this teamwork manifested?
7. Discuss how the following aspects of customer–supplier relationships change when parties move from being adversaries to being partners: criteria for purchase decisions, responsibility for design, number of suppliers, nature of contracts, and working relationships.

8. Discuss significant differences between customer–supplier partnership agreements and traditional purchase contracts with regard to the following: time period of agreement; mutual expectations about price, quality, delivery, and ordering; and expectations about interaction between customer and supplier.
9. What is vendor-managed inventory?
10. Discuss the process of evaluating suppliers for selection and ongoing performance assessment. Besides the usual criterion of price, what should a company look for in selecting among suppliers?
11. Discuss the role of the purchasing function in managing supplier relationships.
12. Virtually all of the lean concepts discussed throughout this book apply not just to manufacturers, but also to their suppliers, and to the carriers and distributors that link customers and suppliers. The section "Lean in the Supply Chain" gave examples. Give other examples of lean applications in the supply chain; consider in particular concepts such as focused factories and cellular manufacturing, level and mixed-model production, and standard operations.
13. Give examples of how the continuous improvement concept can be applied by a team of customers, suppliers, and carriers.

PROBLEMS

1. A manufacturer negotiates with a nearby supplier to deliver parts twice a day. The round trip travel time between the manufacturer and supplier is about 1.5 days. The supplier needs one day to fill any order. The manufacturer uses 600 units of the part per day, and the container size is 50 units.
   a. What number of S-kanbans for the part is needed?
   b. At any given time how many of these containers are at the customer, at the supplier, and en route between them?
   c. In general, as the round-trip customer–supplier travel time increases to three, four, or more days, so does the required number of kanbans. At any given time, where are these additional kanbans?

2. A supplier makes daily deliveries to three customers, similar to the case illustrated in Figure 16.8a. Assume travel time for each leg of the route is 1 hour; loading the truck at the supplier takes 1 hour; unloading time at each customer is 1 hour. With one truck and one driver, one delivery a day to the three customers is possible in an 8-hour day. See Figure 16.14.

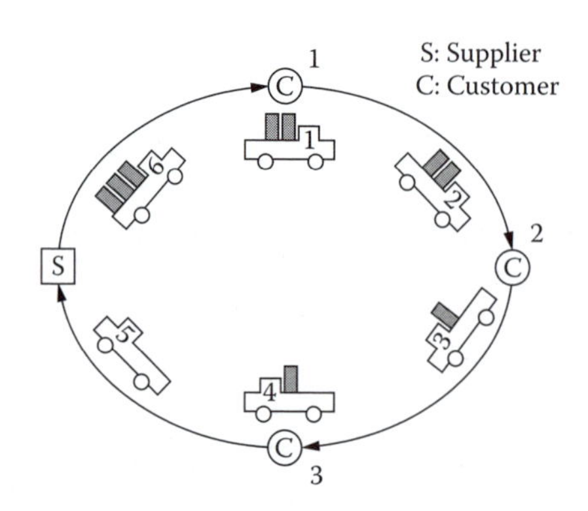

Figure 16.14 Supply route for problem 2.

a. If every customer wants two daily deliveries, how many trucks are needed? Assume one driver who switches trucks at every stop, analogous to the example in Figure 16.9. Draw a diagram like the SOR diagram described in Chapter 11 to illustrate the process. (The four stops along the route are analogous to workstations in a cell. The driver does not have to wait for trucks to be loaded or unloaded, so the load and unload times are analogous to automatic machine times.)
b. Assume two drivers: each morning one reports to the supplier, the other to customer 2. With six trucks, how many deliveries a day can they make to the three customers in 8 hours? Draw an SOR diagram. What happens with five trucks?
c. For part b and six trucks, what is the maximum load and unload time at a customer or supplier (without affecting the number of daily deliveries)? How does reducing the load and unload times to 30 minutes affect the number of daily deliveries?

Index

A

B

C

D

E

G

H

I

J

U

V

W

About the Author

John Nicholas is professor of operations management at Loyola University Chicago where he teaches in the areas of production and operations management, healthcare management, project management, and global operations management. He first introduced a course on lean production at Loyola in 1990. As a management consultant he has conducted productivity improvement projects and training programs in process improvement, quality circles, project management, and teamwork.

He is the author of numerous academic and technical trade publications and four books, including *The Portal to Lean Production: Principles and Practices for Doing More with Less* (coauthored with Avi Soni, 2006, published by Auerbach) and *Project Management for Business, Engineering and Technology: Principles and Practices* (coauthored with Herman Steyn, 2008, published by Elsevier/Butterworth-Heinemann).

Prior to Loyola John held the positions of test engineer and team lead for Lockheed/Martin Corporation, senior business analyst at Bank of America, and research associate at Argonne National Laboratory. He has a BS in aerospace engineering and an MBA in operations research and management, both from the University of Illinois, and a PhD in industrial engineering and applied behavioral science from Northwestern University.